MODERN CHEMISTRY

TEACHER'S SOLUTIONS MANUAL

HOLT, RINEHART AND WINSTON

A Harcourt Classroom Education Company

Austin • **New York** • **Orlando** • **Atlanta** • **San Francisco** • **Boston** • **Dallas** • **Toronto** • **London**

Modern Chemistry
Teacher's Solutions Manual

Copyright © by Holt, Rinehart and Winston

All rights reserved. No part of this publication may be reproduced or transmitted in any form or by any means, electronic or mechanical, including photocopy, recording, or any information storage and retrieval system, without permission in writing from the publisher.

Requests for permission to make copies of any part of the work should be mailed to the following address:

Permissions Department
Holt, Rinehart and Winston
10801 N. MoPac Expressway
Austin, Texas 78759.

Printed in the United States of America

ISBN 0-03-066079-3

    4 5 6 7   862   05 04 03

TABLE OF CONTENTS

Section I Pupil's Edition Solutions

Chapter 2	Measurements and Calculations	**1**
Chapter 3	Atoms: The Building Blocks of Matter	**11**
Chapter 4	Arrangement of Electrons in Atoms	**23**
Chapter 6	Chemical Bonding	**25**
Chapter 7	Chemical Formulas and Chemical Compounds	**49**
Chapter 9	Stoichiometry	**79**
Chapter 10	Physical Characteristics of Gases	**113**
Chapter 11	Molecular Composition of Gases	**125**
Chapter 12	Liquids and Solids	**147**
Chapter 13	Solutions	**153**
Chapter 14	Ions in Aqueous Solutions and Colligative Properties	**161**
Chapter 15	Acids and Bases	**175**
Chapter 16	Acid-Base Titration and pH	**177**
Chapter 17	Reaction Energy and Reaction Kinetics	**193**
Chapter 18	Chemical Equilibrium	**205**
Chapter 19	Oxidation-Reduction Reactions	**217**
Chapter 20	Carbon and Hydrocarbons	**223**
Chapter 21	Other Organic Compounds	**225**
Chapter 22	Nuclear Chemistry	**227**

Teacher's Note: Chapters 1, 5, and 8 contain no quantitative problems, so solutions are not provided for these chapters.

Section II Problem Bank Solutions **237**

CHAPTER 2
Measurements and Calculations

Practice, p. 40

1. Given: $V = 310 \text{ cm}^3$
$m = 853 \text{ g}$
Unknown: D

$D = \dfrac{m}{V} = \dfrac{853 \text{ g}}{310 \text{ cm}^3} = 2.75 \text{ g/cm}^3$

2. Given: $D = 3.26 \text{ g/cm}^3$
$V = 0.350 \text{ cm}^3$
Unknown: m

$D = \dfrac{m}{V}$

$m = DV = (3.26 \text{ g/cm}^3)(0.350 \text{ cm}^3) = 1.14 \text{ g}$

3. Given: $m = 76.2 \text{ g}$
$D = 13.6 \text{ g/mL}$
Unknown: V

$D = \dfrac{m}{V}$

$V = \dfrac{m}{D} = \dfrac{76.2 \text{ g}}{13.6 \text{ g/mL}} = 5.60 \text{ mL}$

ATE, Additional Sample Problems 2-1, p. 40

1. Given: $m = 74.0 \text{ g}$
$V = 20.3 \text{ cm}^3$
Unknown: D

$D = \dfrac{m}{V} = \dfrac{74.0 \text{ g}}{20.3 \text{ cm}^3} = 3.65 \text{ g/cm}^3$

2. Given: $m = 95.1 \text{ g}$
$D = 0.857 \text{ g/cm}^3$
Unknown: V

$D = \dfrac{m}{V}$

$V = \dfrac{m}{D} = \dfrac{95.1 \text{ g}}{0.857 \text{ g/cm}^3} = 111 \text{ cm}^3$

ATE, Additional Sample Problems 2-2, p. 41

1. Given: 1.00 day
Unknown: time in s

1 day = 24 h
1 h = 60 min
1 min = 60 s

$1.00 \text{ day} \times \dfrac{24 \text{ h}}{\text{day}} \times \dfrac{60 \text{ min}}{\text{h}} \times \dfrac{60 \text{ s}}{\text{min}} = 86\,400 \text{ s}$

2. Given: 6.25 kg
Unknown: mass in cg

1 g = 100 cg
1 kg = 1000 g

$6.25 \text{ kg} \times \dfrac{1000 \text{ g}}{\text{kg}} \times \dfrac{100 \text{ cg}}{\text{g}} = 625\,000 \text{ cg}$

Practice, p. 42

1. Given: 16.45 m
Unknown: length in cm and km

1 m = 100 cm

$16.45 \text{ m} \times \dfrac{100 \text{ cm}}{\text{m}} = 1645 \text{ cm}$

1 km = 1000 m

$16.45 \text{ m} \times \dfrac{1 \text{ km}}{1000 \text{ m}} = 0.016\,45 \text{ km}$

2. Given: 0.014 mg
Unknown: mass in g

1 g = 1000 mg

$0.014 \text{ mg} \times \dfrac{1 \text{ g}}{1000 \text{ mg}} = 0.000\,014 \text{ g}$

Section Review, p. 42

3. a. Given: 10.5 g
Unknown: mass in kg

$10.5 \text{ g} \times \dfrac{1 \text{ kg}}{1000 \text{ g}} = 0.0105 \text{ kg}$

b. Given: 1.57 km
Unknown: length in m

$1.57 \text{ km} \times \dfrac{1000 \text{ m}}{\text{km}} = 1570 \text{ m}$

c. Given: 3.54 µg
Unknown: mass in g

$3.54 \text{ µg} \times \dfrac{1 \text{ g}}{1\,000\,000 \text{ µg}} = 0.000\,003\,54 \text{ g}$

d. Given: 3.5 mol
Unknown: amount of substance in µmol

$3.5 \text{ mol} \times 1\,000\,000 \text{ µmol} = 3\,500\,000 \text{ µmol}$

e. Given: 1.2 L
Unknown: volume in mL

$1.2 \text{ L} \times \dfrac{1000 \text{ mL}}{\text{L}} = 1200 \text{ mL}$

f. Given: 358 cm³
Unknown: volume in m³

$358 \text{ cm}^3 \times \dfrac{1 \text{ m}^3}{1\,000\,000 \text{ cm}^3} = 0.000\,358 \text{ m}^3$

g. Given: 548.6 mL
Unknown: volume in cm³

$548.6 \text{ mL} \times \dfrac{1 \text{ cm}^3}{\text{mL}} = 548.6 \text{ cm}^3$

5. a. Given: $m = 84.7$ g
$V = 49.6$ cm³
Unknown: D

$D = \dfrac{m}{V}$

$= \dfrac{84.7 \text{ g}}{49.6 \text{ cm}^3} = 1.71 \text{ g/cm}^3$

b. Given: $m = 7.75$ g
$D = 1.71$ g/cm³
Unknown: V

$D = \dfrac{m}{V}$

$V = \dfrac{m}{D}$

$= \dfrac{7.75 \text{ g}}{1.71 \text{ g/cm}^3} = 4.53 \text{ cm}^3$

Practice, p. 45

1. Given: Value$_{experimental}$ = 17.7 g
Value$_{accepted}$ = 21.2 g
Unknown: Percent error

Percent error = $\dfrac{\text{Value}_{accepted} - \text{Value}_{experimental}}{\text{Value}_{accepted}} \times 100$

Percent error = $\dfrac{21.2 \text{ g} - 17.7 \text{ g}}{21.2 \text{ g}} \times 100 = 17\%$

2. Given: Value$_{experimental}$ = 4.26 mL
Value$_{accepted}$ = 4.15 mL
Unknown: Percent error

Percent error = $\dfrac{\text{Value}_{accepted} - \text{Value}_{experimental}}{\text{Value}_{accepted}} \times 100$

Percent error = $\dfrac{4.15 \text{ mL} - 4.26 \text{ mL}}{4.15 \text{ mL}} \times 100 = -2.7\%$

ATE, Additional Sample Problems 2-3, p. 45

1. Given: Value$_{experimental}$ = 4.25 cm
Value$_{accepted}$ = 4.08 cm
Unknown: Percent error

Percent error = $\dfrac{\text{Value}_{accepted} - \text{Value}_{experimental}}{\text{Value}_{accepted}} \times 100$

Percent error = $\dfrac{4.08 \text{ cm} - 4.25 \text{ cm}}{4.08 \text{ cm}} \times 100 = -4.2\%$

2. Given: Value$_{accepted}$ = 7.44 g/cm^3
Value$_{experimental}$ = 7.30 g/cm^3
Unknown: Percent error

Percent error = $\dfrac{\text{Value}_{accepted} - \text{Value}_{experimental}}{\text{Value}_{accepted}} \times 100$

Percent error = $\dfrac{7.44 \text{ g/cm}^3 - 7.30 \text{ g/cm}^3}{7.44 \text{ g/cm}^3} \times 100 = 1.9\%$

Practice, p. 50

1. Given: 2.099 g
0.05681 g
Unknown: sum of given values

2.099 g + 0.05681 g = 2.156 g

2. Given: 87.3 cm
1.655 cm
Unknown: difference between first and second given value

87.3 cm − 1.655 cm = 85.6 cm

3. Given: length = 1.34 μm
width = 0.7488 μm
Unknown: area

area = length × width

area = 1.34 μm × 0.7488 μm = 1.00 μm^2

4. Given: D = 1.2 g/cm^3
length = 28 cm
width = 22 cm
height = 2 × 3.0 mm = 6.0 mm = 0.60 cm
Unknown: m

$D = \dfrac{m}{V}$

V = length × width × height

$m = DV = D$ × length × width × height

m = 1.2 g/cm^3 × 28 cm × 22 cm × 0.60 cm = 440 g

ATE, Additional Sample Problems 2-5, p. 50

1. Given: length = width = height = 3.23 cm
Unknown: V

$V = \text{length} \times \text{width} \times \text{height}$
$V = (3.23 \text{ cm})^3 = 33.7 \text{ cm}^3$

2. Given: 67.14 kg
8.2 kg
Unknown: sum of given values

$67.14 \text{ kg} + 8.2 \text{ kg} = 75.3 \text{ kg}$

3. Given: $m = 17.982$ g
$V = 4.13$ cm^3
Unknown: D

$D = \dfrac{m}{V} = \dfrac{17.982 \text{ g}}{4.13 \text{ cm}^3} = 4.35 \text{ g/cm}^3$

Practice, p. 54

1. Given: $m = 1.73 \times 10^{-3}$ g
$D = 0.178\,47$ g/L
Unknown: V

$D = \dfrac{m}{V}$

$V = \dfrac{m}{D} = \dfrac{1.73 \times 10^{-3} \text{ g}}{0.178\,47 \text{ g/L}} = 0.00969 \text{ L}$

$V = 9.69 \times 10^{-3} \text{ L} \times \dfrac{1000 \text{ mL}}{\text{L}} = 9.69 \text{ mL}$

2. Given: $m = 6.25 \times 10^5$ g
$V = 92.5$ cm $\times$ 47.3 cm $\times$ 85.4 cm
Unknown: D

$D = \dfrac{m}{V} = \dfrac{6.25 \times 10^5 \text{ g}}{92.5 \text{ cm} \times 47.3 \text{ cm} \times 85.4 \text{ cm}} = 1.67 \text{ g/cm}^3$

3. Given: 5.12×10^5 km
Unknown: length in mm

$5.12 \times 10^5 \text{ km} \times \dfrac{10^6 \text{ mm}}{\text{km}} = 5.12 \times 10^{11} \text{ mm}$

ATE, Additional Sample Problems 2-6, p. 54

1. Given: $V = 41.4$ mL
$m = 58.24$ g
Unknown: D

$D = \dfrac{m}{V} = \dfrac{58.24 \text{ g}}{41.4 \text{ mL}} = 1.41 \text{ g/mL}$

2. Given: 6.2×10^7 cm
Unknown: length in km

$6.2 \times 10^7 \text{ cm} \times \dfrac{1 \text{ m}}{10^2 \text{ cm}} \times \dfrac{1 \text{ km}}{10^3 \text{ m}} = 6.2 \times 10^2 \text{ km}$

3. Given: 3 weeks (exactly)
Unknown: time in hours

$3 \text{ weeks} \times \dfrac{7 \text{ day}}{\text{week}} \times \dfrac{24 \text{ h}}{\text{day}} = 504 \text{ h}$

Section Review, p. 57

4. a. Given: 52.13 g + 1.7502 g
Unknown: sum of values

$52.13 \text{ g} + 1.7502 \text{ g} = 53.88 \text{ g}$

4. b. Given: 12 m × 6.41 m
Unknown: product of values

$12 \text{ m} \times 6.41 \text{ m} = 77 \text{ m}^2$

c. Given: $\dfrac{16.25 \text{ g}}{5.1442 \text{ mL}}$
Unknown: quotient of values

$\dfrac{16.25 \text{ g}}{5.1442 \text{ mL}} = 3.159 \text{ g/mL}$

5. a. Given: $(1.54 \times 10^{-2} \text{ g}) + (2.86 \times 10^{-1} \text{ g})$
Unknown: sum of values

$1.54 \times 10^{-2} \text{ g} = 0.154 \times 10^{-1} \text{ g}$

$(0.154 \times 10^{-1} \text{ g}) + (2.86 \times 10^{-1} \text{ g}) = 3.01 \times 10^{-1} \text{ g}$

b. Given: $(7.023 \times 10^9 \text{ g}) - (6.62 \times 10^7 \text{ g})$
Unknown: difference between values

$6.62 \times 10^7 \text{ g} = 0.0662 \times 10^9 \text{ g}$

$(7.023 \times 10^9 \text{ g}) - (0.0662 \times 10^9 \text{ g}) = 6.957 \times 10^9 \text{ g}$

5. c. Given: $(8.99 \times 10^{-4} \text{ m}) \times (3.57 \times 10^4 \text{ m})$
Unknown: product of values

$(8.99 \times 10^{-4} \text{ m}) \times (3.57 \times 10^4 \text{ m}) = 3.21 \times 10^1 \text{ m}^2$

d. Given: $\dfrac{2.17 \times 10^{-3} \text{ g}}{5.022 \times 10^4 \text{ mL}}$
Unknown: quotient of values

$\dfrac{2.17 \times 10^{-3} \text{ g}}{5.022 \times 10^4 \text{ mL}} = 4.32 \times 10^{-8} \text{ g/mL}$

7. a. Given: $m_{total} = 215.6 \text{ g}$
$m_{beaker} = 110.4 \text{ g}$
Unknown: m_{oil}

$m_{total} = m_{oil} + m_{beaker}$

$m_{oil} = m_{total} - m_{beaker} = 215.6 \text{ g} - 110.4 \text{ g} = 105.2 \text{ g}$

b. Given: $V_{oil} = 114 \text{ cm}^3$
$m_{oil} = 105.2 \text{ g}$
Unknown: D_{oil}

$D = \dfrac{m}{V}$

$D_{oil} = \dfrac{m_{oil}}{V_{oil}} = \dfrac{105.2 \text{ g}}{114 \text{ cm}^3} = 0.923 \text{ g/cm}^3$

8. Given: $V = 5.0 \times 10^{-3} \text{ cm}^3$
$D = 19.3 \text{ g/cm}^3$
Unknown: m

$D = \dfrac{m}{V}$

$m = DV = (19.3 \text{ g/cm}^3)(5.00 \times 10^{-3} \text{ cm}^3) = 9.6 \times 10^{-2} \text{ g}$

Review Problems

27. Given: length = 0.25 m
width = 6.1 m
height = 4.9 m
Unknown: V

$V = \text{length} \times \text{width} \times \text{height}$

$V = 0.25 \text{ m} \times 6.1 \text{ m} \times 4.9 \text{ m} = 7.5 \text{ m}^3$

28. Given: $m = 5.03$ g
$V = 3.24$ mL
Unknown: D

$D = \dfrac{m}{V} = \dfrac{5.03 \text{ g}}{3.24 \text{ mL}} = 1.55$ g/mL

29. Given: $V = 55.1$ cm^3
$D = 6.72$ g/cm^3
Unknown: m

$D = \dfrac{m}{V}$

$m = DV = (6.72 \text{ g/cm}^3)(55.1 \text{ cm}^3) = 3.70 \times 10^2$ g

30. Given: $D = 0.824$ g/mL
$m = 0.451$ g
Unknown: V

$D = \dfrac{m}{V}$

$V = \dfrac{m}{D} = \dfrac{0.451 \text{ g}}{0.824 \text{ g/mL}} = 0.547$ mL

31. Given: 882 µg
Unknown: mass in g

$882 \text{ µg} \times \dfrac{1 \text{ g}}{1\,000\,000 \text{ µg}} = 8.82 \times 10^{-4}$ g

32. Given: 0.603 L
Unknown: volume in mL

$0.603 \text{ L} \times \dfrac{1000 \text{ mL}}{\text{L}} = 603$ mL

33. a. Given: $D = 19.3$ g/cm^3
$m = 0.715$ kg
Unknown: V

$D = \dfrac{m}{V}$

$V = \dfrac{m}{D} = \dfrac{0.715 \text{ kg}}{19.3 \text{ g/cm}^3} \times \dfrac{1000 \text{ g}}{\text{kg}} = 37.0$ cm^3

b. Given: $V = 37.0$ cm^3
Unknown: length of cube's edge

$V = \text{length}^3$

$\text{length} = \sqrt[3]{V} = \sqrt[3]{37.0 \text{ cm}^3} = 3.33$ cm

34. a. Given: 92.25 m
Unknown: length in km

$92.25 \text{ m} \times \dfrac{1 \text{ km}}{10^3 \text{ m}} = 9.225 \times 10^{-2}$ km

b. Given: 9.225×10^{-2} km
Unknown: length in cm

$9.225 \times 10^{-2} \text{ km} \times \dfrac{10^3 \text{ m}}{\text{km}} \times \dfrac{10^2 \text{ cm}}{\text{m}} = 9.225 \times 10^3$ cm

35. Given: Value$_\text{experimental}$ = 9.67 g
Value$_\text{accepted}$ = 9.82 g
Unknown: Percent error

Percent error = $\dfrac{\text{Value}_\text{accepted} - \text{Value}_\text{experimental}}{\text{Value}_\text{accepted}} \times 100$

Percent error = $\dfrac{9.82 \text{ g} - 9.67 \text{ g}}{9.82 \text{ g}} \times 100 = 1.5\%$

36. Given: Value$_\text{accepted}$ = 1.54 g/cm^3
Value$_\text{experimental}$ = 1.25 g/cm^3
Unknown: Percent error

Percent error = $\dfrac{\text{Value}_\text{accepted} - \text{Value}_\text{experimental}}{\text{Value}_\text{accepted}} \times 100$

Percent error = $\dfrac{1.54 \text{ g/cm}^3 - 1.25 \text{ g/cm}^3}{1.54 \text{ g/cm}^3} \times 100 = 19\%$

37. Given: Value$_\text{experimental}$ = 0.229 cm
Value$_\text{accepted}$ = 0.225 cm
Unknown: Percent error

Percent error = $\dfrac{\text{Value}_\text{accepted} - \text{Value}_\text{experimental}}{\text{Value}_\text{accepted}} \times 100$

Percent error = $\dfrac{0.225 \text{ cm} - 0.229 \text{ cm}}{0.225 \text{ cm}} \times 100 = -2\%$

39. Given: 6.078 g
0.3329 g
Unknown: sum of given values

6.078 g + 0.3329 g = 6.411 g

---

40. Given: 7.11 cm
8.2 cm
Unknown: difference between second and first given value

8.2 cm − 7.11 cm = 1.1 cm

---

41. Given: 0.8102 m
3.44 m
Unknown: product of given values

0.8102 m × 3.44 m = 2.79 m

---

42. Given: 94.20 g
3.167 22 mL
Unknown: first given value divided by second given value

$$\frac{94.20 \text{ g}}{3.167\ 22 \text{ mL}} = 29.74 \text{ g/mL}$$

---

45. Given: $\dfrac{6.124\ 33 \times 10^6 \text{ m}^3}{7.15 \times 10^{-3} \text{ m}}$

Unknown: quotient of values in scientific notation

$$\frac{6.124\ 33 \times 10^6 \text{ m}^3}{7.15 \times 10^{-3} \text{ m}} = 8.57 \times 10^8 \text{ m}^2$$

---

46. ANALYZE
Given: $m = 2.03 \times 10^{-3}$ g
$D = 9.133 \times 10^{-1}$ g/cm^3
Unknown: V

PLAN Rearrange the density equation to solve for volume.

$$D = \frac{m}{V}$$

$$V = \frac{m}{D}$$

COMPUTE $V = \dfrac{2.03 \times 10^{-3} \text{ g}}{9.133 \times 10^{-1} \text{ g/cm}^3} = 2.22 \times 10^{-3}$ cm^3

EVALUATE The unit of volume, cm^3, is correct. An order-of-magnitude estimate would put the answer at about 1/1000 cm^3. The correct number of significant digits is three.

Mixed Review

47. Given: $m_1 = 100.6$ kg
$m_2 = 96.4$ kg
Unknown: given values expressed in scientific notation, difference between first and second given values

$m_1 = 1.006 \times 10^2$ kg

$m_2 = 9.64 \times 10^1$ kg

$m_1 - m_2 = 100.6$ kg − 96.41 kg = 4.2 kg

48. Given: length = 1.07×10^2 m
width = 31 m
height = 4.25×10^2 m
Unknown: V

$V = \text{length} \times \text{width} \times \text{height}$
$V = (1.07 \times 10^2 \text{ m})(31 \text{ m})(4.25 \times 10^2 \text{ m})$
$V = 1.4 \times 10^6 \text{ m}^3$

49. Given: $m = 57.6$ g
$V = 40.25$ cm^3
Unknown: D

$D = \dfrac{m}{V} = \dfrac{57.6 \text{ g}}{40.25 \text{ cm}^3} = 1.43 \text{ g/cm}^3$

50. Given: 0.947 mg
Unknown: mass in g and kg

$0.947 \text{ mg} \times \dfrac{1 \text{ g}}{10^3 \text{ mg}} = 9.47 \times 10^{-4} \text{ g}$

$9.47 \times 10^{-4} \text{ g} \times \dfrac{1 \text{ kg}}{10^3 \text{ g}} = 9.47 \times 10^{-7} \text{ kg}$

51. Given: Value$_{\text{experimental}}$ = 6.80 g/cm^3
Value$_{\text{accepted}}$ = 7.86 g/cm^3
Unknown: Percent error

Percent error = $\dfrac{\text{Value}_{\text{accepted}} - \text{Value}_{\text{experimental}}}{\text{Value}_{\text{accepted}}} \times 100$

Percent error = $\dfrac{7.86 \text{ g/cm}^3 - 6.80 \text{ g/cm}^3}{7.86 \text{ g/cm}^3} \times 100 = 13.5\%$

Handbook Search

52. Given: $r_{\text{Li}} = 152$ pm
$r_{\text{Na}} = 186$ pm
$r_{\text{K}} = 232$ pm
$r_{\text{Rb}} = 248$ pm
$r_{\text{Cs}} = 265$ pm
$r_{\text{Fr}} = 270$ pm
Unknown: $V_{\text{Li}}, V_{\text{Na}}, V_{\text{K}}, V_{\text{Rb}}, V_{\text{Cs}}, V_{\text{Fr}}$

$V = \tfrac{4}{3}\pi r^3$

$V_{\text{Li}} = \tfrac{4}{3}\pi(r_{\text{Li}})^3 = \tfrac{4}{3}\pi(152 \text{ pm})^3 = 1.47 \times 10^7 \text{ pm}^3$

$V_{\text{Na}} = \tfrac{4}{3}\pi(r_{\text{Na}})^3 = \tfrac{4}{3}\pi(186 \text{ pm})^3 = 2.70 \times 10^7 \text{ pm}^3$

$V_{\text{K}} = \tfrac{4}{3}\pi(r_{\text{K}})^3 = \tfrac{4}{3}\pi(232 \text{ pm})^3 = 5.23 \times 10^7 \text{ pm}^3$

$V_{\text{Rb}} = \tfrac{4}{3}\pi(r_{\text{Rb}})^3 = \tfrac{4}{3}\pi(248 \text{ pm})^3 = 6.39 \times 10^7 \text{ pm}^3$

$V_{\text{Cs}} = \tfrac{4}{3}\pi(r_{\text{Cs}})^3 = \tfrac{4}{3}\pi(265 \text{ pm})^3 = 7.80 \times 10^7 \text{ pm}^3$

$V_{\text{Fr}} = \tfrac{4}{3}\pi(r_{\text{Fr}})^3 = \tfrac{4}{3}\pi(270 \text{ pm})^3 = 8.24 \times 10^7 \text{ pm}^3$

53. Given: $r_{\text{Na}} = 186$ pm
length = 5.00 cm
Unknown: number of Na atoms

length = (number of Na atoms)(diameter of one Na atom)
= (number of Na atoms)(2 r_{Na})

number of Na atoms = $\dfrac{\text{length}}{2 \, r_{\text{Na}}}$

1 cm = 10^{10} pm

number of Na atoms = $\dfrac{5.00 \text{ cm}}{(2)(186 \text{ pm})} \times \dfrac{10^{10} \text{ pm}}{\text{cm}} = 1.34 \times 10^8$ Na atoms

54. a. Given: $V_{\text{Na}} = 3.00$ cm $\times$ 5.00 cm $\times$ 5.00 cm = 75.0 cm^3
$m_{\text{Na}} = 75.5$ g
Unknown: D_{Na}

$D = \dfrac{m}{V}$

$D_{\text{Na}} = \dfrac{m_{\text{Na}}}{V_{\text{Na}}} = \dfrac{75.5 \text{ g}}{75.0 \text{ cm}^3} = 1.01 \text{ g/cm}^3$

54. b. Given: $\text{Value}_{\text{experimental}} = 1.01 \text{ g/cm}^3$
Value$_{\text{accepted}} = 0.971 \text{ g/cm}^3$
Unknown: Percent error

$$\text{Percent error} = \frac{\text{Value}_{\text{accepted}} - \text{Value}_{\text{experimental}}}{\text{Value}_{\text{accepted}}} \times 100$$

$$\text{Percent error} = \frac{0.971 \text{ g/cm}^3 - 1.01 \text{ g/cm}^3}{0.971 \text{ g/cm}^3} \times 100 = -4\%$$

Alternative Assessment

58. a. Given: Total fat per serving = 2 g
Calories from fat per serving = 15 Calories
Unknown: conversion factor in Calories per gram

15 Cal = 2 g fat

$$\frac{15 \text{ Cal}}{2 \text{ g}} = 8 \text{ Cal/g}$$

b. Given: Serving size = 30 g
Number of servings = 20
Unknown: mass of total in kg

total mass = m_{tot} = Serving size × Number of servings

$$m_{\text{tot}} = (30 \text{ g})(20) \times \frac{1 \text{ kg}}{1000 \text{ g}} = 0.6 \text{ kg}$$

c. Given: 2 g
Unknown: mass in μg

$$2 \text{ g} \times \frac{10^6 \text{ μg}}{\text{g}} = 2 \times 10^6 \text{ μg}$$

CHAPTER 3
Atoms: the Building Blocks of Matter

Practice 3-1, p. 78

1. Given: name and mass number of bromine-80

Unknown: number of protons, electrons, and neutrons

atomic number = number of protons = number of electrons
mass number = number of neutrons + number of protons.

For bromine, the atomic number = 35
For bromine-80, the mass number = 80

number of protons = 35 protons
number of electrons = 35 electrons

number of neutrons = mass number − atomic number = 80 − 35
number of neutrons = 45 neutrons

2. Given: name and mass number of carbon-13

Unknown: nuclear symbol for carbon-13

chemical symbol for carbon: C
atomic number for carbon = 6 (located at lower left of symbol)
mass number for carbon-13 = 13 (located at upper left of symbol)

carbon-13: $^{13}_{6}C$

3. Given: element that has 15 electrons and 15 neutrons

Unknown: hyphen notation for given element nuclide

atomic number = number of protons = number of electrons
mass number = number of protons + number of neutrons

atomic number = 15 (element is phosphorus)

mass number = 15 protons + 15 neutrons = 30

nuclide is phosphorus-30

ATE, Additional Sample Problems 3–1, p. 78

1. Given: name and mass number of carbon-13

Unknown: number of protons, electrons, and neutrons

atomic number = number of protons = number of electrons
mass number = number of neutrons + number of protons

For carbon, the atomic number = 6
For carbon-13, the mass number = 13

number of protons = 6 protons
number of electrons = 6 electrons
number of neutrons = mass number − atomic number = 13 − 6
number of neutrons = 7 neutrons

2. Given: name and mass number of oxygen-16

Unknown: nuclear symbol for oxygen-16

chemical symbol for oxygen: O

atomic number for oxygen = 8 (located at lower left of symbol)
mass number for oxygen-16 = 16 (located at upper left of symbol)

oxygen-16: $^{16}_{8}O$

3. Given: element that has 7 electrons and 9 neutrons
   Unknown: hyphen notation for given nuclide

   atomic number = number of protons = number of electrons
   mass number = number of protons + number of neutrons
   atomic number = 7 (element is nitrogen)
   mass number = 7 protons + 9 neutrons = 16
   nuclide is nitrogen-16

ATE, Additional Sample Problems 3-2, p. 82

1. Given: 3.6 mol C
   Unknown: mass of C in grams

   $3.6 \text{ mol C} \times \dfrac{12.01 \text{ g C}}{\text{mol C}} = 43 \text{ g C}$

2. Given: 0.733 mol Cl
   Unknown: mass of Cl in grams

   $0.733 \text{ mol Cl} \times \dfrac{35.45 \text{ g Cl}}{\text{mol Cl}} = 26.0 \text{ g Cl}$

Practice 3-2, p. 83

1. Given: 2.25 mol Fe
   Unknown: mass of Fe in grams

   $2.25 \text{ mol Fe} \times \dfrac{55.85 \text{ g Fe}}{\text{mol Fe}} = 126 \text{ g Fe}$

2. Given: 0.375 mol K
   Unknown: mass of K in grams

   $0.375 \text{ mol K} \times \dfrac{39.10 \text{ g K}}{\text{mol K}} = 14.7 \text{ g K}$

3. Given: 0.0135 mol Na
   Unknown: mass of Na in grams

   $0.0135 \text{ mol Na} \times \dfrac{22.99 \text{ g Na}}{\text{mol Na}} = 0.310 \text{ g Na}$

4. Given: 16.3 mol Ni
   Unknown: mass of Ni in grams

   $16.3 \text{ mol Ni} \times \dfrac{58.69 \text{ g Ni}}{\text{mol Ni}} = 957 \text{ g Ni}$

Practice 3–3, p. 83

1. Given: 5.00 g Ca
   Unknown: amount of Ca in moles

   $5.00 \text{ g Ca} \times \dfrac{\text{mol Ca}}{40.08 \text{ g Ca}} = 0.125 \text{ mol Ca}$

2. Given: 3.60×10^{-10} g Au
   Unknown: amount of Au in moles

   $3.60 \times 10^{-10} \text{ g Au} \times \dfrac{\text{mol Au}}{196.97 \text{ g Au}} = 1.83 \times 10^{-12} \text{ mol Au}$

ATE, Additional Sample Problems 3-3, p. 83

1. Given: 3.22 g Cu
Unknown: amount of Cu in moles

$$3.22 \text{ g Cu} \times \frac{\text{mol Cu}}{63.55 \text{ g Cu}} = 0.0507 \text{ mol Cu}$$

2. Given: 2.72×10^{-4} g Li
Unknown: amount of Li in moles

$$2.72 \times 10^{-4} \text{ g Li} \times \frac{\text{mol Li}}{6.94 \text{ g Li}} = 3.92 \times 10^{-5} \text{ mol Li}$$

Practice 3–4, p. 84

1. Given: 1.50×10^{12} atoms Pb
Unknown: amount of Pb in moles

$$1.50 \times 10^{12} \text{ Pb atoms} \times \frac{\text{mol Pb}}{6.022 \times 10^{23} \text{ Pb atoms}} = 2.49 \times 10^{-12} \text{ mol Pb}$$

2. Given: 2500 atoms Sn
Unknown: amount of Sn in moles

$$2500 \text{ Sn atoms} \times \frac{\text{mol Sn}}{6.022 \times 10^{23} \text{ Sn atoms}} = 4.2 \times 10^{-21} \text{ mol Sn}$$

3. Given: 2.75 mol Al
Unknown: number of Al atoms

$$2.75 \text{ mol Al} \times \frac{6.022 \times 10^{23} \text{ Al atoms}}{\text{mol Al}} = 1.66 \times 10^{24} \text{ Al atoms}$$

ATE, Additional Sample Problems 3-4, p. 84

1. Given: 2.25×10^{22} atoms C
Unknown: amount of C in moles

$$2.25 \times 10^{22} \text{ C atoms} \times \frac{\text{mol C}}{6.022 \times 10^{23} \text{ C atoms}} = 0.0374 \text{ mol C}$$

2. Given: 2×10^{6} atoms O
Unknown: amount of O in moles

$$2 \times 10^{6} \text{ O atoms} \times \frac{\text{mol O}}{6.022 \times 10^{23} \text{ O atoms}} = 3.321 \times 10^{-18} \text{ mol O}$$

3. Given: 3.80 mol Na
Unknown: number of Na atoms

$$3.80 \text{ mol Na} \times \frac{6.022 \times 10^{23} \text{ Na atoms}}{\text{mol Na}} = 2.29 \times 10^{24} \text{ Na atoms}$$

ATE, Additional Sample Problems 3-5, p. 84

4. Given: 5.0×10^{9} atoms Ne
Unknown: mass of Ne in grams

$$5.0 \times 10^{9} \text{ Ne atoms} \times \frac{\text{mol Ne}}{6.022 \times 10^{23} \text{ Ne atoms}} \times \frac{20.18 \text{ g Ne}}{\text{mol Ne}}$$
$$= 1.7 \times 10^{-13} \text{ g Ne}$$

5. Given: 0.020 g C
Unknown: number of C atoms

$$0.020 \text{ g C} \times \frac{\text{mol C}}{12.01 \text{ g C}} \times \frac{6.022 \times 10^{23} \text{ C atoms}}{\text{mol C}} = 1.0 \times 10^{21} \text{ C atoms}$$

6. Given: 10.0 g B
Unknown: mass of Ag with same number of atoms as 10.0 g B

One mole of boron has the same number of atoms as one mole of silver. Therefore
1 mol B = 1 mol Ag

$$10.0 \text{ g B} \times \frac{\text{mol B}}{10.87 \text{ g B}} \times \frac{\text{mol Ag}}{\text{mol B}} \times \frac{107.87 \text{ g Ag}}{\text{mol Ag}} = 99.8 \text{ g Ag}$$

Practice 3–5, p. 85

1. Given: 7.5×10^{15} atoms Ni
Unknown: mass of Ni in grams

$$7.5 \times 10^{15} \text{ Ni atoms} \times \frac{\text{mol Ni}}{6.022 \times 10^{23} \text{ Ni atoms}} \times \frac{58.69 \text{ g Ni}}{\text{mol Ni}}$$
$$= 7.3 \times 10^{-7} \text{ g Ni}$$

2. Given: 4.00 g S
Unknown: number of S atoms

$$4.00 \text{ g S} \times \frac{\text{mol S}}{32.07 \text{ g S}} \times \frac{6.022 \times 10^{23} \text{ S atoms}}{\text{mol S}} = 7.51 \times 10^{22} \text{ S atoms}$$

3. Given: 9.0 g Al
Unknown: mass of Au with same number of atoms as 9.0 g Al

1 mol Al = 1 mol Au

$$9.0 \text{ g Al} \times \frac{\text{mol Al}}{26.98 \text{ g Al}} \times \frac{\text{mol Au}}{\text{mol Al}} \times \frac{196.97 \text{ g Au}}{\text{mol Au}} = 66 \text{ g Au}$$

Section Review, p. 85

2. a. Given: name and mass number of sodium-23
Unknown: number of protons, electrons, and neutrons

atomic number = number of protons = number of electrons
mass number = number of neutrons + number of protons

For sodium, the atomic number = 11
For sodium-23, the mass number = 23

number of protons = 11 protons
number of electrons = 11 electrons
number of neutrons = mass number − atomic number = 23 − 11
number of neutrons = 12 neutrons

b. Given: name and mass number of calcium-40
Unknown: number of protons, electrons, and neutrons

atomic number = number of protons = number of electrons
mass number = number of neutrons + number of protons

For calcium, the atomic number = 20
For calcium-40, the mass number = 40

number of protons = 20 protons
number of electrons = 20 electrons

number of neutrons = mass number − atomic number = 40 − 20
number of neutrons = 20 neutrons

c. Given: $^{64}_{29}$Cu
Unknown: number of protons, electrons, and neutrons

atomic number = number of protons = number of electrons = 29
mass number = number of neutrons + number of protons = 64

number of neutrons = mass number − atomic number = 64 − 29 = 35
number of protons = 29 protons
number of electrons = 29 electrons
number of neutrons = 35 neutrons

d. Given: $^{108}_{47}$Ag
Unknown: number of protons, electrons, and neutrons

atomic number = number of protons = number of electrons = 47
mass number = number of neutrons + number of protons = 108
number of neutrons = mass number − atomic number = 108 − 47 = 61

number of protons = 47 protons
number of electrons = 47 electrons
number of neutrons = 61 neutrons

3. a. Given: mass number = 28, atomic number = 14
Unknown: nuclear symbol and hyphen notation for given nuclide

atomic number = number of protons
element with 14 protons: silicon, Si

nuclear symbol → $^{28}_{14}$Si

hyphen notation → Silicon-28

b. Given: element that has 26 protons and 30 neutrons
Unknown: nuclear symbol and hyphen notation for given nuclide

atomic number = number of protons = 26
element with 26 protons: iron, Fe
mass number = number of neutrons + number of protons = 30 + 26 = 56

nuclear symbol → $^{56}_{26}$Fe

hyphen notation → iron-56

c. Given: element that has 56 electrons and 82 neutrons
Unknown: nuclear symbol and hyphen notation for given nuclide

atomic number = number of protons = number of electrons
mass number = number of protons + number of neutrons

atomic number = 56 (element is barium, Ba)
mass number = number of neutrons + number of protons = 56 + 82 = 138

nuclear symbol → $^{138}_{56}$Ba

hyphen notation → barium-138

5. a. Given: 2.00 mol N
Unknown: mass of N in grams

$2.00 \text{ mol N} \times \dfrac{14.01 \text{ g N}}{\text{mol N}} = 28.0 \text{ g N}$

b. Given: 3.01 × 10²³ atoms Cl
Unknown: mass of Cl in grams

$3.01 \times 10^{23} \text{ Cl atoms} \times \dfrac{\text{mol Cl}}{6.022 \times 10^{23} \text{ Cl atoms}} \times \dfrac{35.45 \text{ g Cl}}{\text{mol Cl}} = 17.7 \text{ g Cl}$

6. a. Given: 12.15 g Mg
Unknown: amount of Mg in moles

$$12.15 \text{ g Mg} \times \frac{\text{mol Mg}}{24.30 \text{ g Mg}} = 0.5000 \text{ mol Mg}$$

b. Given: 1.50×10^{23} atoms F
Unknown: amount of F in moles

$$1.50 \times 10^{23} \text{ F atoms} \times \frac{\text{mol F}}{6.022 \times 10^{23} \text{ F atoms}} = 0.249 \text{ mol F}$$

7. a. Given: 2.50 mol Zn
Unknown: number of Zn atoms

$$2.50 \text{ mol Zn} \times \frac{6.022 \times 10^{23} \text{ Zn atoms}}{\text{mol Zn}} = 1.51 \times 10^{24} \text{ Zn atoms}$$

b. Given: 1.50 g C
Unknown: number of C atoms

$$1.50 \text{ g C} \times \frac{\text{mol C}}{12.01 \text{ g C}} \times \frac{6.022 \times 10^{23} \text{ C atoms}}{\text{mol C}} = 7.52 \times 10^{22} \text{ C atoms}$$

Reviewing Concepts

2. Given: element A with 2 mass units
element B with 3 mass units
Unknown: masses of compounds AB and A_2B_3

mass of AB = mass A + mass B = 2 mass units + 3 mass units = 5 mass units

mass of A_2B_3 = (2 × mass A) + (3 × mass B)

mass of A_2B_3 = (2)(2 mass units) + (3)(3 mass units) = **13 mass units**

8. Given: isotopes Si-28, Si-29, and Si-30
Unknown: number of protons, electrons, and neutrons

atomic number = number of protons = number of electrons
mass number = number of neutrons + number of protons
number of neutrons = mass number − atomic number
atomic number for Si (silicon) = 14

all Si isotopes have 14 protons and 14 electrons

mass number for Si-28 = 28
number of neutrons = 28 − 14 = 14

mass number for Si-29 = 29
number of neutrons = 29 − 14 = 15

mass number for Si-30 = 30
number of neutrons = 30 − 14 = 16

11. a. Given: atomic number = 2, mass number = 4
Unknown: hyphen notation for given isotope

atomic number = 2 → element is helium

isotope is helium-4

b. Given: atomic number = 8, mass number = 16
Unknown: hyphen notation for given isotope

atomic number = 8 → element is oxygen

isotope is oxygen-16

c. Given: atomic number = 19, mass number = 39

Unknown: hyphen notation for given isotope

atomic number = 19 → element is potassium

isotope is potassium-39

13. a. Given: atom with $\frac{1}{3}$ mass of carbon-12

Unknown: atomic mass of given atom

mass of carbon-12 = 12 amu

mass of given element = $\frac{1}{3}$(12 amu) = 4 amu

b. Given: atom with 4.5 times mass of carbon-12

Unknown: atomic mass of given atom

mass of carbon-12 = 12 amu

mass of given element = (4.5)(12 amu) = 54 amu

Review Problems

17. a. Given: 1.00 mol Li

Unknown: mass of Li in grams

$$1.00 \text{ mol Li} \times \frac{6.94 \text{ g Li}}{\text{mol Li}} = 6.94 \text{ g Li}$$

b. Given: 1.00 mol Al

Unknown: mass of Al in grams

$$1.00 \text{ mol Al} \times \frac{26.98 \text{ g Al}}{\text{mol Al}} = 27.0 \text{ g Al}$$

c. Given: 1.00 molar mass Ca

Unknown: mass of Ca in grams

$$\frac{40.08 \text{ g Ca}}{\text{mol Ca}} \times 1.00 \text{ mol Ca} = 40.1 \text{ g Ca}$$

d. Given: 1.00 molar mass Fe

Unknown: mass of Fe in grams

$$\frac{55.85 \text{ g Fe}}{\text{mol Fe}} \times 1.00 \text{ mol Fe} = 55.8 \text{ g Fe}$$

e. Given: 6.022×10^{23} atoms C

Unknown: mass of C in grams

$$6.022 \times 10^{23} \text{ C atoms} \times \frac{\text{mol C}}{6.022 \times 10^{23} \text{ C atoms}} \times \frac{12.01 \text{ g C}}{\text{mol C}} = 12.01 \text{ g C}$$

f. Given: 6.022×10^{23} atoms Ag

Unknown: mass of Ag in grams

$$6.022 \times 10^{23} \text{ Ag atoms} \times \frac{\text{mol Ag}}{6.022 \times 10^{23} \text{ Ag atoms}} \times \frac{107.87 \text{ g Ag}}{\text{mol Ag}}$$

= 107.9 g Ag

18. a. Given: 6.022×10^{23} atoms Ne

Unknown: amount of Ne in moles

$$6.022 \times 10^{23} \text{ Ne atoms} \times \frac{\text{mol Ne}}{6.022 \times 10^{23} \text{ Ne atoms}} = 1.000 \text{ mol Ne}$$

b. Given: 3.011×10^{23} atoms Mg
Unknown: amount of Mg in moles

3.011×10^{23} Mg atoms $\times \dfrac{\text{mol Mg}}{6.022 \times 10^{23} \text{ Mg atoms}} = 0.5000$ mol Mg

c. Given: 3.25×10^5 g Pb
Unknown: amount of Pb in moles

3.25×10^5 g Pb $\times \dfrac{\text{mol Pb}}{207.2 \text{ g Pb}} = 1.57 \times 10^3$ mol Pb

d. Given: 4.50×10^{-12} g O
Unknown: amount of O in moles

4.50×10^{-12} g O $\times \dfrac{\text{mol O}}{16.00 \text{ g O}} = 2.81 \times 10^{-13}$ mol O

---

19. Given: mass of argon-36 = 35.97 amu, abundance = 0.337 %
mass of argon-38 = 37.96 amu, abundance = 0.063 %
mass of argon-40 = 39.96 amu, abundance = 99.600 %

Unknown: average atomic mass of Ar to two decimal places

average atomic mass Ar = (mass argon-36)(abundance argon-36)
+ (mass argon-38)(abundance argon-38)
+ (mass argon-40)(abundance argon-40)

average atomic mass Ar = (35.97 amu)(0.00337) + (37.96 amu)(0.00063)
+ (39.96 amu)(0.99600)

average atomic mass Ar = 0.121 amu + 0.024 amu + 39.80 amu = 39.95 amu

---

20. Given: mass of boron-11 = 11.01 amu, abundance = 80.20 %
abundance of unknown isotope = 19.80 %
average atomic mass B = 10.81 amu

Unknown: atomic mass of unknown isotope

average atomic mass B = (mass boron-11)(abundance boron-11) + (mass unknown isotope)(abundance unknown isotope)

mass unknown isotope
$= \dfrac{\text{average atomic mass B} - (\text{mass boron-11})(\text{abundance boron-11})}{\text{abundance unknown isotope}}$

mass unknown isotope $= \dfrac{10.81 \text{ amu} - (11.01 \text{ amu})(0.8020)}{0.1980} = \dfrac{1.98 \text{ amu}}{0.1980}$

mass unknown isotope of boron = 10.00 amu

---

21. a. Given: 1.50 mol Na
Unknown: number of Na atoms

1.50 mol Na $\times \dfrac{6.022 \times 10^{23} \text{ Na atoms}}{\text{mol Na}} = 9.03 \times 10^{23}$ Na atoms

b. Given: 6.755 mol Pb
Unknown: number of Pb atoms

6.755 mol Pb $\times \dfrac{6.022 \times 10^{23} \text{ Pb atoms}}{\text{mol Pb}} = 4.068 \times 10^{24}$ Pb atoms

c. Given: 7.02 g Si
Unknown: number of Si atoms

7.02 g Si $\times \dfrac{\text{mol Si}}{28.09 \text{ g Si}} \times \dfrac{6.022 \times 10^{23} \text{ Si atoms}}{\text{mol Si}} = 1.50 \times 10^{23}$ Si atoms

22. a. Given: 3.011×10^{23} atoms F
Unknown: mass of F in grams

3.011×10^{23} F atoms $\times \dfrac{\text{mol F}}{6.022 \times 10^{23} \text{ F atoms}} \times \dfrac{19.00 \text{ g F}}{\text{mol F}} = 9.500$ g F

b. Given: 1.50×10^{23} atoms Mg
Unknown: mass of Mg in grams

1.50×10^{23} Mg atoms $\times \dfrac{\text{mol Mg}}{6.022 \times 10^{23} \text{ Mg atoms}} \times \dfrac{24.30 \text{ g Mg}}{\text{mol Mg}} = 6.05$ g Mg

c. Given: 4.50×10^{12} atoms Cl
Unknown: mass of Cl in grams

4.50×10^{12} Cl atoms $\times \dfrac{\text{mol Cl}}{6.022 \times 10^{23} \text{ Cl atoms}} \times \dfrac{35.45 \text{ g Cl}}{\text{mol Cl}}$

$= 2.65 \times 10^{-10}$ Cl

d. Given: 8.42×10^{18} atoms Br
Unknown: mass of Br in grams

8.42×10^{18} Br atoms $\times \dfrac{\text{mol Br}}{6.022 \times 10^{23} \text{ Br atoms}} \times \dfrac{79.90 \text{ g Br}}{\text{mol Br}}$

$= 1.12 \times 10^{-3}$ g Br

e. Given: 25 atoms W
Unknown: mass of W in grams

25 W atoms $\times \dfrac{\text{mol W}}{6.022 \times 10^{23} \text{ W atoms}} \times \dfrac{183.84 \text{ g W}}{\text{mol W}} = 7.6 \times 10^{-21}$ W

f. Given: 1 atom Au
Unknown: mass of Au in grams

1 Au atom $\times \dfrac{\text{mol Au}}{6.022 \times 10^{23} \text{ Au atoms}} \times \dfrac{196.97 \text{ g Au}}{\text{mol Au}} = 3 \times 10^{-22}$ g Au

23. a. Given: 5.40 g B
Unknown: number of B atoms

5.40 g B $\times \dfrac{\text{mol B}}{10.81 \text{ g B}} \times \dfrac{6.022 \times 10^{23} \text{ B atoms}}{\text{mol B}} = 3.01 \times 10^{23}$ B atoms

b. Given: 0.250 mol S
Unknown: number of S atoms

0.250 mol S $\times \dfrac{6.022 \times 10^{23} \text{ S atoms}}{\text{mol S}} = 1.51 \times 10^{23}$ S atoms

c. Given: 0.0384 mol K
Unknown: number of K atoms

0.0384 mol K $\times \dfrac{6.022 \times 10^{23} \text{ K atoms}}{\text{mol K}} = 2.31 \times 10^{22}$ K atoms

d. Given: 0.025 50 g Pt
Unknown: number of Pt atoms

0.025 50 g Pt $\times \dfrac{\text{mol Pt}}{195.08 \text{ g Pt}} \times \dfrac{6.022 \times 10^{23} \text{ Pt atoms}}{\text{mol Pt}}$

$= 7.871 \times 10^{19}$ Pt atoms

e. Given: 1.00×10^{-10} g Au
Unknown: number of Au atoms

1.00×10^{-10} g Au $\times \dfrac{\text{mol Au}}{196.97 \text{ g Au}} \times \dfrac{6.022 \times 10^{23} \text{ Au atoms}}{\text{mol Au}}$

$= 3.06 \times 10^{11}$ Au atoms

Mixed Review

24. a. Given: 3.00 mol Al
Unknown: mass of Al in grams

3.00 mol Al $\times \dfrac{26.98 \text{ g Al}}{\text{mol Al}} = 80.9$ g Al

b. Given: 2.56×10^{24} atoms Li
Unknown: mass of Li in grams

2.56×10^{24} Li atoms $\times \dfrac{\text{mol Li}}{6.022 \times 10^{23} \text{ Li atoms}} \times \dfrac{6.94 \text{ g Li}}{\text{mol Li}} = 29.5$ g Li

c. Given: 1.38 mol N
Unknown: mass of N in grams

1.38 mol N $\times \dfrac{14.01 \text{ g N}}{\text{mol N}} = 19.3$ g N

d. Given: 4.86×10^{24} atoms Au
Unknown: mass of Au in grams

4.86×10^{24} Au atoms $\times \dfrac{\text{mol Au}}{6.022 \times 10^{23} \text{ Au atoms}} \times \dfrac{196.97 \text{ g Au}}{\text{mol Au}} = 1590$ g Au

e. Given: 6.50 mol Cu
Unknown: mass of Cu in grams

6.50 mol Cu $\times \dfrac{63.55 \text{ g Cu}}{\text{mol Cu}} = 413$ g Cu

f. Given: 2.57×10^{8} mol S
Unknown: mass of S in grams

2.57×10^{8} mol S $\times \dfrac{32.07 \text{ g S}}{\text{mol S}} = 8.24 \times 10^{9}$ g S

g. Given: 1.05×10^{18} atoms Hg
Unknown: mass of Hg in grams

1.05×10^{18} Hg atoms $\times \dfrac{\text{mol Hg}}{6.022 \times 10^{23} \text{ Hg atoms}} \times \dfrac{200.59 \text{ g Hg}}{\text{mol Hg}}$
$= 3.50 \times 10^{-4}$ g Hg

---

28. a. Given: 40.1 g Ca
Unknown: amount of Ca in moles

40.1 g Ca $\times \dfrac{\text{mol Ca}}{40.08 \text{ g Ca}} = 1.00$ mol Ca

b. Given: 11.5 g Na
Unknown: amount of Na in moles

11.5 g Na $\times \dfrac{\text{mol Na}}{22.99 \text{ g Na}} = 0.500$ mol Na

c. Given: 5.87 g Ni
Unknown: amount of Ni in moles

5.87 g Ni $\times \dfrac{\text{mol Ni}}{58.69 \text{ g Ni}} = 0.100$ mol Ni

d. Given: 150 g S
Unknown: amount of S in moles

150 g S $\times \dfrac{\text{mol S}}{32.07 \text{ g S}} = 4.7$ mol S

e. Given: 2.65 g Fe
Unknown: amount of Fe in moles

2.65 g Fe $\times \dfrac{\text{mol Fe}}{55.85 \text{ g Fe}} = 0.0474$ mol Fe

f. Given: 0.007 50 g Ag
Unknown: amount of Ag in moles

0.007 50 g Ag $\times \dfrac{\text{mol Ag}}{107.87 \text{ g Ag}} = 6.95 \times 10^{-5}$ mol Ag

g. Given: 2.25×10^{25} atoms Zn
Unknown: amount of Zn in moles

2.25×10^{25} Zn atoms $\times \dfrac{\text{mol Zn}}{6.022 \times 10^{23} \text{ Zn atoms}} = 37.4$ mol Zn

h. Given: 50.0 atoms Ba
Unknown: amount of Ba in moles

$$50.0 \text{ Ba atoms} \times \frac{\text{mol Ba}}{6.022 \times 10^{23} \text{ Ba atoms}} = 8.30 \times 10^{-23} \text{ mol Ba}$$

30. a. Given: atom with mass 12 times that of carbon-12
Unknown: approximate atomic mass of given atom

atomic mass of carbon-12 = 12.0 amu

unknown mass = (12)(12.0 amu) = 144 amu

b. Given: atom with mass $\frac{1}{2}$ that of carbon-12
Unknown: approximate atomic mass of given atom

atomic mass of carbon-12 = 12.0 amu

unknown mass = $\frac{1}{2}$(12.0 amu) = 6.0 amu

Critical Thinking

34. Given: 92 protons, 143 neutrons, 92 electrons, a $^{235}_{92}$U atom
Unknown: mass converted into energy by formation of atom

mass of uranium-235 = 235.0 amu
mass of proton = 1.0073 amu
mass of neutron = 1.0087 amu

Electrons are not involved in nuclear binding energy.

mass converted = (92)(mass of proton) + (143)(mass of neutron) − mass of uranium-235

mass converted = (92)(1.0073 amu) + (143)(1.0087 amu) − 235.0 amu

mass converted = 236.9 amu − 235.0 amu = 1.9 amu

CHAPTER 4

Arrangement of Electrons in Atoms

Review Problems, pp. 118–121

31. Given: $\lambda = 4.257 \times 10^{-7}$ cm
Unknown: ν

$c = \lambda \nu$

$\nu = \dfrac{c}{\lambda} = \dfrac{3.0 \times 10^8 \text{ m/s}}{(4.257 \times 10^{-7} \text{ cm})(10^{-2} \text{ m/cm})}$

$\nu = 7.0 \times 10^{16}$ Hz

32. Given: $\nu = 3.55 \times 10^{17}$ Hz
Unknown: E_{photon}

$E_{photon} = h\nu = (6.626 \times 10^{-34} \text{ J} \cdot \text{s})(3.55 \times 10^{17} \text{ Hz})$

$E_{photon} = 2.35 \times 10^{-15}$ J

33. Given: $E = h\nu$
$c = \lambda \nu$
Unknown: equation expressing E in terms of h, c, and λ

$c = \lambda \nu$

$\nu = \dfrac{c}{\lambda}$

$E = h\nu = \dfrac{hc}{\lambda}$

$E = \dfrac{hc}{\lambda}$

34. Given: distance = 8.00×10^7 km
Unknown: time that it takes radio wave to travel given distance

$\text{speed} = c = \dfrac{\text{distance}}{\text{time}}$

$\text{time} = \dfrac{\text{distance}}{c} = \dfrac{(8.00 \times 10^7 \text{ km})(10^3 \text{ m/km})}{3.0 \times 10^8 \text{ m/s}}$

$\text{time} = 2.7 \times 10^2$ s

35. Given: $\lambda = 1.00 \times 10^{-3}$ nm
Unknown: E_{photon}

$E_{photon} = h\nu = \dfrac{hc}{\lambda} = \dfrac{(6.626 \times 10^{-34} \text{ J} \cdot \text{s})(3.0 \times 10^8 \text{ m/s})}{(1.00 \times 10^{-3} \text{ nm})(10^{-9} \text{ m/nm})}$

$E_{photon} = 2.0 \times 10^{-13}$ J

41. Given: $c = 3.0 \times 10^8$ m/s
$\nu = 7.500 \times 10^{12}$ Hz
Unknown: λ

$c = \lambda \nu$

$\lambda = \dfrac{c}{\nu} = \dfrac{3.0 \times 10^8 \text{ m/s}}{7.500 \times 10^{12} \text{ Hz}}$

$\lambda = 4.0 \times 10^{-5}$ m

44. Given: $E_{photon} = 1.55 \times 10^{-24}$ J
Unknown: ν

$E_{photon} = h\nu$

$\nu = \dfrac{E_{photon}}{h} = \dfrac{1.55 \times 10^{-24} \text{ J}}{6.626 \times 10^{-34} \text{ J} \cdot \text{s}}$

$\nu = 2.34 \times 10^9$ Hz

47. a. Given: $E_{photon} = 3.37 \times 10^{-19}$ J
Unknown: ν

$E_{photon} = h\nu$

$\nu = \dfrac{E_{photon}}{h} = \dfrac{3.37 \times 10^{-19} \text{ J}}{6.626 \times 10^{-34} \text{ J} \cdot \text{s}}$

$\nu = 5.09 \times 10^{14}$ Hz

b. Given: $\nu = 5.09 \times 10^{14}$ Hz
Unknown: λ

$c = \lambda \nu$

$\lambda = \dfrac{c}{\nu} = \dfrac{3.0 \times 10^{8} \text{ m/s}}{5.09 \times 10^{14} \text{ Hz}}$

$\lambda = 5.9 \times 10^{-7}$ m

---

50. Given: $\lambda_{max} = 284$ nm
Unknown: work function of chromium

work function $= E_{photon,\,min} = h\nu_{min}$

$c = \lambda \nu = \lambda_{max}\, \nu_{min}$

$\nu_{min} = \dfrac{c}{\lambda_{max}}$

work function $= h\nu_{min} = \dfrac{hc}{\lambda_{max}} = \dfrac{(6.626 \times 10^{-34} \text{ J} \cdot \text{s})(3.0 \times 10^{8} \text{ m/s})}{(284 \text{ nm})(10^{-9} \text{ m/nm})}$

CHAPTER 6
Chemical Bonding

Practice, p. 163

Given: bonding between Cl and Ca, Cl and O, and Cl and Br

Unknown: classification of each given bond and indication of the more-negative atom in each pair

Using Figure 5-20, the electronegativities of the elements are:

Cl → 3.0
Ca → 1.0
O → 3.5
Br → 2.8

In each bonded pair, the element with the greater electronegativity is the more-negative atom.

For each bond, the electronegativity difference is:

Cl and Ca	3.0 − 1.0 = 2.0
O and Cl	3.5 − 3.0 = 0.5
Cl and Br	3.0 − 2.8 = 0.2

Bond type can be determined from Figure 6-2.

Bonded atoms	Bond type	More-negative atom
Cl and Ca	ionic	Cl (3.0 > 1.0)
O and Cl	polar-covalent	O (3.5 > 3.0)
Cl and Br	nonpolar-covalent	Cl (3.0 > 2.8)

ATE, Additional Sample Problem 6–1, p. 199A

Given: table of bonded elements and electronegativity differences

Unknown: bond types and more electronegative of the given atoms in each pair

Elements bonded	Electronegativity difference
a. C and H	0.4
b. C and S	0.0
c. O and H	1.4
d. Na and Cl	2.1
e. Cs and S	1.8

Bond type can be determined from Figure 6-2.

Elements from Groups 13 through 17 have higher electronegativities than elements from Groups 1 and 2.

Elements bonded	Electronegativity difference	Bond type	More-negative atom
a. C and H	0.4	polar-covalent	C
b. C and S	0.0	nonpolar-covalent	same electronegativity
c. O and H	1.4	polar-covalent	O
d. Na and Cl	2.1	ionic	Cl
e. Cs and S	1.8	ionic	S

Section Review, p. 163

3a. Given: H and F
Unknown: type of bond between given atoms

According to Figure 5-20, H and F have electronegativities of 2.1 and 4.0, respectively. The difference in the electronegativities is

$$4.0 - 2.1 = 1.9,$$

which, according to Figure 6-2, corresponds to an ionic bond.

b. Given: Cu and S
Unknown: type of bond between given atoms

According to Figure 5-20, Cu and S have electronegativities of 1.9 and 2.5, respectively. The difference in the electronegativities is

$$2.5 - 1.9 = 0.6,$$

which, according to Figure 6-2, corresponds to a polar-covalent bond.

c. Given: I and Br
Unknown: type of bond between given atoms

According to Figure 5-20, I and Br have electronegativities of 2.5 and 2.8, respectively. The difference in the electronegativities is

$$2.8 - 2.5 = 0.3,$$

which, according to Figure 6-2, corresponds to a polar-covalent bond.

ATE, Additional Sample Problem 6-2, p. 170

a. Given: phosphorus
Unknown: electron-dot notation for given element

According to its electron configuration ($[Ne]3s^23p^3$), P has 5 valence electrons (2 + 3). The electron-dot notation for P is therefore

·P:

b. Given: silicon
Unknown: electron-dot notation for given element

According to its electron configuration ($[Ne]3s^23p^2$), Si has 4 valence electrons (2 + 2). The electron-dot notation for Si is therefore

·Si·

c. Given: sulfur
Unknown: electron-dot notation for given element

According to its electron configuration ($[Ne]3s^23p^4$), S has 6 valence electrons (2 + 4). The electron-dot notation for S is therefore

:S:

d. Given: chlorine
Unknown: electron-dot notation for given element

According to its electron configuration ($[Ne]3s^23p^5$), Cl has 7 valence electrons (2 + 5). The electron-dot notation for Cl is therefore

:Cl:

e. Given: xenon
Unknown: electron-dot notation for given element

According to its electron configuration ($[Kr]4d^{10}5s^25p^6$), Xe has 8 valence electrons (2 + 6). The electron-dot notation for Xe is therefore

:Xe:

ATE, Additional Sample Problem 6-3, p. 199A

a. Given: H_2O
Unknown: the Lewis structure for the given molecule

O has 6 valence electrons. H has one valence electron.

:Ö: H·

Total number of valence electrons:

O $1 \times 6e^- = 6e^-$
H $2 \times 1e^- = 2e^-$
                    $\overline{8e^-}$

H:Ö:H

b. Given: CH_4
Unknown: the Lewis structure for the given molecule

C has 4 valence electrons. H has one valence electron.

·Ċ·   H·

Total number of valence electrons:

C $1 \times 4e^- = 4e^-$
H $4 \times 1e^- = 4e^-$
                    $\overline{8e^-}$

```
    H
  ..
H:C:H
  ..
    H
```

c. Given: CH_4O
Unknown: the Lewis structure for the given molecule

C has 4 valence electrons. H has one valence electron. O has 6 valence electrons.

·Ċ·   H·   :Ö:

Total number of valence electrons:

C $1 \times 4e^- = 4e^-$
H $4 \times 1e^- = 4e^-$
O $1 \times 6e^- = 6e^-$
                    $\overline{14e^-}$

```
    H
    ..
H:C:O:H
    ..
    H
```

d. Given: HCl
Unknown: the Lewis structure for the given molecule

Cl has 7 valence electrons. H has one valence electron.

:C̈l: H·

Total number of valence electrons:

Cl $1 \times 7e^- = 7e^-$
H $1 \times 1e^- = 1e^-$
                       $8e^-$

H:C̈l:

Practice, p. 172

1. Given: NH_3
Unknown: the Lewis structure for the given molecule

N has 5 valence electrons. H has one valence electron.

·N̈·   H·

Total number of valence electrons:

N $1 \times 5e^- = 5e^-$
H $3 \times 1e^- = 3e^-$
                     $8e^-$

H:N̈:H
  H

2. Given: H_2S
Unknown: the Lewis structure for the given molecule

S has 6 valence electrons. H has one valence electron.

·S̈·   H·

Total number of valence electrons:

S $1 \times 6e^- = 6e^-$
H $2 \times 1e^- = 2e^-$
                     $8e^-$

H:S̈:H

Practice, p. 174

1. Given: CO_2
Unknown: the Lewis structure for the given molecule

C has 4 valence electrons. O has 6 valence electrons.

·Ċ·   :Ö:

Total number of valence electrons:

C $1 \times 4e^- = 4e^-$
O $2 \times 6e^- = 12e^-$
                     $16e^-$

Ö::C::Ö,   or   Ö=C=Ö

This configuration gives each atom the correct number of valence electrons (C = 4, O = 6).

2. Given: hydrogen cyanide (HCN)

Unknown: the Lewis structure for the given molecule

C has 4 valence electrons. N has 5 valence electrons. H has one valence electron.

·C̈·   ·N̈:   H·

Total number of valence electrons:

C $1 \times 4e^- = 4e^-$
N $1 \times 5e^- = 5e^-$
H $1 \times 1e^- = 1e^-$
                    $\overline{10e^-}$

H:C⋮⋮N:,   or   H—C≡N:

This configuration gives each atom the correct number of valence electrons (H = 1, C = 4, N = 5).

ATE, Additional Sample Problem 6-4, p. 199A

a. Given: O_2

Unknown: the Lewis structure for the given molecule

O has 6 valence electrons.

:Ö·

Total number of valence electrons:

O $2 \times 6e^- = 12e^-$

Ö::Ö,   or   Ö=Ö

This configuration gives each O atom the correct number of valence electrons, 6.

b. Given: C_2H_4

Unknown: the Lewis structure for the given molecule

C has 4 valence electrons. H has one valence electron.

·C̈·   H·

Total number of valence electrons:

C $2 \times 4e^- = 8e^-$
H $4 \times 1e^- = 4e^-$
                    $\overline{12e^-}$

```
H H           H  H
 ··            |  |
C::C,   or    C==C
 ··            |  |
H H           H  H
```

This configuration gives each atom the correct number of valence electrons (C = 4, H = 1).

c. Given: C_2H_2

Unknown: the Lewis structure for the given molecule

C has 4 valence electrons. H has one valence electron.

·C̈·   H·

Total number of valence electrons:

C $2 \times 4e^- = 8e^-$
H $2 \times 1e^- = 2e^-$
                    $\overline{10e^-}$

H:C⋮⋮C:H,   or   H—C≡C—H

This configuration gives each atom the correct number of valence electrons (C = 4, H = 1).

Section Review, p. 175

4a. Given: IBr
Unknown: the Lewis structure for the given molecule

I has 7 valence electrons. Br has 7 valence electrons.

$:\overset{..}{\underset{..}{I}}\cdot \quad :\overset{..}{\underset{.}{Br}}\cdot$

Total number of valence electrons:

I $\quad 1 \times 7e^- = 7e^-$
Br $\quad 1 \times 7e^- = 7e^-$
$\quad\quad\quad\quad\quad\quad\overline{14e^-}$

$:\overset{..}{\underset{..}{I}}:\overset{..}{\underset{..}{Br}}:$

b. Given: CH_3Br
Unknown: the Lewis structure for the given molecule

C has 4 valence electrons. H has one valence electron. Br has 7 valence electrons.

$\cdot\overset{.}{\underset{.}{C}}\cdot \quad H\cdot \quad :\overset{..}{\underset{.}{Br}}:$

Total number of valence electrons:

C $\quad 1 \times 4e^- = 4e^-$
H $\quad 3 \times 1e^- = 3e^-$
Br $\quad 1 \times 7e^- = 7e^-$
$\quad\quad\quad\quad\quad\quad\overline{14e^-}$

$\quad\quad H$
$H:\overset{..}{\underset{..}{C}}:\overset{..}{\underset{..}{Br}}:$
$\quad\quad H$

c. Given: C_2HCl
Unknown: the Lewis structure for the given molecule

C has 4 valence electrons. H has one valence electron. Cl has 7 valence electrons.

$\cdot\overset{.}{\underset{.}{C}}\cdot \quad H\cdot \quad :\overset{..}{\underset{..}{Cl}}:$

Total number of valence electrons:

C $\quad 2 \times 4e^- = 8e^-$
H $\quad 1 \times 1e^- = 1e^-$
Cl $\quad 1 \times 7e^- = 7e^-$
$\quad\quad\quad\quad\quad\quad\overline{16e^-}$

$H:C::C:\overset{..}{\underset{..}{Cl}}:,\quad$ or $\quad H-C\equiv C-Cl$

This configuration gives each atom the correct number of valence electrons (C = 4, H = 1, Cl = 7).

d. Given: $SiCl_4$
Unknown: the Lewis structure for the given molecule

Si has 4 valence electrons. Cl has 7 valence electrons.

·Ṡi·   :C̈l·

Total number of valence electrons:

Si $1 \times 4e^- = 4e^-$
Cl $4 \times 7e^- = \underline{28e^-}$
                         $32e^-$

$$\begin{array}{c} :\ddot{C}l: \\ :\ddot{C}l:Si:\ddot{C}l: \\ :\ddot{C}l: \end{array}$$

e. Given: F_2O
Unknown: the Lewis structure for the given molecule

F has 7 valence electrons. O has 6 valence electrons.

:F̈·   ·Ö·

Total number of valence electrons:

F $2 \times 7e^- = 14e^-$
O $1 \times 6e^- = \underline{6e^-}$
                         $20e^-$

:F̈:Ö:F̈:

Section Review, p. 180

2a. Given: Li and Cl
Unknown: ionic compounds formed, using electron-dot notation

Li· gives up an electron to form the Li^+ cation. :C̈l· acquires an electron to form the :C̈l:$^-$ anion.

Li· + :C̈l· → Li^+ + :C̈l:$^-$ → LiCl

b. Given: Ca and I
Unknown: ionic compounds formed, using electron-dot notation

·Ca· gives up two electrons to form the Ca^{2+} cation. :Ï· acquires an electron to form the :Ï:$^-$ anion.

Two iodine anions combine with one calcium cation to form an electrically neutral compound.

·Ca· + :Ï· + :Ï· → Ca^{2+} + :Ï:$^-$ + :Ï:$^-$ → CaI_2

Practice, p. 185

1a. Given: HI
Unknown: geometry of the given molecule, using VSEPR theory

The molecule has only two atoms, and therefore must be linear.

b. Given: CBr_4
Unknown: geometry of the given molecule, using VSEPR theory

The Lewis structure for CBr_4 is

$$\ddot{:}\ddot{Br}\ddot{:}$$
$$:\!\ddot{Br}\!:\!C\!:\!\ddot{Br}\!:$$
$$:\!\ddot{Br}\!:$$

The octet of electrons around C indicates that the molecule is of the AB_4 type. Its geometry is therefore tetrahedral.

c. Given: $AlBr_3$
Unknown: geometry of the given molecule, using VSEPR theory

The Lewis structure for $AlBr_3$ is

$$:\!\ddot{Br}\!:\!Al\!:\!\ddot{Br}\!:$$
$$:\!\ddot{Br}\!:$$

The molecule is an exception to the octet rule because Al forms only **three** bonds. Therefore, the molecule is of the AB_3 type and its geometry is trigonal-planar.

d. Given: CH_2Cl_2
Unknown: geometry of the given molecule, using VSEPR theory

The Lewis structure for CH_2Cl_2 is

$$H$$
$$:\!\ddot{Cl}\!:\!C\!:\!\ddot{Cl}\!:$$
$$H$$

The octet of electrons around C indicates that the molecule is of the AB_4 type. Its geometry is therefore tetrahedral.

ATE, Additional Sample Problem 6-5, p. 185

a. Given: CCl_4
Unknown: geometry of the given molecule, using VSEPR theory

The Lewis structure for CCl_4 is

$$:\!\ddot{Cl}\!:$$
$$:\!\ddot{Cl}\!:\!C\!:\!\ddot{Cl}\!:$$
$$:\!\ddot{Cl}\!:$$

The octet of electrons around C indicates that the molecule is of the AB_4 type. Its geometry is therefore tetrahedral.

b. Given: HCN
Unknown: geometry of the given molecule, using VSEPR theory

The Lewis structure for HCN is

$$H\!:\!C\!::\!N\!:$$

Although there is an octet of electrons around C, six electrons are combined to form a triple bond. The remaining electron pair in the C–H bond is located opposite the triple bond, so that the molecule is of the AB_2 type. Its geometry is therefore linear.

c. Given: $SiBr_4$
Unknown: geometry of the given molecule, using VSEPR theory

The Lewis structure for $SiBr_4$ is

$$:\!\ddot{Br}\!:$$
$$:\!\ddot{Br}\!:\!Si\!:\!\ddot{Br}\!:$$
$$:\!\ddot{Br}\!:$$

The octet of electrons around Si indicates that the molecule is of the AB_4 type. Its geometry is therefore tetrahedral.

Practice, p. 187

1a. Given: :F̈—S̈—F̈:
Unknown: geometry of the given molecule, using VSEPR theory

The Lewis structure for F_2S indicates that there are two unshared electron pairs on the S atom, as well as two F atoms bonded to the S atom. This arrangement is of the AB_2E_2 type, whose shape is bent or angular.

b. Given: :C̈l—P—C̈l:
      |
     :C̈l:

Unknown: geometry of the given molecule, using VSEPR theory

The Lewis structure for PCl_3 indicates one unshared electron pair and three atoms bonded to the P atom. This arrangement is of the AB_3E type, which has a geometry that is trigonal-pyramidal.

ATE, Additional Sample Problem 6-6, p. 187

a. Given: AsF_5
Unknown: geometry of the given molecule, using VSEPR theory

The Lewis structure for AsF_5 is

The molecule with five single bonds to and no unshared electron pairs on the central atom is of the AB_5 type. Its geometry is trigonal-bipyramidal.

b. Given: SeF_6
Unknown: geometry of the given molecule, using VSEPR theory

The Lewis structure for SeF_6 is

The molecule with six single bonds to and no unshared electron pairs on the central atom is of the AB_6 type. Its geometry is octahedral.

c. Given: CF_4
Unknown: geometry of the given molecule, using VSEPR theory

The Lewis structure for CF_4 is

The molecule with four single bonds to and no unshared electron pairs on the central atom is of the AB_4 type. Its geometry is tetrahedral.

d. Given: NO_3^-
Unknown: geometry of the given ion, using VSEPR theory

The Lewis structure of NO_3^- is

The molecule or ion with three bonds to and no unshared electron pairs on the central atom is of the AB_3 type. Its geometry is trigonal-planar.

MODERN CHEMISTRY CHAPTER 6 SOLUTIONS

Section Review, p. 193

2a. Given: SO_2
Unknown: the Lewis structure and geometry of the given molecule

S and O have 6 valence electrons each. The total number of valence electrons in the molecule is

$6 + (2 \times 6) = 6 + 12 = 18$

:Ö:S::Ö, or :Ö—S=Ö

The molecule with two bonds to and one unshared electron pair on the central atom is of the AB_2E type. Its geometry is bent or angular.

b. Given: CI_4
Unknown: the Lewis structure and geometry of the given molecule

C has 4 valence electrons. I has 7 valence electrons. The total number of valence electrons in the molecule is

$4 + (4 \times 7) = 4 + 28 = 32$

The Lewis structure for CI_4 is

:Ï:
:Ï:C:Ï:
:Ï:

The molecule with four bonds to and no unshared electron pairs on the central atom is of the AB_4 type. Its geometry is tetrahedral.

c. Given: BCl_3
Unknown: the Lewis structure and geometry of the given molecule

B has 3 valence electrons. Cl has 7 valence electrons. The total number of valence electrons in the molecule is

$3 + (3 \times 7) = 3 + 21 = 24$

The Lewis structure for BCl_3 is

:C̈l:
:C̈l:B:C̈l:

The molecule with three bonds to and no unshared electron pairs on the central atom is of the AB_3 type. Its geometry is trigonal-planar.

Reviewing Concepts

11a. Given: H
Unknown: number of valence electrons in given atom

The electron configuration for H is $1s^1$. H therefore has 1 valence electron.

b. Given: F
Unknown: number of valence electrons in given atom

The electron configuration for F is $[He]2s^22p^5$. Counting the outer-shell superscripts, F has 7 valence electrons.

c. Given: Mg
Unknown: number of valence electrons in given atom

The electron configuration for Mg is $[Ne]3s^2$. Counting the outer-shell superscripts, Mg has 2 valence electrons.

d. Given: O
Unknown: number of valence electrons in given atom

The electron configuration for O is $[He]2s^22p^4$. Counting the outer-shell superscripts, O has 6 valence electrons.

e. Given: Al
Unknown: number of valence electrons in given atom

The electron configuration for Al is $[Ne]3s^23p^1$. Counting the outer-shell superscripts, Al has 3 valence electrons.

f. Given: N
Unknown: number of valence electrons in given atom

The electron configuration for N is $[He]2s^22p^3$. Counting the outer-shell superscripts, N has 5 valence electrons.

g. Given: C
Unknown: number of valence electrons in given atom

The electron configuration for C is $[He]2s^22p^2$. Counting the outer-shell superscripts, C has 4 valence electrons.

23b. Given: F_2 and HF
Unknown: geometry of the given molecules, using VSEPR theory

There are only two atoms in each molecule; therefore both molecules are linear.

Review Problems

33. For each pair of atoms, determine their electronegativities from Figure 5-20, and subtract the smaller from the larger to obtain the electronegativity difference. Refer to Figure 6-2 and use the electronegativity difference to determine bond type. The more negative atom is the atom with the higher electronegativity.

Example:

a. H and I

electronegativity difference = 2.5 − 2.1 = 0.4

A bond with an electronegativity difference of 0.4 is of the *polar-covalent* type.

I is the more negative atom since its electronegativity is 2.5, greater than that of H, 2.1.

34. Given: H and I, S and O, K and Br, Si and Cl, H and F, Se and S, C and H

Unknown: order of bonded pairs from least to most covalent

Using the electronegativity differences from item 33, noting that the lower the number the more covalent the bond is, the results are tabulated below.

Bonded atoms	Electronegativity difference
K and Br	2.0
H and F	1.9
Si and Cl	1.2
S and O	1.0
H and I, C and H	0.4
Se and S	0.1

37a. Given: Li
Unknown: electron-dot notation for given element

According to its electron configuration ([He]$2s^1$), Li has 1 valence electron. The electron-dot notation for Li is therefore

Li·

b. Given: Ca
Unknown: electron-dot notation for given element

According to its electron configuration ([Ar]$4s^2$), Ca has 2 valence electrons. The electron-dot notation for Ca is therefore

·Ca·

c. Given: Cl
Unknown: electron-dot notation for given element

According to its electron configuration ([Ne]$3s^23p^5$), Cl has 7 valence electrons (2 + 5). The electron-dot notation for Cl is therefore

:C̈l:

d. Given: O
Unknown: electron-dot notation for given element

According to its electron configuration ([He]$2s^22p^4$), O has 6 valence electrons (2 + 4). The electron-dot notation for O is therefore

:Ö:

e. Given: C
Unknown: electron-dot notation for given element

According to its electron configuration ([He]$2s^22p^2$), C has 4 valence electrons (2 + 2). The electron-dot notation for C is therefore

·C̈·

f. Given: P
Unknown: electron-dot notation for given element

According to its electron configuration ([Ne]$3s^23p^3$), P has 5 valence electrons (2 + 3). The electron-dot notation for P is therefore

·P̈:

g. Given: Al
Unknown: electron-dot notation for given element

According to its electron configuration ([Ne]$3s^23p^1$), Al has 3 valence electrons (2 + 1). The electron-dot notation for Al is therefore

·Äl·

h. Given: S
Unknown: electron-dot notation for given element

According to its electron configuration ([Ne]$3s^23p^4$), S has 6 valence electrons (2 + 4). The electron-dot notation for S is therefore

:S̈:

38a. Given: Na and S
Unknown: ionic compound formed, using electron-dot notation

Na· gives up an electron to form the Na$^+$ cation.

:S̈: acquires two electrons to form the :S̈:$^{2-}$ anion.

Na· + Na· + :S̈: → Na$^+$ + Na$^+$ + :S̈:$^{2-}$ → Na$_2$S

b. Given: Ca and O
Unknown: ionic compound formed, using electron-dot notation

$\cdot$Ca$\cdot$ gives up two electrons to form the Ca^{2+} cation.

:Ö: acquires two electrons to form the :Ö:$^{2-}$ anion.

$\cdot$Ca$\cdot$ + :Ö: → Ca^{2+} + :Ö:$^{2-}$ → CaO

c. Given: Al and S
Unknown: ionic compound formed, using electron-dot notation

$\cdot$Äl$\cdot$ gives up three electrons to form the Al^{3+} cation.

:S̈: acquires two electrons to form the :S̈:$^{2-}$ anion.

$\cdot$Äl$\cdot$ + $\cdot$Äl$\cdot$ + :S̈: + :S̈: + :S̈: → Al^{3+} + Al^{3+} + :S̈:$^{2-}$ + :S̈:$^{2-}$ + :S̈:$^{2-}$ → Al$_2$S$_3$

39a. Given: compound containing one C atom and four F atoms
Unknown: the Lewis structure for the given molecule

C has 4 valence electrons. F has 7 valence electrons.

$\cdot$C̈$\cdot$:F̈:

Total number of valence electrons:

C   1 × 4e$^-$ = 4e$^-$
F   4 × 7e$^-$ = 28e$^-$
                  ―――――
                  32e$^-$

```
      :F:
       ..
:F:C:F:
       ..
      :F:
```

b. Given: compound containing two H atoms and one Se atom
Unknown: the Lewis structure for the given molecule

Se has 6 valence electrons. H has one valence electron.

:S̈e: H$\cdot$

Total number of valence electrons:

Se   1 × 6e$^-$ = 6e$^-$
H   2 × 1e$^-$ = 2e$^-$
                  ―――――
                  8e$^-$

H:S̈e:H

c. Given: compound containing one N atom and three I atoms
Unknown: the Lewis structure for the given molecule

N has 5 valence electrons. I has 7 valence electrons.

:N$\cdot$:Ï:

Total number of valence electrons:

N   1 × 5e$^-$ = 5e$^-$
I   3 × 7e$^-$ = 21e$^-$
                  ―――――
                  26e$^-$

```
:Ï:N̈:Ï:
   :Ï:
```

d. Given: compound containing one Si atom and four Br atoms

Unknown: the Lewis structure for the given molecule

Si has 4 valence electrons. Br has 7 valence electrons.

$\cdot \overset{\cdot}{\underset{\cdot}{Si}} \cdot \qquad : \overset{\cdot \cdot}{\underset{\cdot \cdot}{Br}} :$

Total number of valence electrons:

Si $1 \times 4e^- = 4e^-$
Br $4 \times 7e^- = \underline{28e^-}$
                       $32e^-$

$\qquad : \overset{\cdot \cdot}{\underset{\cdot \cdot}{Br}} :$
$: \overset{\cdot \cdot}{\underset{\cdot \cdot}{Br}} : Si : \overset{\cdot \cdot}{\underset{\cdot \cdot}{Br}} :$
$\qquad : \overset{\cdot \cdot}{\underset{\cdot \cdot}{Br}} :$

e. Given: compound containing one C atom, one Cl atom, and three H atoms

Unknown: the Lewis structure for the given molecule

C has 4 valence electrons. Cl has 7 valence electrons. H has one valence electron.

$\cdot \overset{\cdot}{\underset{\cdot}{C}} \cdot \qquad : \overset{\cdot \cdot}{\underset{\cdot \cdot}{Cl}} : \qquad H \cdot$

Total number of valence electrons:

C $1 \times 4e^- = 4e^-$
Cl $1 \times 7e^- = 7e^-$
H $3 \times 1e^- = \underline{3e^-}$
                       $14e^-$

$\qquad H$
$H : \overset{\cdot \cdot}{\underset{\cdot \cdot}{C}} : \overset{\cdot \cdot}{\underset{\cdot \cdot}{Cl}} :$
$\qquad H$

---

40. Given: BF_3

Unknown: type of hybrid orbitals in given molecule

B has 3 valence electrons. F has 7 valence electrons. The total number of valence electrons in the molecule is

$3 + (3 \times 7) = 3 + 21 = 24$

The Lewis structure for BF_3 is

$: \overset{\cdot \cdot}{\underset{\cdot \cdot}{F}} : B : \overset{\cdot \cdot}{\underset{\cdot \cdot}{F}} :, \quad$ or $\quad \overset{: \overset{\cdot \cdot}{\underset{\cdot \cdot}{F}} \diagdown \quad \diagup \overset{\cdot \cdot}{\underset{\cdot \cdot}{F}} :}{\underset{: \overset{\cdot \cdot}{\underset{\cdot \cdot}{F}} :}{B}}$
$\quad : \overset{\cdot \cdot}{\underset{\cdot \cdot}{F}} :$

According to VSEPR theory, a molecule with three single bonds to and no unshared electron pairs on the central atom is of the AB_3 type, and therefore has trigonal-planar geometry. According to Table 6-6, this geometry is achieved with sp^2 hybridization.

BF_3 therefore has sp^2 hybrid orbitals.

41a. Given: O_2

Unknown: the Lewis structure for the given molecule

O has 6 valence electrons.

$: \overset{\cdot}{\underset{\cdot}{O}} :$

Total number of valence electrons:

O $2 \times 6e^- = 12e^-$

$\overset{\cdot \cdot}{\underset{\cdot \cdot}{O}} :: \overset{\cdot \cdot}{\underset{\cdot \cdot}{O}}, \quad$ or $\quad \overset{\cdot \cdot}{\underset{\cdot \cdot}{O}} = \overset{\cdot \cdot}{\underset{\cdot \cdot}{O}}$

This configuration gives each atom the correct number of valence electrons, 6.

b. Given: N_2
Unknown: the Lewis structure for the given molecule

N has 5 valence electrons.

·N̈:

Total number of valence electrons:

N $2 \times 5e^- = 10e^-$

:N::N:,   or   :N≡N:

This configuration gives each atom a complete octet.

c. Given: CO
Unknown: the Lewis structure for the given molecule

C has 4 valence electrons. O has 6 valence electrons.

·C̈·   ·Ö:

Total number of valence electrons:

C $1 \times 4e^- = 4e^-$
O $1 \times 6e^- = \underline{6e^-}$
                       $10e^-$

:C::O:,   or   :C≡O:

This configuration gives each atom a complete octet.

d. Given: SO_2
Unknown: the Lewis structure for the given molecule

S has 6 valence electrons. O has 6 valence electrons.

·S̈:   ·Ö:

Total number of valence electrons:

S $1 \times 6e^- = 6e^-$
O $2 \times 6e^- = \underline{12e^-}$
                       $18e^-$

A lone electron pair from either oxygen atom can be moved to the bond between that oxygen atom and the sulfur atom, giving the sulfur atom a complete octet. This forms a double bond. Because neither oxygen atom is favored for the double bond, the Lewis structure is represented as two resonance structures.

:Ö:S̈::Ö: ↔ :Ö::S̈:Ö:

or

:Ö—S̈=Ö: ↔ :Ö=S̈—Ö:

42a. Given: OH^-
Unknown: the Lewis structure for the given ion

O has 6 valence electrons. H has one valence electron.

·Ö·   H·

Total number of valence electrons:

O $1 \times 6e^- = 6e^-$
H $1 \times 1e^- = 1e^-$
ion's negative charge $1 \times 1e^- = \underline{1e^-}$
                                                $8e^-$

$\left[\text{:Ö:H}\right]^-$

MODERN CHEMISTRY CHAPTER 6 SOLUTIONS

b. Given: $H_3C_2O_2^-$
Unknown: the Lewis structure for the given ion

C has 4 valence electrons. H has one valence electron. O has 6 valence electrons.

·Ċ·   H·   ·Ö·

Total number of valence electrons:

C	$2 \times 4e^-$ =	$8e^-$
O	$2 \times 6e^-$ =	$12e^-$
H	$3 \times 1e^-$ =	$3e^-$
ion's negative charge	$1 \times 1e^-$ =	$1e^-$
		$24e^-$

A double bond is formed between the carbon atom and one of the oxygen atoms in order to give each atom a complete octet. Because neither oxygen atom is favored for the double bond, the Lewis structure is represented as two resonance structures.

$$\left[\begin{array}{c} H \\ H:\ddot{C}:\ddot{C}::\ddot{O} \\ H:\ddot{O}: \end{array} \right]^- \leftrightarrow \left[\begin{array}{c} H \\ H:\ddot{C}:\ddot{C}:\ddot{O}: \\ H:\ddot{O}: \end{array} \right]^-$$

c. Given: BrO_3^-
Unknown: the Lewis structure for the given ion

Br has 7 valence electrons. O has 6 valence electrons.

:B̈r: ·Ö·

Total number of valence electrons:

Br	$1 \times 7e^-$ =	$7e^-$
O	$3 \times 6e^-$ =	$18e^-$
ion's negative charge	$1 \times 1e^-$ =	$1e^-$
		$26e^-$

$$\left[\begin{array}{c} :\ddot{O}: \\ :\ddot{O}:\ddot{B}r:\ddot{O}: \end{array} \right]^-$$

46. Given: H–H, H–O, H–F, Br–Br, H–Cl, H–N
Unknown: whether bonds in the given molecules are polar or nonpolar

Diatomic molecules have polar bonds if the two atoms are different and nonpolar bonds if the two atoms are the same.

	Molecule	Bond polarity
a.	H–H	nonpolar
b.	H–O	polar
c.	H–F	polar
d.	Br–Br	nonpolar
e.	H–Cl	polar
f.	H–N	polar

47a. Given: H_2O
Unknown: polarity of given molecule

In the molecule there are eight valence electrons in the outer shell of oxygen. Four electrons are used in the covalent bonds with hydrogen, while four exist as unshared electron pairs.

According to Table 6-5, a molecule whose central atom has two bonds and two lone electron pairs is of the AB_2E_2 type, and has bent or angular geometry. The bent shape causes electric charge to be unevenly distributed, and so H_2O is polar.

b. Given: I_2
Unknown: polarity of given molecule

I_2 has two atoms, and therefore is linear. Because the atoms are identical, there is zero difference in electronegativity, so the molecule is nonpolar.

c. Given: CF_4
Unknown: polarity of given molecule

In the molecule there are eight valence electrons in the outer shell of carbon, all of which are used in bonds to the four fluorine atoms.

According to Table 6-5, a molecule with four bonds to the central atom and no unshared pairs is of the type AB_4 which has a tetrahedral geometry. Electric charge is evenly distributed over this molecule when all four bonded atoms are identical, so CF_4 is nonpolar.

d. Given: NH_3
Unknown: polarity of given molecule

In the molecule there are eight valence electrons in the outer shell of nitrogen. Six electrons are used in the covalent bonds with hydrogen, while two exist as an unshared electron pair.

According to Table 6-5, a molecule whose central atom has three bonds and one lone electron pair is of the AB_3E type, and has a trigonal-pyramidal geometry. Electric charge is unevenly distributed over this molecule, so NH_3 is polar.

e. Given: CO_2
Unknown: polarity of given molecule

In the molecule there are eight valence electrons in the outer shell of carbon. All electrons are used to form two double bonds.

According to Table 6-5, a molecule whose central atom has two bonds and no unshared electron pairs is of the AB_2 type, and has linear geometry. Electric charge is evenly distributed over this molecule when the two bonded atoms are identical, so CO_2 is nonpolar.

48a. Given: SCl_2
Unknown: the Lewis structure and molecular geometry of the given molecule

S has 6 valence electrons. Cl has 7 valence electrons. The total number of valence electrons in the molecule is

$6 + (2 \times 7) = 6 + 14 = 20$

The Lewis structure for SCl_2 is

:C̈l:S̈:C̈l:

The molecule with two bonds to and two unshared electron pairs on the central atom is of the AB_2E_2 type. Its geometry is bent or angular.

b. Given: PI_3
Unknown: the Lewis structure and molecular geometry of the given molecule

P has 5 valence electrons. I has 7 valence electrons. The total number of valence electrons in the molecule is

$5 + (3 \times 7) = 5 + 21 = 26$

The Lewis structure for PI_3 is

:Ï:P̈:Ï:
:Ï:

The molecule with three bonds to and one unshared electron pair on the central atom is of the AB_3E type. Its geometry is trigonal-pyramidal.

c. Given: Cl_2O
Unknown: the Lewis structure and molecular geometry of the given molecule

O has 6 valence electrons. Cl has 7 valence electrons. The total number of valence electrons in the molecule is

$6 + (2 \times 7) = 6 + 14 = 20$

The Lewis structure for Cl_2O is

:Cl̈:Ö:Cl̈:

The molecule with two bonds to and two unshared electron pairs on the central atom is of the AB_2E_2 type. Its geometry is bent or angular.

d. Given: NH_2Cl
Unknown: the Lewis structure and molecular geometry of the given molecule

N has 5 valence electrons. H has one valence electron. Cl has 7 valence electrons. The total number of valence electrons in the molecule is

$5 + (2 \times 1) + 7 = 5 + 2 + 7 = 14$

The Lewis structure for NH_2Cl is

:C̈l:
H:N̈:H

The molecule with three bonds to and one unshared electron pair on the central atom is of the AB_3E type. Its geometry is trigonal-pyramidal.

e. Given: $SiCl_3Br$
Unknown: the Lewis structure and molecular geometry of the given molecule

Si has 4 valence electrons. Cl and Br both have 7 valence electrons. The total number of valence electrons in the molecule is

$4 + (4 \times 7) = 4 + 28 = 32$

The Lewis structure for $SiCl_3Br$ is

:C̈l:
:C̈l:Si:C̈l:
:B̈r:

The molecule with four bonds to and no unshared electron pairs on the central atom is of the AB_4 type. Its geometry is tetrahedral.

f. Given: ONCl
Unknown: the Lewis structure and molecular geometry of the given molecule

O has 6 valence electrons. N has 5 valence electrons. Cl has 7 valence electrons. The total number of valence electrons in the molecule is

$6 + 5 + 7 = 18$

The Lewis structure for ONCl is

Ö::N:C̈l:

This configuration gives each atom a complete octet. The molecule with two bonds to and one unshared electron pair on the central atom is of the AB_2E type. Its geometry is bent or angular.

49a. Given: NO_3^-
Unknown: the Lewis structure and geometry of the given ion

N has 5 valence electrons. O has 6 valence electrons. The total number of valence electrons in the ion (plus one to account for the ion's negative charge) is

$5 + (3 \times 6) + 1 = 24$

The Lewis structure for NO_3^- is

$\left[\begin{array}{c} :\ddot{O}:N:\ddot{O}: \\ :\ddot{O}: \end{array} \right]^-$

This configuration gives each atom a complete octet. The molecule or ion with three bonds to and no unshared electron pairs on the central atom is of the AB_3 type. Its geometry is trigonal-planar.

b. Given: NH_4^+
Unknown: the Lewis structure and geometry of the given ion

N has 5 valence electrons. H has one valence electron. The total number of valence electrons in the ion (minus one to account for the ion's positive charge) is

$5 + (4 \times 1) - 1 = 8$

The Lewis structure for NH_4^+ is

$$\left[\begin{array}{c} H \\ H:\!N\!:\!H \\ H \end{array} \right]^+$$

The molecule or ion with four bonds to and no unshared electron pairs on the central atom is of the AB_4 type. Its geometry is tetrahedral.

c. Given: SO_4^{2-}
Unknown: the Lewis structure and geometry of the given ion

S and O both have 6 valence electrons. The total number of valence electrons in the ion (plus two to account for the ion's negative charge) is

$6 + (4 \times 6) + 2 = 32$

The Lewis structure for SO_4^{2-} is

$$\left[\begin{array}{c} \ddot{\text{:O:}} \\ \text{:O:S:O:} \\ \text{:O:} \end{array} \right]^{2-}$$

The molecule or ion with four bonds to and no unshared electron pairs on the central atom is of the AB_4 type. Its geometry is tetrahedral.

d. Given: ClO_2^-
Unknown: the Lewis structure and geometry of the given ion

Cl has 7 valence electrons. O has 6 valence electrons. The total number of valence electrons in the ion (plus one to account for the ion's negative charge) is

$7 + (2 \times 6) + 1 = 20$

The Lewis structure for ClO_2^- is

$$\left[:\!\ddot{\text{O}}\!:\!\ddot{\text{Cl}}\!:\!\ddot{\text{O}}\!: \right]^-$$

The molecule or ion with two bonds to and two unshared electron pairs on the central atom is of the AB_2E_2 type. Its geometry is bent or angular.

Mixed Review

50. Given: list of pairs of molecule types
Unknown: order of molecule pairs from strongest to weakest attraction

a. polar molecule and polar molecule
b. nonpolar molecule and nonpolar molecule
c. polar molecule and ion
d. ion and ion

The stronger the electric charge for each molecule, the stronger the attractive force on another molecule. Ions have the greatest electrical charge, followed by polar molecules. Nonpolar molecules exert the weakest forces. The order of the molecule pairs from strongest attraction to weakest is therefore

d., c., a., b.

51a. Given: CCl_4
Unknown: geometry of given molecule

There are four atoms bonded covalently to the central carbon atom. There are no lone electron pairs on the carbon atom. The molecule is therefore of the AB_4 type and its geometry is tetrahedral.

b. Given: $BeCl_2$
Unknown: geometry of given molecule

There are two atoms bonded covalently to the central beryllium atom, which does not obey the octet rule and has only a total of four valence electrons in its outer shell. There are no unshared electron pairs on the Be atom, so the molecule is of the AB_2 type. Its geometry is therefore linear.

c. Given: PH_3
Unknown: geometry of given molecule

There are three atoms bonded covalently to the central phosphorus atom. There is one lone electron pair on the phosphorus atom. The molecule is therefore of the AB_3E type and its geometry is trigonal-pyramidal.

54. Given: SO_3
Unknown: three resonance structures for given molecule

S has 6 valence electrons. O has 6 valence electrons. The total number of valence electrons in the molecule is

$6 + (3 \times 6) = 24$

The Lewis structure for SO_3 is

:Ö:S:Ö:
:Ö:

To provide an electron octet to the sulfur atom, the sulfur atom can form a double bond with any of the three oxygen atoms. The three possible resonance structures are therefore

[Three resonance structures of SO_3 shown with double bond alternating between the three oxygen positions] ↔ ↔

56a. Given: He
Unknown: electron-dot notation for given element

According to its electron configuration ($1s^2$), He has 2 valence electrons. The electron-dot notation for He is therefore

:He

b. Given: Cl
Unknown: electron-dot notation for given element

According to its electron configuration ($[Ne]3s^23p^5$), Cl has 7 valence electrons (2 + 5). The electron-dot notation for Cl is therefore

:Cl̈:

c. Given: O
Unknown: electron-dot notation for given element

According to its electron configuration ($[He]2s^22p^4$), O has 6 valence electrons (2 + 4). The electron-dot notation for O is therefore

:Ö:

d. Given: P
Unknown: electron-dot notation for given element

According to its electron configuration ($[Ne]3s^23p^3$), P has 5 valence electrons (2 + 3). The electron-dot notation for P is therefore

·P̈:

e. Given: B
Unknown: electron-dot notation for given element

According to its electron configuration ($[He]2s^22p^1$), B has 3 valence electrons. The electron-dot notation for B is therefore

·B·

57. Given: CH₃OH
Unknown: structural formula for the given molecule

C has 4 valence electrons and is the central molecule. H has one valence electron. O has 6 valence electrons. The total number of valence electrons is

4 + 4 + 6 = 14

The Lewis structure for CH₃OH is

```
      H ..
H:C:O:H,
      H
```

which has the correct number of valence electrons and an electron octet around the carbon and oxygen atoms. The structural formula contains single bonds throughout.

```
    H
    |
H—C—O—H
    |
    H
```

63a. Given: Zn and O
Unknown: electronegativity difference, probable bonding type, and more-electronegative atom of given pair of atoms

Using Figure 5-20, the electronegativities of the two given elements

Zn → 1.6
O → 3.5

The more-electronegative atom is O.

The electronegativity difference is 3.5 − 1.6 = 1.9

The bond in ZnO, according to Figure 6-2, is probably ionic.

b. Given: Br and I
Unknown: electronegativity difference, probable bonding type, and more-electronegative atom of given pair of atoms

Using Figure 5-20, the electronegativities of the two given elements are as follows:

Br → 2.8
I → 2.5

The more-electronegative atom is Br.

The electronegativity difference is 2.8 − 2.5 = 0.3

The bond in BrI, according to Figure 6-2, is probably nonpolar-covalent.

c. Given: S and Cl
Unknown: electronegativity difference, probable bonding type, and more-electronegative atom of given pair of atoms

Using Figure 5-20, the electronegativities of the two given elements are as follows:

S → 2.5
Cl → 3.0

The more-electronegative atom is Cl.

The electronegativity difference is 3.0 − 2.5 = 0.5

The bond between the S and Cl atoms, according to Figure 6-2, is probably polar-covalent.

64a. Given: PCl_3
Unknown: the Lewis structure for the given molecule

The electron-dot notations for P, which has 5 valence electrons, and Cl, which has 7 valence electrons, are

$\cdot \ddot{P} \cdot \qquad : \ddot{Cl} :$

The total number of valence electrons in the molecule's atoms is

$1 \times 5e^- = 5e^-$
$3 \times 7e^- = \underline{21e^-}$
$\qquad\qquad\quad 26e^-$

The resulting Lewis structure is:

:Cl:P:Cl:
　:Cl:

b. Given: CCl_2F_2
Unknown: the Lewis structure for the given molecule

The electron-dot notations for C, which has 4 valence electrons, Cl, which has 7 valence electrons, and F, which also has 7 valence electrons, are

$\cdot \dot{C} \cdot \qquad : \ddot{Cl} : \qquad : \ddot{F} :$

The total number of valence electrons in the molecule's atoms is

$1 \times 4e^- = 4e^-$
$2 \times 7e^- = 14e^-$
$2 \times 7e^- = \underline{14e^-}$
$\qquad\qquad\quad 32e^-$

The resulting Lewis structure is:

　:F:
:Cl:C:Cl:
　:F:

c. Given: CH_3NH_2
Unknown: the Lewis structure for the given molecule

The electron-dot notations for C, which has 4 valence electrons, N, which has 5 valence electrons, and H, which has 1 valence electron, are

$\cdot \dot{C} \cdot \qquad \cdot \ddot{N} \cdot \qquad H \cdot$

The total number of valence electrons in the molecule's atoms is

$1 \times 4e^- = 4e^-$
$1 \times 5e^- = 5e^-$
$1 \times 5e^- = \underline{5e^-}$
$\qquad\qquad\quad 14e^-$

The resulting Lewis structure is:

　H
H:C:N:H
　H H

65. Given: BeCl₂
Unknown: the Lewis structure for the given molecule

The electron-dot notations for Be, which has 2 valence electrons, and Cl, which has 7 valence electrons, are

·Be·   :C̈l·

The total number of valence electrons in the molecule's atoms is

$1 \times 2e^- = 2e^-$

$2 \times 7e^- = 14e^-$
_____
$16e^-$

:C̈l:Be:C̈l:

66a. Given: NO₂⁻
Unknown: the Lewis structure and geometry of the given ion

N has 5 valence electrons. O has 6 valence electrons. The total number of valence electrons in the ion (plus one to account for the ion's negative charge) is

$5 + (2 \times 6) + 1 = 18$

The Lewis structure for NO₂⁻ can be found by connecting the oxygen atoms to nitrogen and placing eight electrons around each atom.

:Ö::N̈::Ö:

This configuration gives each atom a complete octet. The molecule or ion with two bonds to and one unshared electron pair on the central atom is of the AB₂E type. Its geometry is bent or angular.

b. Given: NO₃⁻
Unknown: the Lewis structure and geometry of the given ion

N has 5 valence electrons. O has 6 valence electrons. The total number of valence electrons in the ion (plus one to account for the ion's negative charge) is

$5 + (3 \times 6) + 1 = 24$

$\left[\begin{array}{c} :\ddot{O}:N:\ddot{O}: \\ :\ddot{O}: \end{array} \right]^-$

This configuration gives each atom a complete octet. The molecule or ion with three bonds to and no unshared electron pair on the central atom is of the AB₃ type. Its geometry is trigonal-planar.

c. Given: NH₄⁺
Unknown: the Lewis structure and geometry of the given ion

N has 5 valence electrons. H has one valence electron. The total number of valence electrons in the ion (minus one to account for the ion's positive charge) is

$5 + (4 \times 1) - 1 = 8$

The Lewis structure for NH₄⁺ is

$\left[\begin{array}{c} H \\ H:N:H \\ H \end{array} \right]^+$

The molecule or ion with four bonds to and no unshared electron pairs on the central atom is of the AB₄ type. Its geometry is tetrahedral.

CHAPTER 7

Chemical Formulas and Chemical Compounds

Practice, Problem 7-6, p. 222

1. a. Given: H_2SO_4
Unknown: the formula mass of the given compound

$2 \text{ H atoms} \times \dfrac{1.01 \text{ amu}}{\text{H atom}} = 2.02 \text{ amu}$

$1 \text{ S atom} \times \dfrac{32.07 \text{ amu}}{\text{S atom}} = 32.07 \text{ amu}$

$4 \text{ O atoms} \times \dfrac{16.00 \text{ amu}}{\text{O atom}} = 64.00 \text{ amu}$

formula mass of H_2SO_4 = 98.09 amu

b. Given: $Ca(NO_3)_2$
Unknown: the formula mass of the given compound

$1 \text{ Ca atom} \times \dfrac{40.08 \text{ amu}}{\text{Ca atom}} = 40.08 \text{ amu}$

$2 \text{ N atoms} \times \dfrac{14.01 \text{ amu}}{\text{N atom}} = 28.02 \text{ amu}$

$6 \text{ O atoms} \times \dfrac{16.00 \text{ amu}}{\text{O atom}} = 96.00 \text{ amu}$

formula mass of $Ca(NO_3)_2$ = 164.10 amu

c. Given: PO_4^{3-}
Unknown: the formula mass of the given ion

$1 \text{ P atom} \times \dfrac{30.97 \text{ amu}}{\text{P atom}} = 30.97 \text{ amu}$

$4 \text{ O atoms} \times \dfrac{16.00 \text{ amu}}{\text{O atom}} = 64.00 \text{ amu}$

formula mass of PO_4^{3-} = 94.97 amu

d. Given: $MgCl_2$
Unknown: the formula mass of the given compound

$1 \text{ Mg atom} \times \dfrac{24.305 \text{ amu}}{\text{Mg atom}} = 24.305 \text{ amu}$

$2 \text{ Cl atoms} \times \dfrac{35.45 \text{ amu}}{\text{Cl atom}} = 70.90 \text{ amu}$

formula mass of $MgCl_2$ = 95.21 amu

ATE, Additional Sample Problem 7-6, p. 222

a. Given: Na_2SO_3
Unknown: the formula mass of the given compound

$2 \text{ Na atoms} \times \dfrac{22.99 \text{ amu}}{\text{Na atom}} = 45.98 \text{ amu}$

$1 \text{ S atom} \times \dfrac{32.07 \text{ amu}}{\text{S atom}} = 32.07 \text{ amu}$

$3 \text{ O atoms} \times \dfrac{16.00 \text{ amu}}{\text{O atom}} = 48.00 \text{ amu}$

formula mass of Na_2SO_3 = 126.05 amu

b. Given: $HClO_3$
Unknown: the formula mass of the given compound

$1 \text{ H atom} \times \dfrac{1.01 \text{ amu}}{\text{H atom}} = 1.01 \text{ amu}$

$1 \text{ Cl atom} \times \dfrac{35.45 \text{ amu}}{\text{Cl atom}} = 35.45 \text{ amu}$

$3 \text{ O atoms} \times \dfrac{16.00 \text{ amu}}{\text{O atom}} = 48.00 \text{ amu}$

formula mass of $HClO_3$ = 84.46 amu

c. Given: MnO_4^-
Unknown: the formula mass of the given ion

$1 \text{ Mn atom} \times \dfrac{54.94 \text{ amu}}{\text{Mn atom}} = 54.94 \text{ amu}$

$4 \text{ O atoms} \times \dfrac{16.00 \text{ amu}}{\text{O atom}} = 64.00 \text{ amu}$

formula mass of MnO_4^- = 118.94 amu

d. Given: C_2H_6O
Unknown: the formula mass of the given compound

$2 \text{ C atoms} \times \dfrac{12.01 \text{ amu}}{\text{C atom}} = 24.02 \text{ amu}$

$6 \text{ H atoms} \times \dfrac{1.01 \text{ amu}}{\text{H atom}} = 6.06 \text{ amu}$

$1 \text{ O atom} \times \dfrac{16.00 \text{ amu}}{\text{O atom}} = 16.00 \text{ amu}$

formula mass of C_2H_6O = 46.08 amu

Practice, p. 223

2. a. Given: Al_2S_3
Unknown: the molar mass of the given compound

$2 \text{ mol Al} \times \dfrac{26.98 \text{ g Al}}{\text{mol Al}} = 53.96 \text{ g Al}$

$3 \text{ mol S} \times \dfrac{32.07 \text{ g S}}{\text{mol S}} = 96.21 \text{ g S}$

molar mass of Al_2S_3 = 150.17 g/mol

b. Given: $NaNO_3$
Unknown: the molar mass of the given compound

$1 \text{ mol Na} \times \dfrac{22.99 \text{ g Na}}{\text{mol Na}} = 22.99 \text{ g Na}$

$1 \text{ mol N} \times \dfrac{14.01 \text{ g N}}{\text{mol N}} = 14.01 \text{ g N}$

$3 \text{ mol O} \times \dfrac{16.00 \text{ g O}}{\text{mol O}} = 48.00 \text{ g O}$

molar mass of $NaNO_3$ = 85.00 g/mol

c. Given: $Ba(OH)_2$
Unknown: the molar mass of the given compound

$1 \text{ mol Ba} \times \dfrac{137.33 \text{ g Ba}}{\text{mol Ba}} = 137.33 \text{ g Ba}$

$2 \text{ mol O} \times \dfrac{16.00 \text{ g O}}{\text{mol O}} = 32.00 \text{ g O}$

$2 \text{ mol H} \times \dfrac{1.01 \text{ g H}}{\text{mol H}} = 2.02 \text{ g H}$

molar mass of $Ba(OH)_2$ = 171.35 g/mol

ATE, Additional Sample Problems 7-7, p. 223

a. Given: K_2SO_4
Unknown: the molar mass of the given compound

$2 \text{ mol K} \times \dfrac{39.10 \text{ g K}}{\text{mol K}} = 78.20 \text{ g K}$

$1 \text{ mol S} \times \dfrac{32.07 \text{ g S}}{\text{mol S}} = 32.07 \text{ g S}$

$4 \text{ mol O} \times \dfrac{16.00 \text{ g O}}{\text{mol O}} = 64.00 \text{ g O}$

molar mass of K_2SO_4 = 174.27 g/mol

b. Given: $(NH_4)_2CrO_4$
Unknown: the molar mass of the given compound

$2 \text{ mol N} \times \dfrac{14.01 \text{ g N}}{\text{mol N}} = 28.02 \text{ g N}$

$8 \text{ mol H} \times \dfrac{1.01 \text{ g H}}{\text{mol H}} = 8.08 \text{ g H}$

$1 \text{ mol Cr} \times \dfrac{52.00 \text{ g Cr}}{\text{mol Cr}} = 52.00 \text{ g Cr}$

$4 \text{ mol O} \times \dfrac{16.00 \text{ g O}}{\text{mol O}} = 64.00 \text{ g O}$

molar mass of $(NH_4)_2CrO_4$ = 152.10 g/mol

ATE, Additional Sample Problems 7-8, p. 224

1. Given: 3.04 mol NH_3
Unknown: mass of NH_3

$1 \text{ mol N} \times \dfrac{14.01 \text{ g N}}{\text{mol N}} = 14.01 \text{ g N}$

$3 \text{ mol H} \times \dfrac{1.01 \text{ g H}}{\text{mol H}} = 3.03 \text{ g H}$

molar mass of NH_3 = 17.04 g/mol

mass $NH_3 = \dfrac{17.04 \text{ g } NH_3}{\text{mol } NH_3} \times 3.04 \text{ mol } NH_3 = 51.8 \text{ g}$

2. Given: 0.257 mol $Ca(NO_3)_2$
Unknown: mass of $Ca(NO_3)_2$

$1 \text{ mol Ca} \times \dfrac{40.08 \text{ g Ca}}{\text{mol Ca}} = 40.08 \text{ g Ca}$

$2 \text{ mol N} \times \dfrac{14.01 \text{ g N}}{\text{mol N}} = 28.02 \text{ g N}$

$6 \text{ mol O} \times \dfrac{16.00 \text{ g O}}{\text{mol O}} = 96.00 \text{ g O}$

molar mass of $Ca(NO_3)_2$ = 164.10 g/mol

mass $Ca(NO_3)_2 = \dfrac{164.10 \text{ g } Ca(NO_3)_2}{\text{mol } Ca(NO_3)_2} \times 0.257 \text{ mol } Ca(NO_3)_2 = 42.2 \text{ g}$

ATE, Additional Sample Problems 7-9, p. 225

1. a. Given: 3.82 g SO_2
Unknown: number of moles SO_2

$1 \text{ mol S} \times \dfrac{32.07 \text{ g S}}{\text{mol S}} = 32.07 \text{ g S}$

$2 \text{ mol O} \times \dfrac{16.00 \text{ g O}}{\text{mol O}} = 32.00 \text{ g O}$

molar mass of SO_2 = 64.07 g/mol

mol $SO_2 = \dfrac{3.82 \text{ g}}{64.07 \text{ g/mol}} = 0.0596 \text{ mol}$

b. Given: 4.15×10^{-3} g $C_6H_{12}O_6$
Unknown: number of moles $C_6H_{12}O_6$

$6 \text{ mol C} \times \dfrac{12.01 \text{ g C}}{\text{mol C}} = 72.06 \text{ g C}$

$12 \text{ mol H} \times \dfrac{1.01 \text{ g H}}{\text{mol H}} = 12.1 \text{ g H}$

$6 \text{ mol O} \times \dfrac{16.00 \text{ g O}}{\text{mol O}} = 96.00 \text{ g O}$

molar mass of $C_6H_{12}O_6 = 180.2$ g/mol

$\text{mol } C_6H_{12}O_6 = \dfrac{4.15 \times 10^{-3} \text{ g}}{180.2 \text{ g/mol}} = 2.30 \times 10^{-5} \text{ mol}$

c. Given: 77.1 g Cl_2
Unknown: number of moles Cl_2

$2 \text{ mol Cl} \times \dfrac{35.45 \text{ g Cl}}{\text{mol Cl}} = 70.90 \text{ g Cl}$

molar mass of $Cl_2 = 70.90$ g/mol

$\text{mol } Cl_2 = \dfrac{77.1 \text{ g}}{70.90 \text{ g/mol}} = 1.09 \text{ mol}$

---

2. a. Given: 5.96×10^{-2} mol SO_2
Unknown: n; number of molecules SO_2

$n = \dfrac{6.022 \times 10^{23} \text{ molecules}}{\text{mol}} \times (5.96 \times 10^{-2} \text{ mol } SO_2)$

$= 3.59 \times 10^{22}$ molecules

b. Given: 2.30×10^{-5} mol $C_6H_{12}O_6$
Unknown: n; number of molecules $C_6H_{12}O_6$

$n = \dfrac{6.022 \times 10^{23} \text{ molecules}}{\text{mol}} \times (2.30 \times 10^{-5} \text{ mol } C_6H_{12}O_6)$

$= 1.39 \times 10^{19}$ molecules

c. Given: 1.09 mol Cl_2
Unknown: n; number of molecules Cl_2

$n = \dfrac{6.022 \times 10^{23} \text{ molecules}}{\text{mol}} \times 1.09 \text{ mol } Cl_2 = 6.56 \times 10^{23} \text{ molecules}$

Practice, p. 226

1. a. Given: 6.60 g $(NH_4)_2SO_4$
Unknown: number of moles $(NH_4)_2SO_4$

$2 \text{ mol N} \times \dfrac{14.01 \text{ g N}}{\text{mol N}} = 28.02 \text{ g N}$

$8 \text{ mol H} \times \dfrac{1.01 \text{ g H}}{\text{mol H}} = 8.08 \text{ g H}$

$1 \text{ mol S} \times \dfrac{32.07 \text{ g S}}{\text{mol S}} = 32.07 \text{ g S}$

$4 \text{ mol O} \times \dfrac{16.00 \text{ g O}}{\text{mol O}} = 64.00 \text{ g O}$

molar mass of $(NH_4)_2SO_4 = 132.17$ g/mol

$\text{mol } (NH_4)_2SO_4 = \dfrac{6.60 \text{ g } (NH_4)_2SO_4}{132.17 \text{ g/mol } (NH_4)_2SO_4} = 4.99 \times 10^{-2} \text{ mol}$

b. Given: 4.5 kg $Ca(OH)_2$
Unknown: number of moles $Ca(OH)_2$

$1 \text{ mol Ca} \times \dfrac{40.08 \text{ g Ca}}{\text{mol Ca}} = 40.08 \text{ g Ca}$

$2 \text{ mol} \times \dfrac{16.00 \text{ g O}}{\text{mol O}} = 32.00 \text{ g O}$

$2 \text{ mol H} \times \dfrac{1.01 \text{ g H}}{\text{mol H}} = 2.02 \text{ g H}$

molar mass of $Ca(OH)_2$ = 74.10 g/mol

$\text{mol } Ca(OH)_2 = 4.5 \text{ kg } Ca(OH)_2 \times \dfrac{1000 \text{ g}}{\text{kg}} \times \dfrac{1 \text{ mol}}{74.10 \text{ g } Ca(OH)_2} = 61 \text{ mol}$

---

2. a. Given: 25.0 g H_2SO_4
Unknown: n; number of molecules H_2SO_4

$2 \text{ mol H} \times \dfrac{1.01 \text{ g H}}{\text{mol H}} = 2.02 \text{ g H}$

$1 \text{ mol S} \times \dfrac{32.07 \text{ g S}}{\text{mol S}} = 32.07 \text{ g S}$

$4 \text{ mol O} \times \dfrac{16.00 \text{ g O}}{\text{mol O}} = 64.00 \text{ g O}$

molar mass of H_2SO_4 = 98.09 g/mol

$n = 25.0 \text{ g } H_2SO_4 \times \dfrac{1 \text{ mol}}{98.07 \text{ g } H_2SO_4} \times \dfrac{6.022 \times 10^{23} \text{ molecules}}{\text{mol}}$

$= 1.53 \times 10^{23}$ molecules

b. Given: 125 g of $C_{12}H_{22}O_{11}$
Unknown: n; number of molecules $C_{12}H_{22}O_{11}$

$12 \text{ mol C} \times \dfrac{12.01 \text{ g C}}{\text{mol C}} = 144.1 \text{ g C}$

$22 \text{ mol H} \times \dfrac{1.01 \text{ g H}}{\text{mol H}} = 22.2 \text{ g H}$

$11 \text{ mol O} \times \dfrac{16.00 \text{ g O}}{\text{mol O}} = 176.0 \text{ g O}$

molar mass of $C_{12}H_{22}O_{11}$ = 342.3 g/mol

$n = 125 \text{ g } C_{12}H_{22}O_{11} \times \dfrac{1 \text{ mol}}{342.3 \text{ g } C_{12}H_{22}O_{11}} \times \dfrac{6.022 \times 10^{23} \text{ molecules}}{\text{mol}}$

$= 2.20 \times 10^{23}$ molecules

---

3. Given: 6.25 mol of $Cu(NO_3)_2$
Unknown: mass of $Cu(NO_3)_2$

$1 \text{ mol Cu} \times \dfrac{63.55 \text{ g Cu}}{\text{mol Cu}} = 63.55 \text{ g Cu}$

$2 \text{ mol N} \times \dfrac{14.01 \text{ g N}}{\text{mol N}} = 28.02 \text{ g N}$

$6 \text{ mol O} \times \dfrac{16.00 \text{ g O}}{\text{mol O}} = 96.00 \text{ g O}$

molar mass of $Cu(NO_3)_2$ = 187.57 g/mol

$\text{mass } Cu(NO_3)_2 = \dfrac{187.57 \text{ g } Cu(NO_3)_2}{\text{mol } Cu(NO_3)_2} \times 6.25 \text{ mol } Cu(NO_3)_2 = 1170 \text{ g}$

ATE, Additional Sample Problems 7-10, p. 227

1. Given: $NaNO_3$
Unknown: percentage composition of given compound

$$1 \text{ mol Na} \times \frac{22.99 \text{ g Na}}{\text{mol Na}} = 22.99 \text{ g Na}$$

$$1 \text{ mol N} \times \frac{14.01 \text{ g N}}{\text{mol N}} = 14.01 \text{ g N}$$

$$3 \text{ mol O} \times \frac{16.00 \text{ g O}}{\text{mol O}} = 48.00 \text{ g O}$$

molar mass of $NaNO_3$ = 85.00 g

$$\frac{22.99 \text{ g Na}}{85.00 \text{ g NaNO}_3} \times 100 = 27.05\% \text{ Na}$$

$$\frac{14.01 \text{ g N}}{85.00 \text{ g NaNO}_3} \times 100 = 16.48\% \text{ N}$$

$$\frac{48.00 \text{ g O}}{85.00 \text{ g NaNO}_3} \times 100 = 56.47\% \text{ O}$$

2. Given: Ag_2SO_4
Unknown: percentage composition of given compound

$$2 \text{ mol Ag} \times \frac{107.87 \text{ g Ag}}{\text{mol Ag}} = 215.74 \text{ g Ag}$$

$$1 \text{ mol S} \times \frac{32.07 \text{ g S}}{\text{mol S}} = 32.07 \text{ g S}$$

$$4 \text{ mol O} \times \frac{16.00 \text{ g O}}{\text{mol O}} = 64.00 \text{ g O}$$

molar mass of Ag_2SO_4 = 311.81 g

$$\frac{215.74 \text{ g Ag}}{311.81 \text{ g Ag}_2SO_4} \times 100 = 69.19\% \text{ Ag}$$

$$\frac{32.07 \text{ g S}}{311.81 \text{ g Ag}_2SO_4} \times 100 = 10.29\% \text{ S}$$

$$\frac{64.00 \text{ g O}}{311.81 \text{ g Ag}_2SO_4} \times 100 = 20.53\% \text{ O}$$

Practice, p. 228

1. a. Given: $PbCl_2$
Unknown: percentage composition of given compound

$$1 \text{ mol Pb} \times \frac{207.2 \text{ g Pb}}{\text{mol Pb}} = 207.2 \text{ g Pb}$$

$$2 \text{ mol Cl} \times \frac{35.45 \text{ g Cl}}{\text{mol Cl}} = 70.90 \text{ g Cl}$$

molar mass of $PbCl_2$ = 278.1 g

$$\frac{207.2 \text{ g Pb}}{278.1 \text{ g PbCl}_2} \times 100 = 74.51\% \text{ Pb}$$

$$\frac{70.90 \text{ g Cl}}{278.1 \text{ g PbCl}_2} \times 100 = 25.49\% \text{ Cl}$$

b. Given: $Ba(NO_3)_2$
Unknown: percentage composition of given compound

$1 \text{ mol Ba} \times \dfrac{137.33 \text{ g Ba}}{\text{mol Ba}} = 137.33 \text{ g Ba}$

$2 \text{ mol N} \times \dfrac{14.01 \text{ g N}}{\text{mol N}} = 28.02 \text{ g N}$

$6 \text{ mol O} \times \dfrac{16.00 \text{ g O}}{\text{mol O}} = 96.00 \text{ g O}$

molar mass of $Ba(NO_3)_2$ = 261.35 g

$\dfrac{137.33 \text{ g Ba}}{261.35 \text{ g Ba(NO}_3)_2} \times 100 = 52.546\% \text{ Ba}$

$\dfrac{28.02 \text{ g N}}{261.35 \text{ g Ba(NO}_3)_2} \times 100 = 10.72\% \text{ N}$

$\dfrac{96.00 \text{ g O}}{261.35 \text{ g Ba(NO}_3)_2} \times 100 = 36.73\% \text{ O}$

2. Given: $ZnSO_4 \cdot 7H_2O$
Unknown: mass percentage of H_2O in given compound

mass H_2O = $7 \text{ mol H}_2\text{O} \times \dfrac{18.02 \text{ g H}_2\text{O}}{\text{mol H}_2\text{O}} = 126.1 \text{ g H}_2\text{O}$

1 mol $ZnSO_4 \cdot 7H_2O$ contains 1 mol Zn, 1 mol S, and 4 mol O in the zinc sulfate molecule proper.

$1 \text{ mol Zn} \times \dfrac{65.39 \text{ g Zn}}{\text{mol Zn}} = 65.39 \text{ g Zn}$

$1 \text{ mol S} \times \dfrac{32.07 \text{ g S}}{\text{mol S}} = 32.07 \text{ g S}$

$4 \text{ mol O} \times \dfrac{16.00 \text{ g O}}{\text{mol O}} = 64.00 \text{ g O}$

molar mass of $ZnSO_4 \cdot 7H_2O$ = 287.6 g

mass percentage of H_2O in $ZnSO_4 \cdot 7H_2O$ = $\dfrac{126.1 \text{ g H}_2\text{O}}{287.6 \text{ g ZnSO}_4 \cdot 7\text{H}_2\text{O}} \times 100$

= 43.85% H_2O

3. Given: 54.87% O by mass in $Mg(OH)_2$
Unknown: mass of O in 175 g of $Mg(OH)_2$

mass O in $Mg(OH)_2$ = mass percentage of O in $Mg(OH)_2$ × mass $Mg(OH)_2$ = 0.5487 × 175 g

mass O in $Mg(OH)_2$ = 96.0 g O

mol O = $\dfrac{\text{mass O}}{\text{molar mass of O}} = \dfrac{96.0 \text{ g O}}{16.00 \text{ g/mol O}} = 6.00 \text{ mol O}$

ATE, Additional Sample Problems 7-11, p. 228

1. Given: $CuSO_4 \cdot 5H_2O$
Unknown: mass percentage of H_2O in given compound

mass H_2O = 5 mol $H_2O \times \dfrac{18.02 \text{ g } H_2O}{\text{mol } H_2O}$ = 90.10 g H_2O

1 mol Cu $\times \dfrac{63.55 \text{ g Cu}}{\text{mol Cu}}$ = 63.55 g Cu

1 mol S $\times \dfrac{32.07 \text{ g S}}{\text{mol S}}$ = 32.07 g S

4 mol O $\times \dfrac{16.00 \text{ g O}}{\text{mol O}}$ = 64.00 g O

molar mass of $CuSO_4 \cdot 5H_2O$ = 249.72 g

mass percentage of H_2O in $CuSO_4 \cdot 5H_2O$

= $\dfrac{90.10 \text{ g } H_2O}{249.72 \text{ g } CuSO_4 \cdot 5H_2O}$

$\times 100$ = 36.08% H_2O

2. a. Given: 52.02% Cl by mass in $ZnCl_2$
Unknown: mass of Cl in 80.3 g of $ZnCl_2$

mass Cl in $ZnCl_2$ = mass percentage of Cl in $ZnCl_2 \times$ mass $ZnCl_2$
= 0.5202 × 80.3 g

mass Cl in $ZnCl_2$ = 41.8 g Cl

2. b. Given: 41.8 g Cl
Unknown: number of moles in sample

molar mass of Cl = 35.45 g/mol

mol Cl = $\dfrac{\text{mass Cl}}{\text{molar mass of Cl}} = \dfrac{41.08 \text{ g Cl}}{35.45 \text{ g/mol}}$ = 1.18 mol Cl

Section Review, p. 228

1. Given: $(NH_4)_2CO_3$
Unknown: the formula and molar masses of the given compound

2 N atoms $\times \dfrac{14.01 \text{ amu}}{\text{N atom}}$ = 28.02 amu

8 H atoms $\times \dfrac{1.01 \text{ amu}}{\text{H atom}}$ = 8.08 amu

1 C atom $\times \dfrac{12.01 \text{ amu}}{\text{C atom}}$ = 12.01 amu

3 O atoms $\times \dfrac{16.00 \text{ amu}}{\text{O amu}}$ = 48.00 amu

formula mass of $(NH_4)_2CO_3$ = 96.11 amu

molar mass of $(NH_4)_2CO_3$ = 96.11 amu $\times \dfrac{1 \text{ g/mol}}{1 \text{ amu}}$ = 96.11 g/mol

2. (see problem **1.**)

3. Given: 3.25 mol $Fe_2(SO_4)_3$
Unknown: mass of $Fe_2(SO_4)_3$

$2 \text{ mol Fe} \times \dfrac{55.85 \text{ g Fe}}{\text{mol Fe}} = 111.7 \text{ g Fe}$

$3 \text{ mol S} \times \dfrac{32.07 \text{ g S}}{\text{mol S}} = 96.21 \text{ g S}$

$12 \text{ mol O} \times \dfrac{16.00 \text{ g O}}{\text{mol O}} = 192.0 \text{ g O}$

molar mass of $Fe_2(SO_4)_3$ = 399.9 g/mol

$\text{mass } Fe_2(SO_4)_3 = \dfrac{399.9 \text{ g } Fe_2(SO_4)_3}{\text{mol } Fe_2(SO_4)_3} \times 3.25 \text{ mol } Fe_2(SO_4)_3 = 1.30 \times 10^3 \text{ g}$

---

4. Given: 250 g HNO_3
Unknown: number of moles HNO_3

$1 \text{ mol H} \times \dfrac{1.01 \text{ g H}}{\text{mol H}} = 1.01 \text{ g H}$

$1 \text{ mol N} \times \dfrac{14.01 \text{ g N}}{\text{mol N}} = 14.01 \text{ g N}$

$3 \text{ mol O} \times \dfrac{16.00 \text{ g O}}{\text{mol O}} = 48.00 \text{ g O}$

molar mass of HNO_3 = 63.02 g/mol

$\text{mol } HNO_3 = \dfrac{250 \text{ g}}{63.02 \text{ g/mol}}$

mol HNO_3 = 3.97 mol

---

5. Given: 100.0 mg of $C_9H_8O_4$
Unknown: n; number of molecules $C_9H_8O_4$

$9 \text{ mol C} \times \dfrac{12.01 \text{ g C}}{\text{mol C}} = 108.1 \text{ g C}$

$8 \text{ mol H} \times \dfrac{1.01 \text{ g H}}{\text{mol H}} = 8.08 \text{ g H}$

$4 \text{ mol O} \times \dfrac{16.00 \text{ g O}}{\text{mol O}} = 64.00 \text{ g O}$

molar mass of $C_9H_8O_4$ = 180.2 g/mol

$n = 100.0 \text{ mg } C_9H_8O_4 \times \dfrac{1 \text{ g}}{1000 \text{ mg}} \times \dfrac{1 \text{ mol}}{180.2 \text{ g } C_9H_8O_4} \times \dfrac{6.022 \times 10^{23} \text{ molecules}}{\text{mol}}$

$= 3.342 \times 10^{20}$ molecules

6. Given: $(NH_4)_2CO_3$
Unknown: percentage composition of given compound

$2 \text{ mol N} \times \dfrac{14.01 \text{ g N}}{\text{mol N}} = 28.02 \text{ g N}$

$8 \text{ mol H} \times \dfrac{1.01 \text{ g H}}{\text{mol H}} = 8.08 \text{ g H}$

$1 \text{ mol C} \times \dfrac{12.01 \text{ g C}}{\text{mol C}} = 12.01 \text{ g C}$

$3 \text{ mol O} \times \dfrac{16.00 \text{ g O}}{\text{mol O}} = 48.00 \text{ g O}$

molar mass of $(NH_4)_2CO_3$ = 96.11 g

$\dfrac{28.02 \text{ g N}}{96.11 \text{ g } (NH_4)_2CO_3} \times 100 = 29.15\% \text{ N}$

$\dfrac{8.08 \text{ g H}}{96.11 \text{ g } (NH_4)_2CO_3} \times 100 = 8.407\% \text{ H}$

$\dfrac{12.01 \text{ g C}}{96.11 \text{ g } (NH_4)_2CO_3} \times 100 = 12.50\% \text{ C}$

$\dfrac{48.00 \text{ g O}}{96.11 \text{ g } (NH_4)_2CO_3} \times 100 = 49.94\% \text{ O}$

ATE, Additional Sample Problems 7-12, p. 230

1. Given: percentage composition: 36.70% K, 33.27% Cl, and 30.03% O
Unknown: empirical formula

mass composition in 100.00 g sample: 36.70 g K, 33.27 g Cl, 30.03 g O

composition in moles:

$36.70 \text{ g K} \times \dfrac{1 \text{ mol K}}{39.10 \text{ g K}} = 0.9386 \text{ mol K}$

$33.27 \text{ g Cl} \times \dfrac{1 \text{ mol Cl}}{35.45 \text{ g Cl}} = 0.9385 \text{ mol Cl}$

$30.03 \text{ g O} \times \dfrac{1 \text{ mol O}}{16.00 \text{ g O}} = 1.877 \text{ mol O}$

smallest whole-number ratio of atoms:

$\dfrac{0.9386 \text{ mol K}}{0.9385} : \dfrac{0.9385 \text{ mol Cl}}{0.9385} : \dfrac{1.877 \text{ mol O}}{0.9385} = 1 \text{ mol K} : 1 \text{ mol Cl} : 2 \text{ mol O}$

The empirical formula is therefore $KClO_2$.

2. Given: percentage composition: 17.15% C, 1.44% H, and 81.41% F
Unknown: empirical formula

mass composition in 100.00 g sample: 17.15 g C, 1.44 g H, 81.41 g F

composition in moles:

$17.15 \text{ g C} \times \dfrac{1 \text{ mol C}}{12.01 \text{ g C}} = 1.428 \text{ mol C}$

$1.44 \text{ g H} \times \dfrac{1 \text{ mol H}}{1.01 \text{ g H}} = 1.43 \text{ mol H}$

$81.41 \text{ g F} \times \dfrac{1 \text{ mol F}}{19.00 \text{ g F}} = 4.285 \text{ mol F}$

smallest whole-number ratio of atoms:

$\dfrac{1.428 \text{ mol C}}{1.428} : \dfrac{1.43 \text{ mol H}}{1.428} : \dfrac{4.285 \text{ mol F}}{1.428} = 1 \text{ mol C} : 1 \text{ mol H} : 3 \text{ mol F}$

The empirical formula is therefore CHF_3.

Practice 7-12 and 7-13, p. 231

1. Given: percentage composition: 63.52% Fe and 36.48% S
Unknown: empirical formula

mass composition in 100.00 g sample: 63.52 g Fe, 36.48 g S

composition in moles:

$$63.52 \text{ g Fe} \times \frac{1 \text{ mol Fe}}{55.85 \text{ g Fe}} = 1.137 \text{ mol Fe}$$

$$36.48 \text{ g S} \times \frac{1 \text{ mol S}}{32.07 \text{ g S}} = 1.138 \text{ mol S}$$

smallest whole-number ratio of atoms:

$$\frac{1.137 \text{ mol Fe}}{1.137} : \frac{1.138 \text{ mol S}}{1.137} = 1 \text{ mol Fe} : 1 \text{ mol S}$$

The empirical formula is therefore FeS.

2. Given: percentage composition: 26.56% K, 35.41% Cr, and the remainder is O
Unknown: empirical formula

% O = (100.00% − 26.56%) − 35.41% = 38.03%

mass composition in 100.00 g sample: 26.56 g K, 35.41 g Cr, 38.03 g O

composition in moles:

$$26.56 \text{ g K} \times \frac{1 \text{ mol K}}{39.10 \text{ g K}} = 0.6793 \text{ mol K}$$

$$35.41 \text{ g Cr} \times \frac{1 \text{ mol Cr}}{52.00 \text{ g Cr}} = 0.6810 \text{ mol Cr}$$

$$38.03 \text{ g O} \times \frac{1 \text{ mol O}}{16.00 \text{ g O}} = 2.377 \text{ mol O}$$

smallest whole-number ratio of atoms:

$$\frac{0.6793 \text{ mol K}}{0.6793} : \frac{0.6810 \text{ mol Cr}}{0.6793} : \frac{2.377 \text{ mol O}}{0.6793}$$

= 1.000 mol K : 1.003 mol Cr : 3.499 mol O

= 2.000 mol K : 2.006 mol Cr : 6.998 mol O

= 2 mol K : 2 mol Cr : 7 mol O

The empirical formula is therefore $K_2Cr_2O_7$.

3. Given: sample mass = 20.0 g
calcium mass = 4.00 g
Unknown: empirical formula

bromine mass = sample mass − calcium mass = 20.0 g − 4.00 g = 16.0 g

composition in moles:

$$4.00 \text{ g Ca} \times \frac{1 \text{ mol Ca}}{40.08 \text{ g Ca}} = 0.0998 \text{ mol Ca}$$

$$16.0 \text{ g Br} \times \frac{1 \text{ mol Br}}{79.90 \text{ g Br}} = 0.200 \text{ mol Br}$$

smallest whole-number ratio of atoms:

$$\frac{0.0998 \text{ mol Ca}}{0.0998} : \frac{0.200 \text{ mol Br}}{0.0998}$$

= 1 mol Ca : 2 mol Br

The empirical formula is therefore $CaBr_2$.

ATE, Additional Sample Problems 7-13, p. 231

1. Given: sample mass = 60.00 g
lead mass = 38.43 g
carbon mass = 17.83 g
hydrogen mass = 3.74 g
Unknown: empirical formula

composition in moles:

$$38.43 \text{ g Pb} \times \frac{1 \text{ mol Pb}}{207.2 \text{ g Pb}} = 0.1855 \text{ mol Pb}$$

$$17.83 \text{ g C} \times \frac{1 \text{ mol C}}{12.01 \text{ g C}} = 1.485 \text{ mol C}$$

$$3.74 \text{ g H} \times \frac{1 \text{ mol H}}{1.01 \text{ g H}} = 3.70 \text{ mol H}$$

smallest whole-number ratio of atoms:

$$\frac{0.1855 \text{ mol Pb}}{0.1855} : \frac{1.485 \text{ mol C}}{0.1855} : \frac{3.70 \text{ mol H}}{0.1855}$$

= 1 mol Pb : 8.005 mol C : 19.9 mol H

= 1 mol Pb : 8 mol C : 20 mol H

The empirical formula is therefore PbC_8H_{20}.

2. Given: sample mass = 170.00 g
sodium mass = 29.84 g
chromium mass = 67.49 g
oxygen mass = 72.67 g
Unknown: empirical formula

composition in moles:

$$29.84 \text{ g Na} \times \frac{1 \text{ mol Na}}{22.99 \text{ g Na}} = 1.298 \text{ mol Na}$$

$$67.49 \text{ g Cr} \times \frac{1 \text{ mol Cr}}{52.00 \text{ g Cr}} = 1.298 \text{ mol Cr}$$

$$72.67 \text{ g O} \times \frac{1 \text{ mol O}}{16.00 \text{ g O}} = 4.542 \text{ mol O}$$

smallest whole-number ratio of atoms:

$$\frac{1.298 \text{ mol Na}}{1.298} : \frac{1.298 \text{ mol Cr}}{1.298} : \frac{4.542 \text{ mol O}}{1.298}$$

= 1 mol Na : 1 mol Cr : 3.499 mol O

= 2 mol Na : 2 mol Cr : 6.998 mol O

= 2 mol Na : 2 mol Cr : 7 mol O

The empirical formula is therefore $Na_2Cr_2O_7$.

Practice, p. 233

1. Given: empirical formula: CH
molecular formula mass = 78.110 amu
Unknown: molecular formula

$$x = \frac{\text{molecular formula mass}}{\text{empirical formula mass}}$$

empirical formula mass of CH = 12.01 amu + 1.01 amu = 13.02 amu

$$x = \frac{78.110 \text{ amu}}{13.02 \text{ amu}} = 6$$

molecular formula: $C_xH_x = C_6H_6$

2. Given: molecular formula mass = 34.00 amu
hydrogen mass = 0.44 g
oxygen mass = 6.92 g

Unknown: molecular formula

composition in moles:

$$0.44 \text{ g H} \times \frac{1 \text{ mol H}}{1.01 \text{ g H}} = 0.44 \text{ mol H}$$

$$6.92 \text{ g O} \times \frac{1 \text{ mol O}}{16.00 \text{ g O}} = 0.432 \text{ mol O}$$

smallest whole-number ratio of atoms:

$$\frac{0.44 \text{ mol H}}{0.432} : \frac{0.432 \text{ mol O}}{0.432} = 1 \text{ mol H} : 1 \text{ mol O}$$

The empirical formula is therefore HO.

$$x = \frac{\text{molecular formula mass}}{\text{empirical formula mass}}$$

empirical formula mass of HO = 1.01 amu + 16.00 amu = 17.01 amu

$$x = \frac{34.00 \text{ amu}}{17.01 \text{ amu}} = 2$$

molecular formula: $H_xO_x = H_2O_2$

ATE, Additional Sample Problems 7-14, p. 233

1. Given: empirical formula: OCNCl
molar mass = 232.41 g/mol

Unknown: molecular formula

$$x = \frac{\text{molar mass}}{\text{empirical molar mass}}$$

$$1 \text{ mol O} \times \frac{16.00 \text{ g O}}{\text{mol O}} = 16.00 \text{ g O}$$

$$1 \text{ mol C} \times \frac{12.01 \text{ g C}}{\text{mol C}} = 12.01 \text{ g C}$$

$$1 \text{ mol N} \times \frac{14.01 \text{ g N}}{\text{mol N}} = 14.01 \text{ g N}$$

$$1 \text{ mol Cl} \times \frac{35.45 \text{ g Cl}}{\text{mol Cl}} = 35.45 \text{ g Cl}$$

empirical molar mass = 77.47 g/mol

$$x = \frac{232.41 \text{ g/mol}}{77.47 \text{ g/mol}} = 3$$

molecular formula: $O_xC_xN_xCl_x = O_3C_3N_3Cl_3$

2. Given: empirical formula: NH_2
molecular formula mass = 32.06 amu

Unknown: molecular formula

$$x = \frac{\text{molecular formula mass}}{\text{empirical formula mass}}$$

empirical formula mass of NH_2 = 14.01 amu + (2 × 1.01 amu) = 16.03 amu

$$x = \frac{32.06 \text{ amu}}{16.03 \text{ amu}} = 2$$

molecular formula: $N_xH_{2x} = N_2H_4$

Section Review, p. 233

1. Given: percentage composition: 36.48% Na, 25.41% S, and 38.11% O

Unknown: empirical formula

mass composition in 100.00 g sample: 36.48 g Na, 25.41 g S, 38.11 g O

composition in moles:

$$36.48 \text{ g Na} \times \frac{1 \text{ mol Na}}{22.99 \text{ g Na}} = 1.587 \text{ mol Na}$$

$$25.41 \text{ g S} \times \frac{1 \text{ mol S}}{32.07 \text{ g S}} = 0.7923 \text{ mol S}$$

$$38.11 \text{ g O} \times \frac{1 \text{ mol O}}{16.00 \text{ g O}} = 2.382 \text{ mol O}$$

smallest whole-number ratio of atoms:

$$\frac{1.587 \text{ mol Na}}{0.7923} : \frac{0.7923 \text{ mol S}}{0.7923} : \frac{2.382 \text{ mol O}}{0.7923}$$

= 2.003 mol Na : 1.000 mol S : 3.006 mol O

= 2 mol Na : 1 mol S : 3 mol O

The empirical formula is therefore Na_2SO_3.

2. Given: percentage composition: 53.70% Fe and 46.30% S

Unknown: empirical formula

mass composition in 100.00 g sample: 53.70 g Fe, 46.30 g S

composition in moles:

$$53.70 \text{ g Fe} \times \frac{1 \text{ mol Fe}}{55.85 \text{ Fe}} = 0.9615 \text{ mol Fe}$$

$$46.30 \text{ g S} \times \frac{1 \text{ mol S}}{32.07 \text{ g S}} = 1.444 \text{ mol S}$$

smallest whole-number ratio of atoms:

$$\frac{0.9615 \text{ mol Fe}}{0.9615} : \frac{1.444 \text{ mol S}}{0.9615} = 1.000 \text{ mol Fe} : 1.502 \text{ mol S}$$

= 2.000 mol Fe : 3.004 mol S

= 2 mol Fe : 3 mol S

The empirical formula is therefore Fe_2S_3.

3. Given: potassium mass = 1.04 g
chromium mass = 0.70 g
oxygen mass = 0.86 g

Unknown: empirical formula

composition in moles:

$$1.04 \text{ g K} \times \frac{1 \text{ mol K}}{39.10 \text{ g K}} = 0.0266 \text{ mol K}$$

$$0.70 \text{ g Cr} \times \frac{1 \text{ mol Cr}}{52.00 \text{ g Cr}} = 0.013 \text{ mol Cr}$$

$$0.86 \text{ g O} \times \frac{1 \text{ mol O}}{16.00 \text{ g O}} = 0.054 \text{ mol O}$$

smallest whole-number ratio of atoms:

$$\frac{0.0266 \text{ mol K}}{0.013} : \frac{0.013 \text{ mol Cr}}{0.013} : \frac{0.054 \text{ mol O}}{0.013}$$

= 2.0 mol K : 1 mol Cr : 4.2 mol O

= 2 mol K : 1 mol Cr : 4 mol O

The empirical formula is therefore K_2CrO_4.

4. Given: molecular formula mass = 108.0 amu
nitrogen mass = 4.04 g
oxygen mass = 11.46 g

Unknown: molecular formula

composition in moles:

$$4.04 \text{ g N} \times \frac{1 \text{ mol N}}{14.01 \text{ g N}} = 0.288 \text{ mol N}$$

$$11.46 \text{ g O} \times \frac{1 \text{ mol O}}{16.00 \text{ g O}} = 0.7162 \text{ mol O}$$

smallest whole-number ratio of atoms:

$$\frac{0.288 \text{ mol N}}{0.288} : \frac{0.7162 \text{ mol O}}{0.288} = 1.00 \text{ mol N} : 2.49 \text{ mol O}$$

$$= 2.00 \text{ mol N} : 4.98 \text{ mol O}$$

$$= 2 \text{ mol N} : 5 \text{ mol O}$$

The empirical formula is therefore N_2O_5.

$$x = \frac{\text{molecular formula mass}}{\text{empirical formula mass}}$$

empirical formula mass of N_2O_5 = (2 × 14.01 amu) + (5 × 16.00 amu)
= 108.02 amu

$$x = \frac{108.0 \text{ amu}}{108.02 \text{ amu}} = 1$$

molecular formula: $N_{2x}O_{5x} = N_2O_5$

---

5. Given: molar mass = 92 g/mol
nitrogen mass = 0.606 g
oxygen mass = 1.390 g

Unknown: molecular formula

$$x = \frac{\text{molar mass}}{\text{empirical molar mass}}$$

composition in moles:

$$0.606 \text{ g N} \times \frac{1 \text{ mol N}}{14.01 \text{ g N}} = 0.0433 \text{ mol N}$$

$$1.390 \text{ g O} \times \frac{1 \text{ mol O}}{16.00 \text{ g O}} = 0.0869 \text{ mol O}$$

smallest whole-number ratio of atoms:

$$\frac{0.0433 \text{ mol N}}{0.0433} : \frac{0.0869 \text{ mol O}}{0.0433}$$

$$= 1 \text{ mol N} : 2 \text{ mol O}$$

The empirical formula is therefore NO_2.

empirical molar mass of NO_2 = 14.01 g/mol + (2 × 16.00 g/mol)
= 46.01 g/mol

$$x = \frac{92 \text{ g/mol}}{46.01 \text{ g/mol}} = 2$$

molecular formula: $N_xO_{2x} = N_2O_4$

Review Problems

30. a. Given: $C_6H_{12}O_6$
Unknown: the formula mass of the given compound

6 C atoms × $\dfrac{12.01 \text{ amu}}{\text{C atom}}$ = 72.06 amu

12 H atoms × $\dfrac{1.01 \text{ amu}}{\text{H atom}}$ = 12.12 amu

6 O atoms × $\dfrac{16.00 \text{ amu}}{\text{O atom}}$ = 96.00 amu

formula mass of $C_6H_{12}O_6$ = 180.18 amu

b. Given: $Ca(CH_3COO)_2$
Unknown: the formula mass of the given compound

1 Ca atom × $\dfrac{40.08 \text{ amu}}{\text{Ca atom}}$ = 40.08 amu

4 C atoms × $\dfrac{12.01 \text{ amu}}{\text{C atom}}$ = 48.04 amu

6 H atoms × $\dfrac{1.01 \text{ amu}}{\text{H atom}}$ = 6.06 amu

4 O atoms × $\dfrac{16.00 \text{ amu}}{\text{O atom}}$ = 64.00 amu

formula mass of $Ca(CH_3COO)_2$ = 158.18 amu

c. Given: NH_4^+
Unknown: the formula mass of the given ion

1 N atom × $\dfrac{14.01 \text{ amu}}{\text{N atom}}$ = 14.01 amu

4 H atoms × $\dfrac{1.01 \text{ amu}}{\text{H atom}}$ = 4.04 amu

formula mass of NH_4^+ = 18.05 amu

d. Given: ClO_3^-
Unknown: the formula mass of the given ion

1 Cl atom × $\dfrac{35.45 \text{ amu}}{\text{Cl atom}}$ = 35.45 amu

3 O atoms × $\dfrac{16.00 \text{ amu}}{\text{O atom}}$ = 48.00 amu

formula mass of ClO_3^- = 83.45 amu

32. a. Given: KNO_3
Unknown: the molar mass of the given compound

1 mol K × $\dfrac{39.10 \text{ g K}}{\text{mol K}}$ = 39.10 g K

1 mol N × $\dfrac{14.01 \text{ g N}}{\text{mol N}}$ = 14.01 g N

3 mol O × $\dfrac{16.00 \text{ g O}}{\text{mol O}}$ = 48.00 g O

molar mass of KNO_3 = 101.11 g/mol

b. Given: Na_2SO_4
Unknown: the molar mass of the given compound

2 mol Na × $\dfrac{22.99 \text{ g Na}}{\text{mol Na}}$ = 45.98 g Na

1 mol S × $\dfrac{32.07 \text{ g S}}{\text{mol S}}$ = 32.07 g S

4 mol O × $\dfrac{16.00 \text{ g O}}{\text{mol O}}$ = 64.00 g O

molar mass of Na_2SO_4 = 142.05 g/mol

c. Given: $Ca(OH)_2$
Unknown: the molar mass of the given compound

$1 \text{ mol Ca} \times \dfrac{40.08 \text{ g Ca}}{\text{mol Ca}} = 40.08 \text{ g Ca}$

$2 \text{ mol O} \times \dfrac{16.00 \text{ g O}}{\text{mol O}} = 32.00 \text{ g O}$

$2 \text{ mol H} \times \dfrac{1.01 \text{ g H}}{\text{mol H}} = 2.02 \text{ g H}$

molar mass of $Ca(OH)_2$ = 74.10 g/mol

d. Given: $(NH_4)_2SO_3$
Unknown: the molar mass of the given compound

$2 \text{ mol N} \times \dfrac{14.01 \text{ g N}}{\text{mol N}} = 28.02 \text{ g N}$

$8 \text{ mol H} \times \dfrac{1.01 \text{ g H}}{\text{mol H}} = 8.08 \text{ g H}$

$1 \text{ mol S} \times \dfrac{32.07 \text{ g S}}{\text{mol S}} = 32.07 \text{ g S}$

$3 \text{ mol O} \times \dfrac{16.00 \text{ g O}}{\text{mol O}} = 48.00 \text{ g O}$

molar mass of $(NH_4)_2SO_3$ = 116.17 g/mol

e. Given: $Ca_3(PO_4)_2$
Unknown: the molar mass of the given compound

$3 \text{ mol Ca} \times \dfrac{40.08 \text{ g Ca}}{\text{mol Ca}} = 120.24 \text{ g Ca}$

$2 \text{ mol P} \times \dfrac{30.97 \text{ g P}}{\text{mol P}} = 61.94 \text{ g P}$

$8 \text{ mol O} \times \dfrac{16.00 \text{ g O}}{\text{mol O}} = 128.0 \text{ g O}$

molar mass of $Ca_3(PO_4)_2$ = 310.18 g/mol

f. Given: $Al_2(CrO_4)_3$
Unknown: the molar mass of the given compound

$2 \text{ mol Al} \times \dfrac{26.98 \text{ g Al}}{\text{mol Al}} = 53.96 \text{ g Al}$

$3 \text{ mol Cr} \times \dfrac{52.00 \text{ g Cr}}{\text{mol Cr}} = 156.0 \text{ g Cr}$

$12 \text{ mol O} \times \dfrac{16.00 \text{ g O}}{\text{mol O}} = 192.0 \text{ g O}$

molar mass of $Al_2(CrO_4)_3$ = 401.96 g/mol

33. a. Given: 4.50 g H_2O
Unknown: number of moles H_2O

$2 \text{ mol H} \times \dfrac{1.01 \text{ g H}}{\text{mol H}} = 2.02 \text{ g H}$

$1 \text{ mol O} \times \dfrac{16.00 \text{ g O}}{\text{mol O}} = 16.00 \text{ g O}$

molar mass of H_2O = 18.02 g/mol

$\text{mol } H_2O = \dfrac{4.50 \text{ g}}{18.02 \text{ g/mol}} = 0.250 \text{ mol}$

b. Given: 471.6 g Ba(OH)$_2$

Unknown: number of moles Ba(OH)$_2$

$1 \text{ mol Ba} \times \dfrac{137.33 \text{ g Ba}}{\text{mol Ba}} = 137.33 \text{ g Ba}$

$2 \text{ mol O} \times \dfrac{16.00 \text{ g O}}{\text{mol O}} = 32.00 \text{ g O}$

$2 \text{ mol H} \times \dfrac{1.01 \text{ g H}}{\text{mol H}} = 2.02 \text{ g H}$

molar mass of Ba(OH)$_2$ = 171.35 g/mol

$\text{mol Ba(OH)}_2 = \dfrac{471.6 \text{ g}}{171.35 \text{ g/mol}} = 2.752 \text{ mol}$

c. Given: 129.68 g Fe$_3$(PO$_4$)$_2$

Unknown: number of moles Fe$_3$(PO$_4$)$_2$

$3 \text{ mol Fe} \times \dfrac{55.85 \text{ g Fe}}{\text{mol Fe}} = 167.6 \text{ g Fe}$

$2 \text{ mol P} \times \dfrac{30.97 \text{ g P}}{\text{mol P}} = 61.94 \text{ g P}$

$8 \text{ mol O} \times \dfrac{16.00 \text{ g O}}{\text{mol O}} = 128.0 \text{ g O}$

molar mass of Fe$_3$(PO$_4$)$_2$ = 357.54 g/mol

$\text{mol Fe}_3\text{(PO}_4\text{)}_2 = \dfrac{129.68 \text{ g}}{357.54 \text{ g/mol}} = 0.3627 \text{ mol}$

34. a. Given: NaCl

Unknown: percentage composition of given compound

$1 \text{ mol Na} \times \dfrac{22.99 \text{ g Na}}{\text{mol Na}} = 22.99 \text{ g Na}$

$1 \text{ mol Cl} \times \dfrac{35.45 \text{ g Cl}}{\text{mol Cl}} = 35.45 \text{ g Cl}$

molar mass of NaCl = 58.44 g

$\dfrac{22.99 \text{ g Na}}{58.44 \text{ g NaCl}} \times 100 = 39.34\% \text{ Na}$

$\dfrac{35.45 \text{ g Cl}}{58.44 \text{ g NaCl}} \times 100 = 60.66\% \text{ Cl}$

b. Given: AgNO$_3$

Unknown: percentage composition of given compound

$1 \text{ mol Ag} \times \dfrac{107.87 \text{ g Ag}}{\text{mol Ag}} = 107.87 \text{ g Ag}$

$1 \text{ mol N} \times \dfrac{14.01 \text{ g N}}{\text{mol N}} = 14.01 \text{ g N}$

$3 \text{ mol O} \times \dfrac{16.00 \text{ g O}}{\text{mol O}} = 48.00 \text{ g O}$

molar mass of AgNO$_3$ = 169.88 g

$\dfrac{107.87 \text{ g Ag}}{169.88 \text{ g AgNO}_3} \times 100 = 63.50\% \text{ Ag}$

$\dfrac{14.01 \text{ g N}}{169.88 \text{ g AgNO}_3} \times 100 = 8.25\% \text{ N}$

$\dfrac{48.00 \text{ g O}}{169.88 \text{ g AgNO}_3} \times 100 = 28.26\% \text{ O}$

c. Given: Mg(OH)$_2$
Unknown: percentage composition of given compound

$1 \text{ mol Mg} \times \dfrac{24.30 \text{ g Mg}}{\text{mol Mg}} = 24.30 \text{ g Mg}$

$2 \text{ mol O} \times \dfrac{16.00 \text{ g O}}{\text{mol O}} = 32.00 \text{ g O}$

$2 \text{ mol H} \times \dfrac{1.01 \text{ g H}}{\text{mol H}} = 2.02 \text{ g H}$

molar mass of Mg(OH)$_2$ = 58.32 g

$\dfrac{24.30 \text{ g Mg}}{58.32 \text{ g Mg(OH)}_2} \times 100 = 41.67\% \text{ Mg}$

$\dfrac{32.00 \text{ g O}}{58.32 \text{ g Mg(OH)}_2} \times 100 = 54.87\% \text{ O}$

$\dfrac{2.02 \text{ g H}}{58.32 \text{ g Mg(OH)}_2} \times 100 = 3.46\% \text{ H}$

---

35. Given: CuSO$_4 \cdot$ 5H$_2$O
Unknown: mass percentage of H$_2$O in given compound

mass H$_2$O = $5 \text{ mol H}_2\text{O} \times \dfrac{18.02 \text{ g H}_2\text{O}}{\text{mol H}_2\text{O}} = 90.10 \text{ g H}_2\text{O}$

$1 \text{ mol Cu} \times \dfrac{63.55 \text{ g Cu}}{\text{mol Cu}} = 63.55 \text{ g Cu}$

$1 \text{ mol S} \times \dfrac{32.07 \text{ g S}}{\text{mol S}} = 32.07 \text{ g S}$

$4 \text{ mol O} \times \dfrac{16.00 \text{ g O}}{\text{mol O}} = 64.00 \text{ g O}$

molar mass of CuSO$_4 \cdot$ 5H$_2$O = 249.72 g

mass percentage of H$_2$O in CuSO$_4 \cdot$ 5H$_2$O =

$\dfrac{90.10 \text{ g H}_2\text{O}}{249.72 \text{ CuSO}_4 \cdot \text{ 5H}_2\text{O}} \times 100 = 36.08\% \text{ H}_2\text{O}$

---

36. Given: percentage composition: 63.50% Ag, 8.25% N, and the remainder is O
Unknown: empirical formula

% O = 100.00% − 63.50% − 8.25% = 28.25%

mass composition in 100.00 g sample: 63.50 g Ag, 8.25 g N, 28.25 g O

composition in moles:

$63.50 \text{ g Ag} \times \dfrac{1 \text{ mol Ag}}{107.87 \text{ g Ag}} = 0.5887 \text{ mol Ag}$

$8.25 \text{ g N} \times \dfrac{1 \text{ mol N}}{14.01 \text{ g N}} = 0.589 \text{ mol N}$

$28.25 \text{ g O} \times \dfrac{1 \text{ mol O}}{16.00 \text{ g O}} = 1.766 \text{ mol O}$

smallest whole-number ratio of atoms:

$\dfrac{0.5887 \text{ mol Ag}}{0.5887} : \dfrac{0.589 \text{ mol N}}{0.5887} : \dfrac{1.766 \text{ mol O}}{0.5887}$

= 1 mol Ag : 1 mol N : 3 mol O

The empirical formula is therefore AgNO$_3$.

37. Given: percentage composition: 52.11% C, 13.14% H, and 34.75% O

Unknown: empirical formula

mass composition in 100.00 g sample: 52.11 g C, 13.14 g H, 34.75 g O

composition in moles:

$$52.11 \text{ g C} \times \frac{1 \text{ mol C}}{12.01 \text{ g C}} = 4.339 \text{ mol C}$$

$$13.14 \text{ g H} \times \frac{1 \text{ mol H}}{1.01 \text{ g H}} = 13.0 \text{ mol H}$$

$$34.75 \text{ g O} \times \frac{1 \text{ mol O}}{16.00 \text{ g O}} = 2.172 \text{ mol O}$$

smallest whole-number ratio of atoms:

$$\frac{4.339 \text{ mol C}}{2.172} : \frac{13.0 \text{ mol H}}{2.172} : \frac{2.172 \text{ mol O}}{2.172}$$

$$= 2 \text{ mol C} : 6 \text{ mol H} : 1 \text{ mol O}$$

The empirical formula is therefore C_2H_6O.

---

38. Given: empirical formula: CH_2O
molar mass = 120.12 g/mol

Unknown: molecular formula

$$x = \frac{\text{molar mass}}{\text{empirical molar mass}}$$

$$1 \text{ mol C} \times \frac{12.01 \text{ g C}}{\text{mol C}} = 12.01 \text{ g C}$$

$$2 \text{ mol H} \times \frac{1.01 \text{ g H}}{\text{mol H}} = 2.02 \text{ g H}$$

$$1 \text{ mol O} \times \frac{16.00 \text{ g O}}{\text{mol O}} = 16.00 \text{ g O}$$

empirical molar mass = 30.03 g/mol

$$x = \frac{120.12 \text{ g/mol}}{30.03 \text{ g/mol}} = 4.000 \rightarrow 4$$

molecular formula: $C_xH_{2x}O_x = C_4H_8O_4$

---

39. Given: molecular formula mass = 42.08 amu
percentage composition: 85.64% C and 14.36% H

Unknown: molecular formula

$$x = \frac{\text{molecular formula mass}}{\text{empirical formula mass}}$$

mass composition in 100.00 g sample: 85.64 g C, 14.36 g H

composition in moles:

$$85.64 \text{ g C} \times \frac{1 \text{ mol C}}{12.01 \text{ g C}} = 7.131 \text{ mol C}$$

$$14.36 \text{ g H} \times \frac{1 \text{ mol H}}{1.01 \text{ g H}} = 14.2 \text{ mol H}$$

smallest whole-number ratio of atoms:

$$\frac{7.131 \text{ mol C}}{7.131} : \frac{14.2 \text{ mol H}}{7.131} = 1 \text{ mol C} : 2 \text{ mol H}$$

The empirical formula is therefore CH_2.

empirical formula mass = 12.01 amu + 2 × 1.01 amu
= 12.01 amu + 2.02 amu = 14.03 amu

$$x = \frac{42.08 \text{ amu}}{14.03 \text{ amu}} = 3$$

molecular formula: $C_xH_{2x} = C_3H_6$

Mixed Review

40. Given: percentage composition: 37.51% C, 4.20% H, and 58.29% O

Unknown: empirical formula

mass composition in 100.00 g sample: 37.51 g C, 4.20 g H, 58.29 g O

composition in moles:

$$37.51 \text{ g C} \times \frac{1 \text{ mol C}}{12.01 \text{ g C}} = 3.123 \text{ mol C}$$

$$4.20 \text{ g H} \times \frac{1 \text{ mol H}}{1.01 \text{ g H}} = 4.16 \text{ mol H}$$

$$58.29 \text{ g O} \times \frac{1 \text{ mol O}}{16.00 \text{ g O}} = 3.643 \text{ mol O}$$

smallest whole-number ratio of atoms:

$$\frac{3.123 \text{ mol C}}{3.123} : \frac{4.16 \text{ mol H}}{3.123} : \frac{3.643 \text{ mol O}}{3.123}$$

= 1.000 mol C : 1.33 mol H : 1.167 mol O

= 6.000 mol C : 7.98 mol H : 7.002 mol O

= 6 mol C : 8 mol H : 7 mol O

The empirical formula is therefore $C_6H_8O_7$.

---

42. a. Given: 1.000 mol NaCl

Unknown: mass of NaCl

$$1 \text{ mol Na} \times \frac{22.99 \text{ g Na}}{\text{mol Na}} = 22.99 \text{ g Na}$$

$$1 \text{ mol Cl} \times \frac{35.45 \text{ g Cl}}{\text{mol Cl}} = 35.45 \text{ g Cl}$$

molar mass of NaCl = 58.44 g/mol

$$\text{mass NaCl} = \frac{58.44 \text{ g NaCl}}{\text{mol NaCl}} \times 1.000 \text{ mol NaCl} = 58.44 \text{ g}$$

b. Given: 2.000 mol H_2O

Unknown: mass of H_2O

molar mass of H_2O = 18.02 g/mol

$$\text{mass } H_2O = \frac{18.02 \text{ g } H_2O}{\text{mol } H_2O} \times 2.000 \text{ mol } H_2O$$

mass H_2O = 36.04 g

c. Given: 3.500 mol $Ca(OH)_2$

Unknown: mass of $Ca(OH)_2$

$$1 \text{ mol Ca} \times \frac{40.08 \text{ g Ca}}{\text{mol Ca}} = 40.08 \text{ g Ca}$$

$$2 \text{ mol O} \times \frac{16.00 \text{ g O}}{\text{mol O}} = 32.00 \text{ g O}$$

$$2 \text{ mol H} \times \frac{1.01 \text{ g H}}{\text{mol H}} = 2.02 \text{ g H}$$

molar mass of $Ca(OH)_2$ = 74.10 g/mol

$$\text{mass } Ca(OH)_2 = \frac{74.10 \text{ g } Ca(OH)_2}{\text{mol } Ca(OH)_2} \times 3.500 \text{ mol } Ca(OH)_2 = 259.4 \text{ g}$$

d. Given: 0.625 mol Ba(NO$_3$)$_2$
Unknown: mass of sample in grams

$$1 \text{ mol Ba} \times \frac{137.33 \text{ g Ba}}{\text{mol Ba}} = 137.33 \text{ g Ba}$$

$$2 \text{ mol N} \times \frac{14.01 \text{ g N}}{\text{mol N}} = 28.02 \text{ g N}$$

$$6 \text{ mol O} \times \frac{16.00 \text{ g O}}{\text{mol O}} = 96.00 \text{ g O}$$

molar mass of Ba(NO$_3$)$_2$ = 261.35 g/mol

$$\text{mass Ba(NO}_3)_2 = \frac{261.35 \text{ g Ba(NO}_3)_2}{\text{mol Ba(NO}_3)_2} \times 0.625 \text{ mol Ba(NO}_3)_2 = 163 \text{ g}$$

---

43. a. Given: XeF$_4$
Unknown: the formula and molar masses of the given compound

$$1 \text{ Xe atom} \times \frac{131.29 \text{ amu}}{\text{Xe atom}} = 131.29 \text{ amu}$$

$$4 \text{ F atoms} \times \frac{19.00 \text{ amu}}{\text{F atom}} = 76.00 \text{ amu}$$

formula mass of XeF$_4$ = 207.29 amu

$$\text{molar mass of XeF}_4 = 207.29 \text{ amu} \times \frac{1 \text{ g/mol}}{1 \text{ amu}} = 207.29 \text{ g/mol}$$

b. Given: C$_{12}$H$_{24}$O$_6$
Unknown: the formula and molar masses of the given compound

$$12 \text{ C atoms} \times \frac{12.01 \text{ amu}}{\text{C atom}} = 144.12 \text{ amu}$$

$$24 \text{ H atoms} \times \frac{1.01 \text{ amu}}{\text{H atom}} = 24.24 \text{ amu}$$

$$6 \text{ O atoms} \times \frac{16.00 \text{ amu}}{\text{O atom}} = 96.00 \text{ amu}$$

formula mass of C$_{12}$H$_{24}$O$_6$ = 264.36 amu

$$\text{molar mass of C}_{12}\text{H}_{24}\text{O}_6 = 264.36 \text{ amu} \times \frac{1 \text{ g/mol}}{1 \text{ amu}} = 264.36 \text{ g/mol}$$

c. Given: Hg$_2$I$_2$
Unknown: the formula and molar masses of the given compound

$$2 \text{ Hg atoms} \times \frac{200.59 \text{ amu}}{\text{Hg atom}} = 401.18 \text{ amu}$$

$$2 \text{ I atoms} \times \frac{126.90 \text{ amu}}{\text{I atom}} = 253.80 \text{ amu}$$

formula mass of Hg$_2$I$_2$ = 654.98 amu

$$\text{molar mass of Hg}_2\text{I}_2 = 654.98 \text{ amu} \times \frac{1 \text{ g/mol}}{1 \text{ amu}} = 654.98 \text{ g/mol}$$

d. Given: CuCN
Unknown: the formula and molar masses of the given compound

$$1 \text{ Cu atom} \times \frac{63.55 \text{ amu}}{\text{Cu atom}} = 63.55 \text{ amu}$$

$$1 \text{ C atom} \times \frac{12.01 \text{ amu}}{\text{C atom}} = 12.01 \text{ amu}$$

$$1 \text{ N atom} \times \frac{14.01 \text{ amu}}{\text{N atom}} = 14.01 \text{ amu}$$

formula mass of CuCN = 89.57 amu

$$\text{molar mass of CuCN} = 89.57 \text{ amu} \times \frac{1 \text{ g/mol}}{1 \text{ amu}} = 89.57 \text{ g/mol}$$

45. Given: $Fe(CHO_2)_3 \cdot H_2O$

Unknown: number of atoms of each element in one molecule of given compound, and mass percentage of H_2O

1 mol $Fe(CHO_2)_3 \cdot H_2O$ contains 1 mol H_2O. H_2O has a molar mass of 18.02 g/mol.

1 mol $Fe(CHO_2)_3 \cdot H_2O$ contains 1 mol Fe, 3 mol C, 3 mol H, and 6 mol O in the iron (III) formate molecule proper.

$$1 \text{ mol Fe} \times \frac{55.85 \text{ g Fe}}{\text{mol Fe}} = 55.85 \text{ g Fe}$$

$$3 \text{ mol C} \times \frac{12.01 \text{ g C}}{\text{mol C}} = 36.03 \text{ g C}$$

$$3 \text{ mol H} \times \frac{1.01 \text{ H}}{\text{mol H}} = 3.03 \text{ g H}$$

$$6 \text{ mol O} \times \frac{16.00 \text{ g O}}{\text{mol O}} = 96.00 \text{ g O}$$

molar mass of $Fe(CHO_2)_3 \cdot H_2O$ = 208.93 g

mass percentage of H_2O in $Fe(CHO_2)_3 \cdot H_2O$

$$= \frac{18.02 \text{ g } H_2O}{208.93 \text{ g } Fe(CHO_2)_3 \cdot H_2O} \times 100 = 8.62\% \text{ } H_2O$$

46. a. Given: HNO_2

Unknown: oxidation numbers for each atom in the given acid

According to rule 4, O has an oxidation number of –2. H is the least electronegative atom in the molecule, so according to rule 5, H has an oxidation number of +1.

$\overset{+1}{H}\overset{-2}{NO_2}$

total oxidation number of oxygen atoms = $-2 \times 2 = -4$

total oxidation number of hydrogen atom = +1

$\overset{+1}{H}\overset{-2}{NO_2}$
$\overset{+1}{}\overset{-4}{}$

balance of oxidation numbers:

$\overset{+1}{H}\overset{-2}{NO_2} \rightarrow \overset{+1}{H}\overset{+1+3-2}{NO_2}$

N therefore has an oxidation number of +3.

H, +1

N, +3

O, –2

b. Given: H_2SO_3
Unknown: oxidation numbers for each atom in the given acid

According to rule 4, O has an oxidation number of -2. H is the least electronegative atom in the molecule, so according to rule 5, H has an oxidation number of $+1$.

$$\overset{+1}{H_2}\overset{-2}{SO_3}$$

total oxidation number of oxygen atoms = $-2 \times 3 = -6$

total oxidation number of hydrogen atoms = $+1 \times 2 = +2$

$$\underset{+2 \quad -6}{\overset{+1 \quad -2}{H_2SO_3}}$$

balance of oxidation numbers:

$$\underset{+2\ +4\ -6}{\overset{+1\quad -2}{H_2SO_3}} \rightarrow \underset{+2\ +4\ -6}{\overset{+1\ +4\ -2}{H_2SO_3}}$$

S therefore has an oxidation number of $+4$.

H, $+1$
S, $+4$
O, -2

c. Given: H_2CO_3
Unknown: oxidation numbers for each atom in the given acid

According to rule 4, O has an oxidation number of -2. H has the least electronegative atom in the molecule, so according to rule 5, H has an oxidation number of $+1$.

$$\overset{+1}{H_2}\overset{-2}{CO_3}$$

total oxidation number of oxygen atoms = $-2 \times 3 = -6$

total oxidation number of hydrogen atoms = $+1 \times 2 = +2$

$$\underset{+2 \quad -6}{\overset{+1 \quad -2}{H_2CO_3}}$$

balance of oxidation numbers:

$$\underset{+2\ +4\ -6}{\overset{+1\quad -2}{H_2CO_3}} \rightarrow \underset{+2\ +4\ -6}{\overset{+1\ +4\ -2}{H_2CO_3}}$$

C therefore has an oxidation number of $+4$.

H, $+1$
C, $+4$
O, -2

d. Given: HI
Unknown: oxidation numbers for each atom

According to rule 8, I can be treated as an anion with an oxidation number of -1. The oxidation number of H, which has an electronegativity that is smaller than that of I, has an oxidation number of $+1$.

H, $+1$
I, -1

47. a. Given: NaClO
Unknown: percentage composition of given compound

$1 \text{ mol Na} \times \dfrac{22.99 \text{ g Na}}{\text{mol Na}} = 22.99 \text{ g Na}$

$1 \text{ mol Cl} \times \dfrac{35.45 \text{ g Cl}}{\text{mol Cl}} = 35.45 \text{ g Cl}$

$1 \text{ mol O} \times \dfrac{16.00 \text{ g O}}{\text{mol O}} = 16.00 \text{ g O}$

molar mass of NaClO = 74.44 g

$\dfrac{22.99 \text{ g Na}}{74.44 \text{ g NaClO}} \times 100 = 30.88\% \text{ Na}$

$\dfrac{35.45 \text{ g Cl}}{74.44 \text{ g NaClO}} \times 100 = 47.62\% \text{ Cl}$

$\dfrac{16.00 \text{ g O}}{74.44 \text{ g NaClO}} \times 100 = 21.49\% \text{ O}$

b. Given: H_2SO_3
Unknown: percentage composition of given compound

$2 \text{ mol H} \times \dfrac{1.01 \text{ g H}}{\text{mol H}} = 2.02 \text{ g H}$

$1 \text{ mol S} \times \dfrac{32.07 \text{ g S}}{\text{mol S}} = 32.07 \text{ g S}$

$3 \text{ mol O} \times \dfrac{16.00 \text{ g O}}{\text{mol O}} = 48.00 \text{ g O}$

molar mass of H_2SO_3 = 82.09 g

$\dfrac{2.02 \text{ g H}}{82.09 \text{ g } H_2SO_3} \times 100 = 2.46\% \text{ H}$

$\dfrac{32.07 \text{ g S}}{82.09 \text{ g } H_2SO_3} \times 100 = 39.07\% \text{ S}$

$\dfrac{48.00 \text{ g O}}{82.09 \text{ g } H_2SO_3} \times 100 = 58.47\% \text{ O}$

c. Given: C_2H_5COOH
Unknown: percentage composition of given compound

$3 \text{ mol C} \times \dfrac{12.01 \text{ g C}}{\text{mol C}} = 36.03 \text{ g C}$

$6 \text{ mol H} \times \dfrac{1.01 \text{ g H}}{\text{mol H}} = 6.06 \text{ g H}$

$2 \text{ mol O} \times \dfrac{16.00 \text{ g O}}{\text{mol O}} = 32.00 \text{ g O}$

molar mass of C_2H_5COOH = 74.09 g

$\dfrac{36.03 \text{ g C}}{74.09 \text{ g } C_2H_5COOH} \times 100 = 48.63\% \text{ C}$

$\dfrac{6.06 \text{ g H}}{74.09 \text{ g } C_2H_5COOH} \times 100 = 8.18\% \text{ H}$

$\dfrac{32.00 \text{ g O}}{74.09 \text{ g } C_2H_5COOH} \times 100 = 43.19\% \text{ O}$

d. Given: $BeCl_2$
Unknown: percentage composition of given compound

$1 \text{ mol Be} \times \dfrac{9.01 \text{ g Be}}{\text{mol Be}} = 9.01 \text{ g Be}$

$2 \text{ mol Cl} \times \dfrac{35.45 \text{ g Cl}}{\text{mol Cl}} = 70.90 \text{ g Cl}$

molar mass of $BeCl_2$ = 79.91 g

$\dfrac{9.01 \text{ g Be}}{79.91 \text{ g BeCl}_2} \times 100 = 11.28\% \text{ Be}$

$\dfrac{70.90 \text{ g Cl}}{79.91 \text{ g BeCl}_2} \times 100 = 88.72\% \text{ Cl}$

---

49. a. Given: CO_2
Unknown: the oxidation numbers for each atom in the given compound

According to rule 4, O has an oxidation number of –2.

$\overset{-2}{C}O_2$

total oxidation number of oxygen atoms = $-2 \times 2 = -4$

$\underset{-4}{\overset{-2}{C}O_2}$

balance of oxidation numbers:

$\underset{+4-4}{\overset{-2}{C}O_2} \rightarrow \underset{+4-4}{\overset{+4-2}{C}O_2}$

C therefore has an oxidation number of +4.

C, +4
O, –2

b. Given: NH_4^+
Unknown: the oxidation numbers for each atom in the given ion

According to rule 5, H has an oxidation number of +1.

$\overset{+1}{N}H_4^+$

total oxidation number of hydrogen atoms = $+1 \times 4 = +4$

$\underset{+4}{\overset{+1}{N}H_4^+}$

balance of oxidation numbers:

$\underset{-3+4}{\overset{+1}{N}H_4^+} \rightarrow \underset{-3+4}{\overset{-3+1}{N}H_4^+}$

N therefore has an oxidation number of –3.

N, –3
H, +1

c. Given: MnO_4^-
Unknown: the oxidation numbers for each atom in the given ion

According to rule 4, O has an oxidation number of −2.

$Mn\overset{-2}{O_4^-}$

total oxidation number of oxygen atoms = $-2 \times 4 = -8$

$\underset{-8}{Mn\overset{-2}{O_4^-}}$

balance of oxidation numbers:

$\underset{+7\ -8}{Mn\overset{-2}{O_4^-}} \rightarrow \underset{+7\ -8}{\overset{+7\ -2}{MnO_4^-}}$

Mn therefore has an oxidation number of +7.

Mn, +7
O, −2

d. Given: $S_2O_3^{2-}$
Unknown: the oxidation numbers for each atom in the given ion

According to rule 4, O has an oxidation number of −2.

$S_2\overset{-2}{O_3^{2-}}$

total oxidation number of oxygen atoms = $-2 \times 3 = -6$

$\underset{-6}{S_2\overset{-2}{O_3^{2-}}}$

balance of oxidation numbers:

$\underset{+4\ -6}{S_2\overset{-2}{O_3^{2-}}} \rightarrow \underset{+4\ -6}{\overset{+2\ -2}{S_2O_3^{2-}}}$

S therefore has an oxidation number of +2.

S, +2
O, −2

e. Given: H_2O_2
Unknown: the oxidation numbers for each atom in the given compound

According to rule 4, O in peroxides has an oxidation number of −1.

$H_2\overset{-1}{O_2}$

total oxidation number of oxygen atoms = $-1 \times 2 = -2$

$\underset{-2}{H_2\overset{-1}{O_2}}$

balance of oxidation numbers:

$\underset{+2\ -2}{H_2\overset{-1}{O_2}} \rightarrow \underset{+2\ -2}{\overset{+1\ -1}{H_2O_2}}$

H therefore has an oxidation number of +1.

H, +1
O, −1

f. Given: P_4O_{10}
Unknown: the oxidation numbers for each atom in the given compound

According to rule 4, O has an oxidation number of –2.

$$P_4\overset{-2}{O}_{10}$$

total oxidation number of oxygen atoms = $-2 \times 10 = -20$

$$\underset{-20}{P_4\overset{-2}{O}_{10}}$$

balance of oxidation numbers:

$$\underset{+20\ -20}{P_4\overset{-2}{O}_{10}} \to \underset{+20\ -20}{\overset{+5\ -2}{P_4O_{10}}}$$

P therefore has an oxidation number of +5.
P, +5
O, –2

g. Given: OF_2
Unknown: the oxidation numbers for each atom in the given compound

F always has an oxidation number of –1, according to rule 3.

$$O\overset{-1}{F}_2$$

total oxidation number of fluorine atoms = $-1 \times 2 = -2$

$$\underset{-2}{O\overset{-1}{F}_2}$$

balance of oxidation numbers:

$$\underset{+2\ -2}{O\overset{-1}{F}_2} \to \underset{+2\ -2}{\overset{+2\ -1}{O F_2}}$$

O therefore has an oxidation number of +2.
O, +2
F, –1

---

50. Given: sample mass = 175.0 g
carbon mass = 56.15 g
hydrogen mass = 9.43 g
oxygen mass = 74.81 g
nitrogen mass = 13.11 g
sodium mass = 21.49 g

Unknown: empirical formula

composition in moles:

$$56.15 \text{ g C} \times \frac{1 \text{ mol C}}{12.01 \text{ g C}} = 4.675 \text{ mol C}$$

$$9.43 \text{ g H} \times \frac{1 \text{ mol H}}{1.01 \text{ g H}} = 9.34 \text{ mol H}$$

$$74.81 \text{ g O} \times \frac{1 \text{ mol O}}{16.00 \text{ g O}} = 4.676 \text{ mol O}$$

$$13.11 \text{ g N} \times \frac{1 \text{ mol N}}{14.01 \text{ g N}} = 0.9358 \text{ mol N}$$

$$21.49 \text{ g Na} \times \frac{1 \text{ mol Na}}{22.99 \text{ g Na}} = 0.9348 \text{ mol Na}$$

smallest whole-number ratio of atoms:

$$\frac{4.675 \text{ mol C}}{0.9348} : \frac{9.34 \text{ mol H}}{0.9348} : \frac{4.676 \text{ mol O}}{0.9348} : \frac{0.9358 \text{ mol N}}{0.9348} : \frac{0.9348 \text{ mol Na}}{0.9348}$$

= 5 mol C : 10 mol H : 5 mol O : 1 mol N : 1 mol Na

The empirical formula is therefore $C_5H_{10}O_5NNa$.

Critical Thinking

51. Given: sulfur trioxide
Unknown: chemical information available from formula name

sulfur → S × 1 → S

trioxide → O × 3 = O_3

formula: SO_3

According to rule 4, O has an oxidation number of –2.

$\overset{-2}{S}O_3$

total oxidation number of oxygen atoms = –2 × 3 = –6

$\underset{-6}{\overset{-2}{S}O_3}$

balance of oxidation numbers

$\underset{+6-6}{\overset{-2}{S}O_3} \rightarrow \underset{+6-6}{\overset{+6\ -2}{S}O_3}$

S therefore has an oxidation number of +6.

S, +6
O, –2

52. Given: mass of crucible = 30.02 g
mass of nickel and crucible = 31.07 g
mass of nickel oxide and crucible = 31.36 g
Unknown: masses of nickel, nickel oxide, and oxygen, and the empirical formula of nickel oxide

mass of nickel = 31.07 g – 30.02 g = 1.05 g Ni

mass of nickel oxide = 31.36 g – 30.02 g = 1.34 g nickel oxide

mass of oxygen = 1.34 g – 1.05 g = 0.29 g O

composition in moles:

$1.05 \text{ g Ni} \times \dfrac{1 \text{ mol Ni}}{58.69 \text{ g Ni}} = 0.0179 \text{ mol Ni}$

$0.29 \text{ g O} \times \dfrac{1 \text{ mol O}}{16.00 \text{ g O}} = 0.018 \text{ mol O}$

smallest whole-number ratio of atoms:

$\dfrac{0.0179 \text{ mol Ni}}{0.0179} : \dfrac{0.018 \text{ mol O}}{0.0179} = 1 \text{ mol Ni} : 1 \text{ mol O}$

The empirical formula is therefore NiO.

CHAPTER 9
Stoichiometry

ATE, Additional Sample Problems 9-1, p. 281

a. Given: amount of Li = 2 mol
Unknown: amount of Li_2O in moles

balanced equation:
$$4Li + O_2 \rightarrow 2Li_2O$$
mole ratio from balanced equation = $\dfrac{2 \text{ mol } Li_2O}{4 \text{ mol Li}}$

$$\text{mol } Li_2O = 2 \text{ mol Li} \times \dfrac{2 \text{ mol } Li_2O}{4 \text{ mol Li}} = 1 \text{ mol } Li_2O$$

b. Given: amount of H_2O_2 = 5 mol
Unknown: amount of O_2 in moles

balanced equation:
$$2H_2O_2 \rightarrow 2H_2O + O_2$$
mole ratio from balanced equation = $\dfrac{1 \text{ mol } O_2}{2 \text{ mol } H_2O_2}$

$$\text{mol } O_2 = 5 \text{ mol } H_2O_2 \times \dfrac{1 \text{ mol } O_2}{2 \text{ mol } H_2O_2} = 2.5 \text{ mol } O_2$$

Practice, p. 282

1. Given: amount of H_2 = 6 mol
Unknown: amount of NH_3 in moles

balanced equation:
$$3H_2 + N_2 \rightarrow 2NH_3$$
mole ratio from balanced equation = $\dfrac{2 \text{ mol } NH_3}{3 \text{ mol } H_2}$

$$\text{mol } NH_3 = 6 \text{ mol } H_2 \times \dfrac{2 \text{ mol } NH_3}{3 \text{ mol } H_2} = 4 \text{ mol } NH_3$$

2. Given: amount of O_2 = 15 mol
Unknown: amount of $KClO_3$ in moles

balanced equation:
$$2KClO_3 \rightarrow 2KCl + 3O_2$$
mole ratio from balanced equation = $\dfrac{2 \text{ mol } KClO_3}{3 \text{ mol } O_2}$

$$\text{mol } KClO_3 = 15 \text{ mol } O_2 \times \dfrac{2 \text{ mol } KClO_3}{3 \text{ mol } O_2} = 10. \text{ mol } KClO_3$$

Practice, p. 284

1. Given: amount of Mg = 2.00 mol
Unknown: mass of MgO in grams

balanced equation:
$$2Mg(s) + O_2(g) \rightarrow 2MgO(s)$$
mole ratio from balanced equation = $\dfrac{2 \text{ mol MgO}}{2 \text{ mol Mg}}$

$$\text{molar mass of MgO} = 1 \times \dfrac{24.30 \text{ g Mg}}{\text{mol Mg}} + 1 \times \dfrac{16.00 \text{ g O}}{\text{mol O}} = 40.30 \text{ g/mol}$$

$$\text{mass MgO} = 2.00 \text{ mol Mg} \times \dfrac{2 \text{ mol MgO}}{2 \text{ mol Mg}} \times \dfrac{40.30 \text{ g MgO}}{\text{mol MgO}} = 80.6 \text{ g MgO}$$

2. Given: amount of Mg = 2.00 mol
Unknown: mass of O_2 in grams

balanced equation:

$2Mg(s) + O_2(g) \rightarrow 2MgO(s)$

mole ratio from balanced equation = $\dfrac{1 \text{ mol } O_2}{2 \text{ mol Mg}}$

molar mass of $O_2 = 2 \times \dfrac{16.00 \text{ g O}}{\text{mol O}} = 32.00 \text{ g/mol}$

mass $O_2 = 2.00 \text{ mol Mg} \times \dfrac{1 \text{ mol } O_2}{2 \text{ mol Mg}} \times \dfrac{32.00 \text{ g } O_2}{\text{mol } O_2} = 32.0 \text{ g } O_2$

---

3. Given: amount of CO_2 = 10 mol; balanced equation
Unknown: mass of $C_6H_{12}O_6$

mole ratio from balanced equation = $\dfrac{1 \text{ mol } C_6H_{12}O_6}{6 \text{ mol } CO_2}$

molar mass of $C_6H_{12}O_6 = 6 \times \dfrac{12.01 \text{ g C}}{\text{mol C}} + 12 \times \dfrac{1.01 \text{ g H}}{\text{mol H}} + 6 \times \dfrac{16.00 \text{ g O}}{\text{mol O}}$

$= 72.06 \text{ g/mol} + 12.1 \text{ g/mol} + 96.00 \text{ g/mol}$

$= 180.2 \text{ g/mol}$

mass $C_6H_{12}O_6 = 10 \text{ mol } CO_2 \times \dfrac{1 \text{ mol } C_6H_{12}O_6}{6 \text{ mol } CO_2} \times \dfrac{180.2 \text{ g } C_6H_{12}O_6}{\text{mol } C_6H_{12}O_6}$

$= 300 \text{ g } C_6H_{12}O_6$

ATE, Additional Sample Problems 9–2 and 9–3, p. 283

a. Given: amount of NaN_3 = 0.500 mol; balanced equation
Unknown: mass of N in grams

mole ratio from balanced equation = $\dfrac{3 \text{ mol } N_2}{2 \text{ mol } NaN_3}$

molar mass of $N_2 = 2 \times \dfrac{14.01 \text{ g N}}{\text{mol N}} = 28.02 \text{ g/mol}$

mass $N_2 = 0.500 \text{ mol } NaN_3 \times \dfrac{3 \text{ mol } N_2}{2 \text{ mol } NaN_3} \times \dfrac{28.02 \text{ g } N_2}{\text{mol } N_2} = 21.0 \text{ g } N_2$

b. Given: amount of C = 2.00 mol; balanced equation
Unknown: mass of SiC in grams

mole ratio from balanced equation = $\dfrac{1 \text{ mol SiC}}{3 \text{ mol C}}$

molar mass of SiC = $1 \times \dfrac{28.09 \text{ g Si}}{\text{mol Si}} + 1 \times \dfrac{12.01 \text{ g C}}{\text{mol C}} = 40.10 \text{ g/mol}$

mass SiC = $2.00 \text{ mol C} \times \dfrac{1 \text{ mol SiC}}{3 \text{ mol C}} \times \dfrac{40.10 \text{ g SiC}}{\text{mol SiC}} = 26.7 \text{ g SiC}$

c. Given: amount of C = 5.00 mol
Unknown: mass of ZnO in grams

balanced equation:

$2ZnO + C \rightarrow 2Zn + CO_2$

mole ratio from balanced equation = $\dfrac{2 \text{ mol ZnO}}{1 \text{ mol C}}$

molar mass of ZnO = $1 \times \dfrac{65.39 \text{ g Zn}}{\text{mol Zn}} + 1 \times \dfrac{16.00 \text{ g O}}{\text{mol O}} = 81.39 \text{ g/mol}$

mass ZnO = $5.00 \text{ mol C} \times \dfrac{2 \text{ mol ZnO}}{1 \text{ mol C}} \times \dfrac{81.39 \text{ g ZnO}}{\text{mol ZnO}} = 814 \text{ g ZnO}$

d. Given: amount of CH_4 = 1.00 mol; balanced equation

Unknown: mass of C in grams

mole ratio from balanced equation = $\dfrac{2 \text{ mol C}}{1 \text{ mol CH}_4}$

molar mass of C = 12.01 g/mol

mass C = 1.00 mol $CH_4 \times \dfrac{2 \text{ mol C}}{1 \text{ mol CH}_4} \times \dfrac{12.01 \text{ g C}}{\text{mol C}} = 24.0$ g C

e. Given: mass of O_2 = 1200. g; balanced equation

Unknown: mass of SO_2 in grams

mole ratio from balanced equation = $\dfrac{2 \text{ mol SO}_2}{1 \text{ mol O}_2}$

molar mass of SO_2 = $1 \times \dfrac{32.07 \text{ g S}}{\text{mol S}} + 2 \times \dfrac{16.00 \text{ g O}}{\text{mol O}}$

= 32.07 g/mol + 32.00 g/mol = 64.07 g/mol

molar mass of O_2 = $2 \times \dfrac{16.00 \text{ g O}}{\text{mol O}} = 32.00$ g/mol

mass SO_2 = 1200. g $O_2 \times \dfrac{\text{mol O}_2}{32.00 \text{ g O}_2} \times \dfrac{2 \text{ mol SO}_2}{1 \text{ mol O}_2} \times \dfrac{64.07 \text{ g SO}_2}{\text{mol SO}_2}$

= 4805 g SO_2

Practice, p. 285

1. Given: mass of O_2 = 125 g

Unknown: amount of HgO in moles

balanced equation:

$2HgO \rightarrow 2Hg + O_2$

mole ratio from balanced equation = $\dfrac{2 \text{ mol HgO}}{1 \text{ mol O}_2}$

molar mass of O_2 = $2 \times \dfrac{16.00 \text{ g O}}{\text{mol O}} = 32.00$ g/mol

amount HgO = 125 g $O_2 \times \dfrac{1 \text{ mol O}_2}{32.00 \text{ g O}_2} \times \dfrac{2 \text{ mol HgO}}{1 \text{ mol O}_2} = 7.81$ mol HgO

2. Given: mass of O_2 = 125 g

Unknown: amount of Hg in moles

balanced equation:

$2HgO \rightarrow 2Hg + O_2$

mole ratio from balanced equation = $\dfrac{2 \text{ mol Hg}}{1 \text{ mol O}_2}$

molar mass of O_2 = 32.00 g/mol

amount Hg = 125 g $O_2 \times \dfrac{1 \text{ mol O}_2}{32.00 \text{ g O}_2} \times \dfrac{2 \text{ mol Hg}}{1 \text{ mol O}_2} = 7.81$ mol Hg

ATE, Additional Sample Problems 9-4, p. 285

1a. Given: mass of NaCl = 250 g; balanced equation
Unknown: amount of Cl_2 in moles

mole ratio from balanced equation = $\dfrac{1 \text{ mol } Cl_2}{2 \text{ mol NaCl}}$

molar mass of NaCl = $1 \times \dfrac{22.99 \text{ g Na}}{\text{mol Na}} + 1 \times \dfrac{35.45 \text{ g Cl}}{\text{mol Cl}}$ = 58.44 g/mol

amount Cl_2 = 250 g NaCl $\times \dfrac{1 \text{ mol NaCl}}{58.44 \text{ g NaCl}} \times \dfrac{1 \text{ mol } Cl_2}{2 \text{ mol NaCl}}$ = 2.14 mol Cl_2

b. Given: mass of NaCl = 250 g; balanced equation
Unknown: amount of H_2 in moles

mole ratio from balanced equation = $\dfrac{1 \text{ mol } H_2}{2 \text{ mol NaCl}}$

molar mass of NaCl = 58.44 g/mol

amount H_2 = 250 g NaCl $\times \dfrac{1 \text{ mol NaCl}}{58.44 \text{ g NaCl}} \times \dfrac{1 \text{ mol } H_2}{2 \text{ mol NaCl}}$ = 2.14 mol H_2

2a. Given: mass of $PtCl_2(NH_3)_2$ = 30.0 g; balanced equation
Unknown: amount of K_2PtCl_4 in moles

mole ratio from balanced equation = $\dfrac{1 \text{ mol } K_2PtCl_4}{1 \text{ mol } PtCl_2(NH_3)_2}$

molar mass of $PtCl_2(NH_3)_2$ = $1 \times \dfrac{195.08 \text{ g Pt}}{\text{mole Pt}} + 2 \times \dfrac{35.45 \text{ g Cl}}{\text{mol Cl}}$

$+ 2 \times \dfrac{14.01 \text{ g N}}{\text{mol N}} + 6 \times \dfrac{1.01 \text{ g H}}{\text{mol H}}$

= 195.08 g/mol + 70.90 g/mol + 28.02 g/mol + 6.06 g/mol

= 300.06 g/mol

amount K_2PtCl_4 = 30.0 g $PtCl_2(NH_3)_2 \times \dfrac{1 \text{ mol } PtCl_2(NH_3)_2}{300.06 \text{ g } PtCl_2(NH_3)_2}$

$\times \dfrac{1 \text{ mol } K_2PtCl_4}{1 \text{ mol } PtCl_2(NH_3)_2}$ = 0.100 mol K_2PtCl_4

b. Given: mass of $PtCl_2(NH_3)_2$ = 30.0 g; balanced equation
Unknown: amount of NH_3 in moles

mole ratio from balanced equation = $\dfrac{2 \text{ mol } NH_3}{1 \text{ mol } PtCl_2(NH_3)_2}$

molar mass of $PtCl_2(NH_3)_2$ = 300.06 g/mol

amount NH_3 = 30.0 g $PtCl_2(NH_3)_2 \times \dfrac{1 \text{ mol } PtCl_2(NH_3)_2}{300.06 \text{ g } PtCl_2(NH_3)_2}$

$\times \dfrac{2 \text{ mol } NH_3}{1 \text{ mol } PtCl_2(NH_3)_2}$ = 0.200 mol NH_3

Practice, p. 287

1a. Given: mass of N_2O = 33.0 g; balanced equation

Unknown: mass of NH_4NO_3 in grams

mole ratio from balanced equation = $\dfrac{1 \text{ mol } NH_4NO_3}{1 \text{ mol } N_2O}$

molar mass of $NH_4NO_3 = 2 \times \dfrac{14.01 \text{ g N}}{\text{mol N}} + 4 \times \dfrac{1.01 \text{ g H}}{\text{mol H}} + 3 \times \dfrac{16.00 \text{ g O}}{\text{mol O}}$

$= 28.02 \text{ g/mol} + 4.04 \text{ g/mol} + 48.00 \text{ g/mol}$

$= 80.05 \text{ g/mol}$

molar mass of $N_2O = 2 \times \dfrac{14.01 \text{ g N}}{\text{mol N}} + 1 \times \dfrac{16.00 \text{ g O}}{\text{mol O}}$

$= 28.02 \text{ g/mol} + 16.00 \text{ g/mol} = 44.02 \text{ g/mol}$

mass $NH_4NO_3 = 33.0 \text{ g } N_2O \times \dfrac{1 \text{ mol } N_2O}{44.02 \text{ g } N_2O} \times \dfrac{1 \text{ mol } NH_4NO_3}{1 \text{ mol } N_2O}$

$\times \dfrac{80.05 \text{ g } NH_4NO_3}{\text{mol } NH_4NO_3} = 60.0 \text{ g } NH_4NO_3$

b. Given: mass of N_2O = 33.0 g; balanced equation

Unknown: mass of H_2O in grams

mole ratio from balanced equation = $\dfrac{2 \text{ mol } H_2O}{1 \text{ mol } N_2O}$

molar mass of N_2O = 44.02 g/mol

molar mass of H_2O = 18.02 g/mol

mass $H_2O = 33.0 \text{ g } N_2O \times \dfrac{1 \text{ mol } N_2O}{44.02 \text{ g } N_2O} \times \dfrac{2 \text{ mol } H_2O}{1 \text{ mol } N_2O} \times \dfrac{18.02 \text{ g } H_2O}{\text{mol } H_2O}$

$= 27.0 \text{ g } H_2O$

2. Given: mass of Cu = 100. g

Unknown: mass of Ag

balanced equation:

$Cu + 2AgNO_3 \rightarrow 2Ag + Cu(NO_3)_2$

mole ratio from balanced equation = $\dfrac{2 \text{ mol Ag}}{1 \text{ mol Cu}}$

molar mass of Ag = 107.87 g/mol

molar mass of Cu = 63.55 g/mol

mass Ag = $100. \text{ g Cu} \times \dfrac{1 \text{ mol Cu}}{63.55 \text{ g Cu}} \times \dfrac{2 \text{ mol Ag}}{1 \text{ mol Cu}} \times \dfrac{107.87 \text{ g Ag}}{\text{mol Ag}} = 339 \text{ g Ag}$

3. Given: mass of Al_2O_3
= 5.0 kg
Unknown: mass of Al

balanced equation:

$$2Al_2O_3 \rightarrow 4Al + 3O_2$$

mole ratio from balanced equation = $\dfrac{4 \text{ mol Al}}{2 \text{ mol Al}_2\text{O}_3}$

molar mass of Al = 26.98 g/mol

molar mass of $Al_2O_3 = 2 \times \dfrac{26.98 \text{ g Al}}{\text{mol Al}} + 3 \times \dfrac{16.00 \text{ g O}}{\text{mol O}}$

= 53.96 g/mol + 48.00 g/mol = 101.96 g/mol

mass Al = 5.0 kg $Al_2O_3 \times \dfrac{1000 \text{ g}}{\text{kg}} \times \dfrac{1 \text{ mol Al}_2\text{O}_3}{101.96 \text{ g Al}_2\text{O}_3} \times \dfrac{4 \text{ mol Al}}{2 \text{ mol Al}_2\text{O}_3}$

$\times \dfrac{26.98 \text{ g Al}}{\text{mol Al}} = 2.6 \times 10^3$ g Al = 2.6 kg Al

ATE, Additional Sample Problems 9–5, p. 287

1a. Given: mass of Na_2O_2
= 50.0 g
Unknown: mass of O_2
in grams

balanced equation:

$$2Na_2O_2(s) + 2H_2O(l) \rightarrow 4NaOH(aq) + O_2(g)$$

mole ratio from balanced equation = $\dfrac{1 \text{ mol O}_2}{2 \text{ mol Na}_2\text{O}_2}$

molar mass of $Na_2O_2 = 2 \times \dfrac{22.99 \text{ g Na}}{\text{mol Na}} + 2 \times \dfrac{16.00 \text{ g O}}{\text{mol O}}$

= 45.98 g/mol + 32.00 g/mol = 77.98 g/mol

molar mass of $O_2 = 2 \times \dfrac{16.00 \text{ g O}}{\text{mol O}} = 32.00$ g/mol

mass O_2 = 50.0 g $Na_2O_2 \times \dfrac{1 \text{ mol Na}_2\text{O}_2}{77.98 \text{ g Na}_2\text{O}_2} \times \dfrac{1 \text{ mol O}_2}{2 \text{ mol Na}_2\text{O}_2} \times \dfrac{32.00 \text{ g O}_2}{\text{mol O}_2}$

= 10.3 g O_2

b. Given: mass of Na_2O_2
= 50.0 g
Unknown: mass of H_2O
in grams

balanced equation:

$$2Na_2O_2(s) + 2H_2O(l) \rightarrow 4NaOH(aq) + O_2(g)$$

mole ratio from balanced equation = $\dfrac{2 \text{ mol H}_2\text{O}}{2 \text{ mol Na}_2\text{O}_2}$

molar mass of Na_2O_2 = 77.98 g/mol

molar mass of H_2O = 18.02 g/mol

mass H_2O = 50.0 g $Na_2O_2 \times \dfrac{1 \text{ mol Na}_2\text{O}_2}{77.98 \text{ g Na}_2\text{O}_2} \times \dfrac{2 \text{ mol H}_2\text{O}}{2 \text{ mol Na}_2\text{O}_2} \times \dfrac{18.02 \text{ g H}_2\text{O}}{\text{mol H}_2\text{O}}$

= 11.6 g H_2O

2a. Given: mass of Mg(OH)$_2$ = 3.00 g; balanced equation

Unknown: mass of MgCl$_2$ in grams

mole ratio from balanced equation = $\dfrac{1 \text{ mol MgCl}_2}{1 \text{ mol Mg(OH)}_2}$

molar mass of MgCl$_2$ = $1 \times \dfrac{24.30 \text{ g Mg}}{\text{mol Mg}} + 2 \times \dfrac{35.45 \text{ g Cl}}{\text{mol Cl}}$

= 24.30 g/mol + 70.90 g/mol = 95.20 g/mol

molar mass of Mg(OH)$_2$ = $1 \times \dfrac{24.30 \text{ g Mg}}{\text{mol Mg}} + 2 \times \dfrac{16.00 \text{ g O}}{\text{mol O}} + 2 \times \dfrac{1.01 \text{ g H}}{\text{mol H}}$

= 24.30 g/mol + 32.00 g/mol + 2.02 g/mol

= 58.32 g/mol

mass MgCl$_2$ = 3.00 g Mg(OH)$_2$ × $\dfrac{\text{mol Mg(OH)}_2}{53.32 \text{ g Mg(OH)}_2}$ × $\dfrac{1 \text{ mol MgCl}_2}{1 \text{ mol Mg(OH)}_2}$

× $\dfrac{95.20 \text{ g MgCl}_2}{\text{mol MgCl}_2}$ = 4.90 g MgCl$_2$

b. Given: mass of Mg(OH)$_2$ = 3.00 g

Unknown: mass of HCl in grams

mole ratio from balanced equation = $\dfrac{2 \text{ mol HCl}}{1 \text{ mol Mg(OH)}_2}$

molar mass of HCl = $1 \times \dfrac{1.01 \text{ g H}}{\text{mol H}} + 1 \times \dfrac{35.45 \text{ g Cl}}{\text{mol Cl}}$ = 36.46 g/mol

molar mass of Mg(OH)$_2$ = 58.32 g/mol

mass HCl = 3.00 g Mg(OH)$_2$ × $\dfrac{1 \text{ mol Mg(OH)}_2}{58.32 \text{ g Mg(OH)}_2}$ × $\dfrac{2 \text{ mol HCl}}{1 \text{ mol Mg(OH)}_2}$

× $\dfrac{36.46 \text{ g HCl}}{\text{mol HCl}}$ = 3.75 g HCl

Section Review, p. 287

1a. Given: amount of NH$_3$ = 4 mol

Unknown: amounts of O$_2$, N$_2$, and H$_2$O in moles

balanced equation:

4NH$_3$ + 3O$_2$ → 2N$_2$ + 6H$_2$O

mole ratios from balanced equation = $\dfrac{3 \text{ mol O}_2}{4 \text{ mol NH}_3}, \dfrac{2 \text{ mol N}_2}{4 \text{ mol NH}_3}, \dfrac{6 \text{ mol H}_2\text{O}}{4 \text{ mol NH}_3}$

mol O$_2$ = 4 mol NH$_3$ × $\dfrac{3 \text{ mol O}_2}{4 \text{ mol NH}_3}$ = 3 mol O$_2$

mol N$_2$ = 4 mol NH$_3$ × $\dfrac{2 \text{ mol N}_2}{4 \text{ mol NH}_3}$ = 2 mol N$_2$

mol H$_2$O = 4 mol NH$_3$ × $\dfrac{6 \text{ mol H}_2\text{O}}{4 \text{ mol NH}_3}$ = 6 mol H$_2$O

b. Given: amount of N$_2$ = 4 mol

Unknown: amounts of NH$_3$, O$_2$, and H$_2$O in moles

mole ratios from balanced equation = $\dfrac{4 \text{ mol NH}_3}{2 \text{ mol N}_2}, \dfrac{3 \text{ mol O}_2}{2 \text{ mol N}_2}, \dfrac{6 \text{ mol H}_2\text{O}}{2 \text{ mol N}_2}$

mol NH$_3$ = 4 mol N$_2$ × $\dfrac{4 \text{ mol NH}_3}{2 \text{ mol N}_2}$ = 8 mol NH$_3$

mol O$_2$ = 4 mol N$_2$ × $\dfrac{3 \text{ mol O}_2}{2 \text{ mol N}_2}$ = 6 mol O$_2$

mol H$_2$O = 4 mol N$_2$ × $\dfrac{6 \text{ mol H}_2\text{O}}{2 \text{ mol N}_2}$ = 12 mol H$_2$O

c. Given: amount of O_2 = 4.5 mol

Unknown: amounts of NH_3, N_2, and H_2O in moles

mole ratios from balanced equation = $\dfrac{4 \text{ mol } NH_3}{3 \text{ mol } O_2}, \dfrac{2 \text{ mol } NH_3}{3 \text{ mol } O_2}, \dfrac{6 \text{ mol } H_2O}{3 \text{ mol } O_2}$

mol NH_3 = 4.5 mol $O_2 \times \dfrac{4 \text{ mol } NH_3}{3 \text{ mol } O_2}$ = 6.0 mol NH_3

mol N_2 = 4.5 mol $O_2 \times \dfrac{2 \text{ mol } N_2}{3 \text{ mol } O_2}$ = 3.0 mol N_2

mol H_2O = 4.5 mol $O_2 \times \dfrac{6 \text{ mol } H_2O}{3 \text{ mol } O_2}$ = 9.0 mol H_2O

2a. Given: amount of Mg = 2.50 mol

Unknown: mass of HCl

balanced equation:

$Mg(s) + 2HCl(aq) \rightarrow MgCl_2(aq) + H_2(g)$

mole ratio from balanced equation = $\dfrac{2 \text{ mol HCl}}{1 \text{ mol Mg}}$

molar mass of HCl = $1 \times \dfrac{1.01 \text{ g H}}{\text{mol H}} + 1 \times \dfrac{35.45 \text{ g HCl}}{\text{mol HCl}}$ = 36.46 g/mol

mass HCl = 2.50 mol Mg $\times \dfrac{2 \text{ mol HCl}}{1 \text{ mol Mg}} \times \dfrac{36.46 \text{ g HCl}}{\text{mol HCl}}$ = 182 g HCl

b. Given: amount of Mg = 2.50 mol

Unknown: masses of $MgCl_2$ and H_2

balanced equation:

$Mg(s) + 2HCl(aq) \rightarrow MgCl_2(aq) + H_2(g)$

mole ratios from balanced equation = $\dfrac{1 \text{ mol } MgCl_2}{1 \text{ mol Mg}}, \dfrac{1 \text{ mol } H_2}{1 \text{ mol Mg}}$

molar mass of $MgCl_2$ = $1 \times \dfrac{24.30 \text{ g Mg}}{\text{mol Mg}} + 2 \times \dfrac{35.45 \text{ g Cl}}{\text{mol Cl}}$ = 95.20 g/mol

mass $MgCl_2$ = 2.50 mol Mg $\times \dfrac{1 \text{ mol } MgCl_2}{1 \text{ mol Mg}} \times \dfrac{95.20 \text{ g } MgCl_2}{\text{mol } MgCl_2}$

= 238 g $MgCl_2$

molar mass of H_2 = $2 \times \dfrac{1.01 \text{ g H}}{\text{mol H}}$ = 2.02 g/mol

mass H_2 = 2.50 mol Mg $\times \dfrac{1 \text{ mol } H_2}{1 \text{ mol Mg}} \times \dfrac{2.02 \text{ g } H_2}{\text{mol } H_2}$ = 5.05 g H_2

3a. Given: mass of CaC_2 = 32.0 g; balanced equation

Unknown: amount of H_2O in moles

mole ratio from balanced equation = $\dfrac{2 \text{ mol } H_2O}{1 \text{ mol } CaC_2}$

molar mass of CaC_2 = $1 \times \dfrac{40.08 \text{ g Ca}}{\text{mol Ca}} + 2 \times \dfrac{12.01 \text{ g C}}{\text{mol C}}$

= 40.08 g/mol + 24.02 g/mol = 64.10 g/mol

amount H_2O = 32 g $CaC_2 \times \dfrac{1 \text{ mol } CaC_2}{64.10 \text{ g } CaC_2} \times \dfrac{2 \text{ mol } H_2O}{1 \text{ mol } CaC_2}$ = 0.998 mol H_2O

b. Given: mass of CaC_2 = 32.0 g

Unknown: amounts of C_2H_2 and $Ca(OH)_2$ in moles

mole ratios from balanced equation = $\dfrac{1 \text{ mol } C_2H_2}{1 \text{ mol } CaC_2}, \dfrac{1 \text{ mol } Ca(OH)_2}{1 \text{ mol } CaC_2}$

molar mass of CaC_2 = 64.10 g/mol

amount C_2H_2 = 32.0 g $CaC_2 \times \dfrac{1 \text{ mol } CaC_2}{64.10 \text{ g } CaC_2} \times \dfrac{1 \text{ mol } C_2H_2}{1 \text{ mol } CaC_2}$

= 0.499 mol C_2H_2

amount $Ca(OH)_2$ = 32.0 g $CaC_2 \times \dfrac{1 \text{ mol } CaC_2}{64.10 \text{ g } CaC_2} \times \dfrac{1 \text{ mol } Ca(OH)_2}{1 \text{ mol } CaC_2}$

= 0.499 mol $Ca(OH)_2$

4. Given: mass of $AgNO_3$ = 75.0 g

Unknown: mass of AgCl

balanced equation:

$NaCl + AgNO_3 \rightarrow AgCl + NaNO_3$

mole ratio from balanced equation = $\dfrac{1 \text{ mol } AgCl}{1 \text{ mol } AgNO_3}$

molar mass of $AgNO_3$ = $1 \times \dfrac{107.87 \text{ g Ag}}{\text{mol Ag}} + 1 \times \dfrac{14.01 \text{ g N}}{\text{mol N}} + 3 \times \dfrac{16.00 \text{ g O}}{\text{mol O}}$

= 107.87 g/mol + 14.01 g/mol + 48.00 g/mol

= 169.88 g/mol

molar mass of AgCl = $1 \times \dfrac{107.87 \text{ g Ag}}{\text{mol Ag}} + 1 \times \dfrac{35.45 \text{ g Cl}}{\text{mol Cl}}$ = 143.32 g/mol

mass AgCl = 75.0 g $AgNO_3 \times \dfrac{1 \text{ mol } AgNO_3}{169.88 \text{ g } AgNO_3} \times \dfrac{1 \text{ mol AgCl}}{1 \text{ mol } AgNO_3}$

$\times \dfrac{143.32 \text{ g AgCl}}{\text{mol AgCl}}$ = 63.3 g AgCl

5. Given: mass of C_2H_2 = 2.50×10^4 g; balanced equation

Unknown: masses of CO_2 and H_2O in grams

balanced equation:

$2C_2H_2(g) + 5O_2(g) \rightarrow 4CO_2(g) + 2H_2O(g)$

mole ratios from balanced equation = $\dfrac{4 \text{ mol } CO_2}{2 \text{ mol } C_2H_2}, \dfrac{2 \text{ mol } H_2O}{2 \text{ mol } C_2H_2}$

molar mass of C_2H_2 = $2 \times \dfrac{12.01 \text{ g C}}{\text{mol C}} + 2 \times \dfrac{1.01 \text{ g H}}{\text{mol H}}$

= 24.02 g/mol + 2.02 g/mol = 26.04 g/mol

molar mass of CO_2 = $1 \times \dfrac{12.01 \text{ g C}}{\text{mol C}} + 2 \times \dfrac{16.00 \text{ g O}}{\text{mol O}}$ = 44.01 g/mol

mass CO_2 = $(2.50 \times 10^4 \text{ g } C_2H_2) \times \dfrac{1 \text{ mol } C_2H_2}{26.04 \text{ g } C_2H_2} \times \dfrac{4 \text{ mol } CO_2}{2 \text{ mol } C_2H_2}$

$\times \dfrac{44.01 \text{ g } CO_2}{\text{mol } CO_2}$ = 8.45×10^4 g CO_2

molar mass of H_2O = 18.02 g/mol

mass H_2O = $(2.50 \times 10^4 \text{ g } C_2H_2) \times \dfrac{1 \text{ mol } C_2H_2}{26.04 \text{ g } C_2H_2} \times \dfrac{2 \text{ mol } H_2O}{2 \text{ mol } C_2H_2}$

$\times \dfrac{18.02 \text{ g } H_2O}{\text{mol } H_2O}$ = 1.73×10^4 g H_2O

Practice, p. 289

1a. Given: amount of N_2H_4 = 1.750 mol; balanced equation amount of H_2O_2 = 0.500 mol

Unknown: limiting reactant

mole ratio of reactants = $\dfrac{2 \text{ mol } H_2O_2}{1 \text{ mol } N_2H_4}$

amount H_2O_2 required = 0.750 mol $N_2H_4 \times \dfrac{2 \text{ mol } H_2O_2}{1 \text{ mol } N_2H_4}$ = 1.50 mol H_2O_2

Because there is only 0.500 mol H_2O_2 available, the limiting reactant is H_2O_2.

b. Given: amount of N_2H_4 = 0.750 mol amount of H_2O_2 = 0.500 mol (limiting reactant)

Unknown: remaining excess reactant in moles

amount N_2H_4 reacting = $\dfrac{1 \text{ mol } N_2H_4}{2 \text{ mol } H_2O_2} \times 0.500$ mol H_2O_2 = 0.250 mol N_2H_4

remaining excess reactant = 0.750 mol N_2H_4 − 0.250 mol N_2H_4
= 0.500 mol N_2H_4

c. Given: limiting reactant amount = 0.500 mol H_2O_2
mole ratios:
$\dfrac{1 \text{ mol } N_2}{2 \text{ mol } H_2O_2}$,
$\dfrac{4 \text{ mol } H_2O}{2 \text{ mol } H_2O_2}$

Unknown: amounts of N_2 and H_2O in moles

amount N_2 = $\dfrac{1 \text{ mol } N_2}{2 \text{ mol } H_2O_2} \times 0.500$ mol H_2O_2 = 0.250 mol N_2

amount H_2O = $\dfrac{4 \text{ mol } H_2O}{2 \text{ mol } H_2O_2} \times 0.500$ mol H_2O_2 = 1.00 mol H_2O

2. Given: mass of Cl = 20.5 g mass of Na = 20.5 g

Unknown: excess reactant

amount Cl = $\dfrac{20.5 \text{ g}}{35.45 \text{ g/mol}}$ = 0.578 mol Cl

amount Na = $\dfrac{20.5 \text{ g}}{22.99 \text{ g/mol}}$ = 0.892 mol Na

One mole of Na reacts with one mole of Cl. Because there is more Na (0.892 mol) than Cl (0.578 mol), only 0.578 mol Na reacts. Na is therefore the excess reactant.

ATE, Additional Sample Problems 9–6, p. 289

1a. Given: amount of CO = 500. mol amount of H_2 = 750. mol balanced equation

Unknown: limiting reactant

mole ratio of reactants = $\dfrac{2 \text{ mol } H_2}{1 \text{ mol CO}}$

amount H_2 required = 500. mol CO $\times \dfrac{2 \text{ mol } H_2}{1 \text{ mol CO}}$ = 1.00×10^3 mol H_2

Because there are only 750. mol H_2 available, the limiting reactant is H_2.

b. Given: amount of CO = 500. mol
amount of H_2 = 750. mol (limiting reactant)
Unknown: remaining excess reactant in moles

amount CO reacting = $\dfrac{1 \text{ mol CO}}{2 \text{ mol H}_2} \times 750.$ mol H_2 = 375 mol CO

remaining excess reactant = 500. mol CO − 375 mol CO = 125 mol CO

c. Given: limiting reactant amount = 750.0 mol H_2
mole ratio: $\dfrac{1 \text{ mol CH}_3\text{OH}}{2 \text{ mol H}_2}$
Unknown: amount of CH_3OH in moles

amount CH_3OH = $\dfrac{1 \text{ mol CH}_3\text{OH}}{2 \text{ mol H}_2} \times 750.$ mol H_2 = 375 mol CH_3OH

2a. Given: amount of $ZnCO_3$ = 1 mol
amount of $C_6H_8O_7$ = 1 mol
balanced equation
Unknown: limiting reactant

mole ratio of reactants = $\dfrac{2 \text{ mol C}_6\text{H}_8\text{O}_7}{3 \text{ mol ZnCO}_3}$

amount $C_6H_8O_7$ required = 1 mol $ZnCO_3 \times \dfrac{2 \text{ mol C}_6\text{H}_8\text{O}_7}{3 \text{ mol ZnCO}_3}$

= 0.7 mol $C_6H_8O_7$

Because only 0.7 mol $C_6H_8O_7$ is needed to react with 1 mol $ZnCO_3$, the limiting reactant is $ZnCO_3$.

b. Given: amount of $ZnCO_3$ = 6 mol
amount of $C_6H_8O_7$ = 10 mol
balanced equation
Unknown: reactant in excess

mole ratio of reactants = $\dfrac{2 \text{ mol C}_6\text{H}_8\text{O}_7}{3 \text{ mol ZnCO}_3}$

amount $C_6H_8O_7$ required = 6 mol $ZnCO_3 \times \dfrac{2 \text{ mol C}_6\text{H}_8\text{O}_7}{3 \text{ mol ZnCO}_3}$ = 4 mol $C_6H_8O_7$

Because there are 10 mol $C_6H_8O_7$ available and only 4 mol needed, the reactant in excess is $C_6H_8O_7$.

c. Given: limiting reactant amount = 6 mol $ZnCO_3$
mole ratio: $\dfrac{1 \text{ mol Zn}_3(\text{C}_6\text{H}_5\text{O}_7)_2}{3 \text{ mol ZnCO}_3}$
Unknown: amount of $Zn_3(C_6H_5O_7)_2$ in moles

amount $Zn_3(C_6H_5O_7)_2$ = 6 mol $ZnCO_3 \times \dfrac{1 \text{ mol Zn}_3(\text{C}_6\text{H}_5\text{O}_7)_2}{3 \text{ mol ZnCO}_3}$

= 2 mol $Zn_3(C_6H_5O_7)_2$

Practice, p. 291

1a. Given: amount of Zn = 2.00 mol
amount of S_8 = 1.00 mol;
balanced equation
Unknown: limiting reactant

mole ratio of reactants = $\dfrac{1 \text{ mol S}_8}{8 \text{ mol Zn}}$

amount S_8 required = 2.00 mol Zn $\times \dfrac{1 \text{ mol S}_8}{1 \text{ mol Zn}}$ = 0.250 mol S_8

Because only 0.250 mol S_8 is needed to react with 2.00 mol Zn, the limiting reactant is Zn.

b. Given: amount of S_8 = 1.00 mol
amount of S_8 reacting with Zn = 0.250 mol S_8
Unknown: amount of reactant in excess in moles

remaining excess reactant = 1.00 mol S_8 − 0.250 mol S_8 = 0.75 mol S_8

c. Given: limiting reactant amount = 2.00 mol Zn
mole ratio: $\dfrac{8 \text{ mol ZnS}}{8 \text{ mol Zn}}$
Unknown: amount of ZnS in moles

amount ZnS = 2.00 mol Zn × $\dfrac{8 \text{ mol ZnS}}{8 \text{ mol Zn}}$ = 2.00 mol ZnS

2a. Given: amount of C = 2.40 mol
amount of H_2O = 3.10 mol
Unknown: limiting reactant

balanced equation:

$C(s) + H_2O(g) \rightarrow H_2(g) + CO(g)$

mole ratio of reactants = $\dfrac{1 \text{ mol } H_2O}{1 \text{ mol C}}$

amount H_2O required = 2.40 mol C × $\dfrac{1 \text{ mol } H_2O}{1 \text{ mol C}}$ = 2.40 mol H_2O

Because only 2.40 mol H_2O are needed to react with 2.40 mol C, the limiting reactant is C.

b. Given: limiting reactant amount = 2.40 mol C
mole ratios:
$\dfrac{1 \text{ mol } H_2}{1 \text{ mol C}}$,
$\dfrac{1 \text{ mol CO}}{1 \text{ mol C}}$
Unknown: amounts of H_2 and CO in moles

amount H_2 = 2.40 mol C × $\dfrac{1 \text{ mol } H_2}{1 \text{ mol C}}$ = 2.40 mol H_2

amount CO = 2.40 mol C × $\dfrac{1 \text{ mol CO}}{1 \text{ mol C}}$ = 2.40 mol CO

c. Given: amount of H_2 = 2.40 mol
amount of CO = 2.40 mol
Unknown: masses of H_2 and CO in grams

molar mass of H_2 = 2 × $\dfrac{1.01 \text{ g H}}{\text{mol H}}$ = 2.02 g/mol

mass H_2 = 2.40 mol H_2 × $\dfrac{2.02 \text{ g } H_2}{\text{mol } H_2}$ = 4.85 g H_2

molar mass of CO = 1 × $\dfrac{12.01 \text{ g CO}}{\text{mol CO}}$ + 1 × $\dfrac{16.00 \text{ g O}}{\text{mol O}}$

= 12.01 g/mol + 16.00 g/mol = 28.01 g/mol

mass CO = 2.40 mol CO × $\dfrac{28.01 \text{ g CO}}{\text{mol CO}}$ = 67.2 g CO

ATE, Additional Sample Problems 9–7, p. 290

1a. Given: mass of $C_7H_6O_3$ = 20.0 g, mass of $C_4H_6O_3$ = 20.0 g, balanced equation

Unknown: limiting reactant and moles of excess reactant needed to complete reaction

molar mass of $C_7H_6O_3 = 7 \times \dfrac{12.01 \text{ g C}}{\text{mol C}} + 6 \times \dfrac{1.01 \text{ g H}}{\text{mol H}} + 3 \times \dfrac{16.00 \text{ g O}}{\text{mol O}}$

$= 138.13 \text{ g/mol}$

amount $C_7H_6O_3 = \dfrac{20.0 \text{ g}}{138.13 \text{ g/mol}} = 0.145 \text{ mol } C_7H_6O_3$

molar mass of $C_4H_6O_3 = 4 \times \dfrac{12.01 \text{ g C}}{\text{mol C}} + 6 \times \dfrac{1.01 \text{ g H}}{\text{mol H}} + 3 \times \dfrac{16.00 \text{ g O}}{\text{mol O}}$

$= 102.10 \text{ g/mol}$

amount $C_4H_6O_3 = \dfrac{20.0 \text{ g}}{102.10 \text{ g/mol}} = 0.196 \text{ mol } C_4H_6O_3$

mole ratio of reactants $= \dfrac{1 \text{ mol } C_4H_6O_3}{2 \text{ mol } C_7H_6O_3}$

amount $C_4H_6O_3$ required $= 0.145 \text{ mol } C_7H_6O_3 \times \dfrac{1 \text{ mol } C_4H_6O_3}{2 \text{ mol } C_7H_6O_3}$

$= 0.0725 \text{ mol } C_4H_6O_3$

Because only 0.0725 mol $C_4H_6O_3$ is needed to react with all of the 0.145 mol $C_7H_6O_3$, the limiting reactant is $C_7H_6O_3$.

amount of excess reactant used = mol $C_4H_6O_3$ required = 0.0725 mol $C_4H_6O_3$

b. Given: limiting reactant amount = 0.145 mol $C_7H_6O_3$

mole ratio: $\dfrac{2 \text{ mol } C_9H_8O_4}{2 \text{ mol } C_7H_6O_3}$

Unknown: mass of $C_9H_8O_4$ in grams

molar mass of $C_9H_8O_5 = 9 \times \dfrac{12.01 \text{ g C}}{\text{mol C}} + 8 \times \dfrac{1.01 \text{ g H}}{\text{mol H}} + 4 \times \dfrac{16.00 \text{ g O}}{\text{mol O}}$

$= 180.2 \text{ g/mol}$

mass $C_9H_8O_4 = 0.145 \text{ mol } C_7H_6O_3 \times \dfrac{2 \text{ mol } C_9H_8O_4}{2 \text{ mol } C_7H_6O_3} \times \dfrac{180.2 \text{ g } C_9H_8O_4}{\text{mol } C_9H_8O_4}$

$= 26.1 \text{ g } C_9H_8O_4$

2. Given: mass of $Ca_3(PO_4)_2$ = 250 g
amount of H_2SO_4 = 3 mol
Unknown: limiting reactant, amount in moles and mass in grams of $CaSO_4$

balanced equation:
$$Ca_3(PO_4)_2(s) + 3H_2SO_4(aq) \rightarrow 3CaSO_4(s) + 2H_3PO_4(aq)$$

mole ratio of reactants = $\dfrac{3 \text{ mol } H_2SO_4}{1 \text{ mol } Ca_3(PO_4)_2}$

molar mass of $Ca_3(PO_4)_2$ = $3 \times \dfrac{40.08 \text{ g Ca}}{\text{mol Ca}} + 2 \times \dfrac{30.97 \text{ g P}}{\text{mol P}}$

$+ 8 \times \dfrac{16.00 \text{ g O}}{\text{mol O}} = 310.2$ g/mol

amount $Ca_3(PO_4)_2 = \dfrac{250 \text{ g}}{310.2 \text{ g/mol}} = 0.81$ mol $Ca_3(PO_4)_2$

amount H_2SO_4 required = 0.81 mol $Ca_3(PO_4)_2 \times \dfrac{3 \text{ mol } H_2SO_4}{1 \text{ mol } Ca_3(PO_4)_2}$

$= 2.4$ mol H_2SO_4

This is less than the 3 mol H_2SO_4 available for the reaction, so the limiting reactant is $Ca_3(PO_4)_2$.

mole ratio of $CaSO_4$ to limiting reactant $Ca_3(PO_4)_2 = \dfrac{3 \text{ mol } CaSO_4}{1 \text{ mol } Ca_3(PO_4)_2}$

amount $CaSO_4 = 0.81$ mol $Ca_3(PO_4)_2 \times \dfrac{3 \text{ mol } CaSO_4}{1 \text{ mol } Ca_3(PO_4)_2} = 2.42$ mol $CaSO_4$

This is less than 3 mol $CaSO_4$, so 3 mol $CaSO_4$ cannot be produced.

molar mass of $CaSO_4 = 1 \times \dfrac{40.08 \text{ g Ca}}{\text{mol Ca}} + 1 \times \dfrac{32.07 \text{ g S}}{\text{mol S}} + 4 \times \dfrac{16.00 \text{ g O}}{\text{mol O}}$

$= 136.15$ g/mol

mass $CaSO_4 = 2.4$ mol $CaSO_4 \times \dfrac{136.15 \text{ g } CaSO_4}{\text{mol } CaSO_4} = 329$ g $CaSO_4$

Practice, p. 294

1. Given: mass of CO = 75.0 g,
balanced equation,
actual yield of CH_3OH = 68.4 g
Unknown: percent yield of CH_3OH

mole ratio from balanced equation = $\dfrac{1 \text{ mol } CH_3OH}{1 \text{ mol CO}}$

molar mass of $CH_3OH = 1 \times \dfrac{12.01 \text{ g C}}{\text{mol C}} + 4 \times \dfrac{1.01 \text{ g H}}{\text{mol H}} + 1 \times \dfrac{16.00 \text{ g O}}{\text{mol O}}$

$= 32.05$ g/mol

molar mass of CO = $1 \times \dfrac{12.01 \text{ g C}}{\text{mol C}} + 1 \times \dfrac{16.00 \text{ g O}}{\text{mol O}} = 28.01$ g/mol

mass CH_3OH(theoretical) = 75.0 g CO $\times \dfrac{1 \text{ mol CO}}{28.01 \text{ g CO}} \times \dfrac{1 \text{ mol } CH_3OH}{1 \text{ mol CO}}$

$\times \dfrac{32.05 \text{ g } CH_3OH}{\text{mol } CH_3OH} = 85.9$ g CH_3OH

percent yield of $CH_3OH = \dfrac{68.4 \text{ g } CH_3OH}{85.9 \text{ g } CH_3OH} \times 100 = 79.6\%$

2. Given: mass of Al = 1.85 g
percent yield of Cu = 56.6%

Unknown: mass of Cu (actual yield)

balanced equation:

$$2Al(s) + 3CuSO_4(aq) \rightarrow Al_2(SO_4)_3(aq) + 3Cu(s)$$

$$\text{mole ratio from balanced equation} = \frac{3 \text{ mol Cu}}{2 \text{ mol Al}}$$

$$\text{mass Cu (theoretical)} = 1.85 \text{ g Al} \times \frac{\text{mol Al}}{26.98 \text{ g. Al}} \times \frac{3 \text{ mol Cu}}{2 \text{ mol Al}} \times \frac{63.55 \text{ g Cu}}{\text{mol Cu}}$$

$$= 6.54 \text{ g Cu}$$

$$\text{mass Cu (actual)} = \frac{56.6 \times 6.54 \text{ g Cu}}{100} = 3.70 \text{ g Cu}$$

ATE, Additional Sample Problem 9–8, p. 294

Given: percent yield of SO_2 = 86.78%
mass of ZnS = 4897 g
balanced equation

Unknown: mass of SO_2 (actual yield)

$$\text{mole ratio from balanced equation} = \frac{2 \text{ mol } SO_2}{2 \text{ mol ZnS}}$$

$$\text{molar mass of } SO_2 = 1 \times \frac{32.07 \text{ g S}}{\text{mol S}} + 2 \times \frac{16.00 \text{ g O}}{\text{mol O}} = 64.07 \text{ g/mol}$$

$$\text{molar mass of ZnS} = 1 \times \frac{65.39 \text{ g Zn}}{\text{mol Zn}} + 1 \times \frac{32.07 \text{ g S}}{\text{mol S}} = 97.46 \text{ g/mol}$$

$$\text{mass } SO_2 \text{ (theoretical)} = 4897 \text{ g ZnS} \times \frac{1 \text{ mol ZnS}}{97.46 \text{ g ZnS}} \times \frac{2 \text{ mol } SO_2}{2 \text{ mol ZnS}}$$

$$\times \frac{64.07 \text{ g } SO_2}{\text{mol } SO_2} = 3219 \text{ g } SO_2$$

$$\text{mass } SO_2 \text{ (actual)} = \frac{86.78}{100} \times 3219 \text{ g } SO_2 = 2794 \text{ g } SO_2$$

Section Review, p. 294

1a. Given: amount of CS_2 = 1.00 mol,
amount of O_2 = 1.00 mol,
balanced equation

Unknown: limiting reactant

$$\text{mole ratio of reactants} = \frac{3 \text{ mol } O_2}{1 \text{ mol } CS_2}$$

$$\text{amount } O_2 \text{ required} = 1.00 \text{ mol } CS_2 \times \frac{3 \text{ mol } O_2}{1 \text{ mol } CS_2} = 3.00 \text{ mol } O_2$$

Because there is only 1.00 mol O_2 available, the limiting reactant is O_2.

b. Given: amount of CS_2 = 1.00 mol
amount of O_2 = 1.00 mol (limiting reactant)

Unknown: remaining excess reactant in moles

$$\text{amount } CS_2 \text{ reacting} = \frac{1 \text{ mol } CS_2}{3 \text{ mol } O_2} \times 1.00 \text{ mol } O_2 = 0.333 \text{ mol } CS_2$$

remaining excess reactant = 1.00 mol CS_2 – 0.333 mol CS_2 = 0.667 mol CS_2

c. Given: limiting reactant amount = 1.00 mol O_2
mole ratios:
$\dfrac{1 \text{ mol } CO_2}{3 \text{ mol } O_2}$,
$\dfrac{2 \text{ mol } SO_2}{3 \text{ mol } O_2}$

Unknown: amounts of CO_2 and SO_2 in moles

amount $CO_2 = \dfrac{1 \text{ mol } CO_2}{3 \text{ mol } O_2} \times 1.00 \text{ mol } O_2 = 0.333 \text{ mol } CO_2$

amount $SO_2 = \dfrac{2 \text{ mol } SO_2}{3 \text{ mol } O_2} \times 1.00 \text{ mol } O_2 = 0.667 \text{ mol } SO_2$

2a. Given: mass of Mg = 16.2 g
mass of H_2O = 12.0 g

Unknown: limiting reactant

balanced equation:

$2Mg(s) + 4H_2O(g) \rightarrow 2Mg(OH)_2(s) + 2H_2(g)$

mole ratio of reactants = $\dfrac{4 \text{ mol } H_2O}{2 \text{ mol } Mg}$

amount $H_2O = 12.0 \text{ g } H_2O \times \dfrac{1 \text{ mol } H_2O}{18.02 \text{ g } H_2O} = 0.666 \text{ mol } H_2O$

amount $Mg = 16.2 \text{ g Mg} \times \dfrac{1 \text{ mol Mg}}{24.30 \text{ g Mg}} = 0.667 \text{ mol Mg}$

amount H_2O required $= 0.667 \text{ mol Mg} \times \dfrac{4 \text{ mol } H_2O}{2 \text{ mol Mg}} = 1.33 \text{ mol } H_2O$

Because there is only 0.666 mol H_2O available, the limiting reactant is H_2O.

b. Given: amount of Mg = 0.667 mol
amount of H_2O = 0.666 mol (limiting reactant)

Unknown: remaining excess reactant in moles

amount Mg reacting = $\dfrac{2 \text{ mol Mg}}{4 \text{ mol } H_2O} \times 0.666 \text{ mol } H_2O = 0.333 \text{ mol Mg}$

remaining excess reactant = 0.667 mol Mg − 0.333 mol Mg = 0.334 mol Mg

c. Given: limiting reactant amount = 0.666 mol H_2O
mole ratios:
$\dfrac{2 \text{ mol } Mg(OH)_2}{4 \text{ mol } H_2O}$,
$\dfrac{2 \text{ mol } H_2}{4 \text{ mol } H_2O}$

Unknown: masses of $Mg(OH)_2$ and H_2 in grams

molar mass of $Mg(OH)_2 = 1 \times \dfrac{24.30 \text{ g Mg}}{\text{mol Mg}} + 2 \times \dfrac{16.00 \text{ g O}}{\text{mol O}}$
$+ 2 \times \dfrac{1.01 \text{ g H}}{\text{mol H}} = 58.32 \text{ g/mol}$

molar mass of $H_2 = 2 \times \dfrac{1.01 \text{ g H}}{\text{mol H}} = 2.02 \text{ g/mol}$

mass $Mg(OH)_2 = 0.666 \text{ mol } H_2O \times \dfrac{2 \text{ mol } Mg(OH)_2}{4 \text{ mol } H_2O} \times \dfrac{58.32 \text{ g } Mg(OH)_2}{\text{mol } Mg(OH)_2}$

$= 19.4 \text{ g } Mg(OH)_2$

mass $H_2 = 0.666 \text{ mol } H_2O \times \dfrac{2 \text{ mol } H_2}{4 \text{ mol } H_2O} \times \dfrac{2.02 \text{ g } H_2}{\text{mol } H_2} = 0.673 \text{ g } H_2$

3a. Given: mass of CuO = 19.9 g
mass of H_2 = 2.02 g
balanced equation
Unknown: limiting reactant

mole ratio of reactants = $\dfrac{1 \text{ mol } H_2}{1 \text{ mol CuO}}$

molar mass of CuO = $1 \times \dfrac{63.55 \text{ g Cu}}{\text{mol Cu}} + 1 \times \dfrac{16.00 \text{ g O}}{\text{mol O}} = 79.55$ g/mol

molar mass of H_2 = $2 \times \dfrac{1.01 \text{ g H}}{\text{mol}} = 2.02$ g/mol

amount CuO = 19.9 g CuO $\times \dfrac{1 \text{ mol CuO}}{79.55 \text{ g CuO}} = 0.250$ mol CuO

amount H_2 = 2.02 g $H_2 \times \dfrac{1 \text{ mol } H_2}{2.02 \text{ g } H_2} = 1.00$ mol H_2

amount H_2 required = 0.250 mol CuO $\times \dfrac{1 \text{ mol } H_2}{1 \text{ mol CuO}} = 0.250$ mol H_2

Because only 0.250 mol H_2 is required to react with 0.250 mol CuO, the limiting reactant is CuO.

b. Given: limiting reactant amount = 0.250 mol CuO
mole ratio: $\dfrac{1 \text{ mol Cu}}{1 \text{ mol CuO}}$
Unknown: mass of Cu in grams

molar mass of Cu = 63.55 g/mol

mass Cu = 0.250 mol CuO $\times \dfrac{1 \text{ mol Cu}}{1 \text{ mol CuO}} \times \dfrac{63.55 \text{ g Cu}}{\text{mol Cu}} = 15.9$ g Cu

4. Given: mass of $CaCO_3$ = 2.00×10^3 g, actual yield of CaO = 1.05×10^3 g, balanced reaction
Unknown: percent yield of CaO

mole ratio from balanced equation = $\dfrac{1 \text{ mol CaO}}{1 \text{ mol CaCO}_3}$

molar mass of $CaCO_3$ = $1 \times \dfrac{40.08 \text{ g Ca}}{\text{mol Ca}} + 1 \times \dfrac{12.01 \text{ g C}}{\text{mol C}} + 3 \times \dfrac{16.00 \text{ g O}}{\text{mol O}}$

$= 100.09$ g/mol

molar mass of CaO = $1 \times \dfrac{40.08 \text{ g Ca}}{\text{mol Ca}} + 1 \times \dfrac{16.00 \text{ g O}}{\text{mol O}} = 56.08$ g/mol

mass CaO (theoretical) = $(2.00 \times 10^3 \text{ g CaCO}_3) \times \dfrac{1 \text{ mol CaCO}_3}{100.09 \text{ g CaCO}_3}$

$\times \dfrac{1 \text{ mol CaO}}{1 \text{ mol CaCO}_3} \times \dfrac{56.08 \text{ g CaO}}{\text{mol CaO}}$

$= 1.12 \times 10^3$ g CaO

percent yield of CaO = $\dfrac{1.05 \times 10^3 \text{ g CaO}}{1.12 \times 10^3 \text{ g CaO}} \times 100 = 93.8\%$

Review Problems

10a. Given: amount of H_2O = 5.0 mol; balanced equation
Unknown: amount of H_2 in moles

mole ratio from balanced equation = $\dfrac{2 \text{ mol } H_2}{2 \text{ mol } H_2O}$

amount H_2 = 5.0 mol $H_2O \times \dfrac{2 \text{ mol } H_2}{2 \text{ mol } H_2O} = 5.0$ mol H_2

b. Given: amount of H_2O = 5.0 mol

Unknown: amount of O_2 in moles

mole ratio from balanced equation = $\dfrac{1 \text{ mol } O_2}{2 \text{ mol } H_2O}$

amount O_2 = 5.0 mol $H_2O \times \dfrac{1 \text{ mol } O_2}{2 \text{ mol } H_2O}$ = 2.5 mol O_2

11a. Given: amount of C_2H_6 = 4.50 mol

Unknown: amount of O_2 in moles

balanced equation:

$2C_2H_6 + 7O_2 \rightarrow 4CO_2 + 6H_2O$

mole ratio from balanced equation = $\dfrac{7 \text{ mol } O_2}{2 \text{ mol } C_2H_6}$

amount O_2 = 4.50 mol $C_2H_6 \times \dfrac{7 \text{ mol } O_2}{2 \text{ mol } C_2H_6}$ = 15.8 mol O_2

b. Given: amount of C_2H_6 = 4.50 mol

Unknown: amounts of CO_2 and H_2O in moles

mole ratios from balanced equation = $\dfrac{4 \text{ mol } CO_2}{2 \text{ mol } C_2H_6}, \dfrac{6 \text{ mol } H_2O}{2 \text{ mol } C_2H_6}$

amount CO_2 = 4.50 mol $C_2H_6 \times \dfrac{4 \text{ mol } CO_2}{2 \text{ mol } C_2H_6}$ = 9.00 mol CO_2

amount H_2O = 4.50 mol $C_2H_6 \times \dfrac{6 \text{ mol } H_2O}{2 \text{ mol } C_2H_6}$ = 13.5 mol H_2O

12. Given: amount of NaCl = 25.0 mol

Unknown: masses of Na and Cl

balanced equation:

$2Na + Cl_2 \rightarrow 2NaCl$

mole ratios from balanced equation = $\dfrac{2 \text{ mol Na}}{2 \text{ mol NaCl}}, \dfrac{1 \text{ mol } Cl_2}{2 \text{ mol NaCl}}$

mass Na = 25.0 mol NaCl $\times \dfrac{2 \text{ mol Na}}{2 \text{ mol NaCl}} \times \dfrac{22.99 \text{ g Na}}{\text{mol Na}}$ = 575 g Na

mass Cl_2 = 25.0 mol NaCl $\times \dfrac{1 \text{ mol } Cl_2}{2 \text{ mol NaCl}} \times \dfrac{70.90 \text{ g } Cl_2}{\text{mol } Cl_2}$ = 886 g Cl_2

13a. Given: mass of Fe_2O_3 = 4.00 kg

Unknown: amount of CO in moles

balanced equation:

$2Fe_2O_3(s) + 6CO(g) \rightarrow 4Fe(s) + 6CO_2(g)$

mole ratio from balanced equation = $\dfrac{6 \text{ mol CO}}{2 \text{ mol } Fe_2O_3}$

molar mass of Fe_2O_3 = $2 \times \dfrac{55.85 \text{ g Fe}}{\text{mol Fe}} + 3 \times \dfrac{16.00 \text{ g O}}{\text{mol O}}$ = 159.7 g/mol

amount CO = 4.00 kg $Fe_2O_3 \times \dfrac{1000 \text{ g}}{\text{kg}} \times \dfrac{1 \text{ mol } Fe_2O_3}{159.7 \text{ g } Fe_2O_3} \times \dfrac{6 \text{ mol CO}}{2 \text{ mol } Fe_2O_3}$

= 75.1 mol CO

13b. Given: mass of Fe_2O_3 = 4.00 kg
Unknown: amounts of Fe and CO_2 in moles

mole ratios from balanced equation = $\dfrac{4 \text{ mol Fe}}{2 \text{ mol Fe}_2\text{O}_3}, \dfrac{6 \text{ mol CO}_2}{2 \text{ mol Fe}_2\text{O}_3}$

molar mass of Fe_2O_3 = 159.7 g/mol

amount Fe = 4.00 kg $Fe_2O_3 \times \dfrac{1000 \text{ g}}{\text{kg}} \times \dfrac{1 \text{ mol Fe}_2\text{O}_3}{159.7 \text{ g Fe}_2\text{O}_3} \times \dfrac{4 \text{ mol Fe}}{2 \text{ mol Fe}_2\text{O}_3}$

= 50.1 mol Fe

amount CO_2 = 4.00 kg $Fe_2O_3 \times \dfrac{1000 \text{ g}}{\text{kg}} \times \dfrac{1 \text{ mol Fe}_2\text{O}_3}{159.7 \text{ g Fe}_2\text{O}_3} \times \dfrac{6 \text{ mol CO}_2}{2 \text{ mol Fe}_2\text{O}_3}$

= 75.1 mol CO_2

---

14. Given: mass of CH_3OH = 100.0 kg
Unknown: masses of CO and H_2

balanced equation:

$CO(g) + 2H_2(g) \rightarrow CH_3OH$

mole ratios from balanced equation = $\dfrac{1 \text{ mol CO}}{1 \text{ mol CH}_3\text{OH}}, \dfrac{2 \text{ mol H}_2}{1 \text{ mol CH}_3\text{OH}}$

molar mass of CH_3OH = $1 \times \dfrac{12.01 \text{ g C}}{\text{mol C}} + 4 \times \dfrac{1.01 \text{ g H}}{\text{mol H}} + 1 \times \dfrac{16.00 \text{ g O}}{\text{mol O}}$

= 32.05 g/mol

molar mass of CO = $1 \times \dfrac{12.01 \text{ g C}}{\text{mol C}} + 1 \times \dfrac{16.00 \text{ g O}}{\text{mol O}}$ = 28.01 g/mol

mass CO = 100.0 kg $CH_3OH \times \dfrac{1000 \text{ g}}{\text{kg}} \times \dfrac{1 \text{ mol CH}_3\text{OH}}{32.05 \text{ g CH}_3\text{OH}} \times \dfrac{1 \text{ mol CO}}{1 \text{ mol CH}_3\text{OH}}$

$\times \dfrac{28.01 \text{ g CO}}{\text{mol CO}}$ = 8.739×10^4 g CO

molar mass of H_2 = $2 \times \dfrac{1.01 \text{ g H}}{\text{mol H}}$ = 2.02 g/mol

mass H_2 = 100.0 kg $H_2 \times \dfrac{1000 \text{ g}}{\text{kg}} \times \dfrac{1 \text{ mol CH}_3\text{OH}}{32.05 \text{ g CH}_3\text{OH}} \times \dfrac{2 \text{ mol H}_2}{1 \text{ mol CH}_3\text{OH}}$

$\times \dfrac{2.02 \text{ g H}_2}{\text{mol H}_2}$ = 1.261×10^4 g H_2

15a. Given: mass of O_2 = 384 g
Unknown: mass of NO_2

balanced equation:

$2NO(g) + O_2(g) \rightarrow 2NO_2(g)$

mole ratio from balanced equation = $\dfrac{2 \text{ mol NO}_2}{1 \text{ mol O}_2}$

molar mass O_2 = $2 \times \dfrac{16.00 \text{ g O}}{\text{mol O}}$ = 32.00 g/mol

molar mass of NO_2 = $1 \times \dfrac{14.01 \text{ g N}}{\text{mol N}} + 2 \times \dfrac{16.00 \text{ g O}}{\text{mol O}}$ = 46.01 g/mol

mass NO_2 = 384 g $O_2 \times \dfrac{1 \text{ mol O}_2}{32.00 \text{ g O}_2} \times \dfrac{2 \text{ mol NO}_2}{1 \text{ mol O}_2} \times \dfrac{46.01 \text{ g NO}_2}{\text{mol NO}_2}$

= 1.10×10^3 g NO_2

b. Given: mass of O_2 = 384 g
Unknown: mass of NO in grams

mole ratio from balanced equation = $\dfrac{2 \text{ mol NO}}{1 \text{ mol } O_2}$

molar mass of NO = $1 \times \dfrac{14.01 \text{ g N}}{\text{mol N}} + 1 \times \dfrac{16.00 \text{ g O}}{\text{mol O}} = 30.01$ g/mol

molar mass of O_2 = 32.00 g/mol

mass NO = 384 g $\times \dfrac{1 \text{ mol } O_2}{32.00 \text{ g } O_2} \times \dfrac{2 \text{ mol NO}}{1 \text{ mol } O_2} \times \dfrac{30.01 \text{ g NO}}{\text{mol NO}} = 720.$ g NO

16a. Given: mass of CO_2 = 925.0 g
Unknown: amount of NaOH in moles

balanced equation:

$2NaOH + CO_2 \rightarrow Na_2CO_3 + H_2O$

mole ratio from balanced equation = $\dfrac{2 \text{ mol NaOH}}{1 \text{ mol } CO_2}$

molar mass of CO_2 = $1 \times \dfrac{12.01 \text{ g C}}{\text{mol C}} + 2 \times \dfrac{16.00 \text{ g O}}{\text{mol O}} = 44.01$ g/mol

amount NaOH = 925.0 g $CO_2 \times \dfrac{1 \text{ mol } CO_2}{44.01 \text{ g } CO_2} \times \dfrac{2 \text{ mol NaOH}}{1 \text{ mol } CO_2}$

= 42.04 mol NaOH

b. Given: mass of CO_2 = 925.0 g
Unknown: amounts of Na_2CO_3 and H_2O in moles

mole ratios from balanced equation = $\dfrac{1 \text{ mol } Na_2CO_3}{1 \text{ mol } CO_2}, \dfrac{1 \text{ mol } H_2O}{1 \text{ mol } CO_2}$

molar mass of CO_2 = 44.01 g/mol

amount Na_2CO_3 = 925.0 g $CO_2 \times \dfrac{1 \text{ mol } CO_2}{44.01 \text{ g } CO_2} \times \dfrac{1 \text{ mol } Na_2CO_3}{1 \text{ mol } CO_2}$

= 21.02 mol Na_2CO_3

amount H_2O = 925.0 g $CO_2 \times \dfrac{1 \text{ mol } CO_2}{44.01 \text{ g } CO_2} \times \dfrac{1 \text{ mol } H_2O}{1 \text{ mol } CO_2} = 21.02$ mol H_2O

17a. Given: amount of $AgNO_3$ = 4.50 mol
Unknown: mass of NaBr

balanced equation:

$AgNO_3 + NaBr \rightarrow AgBr + NaNO_3$

mole ratio from balanced equation = $\dfrac{1 \text{ mol NaBr}}{1 \text{ mol } AgNO_3}$

molar mass of NaBr = $1 \times \dfrac{22.99 \text{ g Na}}{\text{mol Na}} + 1 \times \dfrac{79.90 \text{ g Br}}{\text{mol Br}} = 102.89$ g/mol

mass NaBr = 4.50 mol $AgNO_3 \times \dfrac{1 \text{ mol NaBr}}{1 \text{ mol } AgNO_3} \times \dfrac{102.89 \text{ g NaBr}}{\text{mol NaBr}}$

= 463 g NaBr

b. Given: amount of $AgNO_3$ = 4.50 mol
Unknown: mass of AgBr

mole ratio from balanced equation = $\dfrac{1 \text{ mol AgBr}}{1 \text{ mol } AgNO_3}$

molar mass of AgBr = $1 \times \dfrac{107.87 \text{ g Ag}}{\text{mol Ag}} + 1 \times \dfrac{79.90 \text{ g Br}}{\text{mol Br}} = 187.77$ g/mol

mass AgBr = 4.50 mol $AgNO_3 \times \dfrac{1 \text{ mol AgBr}}{1 \text{ mol } AgNO_3} \times \dfrac{187.77 \text{ g AgBr}}{\text{mol AgBr}}$

= 845 g AgBr

18a. Given: mass of H_2SO_4 = 150.0 g
Unknown: amount of $NaHCO_3$ in moles

balanced equation:

$2NaHCO_3 + H_2SO_4 \rightarrow 2CO_2 + Na_2SO_4 + 2H_2O$

mole ratio from balanced equation = $\dfrac{2 \text{ mol } NaHCO_3}{1 \text{ mol } H_2SO_4}$

molar mass of $H_2SO_4 = 2 \times \dfrac{1.01 \text{ g H}}{\text{mol H}} + 1 \times \dfrac{32.07 \text{ g S}}{\text{mol S}} + 4 \times \dfrac{16.00 \text{ g O}}{\text{mol O}}$

= 98.09 g/mol

amount $NaHCO_3$ = 150.0 g $H_2SO_4 \times \dfrac{1 \text{ mol } H_2SO_4}{98.09 \text{ g } H_2SO_4} \times \dfrac{2 \text{ mol } NaHCO_3}{1 \text{ mol } H_2SO_4}$

= 3.058 mol $NaHCO_3$

b. Given: mass of H_2SO_4 = 150.0 g
Unknown: amounts of CO_2, Na_2SO_4, and H_2O in moles

mole ratios from balanced equation = $\dfrac{2 \text{ mol } CO_2}{1 \text{ mol } H_2SO_4}, \dfrac{1 \text{ mol } Na_2SO_4}{1 \text{ mol } H_2SO_4},$

$\dfrac{2 \text{ mol } H_2O}{1 \text{ mol } H_2SO_4}$

molar mass of H_2SO_4 = 98.09 g/mol

amount CO_2 = 150.0 g $H_2SO_4 \times \dfrac{1 \text{ mol } H_2SO_4}{98.09 \text{ g } H_2SO_4} \times \dfrac{2 \text{ mol } CO_2}{1 \text{ mol } H_2SO_4}$

= 3.058 mol CO_2

amount Na_2SO_4 = 150.0 g $H_2SO_4 \times \dfrac{1 \text{ mol } H_2SO_4}{98.09 \text{ g } H_2SO_4} \times \dfrac{1 \text{ mol } Na_2SO_4}{1 \text{ mol } H_2SO_4}$

= 1.529 mol Na_2SO_4

amount H_2O = 150.0 g $H_2SO_4 \times \dfrac{1 \text{ mol } H_2SO_4}{98.09 \text{ g } H_2SO_4} \times \dfrac{2 \text{ mol } H_2O}{1 \text{ mol } H_2SO_4}$

= 3.058 mol H_2O

19b. Given: amount of NaOH = 0.75 mol
Unknown: mass of H_2SO_4

mole ratio from balanced equation = $\dfrac{1 \text{ mol } H_2SO_4}{2 \text{ mol NaOH}}$

molar mass of $H_2SO_4 = 2 \times \dfrac{1.01 \text{ g H}}{\text{mol H}} + 1 \times \dfrac{32.07 \text{ g S}}{\text{mol S}} + 4 \times \dfrac{16.00 \text{ g O}}{\text{mol O}}$

= 98.09 g/mol

mass H_2SO_4 = 0.75 mol NaOH $\times \dfrac{1 \text{ mol } H_2SO_4}{2 \text{ mol NaOH}} \times \dfrac{98.09 \text{ g } H_2SO_4}{\text{mol } H_2SO_4}$

= 37 g H_2SO_4

c. Given: amount of NaOH = 0.75 mol
Unknown: masses of Na_2SO_4 and H_2O

mole ratios from balanced equations = $\dfrac{1 \text{ mol } Na_2SO_4}{2 \text{ mol NaOH}}, \dfrac{2 \text{ mol } H_2O}{2 \text{ mol NaOH}}$

molar mass of $Na_2SO_4 = 2 \times \dfrac{22.99 \text{ g Na}}{\text{mol Na}} + 1 \times \dfrac{32.07 \text{ g S}}{\text{mol S}} + 4 \times \dfrac{16.00 \text{ g O}}{\text{mol O}}$

= 142.05 g/mol

mass Na_2SO_4 = 0.75 mol NaOH $\times \dfrac{1 \text{ mol } Na_2SO_4}{2 \text{ mol NaOH}} \times \dfrac{142.05 \text{ g } Na_2SO_4}{\text{mol } Na_2SO_4}$

= 53 g Na_2SO_4

molar mass of H_2O = 18.02 g/mol

mass H_2O = 0.75 mol NaOH $\times \dfrac{2 \text{ mol } H_2O}{2 \text{ mol NaOH}} \times \dfrac{18.02 \text{ g } H_2O}{\text{mol } H_2O}$ = 14 g H_2O

20a. Given: mass of Ag = 2.25 g
Unknown: amount of $Cu(NO_3)_2$ in moles

balanced equation:

$Cu + 2AgNO_3 \rightarrow 2Ag + Cu(NO_3)_2$

mole ratio from balanced equation = $\dfrac{1 \text{ mol } Cu(NO_3)_2}{2 \text{ mol Ag}}$

molar mass of Ag = 107.87 g/mol

amount $Cu(NO_3)_2$ = 2.25 g Ag $\times \dfrac{1 \text{ mol Ag}}{107.87 \text{ g Ag}} \times \dfrac{1 \text{ mol } Cu(NO_3)_2}{2 \text{ mol Ag}}$

= 0.0104 mol $Cu(NO_3)_2$

b. Given: mass of Ag = 2.25 g
Unknown: amounts of Cu and $AgNO_3$ in moles

mole ratios from balanced equation = $\dfrac{1 \text{ mol Cu}}{2 \text{ mol Ag}}, \dfrac{2 \text{ mol } AgNO_3}{2 \text{ mol Ag}}$

molar mass of Ag = 107.87 g/mol

amount Cu = 2.25 g Ag $\times \dfrac{1 \text{ mol Ag}}{107.87 \text{ g Ag}} \times \dfrac{1 \text{ mol Cu}}{2 \text{ mol Ag}}$ = 0.0104 mol Cu

amount $AgNO_3$ = 2.25 g Ag $\times \dfrac{1 \text{ mol Ag}}{107.87 \text{ g Ag}} \times \dfrac{2 \text{ mol } AgNO_3}{2 \text{ mol Ag}}$

= 0.0209 mol $AgNO_3$

21a. Given: amount of $C_7H_6O_3$ = 75.0 mol
Unknown: mass of $C_9H_8O_4$ in kg

balanced equation:

$C_7H_6O_3(s) + C_4H_6O_3(l) \rightarrow C_9H_8O_4(s) + HC_2H_3O_2(l)$

mole ratio from balanced equation = $\dfrac{1 \text{ mol } C_9H_8O_4}{1 \text{ mol } C_7H_6O_3}$

molar mass of $C_9H_8O_4 = 9 \times \dfrac{12.01 \text{ g C}}{\text{mol C}} + 8 \times \dfrac{1.01 \text{ g H}}{\text{mol H}} + 4 \times \dfrac{16.00 \text{ g O}}{\text{mol O}}$

= 180.2 g/mol

mass $C_9H_8O_4$ = 75.0 mol $C_7H_6O_3 \times \dfrac{1 \text{ mol } C_9H_8O_4}{1 \text{ mol } C_7H_6O_3} \times \dfrac{180.2 \text{ g } C_9H_8O_4}{\text{mol } C_9H_8O_4}$

$\times \dfrac{1 \text{ kg}}{1000 \text{ g}}$ = 13.5 kg $C_9H_8O_4$

21b. Given: amount of $C_7H_6O_3$ = 75.0 mol

Unknown: mass of $C_4H_6O_3$ in kg

mole ratio from balanced equation = $\dfrac{1 \text{ mol } C_4H_6O_3}{1 \text{ mol } C_7H_6O_3}$

molar mass of $C_4H_6O_3$ = $4 \times \dfrac{12.01 \text{ g C}}{\text{mol C}} + 6 \times \dfrac{1.01 \text{ g H}}{\text{mol H}} + 3 \times \dfrac{16.00 \text{ g O}}{\text{mol O}}$

= 102.10 g/mol

mass $C_4H_6O_3$ = 75.0 mol $C_7H_6O_3 \times \dfrac{1 \text{ mol } C_4H_6O_3}{1 \text{ mol } C_7H_6O_3} \times \dfrac{102.10 \text{ g } C_4H_6O_3}{\text{mol } C_4H_6O_3}$

$\times \dfrac{1 \text{ kg}}{1000 \text{ g}}$ = 7.66 kg $C_4H_6O_3$

c. Given: amount of $C_7H_6O_3$ = 75.0 mol; density of $HC_2H_3O_2$ = 1.05 g/cm³

Unknown: volume of $HC_2H_3O_2$ in liters

mole ratio from balanced equation = $\dfrac{1 \text{ mol } HC_2H_3O_2}{1 \text{ mol } C_7H_6O_3}$

molar mass of $HC_2H_3O_2$ = $4 \times \dfrac{1.01 \text{ g H}}{\text{mol H}} + 2 \times \dfrac{12.01 \text{ g C}}{\text{mol C}} + 2 \times \dfrac{16.00 \text{ g O}}{\text{mol O}}$

= 60.06 g/mol

V $HC_2H_3O_2$ = 75.0 mol $C_7H_6O_3 \times \dfrac{1 \text{ mol } HC_2H_3O_2}{1 \text{ mol } C_7H_6O_3} \times \dfrac{60.06 \text{ g } HC_2H_3O_2}{\text{mol } HC_2H_3O_2}$

$\times \dfrac{1 \text{ cm}^3 \text{ } HC_2H_3O_2}{1.05 \text{ g } HC_2H_3O_2} \times \dfrac{1 \text{ L}}{1000 \text{ cm}^3}$ = 4.29 L $HC_2H_3O_2$

22a. Given: amount of HCl = 2.0 mol, amount of NaOH = 2.5 mol, balanced equation

Unknown: limiting reactant

mole ratio of reactants = $\dfrac{1 \text{ mol NaOH}}{1 \text{ mol HCl}}$

amount NaOH required = 2.0 mol HCl $\times \dfrac{1 \text{ mol NaOH}}{1 \text{ mol HCl}}$ = 2.0 mol NaOH

Because only 2.0 mol NaOH are needed to react with 2.0 mol HCl, the limiting reactant is HCl.

b. Given: amount of Zn = 2.5 mol, amount of HCl = 6.0 mol, balanced equation

Unknown: limiting reactant

balanced equation:

Zn + 2HCl → $ZnCl_2$ + H_2

mole ratio of reactants = $\dfrac{2 \text{ mol HCl}}{1 \text{ mol Zn}}$

amount HCl required = 2.5 mol Zn $\times \dfrac{2 \text{ mol HCl}}{1 \text{ mol Zn}}$ = 5.0 mol HCl

Because only 5.0 mol HCl are needed to react with 2.5 mol Zn, the limiting reactant is Zn.

c. Given: amount of $Fe(OH)_3$ = 4.0 mol, amount of H_2SO_4 = 6.5 mol, balanced equation

Unknown: limiting reactant

mole ratio of reactants = $\dfrac{3 \text{ mol } H_2SO_4}{2 \text{ mol } Fe(OH)_3}$

amount H_2SO_4 required = 4.0 mol $Fe(OH)_3 \times \dfrac{3 \text{ mol } H_2SO_4}{2 \text{ mol } Fe(OH)_3}$

= 6.0 mol H_2SO_4

Because only 6.0 mol H_2SO_4 are needed to react with 4.0 mol $Fe(OH)_3$, the limiting reactant is $Fe(OH)_3$.

23a. Given: amount of HCl = 2.0 mol (limiting reactant); amount of NaOH = 2.5 mol

Unknown: remaining excess reactant in moles

From problem **22.a**, 2.0 mol NaOH react with 2.0 mol HCl.

remaining excess reactant = 2.5 mol NaOH − 2.0 mol NaOH = 0.5 mol NaOH

b. Given: amount of Zn = 2.5 mol (limiting reactant); amount of HCl = 6.0 mol

Unknown: remaining excess reactant in moles

From problem **22.b**, 5.0 mol HCl react with 2.5 mol Zn.

remaining excess reactant = 6.0 mol HCl − 5.0 mol HCl = 1.0 mol HCl

c. Given: amount of $Fe(OH)_3$ = 4.0 mol (limiting reactant); amount of H_2SO_4 = 6.5 mol

Unknown: remaining excess reactant in moles

From problem **22.c**, 6.0 mol H_2SO_4 react with 4.0 mol $Fe(OH)_3$.

remaining excess reactant = 6.5 mol H_2SO_4 − 6.0 mol H_2SO_4 = 0.5 mol H_2SO_4

24a. Given: limiting reactant amount = 2.0 mol HCl; mole ratios: $\dfrac{1 \text{ mol NaCl}}{1 \text{ mol HCl}}$, $\dfrac{1 \text{ mol H}_2\text{O}}{1 \text{ mol HCl}}$

Unknown: amounts of NaCl and H_2O in moles

amount NaCl = 2.0 mol HCl × $\dfrac{1 \text{ mol NaCl}}{1 \text{ mol HCl}}$ = 2.0 mol NaCl

amount H_2O = 2.0 mol HCl × $\dfrac{1 \text{ mol H}_2\text{O}}{1 \text{ mol HCl}}$ = 2.0 mol H_2O

b. Given: limiting reactant amount = 2.5 mol Zn; mole ratios: $\dfrac{1 \text{ mol ZnCl}_2}{1 \text{ mol Zn}}$, $\dfrac{1 \text{ mol H}_2}{1 \text{ mol Zn}}$

Unknown: amounts of $ZnCl_2$ and H_2 in moles

amount $ZnCl_2$ = 2.5 mol Zn × $\dfrac{1 \text{ mol ZnCl}_2}{1 \text{ mol Zn}}$ = 2.5 mol $ZnCl_2$

amount H_2 = 2.5 mol Zn × $\dfrac{1 \text{ mol H}_2}{1 \text{ mol Zn}}$ = 2.5 mol H_2

c. Given: limiting reactant amount = 4.0 mol $Fe(OH)_3$

mole ratios:
$$\frac{1 \text{ mol } Fe_2(SO_4)_3}{2 \text{ mol } Fe(OH)_3}, \frac{6 \text{ mol } H_2O}{2 \text{ mol } Fe(OH)_3}$$

Unknown: amounts of H_2O and $Fe_2(SO_4)_3$ in moles

amount $Fe_2(SO_4)_3$ = 4.0 mol $Fe(OH)_3 \times \dfrac{1 \text{ mol } Fe_2(SO_4)_3}{2 \text{ mol } Fe(OH)_3}$

= 2.0 mol $Fe_2(SO_4)_3$

amount H_2O = 4.0 mol $Fe(OH)_3 \times \dfrac{6 \text{ mol } H_2O}{2 \text{ mol } Fe(OH)_3}$ = 12 mol H_2O

25a. Given: amount of Cu = 2.50 mol, amount of $AgNO_3$ = 5.50 mol

Unknown: limiting reactant

balanced equation:

$Cu + 2AgNO_3 \rightarrow Cu(NO_3)_2 + 2Ag$

mole ratio of reactants = $\dfrac{2 \text{ mol } AgNO_3}{1 \text{ mol } Cu}$

amount $AgNO_3$ required = 2.50 mol Cu $\times \dfrac{2 \text{ mol } AgNO_3}{1 \text{ mol } Cu}$ = 5.00 mol $AgNO_3$

Because only 5.00 mol $AgNO_3$ are needed to react with 2.50 mol Cu, the limiting reactant is Cu.

b. Unknown: remaining excess reactant in moles

5.00 mol $AgNO_3$ reacts with 2.50 mol Cu

remaining excess reactant = 5.50 mol $AgNO_3$ − 5.00 mol $AgNO_3$
= 0.50 mol $AgNO_3$

c. Given: limiting reactant amount = 2.50 mol Cu,

mole ratios:
$$\frac{1 \text{ mol } Cu(NO_3)_2}{1 \text{ mol } Cu}, \frac{2 \text{ mol } Ag}{1 \text{ mol } Cu}$$

Unknown: amounts of $Cu(NO_3)_2$ and Ag in moles

amount $Cu(NO_3)_2$ = 2.50 mol Cu $\times \dfrac{1 \text{ mol } Cu(NO_3)_2}{1 \text{ mol } Cu}$ = 2.50 mol $Cu(NO_3)_2$

amount Ag = 2.50 mol Cu $\times \dfrac{2 \text{ mol } Ag}{1 \text{ mol } Cu}$ = 5.00 mol Ag

d. Given: amount of $Cu(NO_3)_2$ = 2.50 mol, amount of Ag = 5.00 mol

Unknown: masses of $Cu(NO_3)_2$ and Ag

molar mass of $Cu(NO_3)_2$ = $1 \times \dfrac{63.55 \text{ g Cu}}{\text{mol Cu}} + 2 \times \dfrac{14.01 \text{ g N}}{\text{mol N}} + 6 \times \dfrac{16.00 \text{ g O}}{\text{mol O}}$

= 187.57 g/mol

mass $Cu(NO_3)_2$ = 2.50 mol $Cu(NO_3)_2 \times \dfrac{187.57 \text{ g } Cu(NO_3)_2}{\text{mol } Cu(NO_3)_2}$

= 469 g $Cu(NO_3)_2$

molar mass of Ag = 107.87 g/mol

mass Ag = 5.00 mol Ag $\times \dfrac{107.87 \text{ g Ag}}{\text{mol Ag}}$ = 539 g Ag

26a. Given: mass of H_2SO_4 = 30.0 g
mass of $Al(OH)_3$ = 25.0 g
Unknown: limiting reactant

balanced equation:

$$3H_2SO_4 + 2Al(OH)_3 \rightarrow Al_2(SO_4)_3 + 6H_2O$$

mole ratio of reactants = $\dfrac{2 \text{ mol Al(OH)}_3}{3 \text{ mol H}_2\text{SO}_4}$

molar mass of H_2SO_4 = $2 \times \dfrac{1.01 \text{ g H}}{\text{mol H}} + 1 \times \dfrac{32.07 \text{ g S}}{\text{mol S}} + 4 \times \dfrac{16.00 \text{ g O}}{\text{mol O}}$

= 98.09 g/mol

amount H_2SO_4 = 30.0 g $H_2SO_4 \times \dfrac{1 \text{ mol H}_2\text{SO}_4}{98.09 \text{ g H}_2\text{SO}_4}$ = 0.306 mol H_2SO_4

molar mass of $Al(OH)_3$ = $1 \times \dfrac{26.98 \text{ g Al}}{\text{mol Al}} + 3 \times \dfrac{16.00 \text{ g O}}{\text{mol O}} + 3 \times \dfrac{1.01 \text{ g H}}{\text{mol H}}$

= 78.01 g/mol

amount $Al(OH)_3$ = 25.0 g $Al(OH)_3 \times \dfrac{1 \text{ mol Al(OH)}_3}{78.01 \text{ g Al(OH)}_3}$ = 0.320 mol $Al(OH)_3$

amount $Al(OH)_3$ required = 0.306 mol $H_2SO_4 \times \dfrac{2 \text{ mol Al(OH)}_3}{3 \text{ mol H}_2\text{SO}_4}$

= 0.204 mol $Al(OH)_3$

Because only 0.204 mol $Al(OH)_3$ is needed to react with 0.306 mol H_2SO_4, the limiting reactant is H_2SO_4.

b. Given: amount of H_2SO_4 = 0.306 mol (limiting reactant)
amount of $Al(OH)_3$ = 0.320 mol
Unknown: remaining excess reactant in grams

molar mass of $Al(OH)_3$ = 78.01 g/mol

0.204 mol $Al(OH)_3$ reacts with 0.306 mol H_2SO_4

remaining excess reactant = 0.320 mol − 0.204 mol

= 0.116 mol $Al(OH)_3$

mass $Al(OH)_3$ remaining = 0.116 mol $Al(OH)_3 \times \dfrac{78.01 \text{ g Al(OH)}_3}{\text{mol Al(OH)}_3}$

= 9.05 g $Al(OH)_3$

c. Given: limiting reactant amount = 0.306 mol H_2SO_4;
mole ratios:
$\dfrac{1 \text{ mol Al}_2(\text{SO}_4)_3}{3 \text{ mol H}_2\text{SO}_4}$,
$\dfrac{6 \text{ mol H}_2\text{O}}{3 \text{ mol H}_2\text{SO}_4}$
Unknown: masses of $Al_2(SO_4)_3$ and H_2O

molar mass of $Al_2(SO_4)_3$ = $2 \times \dfrac{26.98 \text{ g Al}}{\text{mol Al}} + 3 \times \dfrac{32.07 \text{ g S}}{\text{mol S}}$

$+ 12 \times \dfrac{16.00 \text{ g O}}{\text{mol O}}$ = 342.2 g/mol

mass $Al_2(SO_4)_3$ = 0.306 mol $H_2SO_4 \times \dfrac{1 \text{ mol Al}_2(\text{SO}_4)_3}{3 \text{ mol H}_2\text{SO}_4} \times \dfrac{342.2 \text{ g Al}_2(\text{SO}_4)_3}{\text{mol Al}_2(\text{SO}_4)_3}$

= 34.9 g $Al_2(SO_4)_3$

molar mass of H_2O = 18.02 g/mol

mass H_2O = 0.306 mol $H_2SO_4 \times \dfrac{6 \text{ mol H}_2\text{O}}{3 \text{ mol H}_2\text{SO}_4} \times \dfrac{18.02 \text{ g H}_2\text{O}}{\text{mol H}_2\text{O}}$

= 11.0 g H_2O

27a. Given: mass of N_2H_4 = 1200. kg, mass of $(CH_3)_2N_2H_2$ = 1000. kg, mass of N_2O_4 = 4500. kg, balanced equation

Unknown: limiting reactant (reactant first used up)

mole ratios of reactants = $\dfrac{1 \text{ mol } (CH_3)_2N_2H_2}{2 \text{ mol } N_2H_4}, \dfrac{3 \text{ mol } N_2O_4}{2 \text{ mol } N_2H_4}$

molar mass of $N_2H_4 = 2 \times \dfrac{14.01 \text{ g N}}{\text{mol N}} + 4 \times \dfrac{1.01 \text{ g H}}{\text{mol H}} = 32.06$ g/mol

molar mass of $(CH_3)_2N_2H_2 = 2 \times \dfrac{12.01 \text{ g C}}{\text{mol C}} + 8 \times \dfrac{1.01 \text{ g H}}{\text{mol H}}$
$+ 2 \times \dfrac{14.01 \text{ g N}}{\text{mol N}} = 60.12$ g/mol

molar mass of $N_2O_4 = 2 \times \dfrac{14.01 \text{ g N}}{\text{mol N}} + 4 \times \dfrac{16.00 \text{ g O}}{\text{mol O}} = 92.02$ g/mol

amount N_2H_4 = 1200. kg $N_2H_4 \times \dfrac{1000 \text{ g}}{\text{kg}} \times \dfrac{1 \text{ mol } N_2H_4}{32.06 \text{ g } N_2H_4}$

= 3.743×10^4 mol N_2H_4

amount $(CH_3)_2N_2H_2$ = 1000. kg $(CH_3)_2N_2H_2 \times \dfrac{1000 \text{ g}}{\text{kg}}$
$\times \dfrac{1 \text{ mol } (CH_3)_2N_2H_2}{60.12 \text{ g } (CH_3)_2N_2H_2} = 1.663 \times 10^4$ mol $(CH_3)_2N_2H_2$

amount N_2O_4 = 4500. kg $N_2O_4 \times \dfrac{1000 \text{ g}}{\text{kg}} \times \dfrac{1 \text{ mol } N_2O_4}{92.02 \text{ g } N_2O_4}$

= 4.890×10^4 mol N_2O_4

amount $(CH_3)_2N_2H_2$ required = $(3.743 \times 10^4$ mol $N_2H_4)$
$\times \dfrac{1 \text{ mol } (CH_3)_2N_2H_2}{2 \text{ mol } N_2H_4} = 1.872 \times 10^4$ mol $(CH_3)_2N_2H_2$

amount N_2O_4 required = $(3.743 \times 10^4$ mol $N_2H_4) \times \dfrac{3 \text{ mol } N_2O_4}{2 \text{ mol } N_2H_4}$

= 5.614×10^4 mol N_2O_4

Both $(CH_3)_2N_2H_2$ and N_2O_4 are used up before all of the N_2H_4 is used up. To determine which is the limiting reactant, use the mole ratio:

$\dfrac{1 \text{ mol } (CH_3)_2N_2H_2}{3 \text{ mol } N_2O_4}$

amount $(CH_3)_2N_2H_2$ required = $(4.890 \times 10^4$ mol $N_2O_4) \times \dfrac{1 \text{ mol } (CH_3)_2N_2H_2}{3 \text{ mol } N_2O_4}$

= 1.630×10^4 mol $(CH_3)_2N_2H_2$

There is an excess of $(CH_3)_2N_2H_2$ for the reaction, so the reactant that is used first is N_2O_4.

b. Given: limiting reactant amount = 4.890×10^4 mol N_2O_4, mole ratio: $\dfrac{8 \text{ mol } H_2O}{3 \text{ mol } N_2O_4}$

Unknown: mass of H_2O in kilograms

molar mass of H_2O = 18.02 g/mol

mass H_2O = $(4.890 \times 10^4 \text{ mol } N_2O_4) \times \dfrac{8 \text{ mol } H_2O}{3 \text{ mol } N_2O_4} \times \dfrac{18.02 \text{ g } H_2O}{\text{mol } H_2O}$

$\times \dfrac{1 \text{ kg}}{1000 \text{ g}} = 2.350 \times 10^3$ kg H_2O

28a. Given: theoretical yield = 20.0 g
actual yield = 15.0 g

Unknown: percent yield

percent yield = $\dfrac{15.0 \text{ g}}{20.0 \text{ g}} \times 100 = 75.0\%$

b. Given: theoretical yield = 1.0 g
percent yield = 90.0%

Unknown: actual yield

percent yield = $\dfrac{\text{actual yield}}{\text{theoretical yield}} \times 100$

actual yield = $\dfrac{\text{percent yield} \times \text{theoretical yield}}{100} = \dfrac{90.0 \times 1.0 \text{ g}}{100}$

actual yield = 0.90 g

c. Given: theoretical yield = 5.00 g
actual yield = 4.75 g

Unknown: percent yield

percent yield = $\dfrac{4.75 \text{ g}}{5.00 \text{ g}} \times 100 = 95.0\%$

d. Given: theoretical yield = 3.45 g
percent yield = 48.0%

Unknown: actual yield

percent yield = $\dfrac{\text{actual yield}}{\text{theoretical yield}} \times 100$

actual yield = $\dfrac{\text{percent yield} \times \text{theoretical yield}}{100} = \dfrac{48.0 \times 3.45 \text{ g}}{100}$

actual yield = 1.66 g

29. Given: percent yield = 83.2%
mass of PCl_3 = 73.7 g
balanced equation

Unknown: mass of PCl_5 (actual yield)

mole ratio from balanced equation = $\dfrac{1 \text{ mol } PCl_5}{1 \text{ mol } PCl_3}$

molar mass of PCl_5 = $1 \times \dfrac{30.97 \text{ g P}}{\text{mol P}} + 5 \times \dfrac{35.45 \text{ g Cl}}{\text{mol Cl}} = 208.2$ g/mol

molar mass of PCl_3 = $1 \times \dfrac{30.97 \text{ g P}}{\text{mol P}} + 3 \times \dfrac{35.45 \text{ g Cl}}{\text{mol Cl}} = 137.4$ g/mol

mass PCl_5 (theoretical) = 73.7 g $PCl_3 \times \dfrac{1 \text{ mol } PCl_3}{137.4 \text{ g } PCl_3} \times \dfrac{1 \text{ mol } PCl_5}{1 \text{ mol } PCl_3}$

$\times \dfrac{208.2 \text{ g } PCl_5}{\text{mol } PCl_5} = 112$ g PCl_5

mass PCl_5 (actual) = $\dfrac{\text{percent yield} \times \text{g } PCl_5 \text{ (theoretical)}}{100}$

$= \dfrac{83.2 \times 112 \text{ g } PCl_5}{100} = 93.1$ g PCl_5

30. Given: mass of NH_3 = 5.00 kg
percent yield for each step = 94.0%
balanced equation

Unknown: mass of HNO_3 (actual yield)

mole ratio of HNO_3 to NH_3 = $\dfrac{2 \text{ mol } HNO_3}{3 \text{ mol } NO_2} \times \dfrac{2 \text{ mol } NO_2}{2 \text{ mol } NO} \times \dfrac{4 \text{ mol } NO}{4 \text{ mol } NH_3}$

$= \dfrac{2 \text{ mol } HNO_3}{3 \text{ mol } NH_3}$

molar mass of $HNO_3 = 1 \times \dfrac{1.01 \text{ g H}}{\text{mol H}} + 1 \times \dfrac{14.01 \text{ g N}}{\text{mol N}} + 3 \times \dfrac{16.00 \text{ g O}}{\text{mol O}}$

$= 63.02 \text{ g/mol}$

molar mass of $NH_3 = 1 \times \dfrac{14.01 \text{ g N}}{\text{mol N}} + 3 \times \dfrac{1.01 \text{ g H}}{\text{mol H}} = 17.04 \text{ g/mol}$

mass HNO_3 (theoretical) = $5.00 \text{ kg } NH_3 \times \dfrac{1000 \text{ g}}{\text{kg}} \times \dfrac{1 \text{ mol } NH_3}{17.04 \text{ g } NH_3}$

$\times \dfrac{2 \text{ mol } HNO_3}{3 \text{ mol } NH_3} \times \dfrac{63.02 \text{ g } HNO_3}{\text{mol } HNO_3} = 1.23 \times 10^4 \text{ g } HNO_3$

The total percent yield equals the product of the percent yields for each step.

total percent yield = $94.0 \times 94.0 \times 94.0 = (94.0)^3$

mass HNO_3 (actual) = $\dfrac{\text{total percent yield} \times \text{g } HNO_3 \text{ (theoretical)}}{(100)^3}$

$= (0.94)^3 \times (1.23 \times 10^4 \text{ g } HNO_3) = 1.02 \times 10^4 \text{ g } HNO_3$

Mixed Review

31. Given: mass of Mg = 185.0 g
mass of $MgCl_2$ = 1000. g

Unknown: percent yield

balanced equation:

$MgCl_2 \rightarrow Mg + Cl_2$

mole ratio from balanced equation = $\dfrac{1 \text{ mol Mg}}{1 \text{ mol } MgCl_2}$

molar mass of Mg = 24.30 g/mol

molar mass of $MgCl_2 = 1 \times \dfrac{24.30 \text{ g Mg}}{\text{mol Mg}} + 2 \times \dfrac{35.45 \text{ g Cl}}{\text{mol Cl}} = 95.20 \text{ g/mol}$

mass Mg (theoretical) = $1000. \text{ g } MgCl_2 \times \dfrac{1 \text{ mol } MgCl_2}{95.20 \text{ g } MgCl_2} \times \dfrac{1 \text{ mol Mg}}{1 \text{ mol } MgCl_2}$

$\times \dfrac{24.30 \text{ g Mg}}{\text{mol Mg}} = 255.3 \text{ g Mg}$

percent yield = $\dfrac{\text{g Mg (actual)}}{\text{g Mg (theoretical)}} \times 100 = \dfrac{185.0 \text{ g Mg}}{255.3 \text{ g Mg}} \times 100$

percent yield = 72.46%

32. Given: volume of CO_2 = 0.750 L
density of CO_2 = 1.20 g/L
mass of $NaHCO_3$ in baking powder = 168 g/kg
balanced equation

Unknown: mass of baking powder

mass of CO_2 = $\dfrac{1.20 \text{ g } CO_2}{\text{L } CO_2} \times 0.750 \text{ L } CO_2 = 0.900 \text{ g } CO_2$

mole ratio from balanced equation = $\dfrac{2 \text{ mol } NaHCO_3}{2 \text{ mol } CO_2}$

molar mass of $NaHCO_3$ = $1 \times \dfrac{22.99 \text{ g Na}}{\text{mol Na}} + 1 \times \dfrac{1.01 \text{ g H}}{\text{mol H}} + 1 \times \dfrac{12.01 \text{ g C}}{\text{mol C}}$

$+ 3 \times \dfrac{16.00 \text{ g O}}{\text{mol O}} = 84.01 \text{ g/mol}$

molar mass of CO_2 = $1 \times \dfrac{12.01 \text{ g C}}{\text{mol C}} + 2 \times \dfrac{16.00 \text{ g O}}{\text{mol O}} = 44.01 \text{ g/mol}$

mass of baking powder = $0.900 \text{ g } CO_2 \times \dfrac{1 \text{ mol } CO_2}{44.01 \text{ g } CO_2} \times \dfrac{2 \text{ mol } NaHCO_3}{2 \text{ mol } CO_2}$

$\times \dfrac{84.01 \text{ g } NaHCO_3}{\text{mol } NaHCO_3} \times \dfrac{1 \text{ kg baking powder}}{168 \text{ g } NaHCO_3}$

= 0.0102 kg baking powder = 10.2 g baking powder

---

33. Given: percent yield = 85.0%
mass of C = 1250 g
balanced equation

Unknown: mass of CH_4 (actual yield)

mole ratio from balanced equation = $\dfrac{1 \text{ mol } CH_4}{2 \text{ mol C}}$

molar mass of C = 12.01 g/mol

molar mass of CH_4 = $1 \times \dfrac{12.01 \text{ g C}}{\text{mol C}} + 4 \times \dfrac{1.01 \text{ g H}}{\text{mol H}} = 16.05 \text{ g/mol}$

mass CH_4 (theoretical) = $1250 \text{ g C} \times \dfrac{1 \text{ mol C}}{12.01 \text{ g C}} \times \dfrac{1 \text{ mol } CH_4}{2 \text{ mol C}}$

$\times \dfrac{16.05 \text{ g } CH_4}{\text{mol } CH_4} = 835 \text{ g } CH_4$

mass CH_4 (actual) = $\dfrac{\text{percent yield} \times \text{g } CH_4 \text{ (theoretical)}}{100}$

$= \dfrac{85.0 \times 835 \text{ g } CH_4}{100} = 710. \text{ g } CH_4$

---

34. Given: percent yield = 95%
mass of C = 2750 g

Unknown: mass of CH_4 (actual yield)

From problem **33**, the mole ratio = $\dfrac{1 \text{ mol } CH_4}{2 \text{ mol C}}$, the molar mass of C = 12.01 g/mol, and the molar mass of CH_4 = 16.05 g/mol.

mass CH_4 (theoretical) = $2750 \text{ g C} \times \dfrac{1 \text{ mol C}}{12.01 \text{ g C}} \times \dfrac{1 \text{ mol } CH_4}{2 \text{ mol C}}$

$\times \dfrac{16.05 \text{ g } CH_4}{\text{mol } CH_4} = 1850 \text{ g } CH_4$

mass CH_4 (actual) = $\dfrac{\text{percent yield} \times \text{g } CH_4 \text{ (theoretical)}}{100}$

$= \dfrac{95 \times 1850 \text{ g } CH_4}{100} = 1760 \text{ g } CH_4$

35. Given: volume of $CaSO_4 \cdot 2H_2O$ = 5.00 L
density of $CaSO_4 \cdot 2H_2O$ = 2.32 g/mL
density of water vapor = 0.581 g/mL

Unknown: volume of water vapor in liters

balanced equation:

$$2CaSO_4 \cdot 2H_2O \rightarrow 2CaSO_4 \cdot \frac{1}{2}H_2O + 3H_2O$$

mole ratio from balanced equation = $\dfrac{3 \text{ mol } H_2O}{2 \text{ mol } CaSO_4 \cdot 2 H_2O}$

molar mass of $CaSO_4 \cdot 2 H_2O = 1 \times \dfrac{40.08 \text{ g Ca}}{\text{mol Ca}} + 1 \times \dfrac{32.07 \text{ g S}}{\text{mol S}}$

$+ 4 \times \dfrac{1.01 \text{ g H}}{\text{mol H}} + 6 \times \dfrac{16.00 \text{ g O}}{\text{mol O}} = 172.19 \text{ g/mol}$

molar mass of H_2O = 18.02 g/mol

mass $CaSO_4 \cdot 2H_2O = \dfrac{2.32 \text{ g}}{\text{mL}} \times 5.00 \text{ L} \times \dfrac{1000 \text{ mL}}{\text{L}}$

$= 1.16 \times 10^4 \text{ g } CaSO_4 \cdot 2H_2O$

mass $H_2O = (1.16 \times 10^4 \text{ g } CaSO_4 \cdot 2H_2O) \times \dfrac{1 \text{ mol } CaSO_4 \cdot 2H_2O}{172.19 \text{ g } CaSO_4 \cdot 2H_2O}$

$\times \dfrac{3 \text{ mol } H_2O}{2 \text{ mol } CaSO_4 \cdot 2H_2O} \times \dfrac{18.02 \text{ g } H_2O}{\text{mol } H_2O} = 1.82 \times 10^3 \text{ g } H_2O$

$V\ H_2O = \dfrac{1.82 \times 10^3 \text{ g } H_2O}{\left(\dfrac{0.581 \text{ g } H_2O}{\text{mL } H_2O}\right)} \times \dfrac{1 \text{ L}}{1000 \text{ mL}} = 3.13 \text{ L water vapor}$

36. Given: mass of ZnO = 2.00 g
balanced equation

Unknown: mass of Au

mole ratio of Au to ZnO = $\dfrac{2 \text{ mol Zn}}{2 \text{ mol ZnO}} \times \dfrac{2 \text{ mol Au}}{3 \text{ mol Zn}} = \dfrac{2 \text{ mol Au}}{3 \text{ mol ZnO}}$

molar mass of ZnO = $1 \times \dfrac{65.39 \text{ g Zn}}{\text{mol Zn}} + 1 \times \dfrac{16.00 \text{ g O}}{\text{mol O}} = 81.39 \text{ g/mol}$

molar mass of Au = 196.97 g/mol

mass Au = $2.00 \text{ g ZnO} \times \dfrac{1 \text{ mol ZnO}}{81.39 \text{ g ZnO}} \times \dfrac{2 \text{ mol Au}}{3 \text{ mol ZnO}} \times \dfrac{196.97 \text{ g Au}}{\text{mol Au}} = 3.23 \text{ g Au}$

Handbook Search

42a. Given: mass of Fe = 3.65×10^3 kg, balanced equation

Unknown: minimum mass of C

mole ratio from balanced equation = $\dfrac{1 \text{ mol C}}{3 \text{ mol Fe}}$

molar mass of Fe = 55.85 g/mol

molar mass of C = 12.01 g/mol

mass C = $(3.65 \times 10^3 \text{ kg Fe}) \times \dfrac{1000 \text{ g}}{\text{kg}} \times \dfrac{1 \text{ mol Fe}}{55.85 \text{ g Fe}} \times \dfrac{1 \text{ mol C}}{3 \text{ mol Fe}}$

$\times \dfrac{12.01 \text{ g C}}{\text{mol C}} \times \dfrac{1 \text{ kg}}{1000 \text{ g}} = 262 \text{ kg C}$

b. Given: mass of Fe = 3.65×10^3 kg
Unknown: mass of Fe_3C

mole ratio from balanced equation = $\dfrac{1 \text{ mol } Fe_3C}{3 \text{ mol Fe}}$

molar mass of $Fe_3C = 3 \times \dfrac{55.85 \text{ g Fe}}{\text{mol Fe}} + 1 \times \dfrac{12.01 \text{ g C}}{\text{mol C}} = 179.6$ g/mol

molar mass of Fe = 55.85 g/mol

mass $Fe_3C = (3.65 \times 10^3$ kg Fe$) \times \dfrac{1000 \text{ g}}{\text{kg}} \times \dfrac{1 \text{ mol Fe}}{55.85 \text{ g Fe}} \times \dfrac{1 \text{ mol } Fe_3C}{3 \text{ mol Fe}}$

$\times \dfrac{179.6 \text{ g } Fe_3C}{\text{mol } Fe_3C} \times \dfrac{1 \text{ kg}}{1000 \text{ g}} = 3.92 \times 10^3$ kg Fe_3C

43a. Given: mass of Al = 30.0 g
Unknown: mass of Al_2O_3

mole ratio from balanced equation = $\dfrac{2 \text{ mol } Al_2O_3}{4 \text{ mol Al}}$

molar mass of Al = 26.98 g/mol

molar mass of $Al_2O_3 = 2 \times \dfrac{26.98 \text{ g Al}}{\text{mol Al}} + 3 \times \dfrac{16.00 \text{ g O}}{\text{mol O}} = 101.96$ g/mol

mass $Al_2O_3 = 30.0$ g Al $\times \dfrac{1 \text{ mol Al}}{26.98 \text{ g Al}} \times \dfrac{2 \text{ mol } Al_2O_3}{4 \text{ mol Al}} \times \dfrac{101.96 \text{ g } Al_2O_3}{\text{mol } Al_2O_3}$

$= 56.6$ g Al_2O_3

44a. Given: mass of MgO = 154.6 g, balanced equation
Unknown: mass of CO_2

mole ratio from balanced equation = $\dfrac{1 \text{ mol } CO_2}{1 \text{ mol MgO}}$

molar mass of MgO = $1 \times \dfrac{24.30 \text{ g Mg}}{\text{mol Mg}} + 1 \times \dfrac{16.00 \text{ g O}}{\text{mol O}} = 40.30$ g/mol

molar mass of $CO_2 = 1 \times \dfrac{12.01 \text{ g C}}{\text{mol C}} + 2 \times \dfrac{16.00 \text{ g O}}{\text{mol O}} = 44.01$ g/mol

mass $CO_2 = 154.6$ g MgO $\times \dfrac{1 \text{ mol MgO}}{40.30 \text{ g MgO}} \times \dfrac{1 \text{ mol } CO_2}{1 \text{ mol MgO}} \times \dfrac{44.01 \text{ g } CO_2}{\text{mol } CO_2}$

$= 168.8$ g CO_2

b. Given: mass of MgO = 154.6 g
Unknown: mass of $MgCO_3$

mole ratio from balanced equation = $\dfrac{1 \text{ mol } MgCO_3}{1 \text{ mol MgO}}$

molar mass of $MgCO_3 = 1 \times \dfrac{24.30 \text{ g Mg}}{\text{mol Mg}} + 1 \times \dfrac{12.01 \text{ g C}}{\text{mol C}} + 3 \times \dfrac{16.00 \text{ g O}}{\text{mol O}}$

$= 84.31$ g/mol

molar mass of MgO = 40.30 g/mol

mass $MgCO_3 = 154.6$ g MgO $\times \dfrac{1 \text{ mol MgO}}{40.30 \text{ g MgO}} \times \dfrac{1 \text{ mol } MgCO_3}{1 \text{ mol MgO}}$

$\times \dfrac{84.31 \text{ g } MgCO_3}{\text{mol } MgCO_3} = 323.4$ g $MgCO_3$

c. Given: mass of P_4O_{10} = 45.7 g, balanced equation

Unknown: mass of $Ca_3(PO_4)_2$

mole ratio from balanced equation = $\dfrac{2 \text{ mol } Ca_3(PO_4)_2}{1 \text{ mol } P_4O_{10}}$

molar mass of P_4O_{10} = $4 \times \dfrac{30.97 \text{ g P}}{\text{mol P}} + 10 \times \dfrac{16.00 \text{ g O}}{\text{mol O}}$ = 283.9 g/mol

molar mass of $Ca_3(PO_4)_2$ = $3 \times \dfrac{40.08 \text{ g Ca}}{\text{mol Ca}} + 2 \times \dfrac{30.97 \text{ g P}}{\text{mol P}}$
$+ 8 \times \dfrac{16.00 \text{ g O}}{\text{mol O}}$ = 310.1 g/mol

mass $Ca_3(PO_4)_2$ = 45.7 g $P_4O_{10} \times \dfrac{1 \text{ mol } P_4O_{10}}{283.9 \text{ g } P_4O_{10}} \times \dfrac{2 \text{ mol } Ca_3(PO_4)_2}{1 \text{ mol } P_4O_{10}}$
$\times \dfrac{310.1 \text{ g } Ca_3(PO_4)_2}{\text{mol } Ca_3(PO_4)_2}$ = 99.9 g $Ca_3(PO_4)_2$

CHAPTER 10
Physical Characteristics of Gases

Practice, p. 312

1. Given: $P = 1.75$ atm

a. $1.75 \text{ atm} \times \dfrac{101.325 \text{ kPa}}{\text{atm}} = 177$ kPa

b. $1.75 \text{ atm} \times \dfrac{760 \text{ mm Hg}}{\text{atm}} = 1330$ mm Hg

2. Given: $P = 570$ torr

a. $570 \text{ torr} \times \dfrac{1 \text{ atm}}{760 \text{ torr}} = 0.750$ atm

b. $0.750 \text{ atm} \times \dfrac{101.325 \text{ kPa}}{\text{atm}} = 76.0$ kPa

Section Review, p. 312

4. a. Given: $P = 151.98$ kPa

$151.98 \text{ kPa} \times \dfrac{1 \text{ atm}}{101.325 \text{ kPa}} = 1.4999$ atm

b. Given: $P = 456$ torr

$456 \text{ torr} \times \dfrac{1 \text{ atm}}{760 \text{ torr}} = 0.600$ atm

c. Given: $P = 912$ mm Hg

$912 \text{ mm Hg} \times \dfrac{1 \text{ atm}}{760 \text{ mm Hg}} = 1.20$ atm

ATE, Additional Sample Problem 10–1, p. 312

Given: $P = 745.8$ mm Hg

a. $745.8 \text{ mm} \times \dfrac{1 \text{ atm}}{760 \text{ mm}} = 0.9813$ atm

b. $0.9813 \text{ atm} \times \dfrac{760 \text{ torr}}{1 \text{ atm}} = 745.8$ torr

c. $0.9813 \text{ atm} \times \dfrac{101.325 \text{ kPa}}{\text{atm}} = 99.43$ kPa

Practice, p. 315

1. Given: $V_1 = 500$ mL He
$P_1 = 1$ atm
$P_2 = 0.5$ atm

Unknown: V_2

$P_1 V_1 = P_2 V_2$

$V_2 = \dfrac{P_1 V_1}{P_2} = \dfrac{(1 \text{ atm})(500 \text{ mL He})}{0.5 \text{ atm}} = 1000$ mL He

2. Given: $P_1 = 1.26$ atm
$V_1 = 7.40$ L
$V_2 = 2.93$ L

Unknown: P_2

$P_1 V_1 = P_2 V_2$

$P_2 = \dfrac{P_1 V_1}{V_2} = \dfrac{(1.26 \text{ atm})(7.40 \text{ L})}{2.93 \text{ L}} = 3.18$ atm

3. Given: $V_1 = 3.5$ L

$P_1 = 1$ atm

Unknowns: P_2, V_2

101.325 kPa = 1 atm

$$\frac{10.2 \text{ m}}{100 \text{ kPa}} = \frac{51 \text{ m}}{x}$$

$10.2 \, x = (51)(100)$

$$x = \frac{(51)(100)}{10.2} = 500 \text{ kPa}$$

$$P_2 = 101 + 500 = 601 \text{ kPa} \times \frac{1 \text{ atm}}{101.325 \text{ kPa}} = 5.93 \text{ atm}$$

$P_1 V_1 = P_2 V_2$

$$V_2 = \frac{P_1 V_1}{P_2} = \frac{(1 \text{ atm})(3.5 \text{ L})}{5.93 \text{ atm}} = 0.59 \text{ L}$$

ATE, Additional Sample Problems 10–2, p. 315

a. Given: $V_1 = 450$ mL

$P_1 = 1$

$P_2 = 15$

Unknown: V_2

$P_1 V_1 = P_2 V_2$

$$P_2 = \frac{P_1 V_1}{V_2} = \frac{(1)(450 \text{ mL})}{15} = 30.0 \text{ mL}$$

b. Given: V_1 of helium = 125 mL

$P_1 = 0.974$ atm

$P_2 = 1.000$ atm

Unknown: V_2

$P_1 V_1 = P_2 V_2$

$$V_2 = \frac{P_1 V_1}{P_2} = \frac{(0.974 \text{ atm})(125 \text{ mL})}{1.000 \text{ atm}} = 122 \text{ mL He}$$

c. Given: $V_1 = 1.375$ L

$P_1 = 1.000$ atm = 101.325 kPa

$P_2 = 10.0$ kPa

Unknown: V_2

$P_1 V_1 = P_2 V_2$

$$V_2 = \frac{P_1 V_1}{P_2} = \frac{(101.325 \text{ kPa})(1.375 \text{ L})}{10.0 \text{ kPa}} = 13.9 \text{ L}$$

Practice, p. 319

1. Given: $V_1 = 2.75$ L

$T_1 = 20°C = 293$ K

$V_2 = 2.46$ L

Unknown: T_2 (in K & °C)

$$\frac{V_1}{T_1} = \frac{V_2}{T_2}$$

$$T_2 = \frac{T_1 V_2}{V_1} = \frac{(293 \text{ K})(2.46 \text{ L})}{2.75 \text{ L}} = 262 \text{ K}$$

K = °C + 273

$T_2 = 262 - 273 = -11°C$

2. Given: $V_1 = 4.22$ L
$T_1 = 65°C = 338$ K
$V_2 = 3.87$ L
Unknown: T_2

$\dfrac{V_1}{T_1} = \dfrac{V_2}{T_2}$

$T_2 = \dfrac{T_1 V_2}{V_1} = \dfrac{(338 \text{ K})(3.87 \text{ L})}{4.22 \text{ L}} = 310$ K

K = °C + 273

$T_2 = 310 - 273 = 37°C$

ATE, Additional Sample Problems 10–3, p. 319

a. Given: $V_1 = 5.5$ L
$T_1 = 25°C = 298$ K
$T_2 = 100°C = 373$ K
Unknown: V_2

$\dfrac{V_1}{T_1} = \dfrac{V_2}{T_2}$

$V_2 = \dfrac{V_1 T_2}{T_1} = \dfrac{(5.5 \text{ L})(373 \text{ K})}{298 \text{ K}} = 6.9$ L

b. Given: $V_1 = 375$ mL
$T_1 = 0.0°C = 273$ K
$V_2 = 500$ mL
Unknown: T_2

$\dfrac{V_1}{T_1} = \dfrac{V_2}{T_2}$

$T_2 = \dfrac{T_1 V_2}{V_1} = \dfrac{(273 \text{ K})(500 \text{ mL})}{375 \text{ mL}} = 364$ K

364 K − 273 = 91°C

Practice, p. 320

1. Given: $P_1 = 1.8$ atm
$T_1 = 20°C = 293$ K
$P_2 = 1.9$ atm
Unknown: T_2

$\dfrac{P_1}{T_1} = \dfrac{P_2}{T_2}$

$T_2 = \dfrac{T_1 P_2}{P_1} = \dfrac{(293 \text{ K})(1.9 \text{ atm})}{1.8 \text{ atm}} = 309$ K

309 K − 273 = 36°C

2. Given: $T_1 = 120°C = 393$ K
$P_1 = 1.07$ atm
$T_2 = 205°C = 478$ K
Unknown: P_2

$\dfrac{P_1}{T_1} = \dfrac{P_2}{T_2}$

$P_2 = \dfrac{P_1 T_2}{T_1} = \dfrac{(1.07 \text{ atm})(478 \text{ K})}{393 \text{ K}} = 1.30$ atm

3. Given: $P_1 = 1.20$ atm
$T_1 = 22°C = 295$ K
$P_2 = 2.00$ atm
Unknown: T_2

$\dfrac{P_1}{T_1} = \dfrac{P_2}{T_2}$

$T_2 = \dfrac{T_1 P_2}{P_1} = \dfrac{(295 \text{ K})(2.00 \text{ atm})}{1.20 \text{ atm}} = 492$ K

492 K − 273 = 219°C

ATE, Additional Sample Problems 10-4, p. 320

a. Given: $T_1 = 20°C$
$\quad\quad\quad\quad = 293\ K$
$\quad\quad P_1 = 1.0\ atm$
$\quad\quad T_2 = 500°C$
$\quad\quad\quad\quad = 773\ K$

Unknown: P_2

$\dfrac{P_1}{T_1} = \dfrac{P_2}{T_2}$

$P_2 = \dfrac{P_1 T_2}{T_1} = \dfrac{(1.0\ atm)(773\ K)}{293\ K} = 2.6\ atm$

b. Given: $T_1 = 25°C$
$\quad\quad\quad\quad = 298\ K$
$\quad\quad P_2 = 1.80\ atm$
$\quad\quad P_1 = 1.75\ atm$

Unknown: T_2

$\dfrac{P_1}{T_1} = \dfrac{P_2}{T_2}$

$T_2 = \dfrac{T_1 P_2}{P_1} = \dfrac{(298\ K)(1.80\ atm)}{1.75\ atm} = 306.5\ K$

$306.5\ K - 273 = 33.5°C$

c. Given: $T_1 = 100°C$
$\quad\quad\quad\quad = 373\ K$
$\quad\quad P_1 = 3.0\ atm$
$\quad\quad T_2 = 300°C$
$\quad\quad\quad\quad = 573\ K$

Unknown: P_2

$\dfrac{P_1}{T_1} = \dfrac{P_2}{T_2}$

$P_2 = \dfrac{P_1 T_2}{T_1} = \dfrac{(3.0\ atm)(573\ K)}{373\ K} = 4.6\ atm$

Practice, p. 322

1. Given: $V_1 = 27.5\ mL$
$\quad\quad T_1 = 22.0°C$
$\quad\quad\quad\quad = 295\ K$
$\quad\quad P_1 = 0.974\ atm$
$\quad\quad T_2 = 15.0°C$
$\quad\quad\quad\quad = 288\ K$
$\quad\quad P_2 = 0.993\ atm$

Unknown: V_2

$\dfrac{P_1 V_1}{T_1} = \dfrac{P_2 V_2}{T_2}$

$V_2 = \dfrac{P_1 V_1 T_2}{P_2 T_1} = \dfrac{(0.974\ atm)(27.5\ mL)(288\ K)}{(0.993\ atm)(295\ K)} = 26.3\ mL$

2. Given: $V_1 = 700\ mL$
$\quad\quad T_1 = 0°C = 273\ K$
$\quad\quad P_1 = 1.00\ atm$
$\quad\quad V_2 = 200\ mL$
$\quad\quad T_2 = 30.0°C$
$\quad\quad\quad\quad = 303\ K$

Unknown: P_2 (in Pa)

$\dfrac{P_1 V_1}{T_1} = \dfrac{P_2 V_2}{T_2}$

$P_2 = \dfrac{P_1 V_1 T_2}{T_1 V_2} = \dfrac{(1.00\ atm)(700\ mL)(303\ K)}{(273\ K)(200\ mL)} \times \dfrac{101.325\ kPa}{atm} = 394\ kPa$

$= 3.94 \times 10^5\ Pa$

ATE, Additional Sample Problems 10–5, p. 322

a. Given: $T_1 = 27.0°C$
$\quad\quad\quad\quad = 300\text{ K}$
$\quad\quad P_1 = 0.200\text{ atm}$
$\quad\quad V_1 = 80.0\text{ mL}$
$\quad\quad T_2 = 0°C = 273\text{ K}$
$\quad\quad P_2 = 1.00\text{ atm}$

Unknown: V_2

$$\frac{P_1 V_1}{T_1} = \frac{P_2 V_2}{T_2}$$

$$V_2 = \frac{P_1 V_1 T_2}{P_2 T_1} = \frac{(0.200\text{ atm})(80.0\text{ mL})(273\text{ K})}{(1.00\text{ atm})(300\text{ K})} = 14.6\text{ mL}$$

b. Given: $V_1 = 75\text{ mL}$
$\quad\quad T_1 = 0°C = 273\text{ K}$
$\quad\quad P_1 = 1.00\text{ atm}$
$\quad\quad T_2 = 17°C = 290\text{ K}$
$\quad\quad P_2 = 0.97\text{ atm}$

Unknown: V_2

$$\frac{P_1 V_1}{T_1} = \frac{P_2 V_2}{T_2}$$

$$V_2 = \frac{P_1 V_1 T_2}{P_2 T_1} = \frac{(1.00\text{ atm})(75\text{ mL})(290\text{ K})}{(0.97\text{ atm})(273\text{ K})} = 82\text{ mL}$$

c. Given: $V_1 = 60.0\text{ mL}$
$\quad\quad T_1 = 273\text{ K}$
$\quad\quad P_1 = 1.00\text{ atm}$
$\quad\quad V_2 = 10.0\text{ mL}$
$\quad\quad T_2 = 25.0°C$
$\quad\quad\quad\quad = 298\text{ K}$

Unknown: P_2

$$\frac{P_1 V_1}{T_1} = \frac{P_2 V_2}{T_2}$$

$$P_2 = \frac{P_1 V_1 T_2}{T_2 V_2} = \frac{(1.00\text{ atm})(60.0\text{ mL})(298\text{ K})}{(273\text{ K})(10.0\text{ mL})} = 6.55\text{ atm}$$

Practice, p. 325

1. Given: $T = 20.0°C$
$\quad\quad P_{H_2} = 742.5\text{ torr}$
$\quad\quad P_{H_2O}$ at $20°C$
$\quad\quad = 17.5\text{ torr}$

Unknown: P_T

$P_T = P_{H_2} + P_{H_2O} = 742.5 + 17.5 = 760.0\text{ torr}$

2. Given: $T = 25°C$
$\quad\quad P_{atm} = 750.0\text{ mm Hg}$
$\quad\quad P_{H_2O}$ at $25°C = 23.8\text{ mm}$

Unknown: P_{He}

$P_T = P_{He} + P_{H_2O}$

$P_{He} = P_T - P_{H_2O} = 750.0 - 23.8 = 726.2\text{ mm Hg}$

Section Review, p. 325

2. Given: $V_1 = 200.0$ mL
$P_1 = 0.960$ atm
$V_2 = 50.0$ mL
Unknown: P_2

$P_1V_1 = P_2V_2$

$P_2 = \dfrac{P_1V_1}{V_2} = \dfrac{(0.960 \text{ atm})(200.0 \text{ mL})}{50.0 \text{ mL}} = 3.84$ atm

3. Given: $V_1 = 0.750$ L
$T_1 = 298$ K
$V_2 = 0.500$ L
Unknown: T_2 (in °C)

$\dfrac{V_1}{T_1} = \dfrac{V_2}{T_2}$

$T_2 = \dfrac{T_1V_2}{V_1} = \dfrac{(298 \text{ K})(0.500 \text{ L})}{0.750 \text{ L}} = 199$ K

$199 \text{ K} - 273 = -74°C$

4. Given: $P_1 = 4.50$ atm
$T_1 = 20.0°C$
$\quad = 293$ K
$P_2 = 4.80$ atm
Unknown: T_2 (in °C)

$\dfrac{P_1}{T_1} = \dfrac{P_2}{T_2}$

$T_2 = \dfrac{T_1P_2}{P_1} = \dfrac{(293 \text{ K})(4.80 \text{ atm})}{4.50 \text{ atm}} = 312.5$ K

$312.5 \text{ K} - 273 = 39.5°C$

6. Given: $V_1 = 720.0$ mL
$T_1 = 25.0°C = 298$ K
$P_{atm} = 755$ torr
$T_2 = 0°C = 273$ K
$P_2 = 760.0$ torr
Unknown: V_2

$\dfrac{P_1V_1}{T_1} = \dfrac{P_2V_2}{T_2}$

P_{H_2O} at $25.0°C = 23.8$ torr (from Appendix A-8)

$P_T = P_{O_2} + P_{H_2O}$

$P_{O_2} = P_T - P_{H_2O} = 755 \text{ torr} - 23.8 \text{ torr} = 731.2 \text{ torr} = P_1$

$V_2 = \dfrac{P_1V_1T_2}{P_2T_1} = \dfrac{(731.2 \text{ torr})(720 \text{ mL})(273 \text{ K})}{(760.0 \text{ torr})(298 \text{ K})} = 635$ mL

ATE, Additional Sample Problems 10–6, p. 325

a. Given: $T = 23.0°C$
$P_T = P_{atm}$
$\quad = 785$ mm Hg
Unknown: P_{N_2}

$P_T = P_{N_2} + P_{H_2O}$

P_{H_2O} at $23°C = 21.1$ mm Hg (from Appendix A-8)

$P_{N_2} = P_T - P_{H_2O} = 785 - 21 = 764$ mm Hg

b. Given: $T = 27.0°C$
$P_T = $ 743.3 mm Hg
Unknown: P_{Ne}

$P_T = P_{Ne} + P_{H_2O}$

P_{H_2O} at $27.0°C = 26.7$ mm Hg (from Appendix A-8)

$P_{Ne} = P_T - P_{H_2O} = 743.3 - 26.7 = 716.6$ mm Hg

Review Problems

15. a. Given: $P_1 = 760$ mm Hg; $V_1 = 1$; $V_2 = \left(\frac{1}{14}\right)(V_1)$

Unknown: P_2

$D = M/V \quad V = M/D \quad P_1V_1 = P_2V_2$

$P_2 = \dfrac{P_1V_1}{V_2} = \dfrac{(760 \text{ mm})(1)}{\frac{1}{14}} = 11\,000$ mm Hg

b. Given: $V_2 = (1.40)(V_1)$

Unknown: P_2

$P_2 = \dfrac{P_1V_1}{V_2} = \dfrac{(760 \text{ mm})(1)}{1.40} = 543$ mm Hg

16. a. Given: $P = 1.25$ atm

Unknown: P in torr

$P = (1.25 \text{ atm})\left(\dfrac{760 \text{ torr}}{\text{atm}}\right) = 950$ torr

b. Given: $P = 2.48 \times 10^{-3}$ atm

$P = (2.48 \times 10^{-3} \text{ atm})\left(\dfrac{760 \text{ torr}}{\text{atm}}\right) = 1.88$ torr

c. Given: $P = 4.75 \times 10^4$ atm

$P = (4.75 \times 10^4 \text{ atm})\left(\dfrac{760 \text{ torr}}{\text{atm}}\right) = 3.61 \times 10^7$ torr

d. Given: $P = 7.60 \times 10^6$ atm

$P = (7.60 \times 10^6 \text{ atm})\left(\dfrac{760 \text{ torr}}{\text{atm}}\right) = 5.78 \times 10^9$ torr

17. a. Given: $P = 125$ mm

Unknown: P in atm

$P = (125 \text{ mm})\left(\dfrac{1 \text{ atm}}{760 \text{ mm}}\right) = 0.164$ atm

b. Given: $P = 3.20$ atm

Unknown: P in Pa

$P = (3.20 \text{ atm})\left(\dfrac{101.325 \text{ kPa}}{\text{atm}}\right) = 324.24 \text{ kPa} = 3.24 \times 10^5$ Pa

c. Given: $P = 5.38$ kPa

Unknown: P in torr

$P = (5.38 \text{ kPa})\left(\dfrac{1 \text{ atm}}{101.325 \text{ kPa}}\right)\left(\dfrac{760 \text{ torr}}{\text{atm}}\right) = 40.4$ torr

18. Given: T in °C

Unknown: T in K

$K = °C + 273$

a. $0.°C + 273 = 273$ K

b. $27°C + 273 = 300$ K

c. $-50.°C + 273 = 223$ K

d. $-273°C + 273 = 0$ K

19. Given: T in K

Unknown: T in °C

$°C = K - 273$

a. $273 \text{ K} - 273 = 0°C$

b. $350. \text{ K} - 273 = 77°C$

c. $100. \text{ K} - 273 = -173°C$

d. $20. \text{ K} - 273 = -253°C$

20. a. Given: $P_1 = 350.$ torr
$V_1 = 200.$ mL
$P_2 = 700.$ torr
Unknown: V_2

$P_1V_1 = P_2V_2$

$V_2 = \dfrac{P_1V_1}{P_2} = \dfrac{(350 \text{ torr})(200 \text{ mL})}{700 \text{ torr}} = 100 \text{ mL}$

b. Given: $P_1 = 0.75$ atm
$V_2 = 435$ mL
$P_2 = 0.48$ atm
Unknown: V_1

$P_1V_1 = P_2V_2$

$V_1 = \dfrac{P_2V_2}{P_1} = \dfrac{(0.48 \text{ atm})(435 \text{ mL})}{0.75 \text{ atm}} = 280 \text{ mL}$

c. Given: $V_1 = 2.4 \times 10^5$ L
$P_2 = 180$ mm Hg
$V_2 = 1.8 \times 10^3$ L
Unknown: P_1

$P_1V_1 = P_2V_2$

$P_1 = \dfrac{P_2V_2}{V_1} = \dfrac{(180 \text{ mm})(1.8 \times 10^3 \text{ L})}{2.4 \times 10^5 \text{ L}} = 1.4 \text{ mm}$

21. Given: $V_1 = 240.$ mL
$P_1 = 0.428$ atm
$P_2 = 0.724$ atm
Unknown: V_2

$P_1V_1 = P_2V_2$

$V_2 = \dfrac{P_1V_1}{P_2} = \dfrac{(0.428 \text{ atm})(240 \text{ mL})}{(0.724 \text{ atm})} = 142 \text{ mL}$

22. Given: $V_1 = 155$ cm^3
$P_1 = 22.5$ kPa
$V_2 = 90.0$ cm^3
Unknown: P_2

$P_1V_1 = P_2V_2$

$P_2 = \dfrac{P_1V_1}{V_2} = \dfrac{(22.5 \text{ kPa})(155 \text{ cm}^3)}{(90.0 \text{ cm}^3)} = 38.8 \text{ kPa}$

23. Given: $V_1 = 450.0$ mL
Unknown: V_2

a. $P_2 = 2P_1$

b. $P_2 = \tfrac{1}{4}P_1$

$P_1V_1 = P_2V_2$

$V_2 = \dfrac{P_1V_1}{P_2} = \dfrac{(1)(450.0 \text{ mL})}{2} = 225.0 \text{ mL}$

$V_2 = \dfrac{(1)(450.0 \text{ mL})}{\tfrac{1}{4}} = 1800 \text{ mL}$

24. Given: $V_1 = 1.00 \times 10^6$ mL
$P_1 = 575$ mm Hg
$P_2 = 1.25$ atm
Unknown: V_2

$P_1V_1 = P_2V_2, \; V_2 = \dfrac{P_1V_1}{P_2}$

$P_1 = (575 \text{ mm})\left(\dfrac{1 \text{ atm}}{760 \text{ mm}}\right) = 0.756 \text{ atm}$

$V_2 = \dfrac{(0.756 \text{ atm})(1.00 \times 10^6 \text{ mL})}{1.25 \text{ atm}} = 6.05 \times 10^5 \text{ mL}$

25. a. Given: $V_1 = 80.0$ mL
$T_1 = 27°C$
    $= 300$ K
$T_2 = 77°C$
    $= 350$ K

Unknown: V_2

$$\frac{V_1}{T_1} = \frac{V_2}{T_2}$$

$$V_2 = \frac{V_1 T_2}{T_1} = \frac{(80.0 \text{ mL})(350 \text{ K})}{300 \text{ K}} = 93.3 \text{ mL}$$

b. Given: $V_1 = 125$ L
$V_2 = 85.0$ L
$T_2 = 127°C$
    $= 400$ K

Unknown: T_1

$$\frac{V_1}{T_1} = \frac{V_2}{T_2}$$

$$T_1 = \frac{V_1 T_2}{V_2} = \frac{(125 \text{ L})(400 \text{ K})}{85.0 \text{ L}} = 588 \text{ K} = 315°C$$

c. Given: $T_1 = -33°C$
    $= 240$ K
$V_2 = 54.0$ mL
$T_2 = 160°C$
    $= 433$ K

Unknown: V_1

$$\frac{V_1}{T_1} = \frac{V_2}{T_2}$$

$$V_1 = \frac{V_2 T_1}{T_2} = \frac{(54.0 \text{ mL})(2.40 \text{ K})}{433 \text{ K}} = 29.9 \text{ mL}$$

26. Given: $V_1 = 140.0$ mL
$T_1 = 67°C = 340$ K
$V_2 = 50.0$ mL

Unknown: T_2

$$\frac{V_1}{T_1} = \frac{V_2}{T_2}$$

$$T_2 = \frac{V_2 T_1}{V_1} = \frac{(50.0 \text{ mL})(340 \text{ K})}{140.0 \text{ mL}} = 121 \text{ K} = -152°C$$

27. Given: $V_1 = 275$ mL
$T_1 = 0°C = 273$ K
$T_2 = 130.$ °C
    $= 403$ K

Unknown: V_2

$$\frac{V_1}{T_1} = \frac{V_2}{T_2}$$

$$V_2 = \frac{V_1 T_2}{T_1} = \frac{(275 \text{ mL})(403 \text{ K})}{273 \text{ K}} = 406 \text{ mL}$$

28. Given: $T_1 = 47°C = 320$ K
$P_1 = 0.329$ atm
$T_2 = 77°C = 350$ K

Unknown: P_2

$$\frac{P_1}{T_1} = \frac{P_2}{T_2}$$

$$P_2 = \frac{P_1 T_2}{T_1} = \frac{(0.329 \text{ atm})(350 \text{ K})}{320 \text{ K}} = 0.360 \text{ atm}$$

29. Given: $T_1 = 27°C = 300$ K
$P_1 = 0.625$ atm
$P_2 = 1.125$ atm

Unknown: T_2

$$\frac{P_1}{T_1} = \frac{P_2}{T_2}$$

$$T_2 = \frac{P_2 T_1}{P_1} = \frac{(1.125 \text{ atm})(300 \text{ K})}{0.625 \text{ atm}} = 540 \text{ K} = 267°C$$

30. Given: $T_1 = -73°C = 200$ K
$P_1 = 1$
$P_2 = 2$

Unknown $= T_2$

$$T_2 = \frac{P_2 T_1}{P_1} = \frac{(2)(200)}{1} = 400 \text{ K} = 127°C$$

31. Given: $T_1 = 47°C = 320$ K
$P_1 = 1.03$ atm
$V_1 = 2.20$ L
$T_2 = 107°C = 380$ K
$P_2 = 0.789$ atm

Unknown: V_2

$$\frac{P_1V_1}{T_1} = \frac{P_2V_2}{T_2}$$

$$V_2 = \frac{P_1V_1T_2}{P_2T_1} = \frac{(1.03 \text{ atm})(2.20 \text{ L})(380 \text{ K})}{(0.789 \text{ atm})(320 \text{ K})} = 3.41 \text{ L}$$

---

32. Given: $V_1 = 350.$ mL
$T_1 = 35°C = 308$ K
$P_1 = 550.$ torr
$V_2 = 425$ mL
$T_2 = 57°C = 330$ K

Unknown: P_2

$$\frac{P_1V_1}{T_1} = \frac{P_2V_2}{T_2}$$

$$P_2 = \frac{P_1V_1T_2}{T_1V_2} = \frac{(550 \text{ torr})(350 \text{ mL})(330 \text{ K})}{(308 \text{ K})(425 \text{ mL})} = 485 \text{ torr}$$

---

33. Given: $V_1 = 1.75$ L
$T_1 = -23°C = 250$ K
$P_1 = 150$ kPa
$V_2 = 1.30$ L
$P_2 = 210$ kPa

Unknown: T_2

$$\frac{P_1V_1}{T_1} = \frac{P_2V_2}{T_2}$$

$$T_2 = \frac{P_2V_2T_1}{P_1V_1} = \frac{(210 \text{ kPa})(1.30 \text{ L})(250 \text{ K})}{(150 \text{ kPa})(1.75 \text{ L})} = 260 \text{ K} = -13°C$$

---

34. Given: $T_1 = 40.°C = 313$ K
$V_1 = 820.$ mL
$T_2 = 60.°C = 333$ K
$V_2 = 1250$ mL
$P_2 = 1.40$ atm

Unknown: P_1

$$\frac{P_1V_1}{T_1} = \frac{P_2V_2}{T_2}$$

$$P_1 = \frac{P_2V_2T_1}{T_2V_1} = \frac{(1.40 \text{ atm})(1250 \text{ mL})(313 \text{ K})}{(333 \text{ K})(820 \text{ mL})} = 2.01 \text{ atm}$$

---

35. Given: $P_1 = 7.75 \times 10^4$ Pa
$T_1 = 17°C = 290$ K
$V_1 = 850.$ cm^3
$V_2 = 720.$ cm^3
$P_2 = 8.10 \times 10^4$ Pa

Unknown: T_2

$$\frac{P_1V_1}{T_1} = \frac{P_2V_2}{T_2}$$

$$T_2 = \frac{P_2V_2T_1}{P_1V_1} = \frac{(8.10 \times 10^4 \text{ Pa})(720 \text{ cm}^3)(290 \text{ K})}{(7.75 \times 10^4 \text{ Pa})(850 \text{ cm}^3)} = 257 \text{ K} = -17°C$$

---

36. Given: $V_1 = 250$ L
$T_1 = 22°C = 295$ K
$P_1 = 0.974$ atm
$T_2 = -52°C = 221$ K
$P_2 = 0.750$ atm

Unknown: V_2

$$\frac{P_1V_1}{T_1} = \frac{P_2V_2}{T_2}$$

$$V_2 = \frac{P_1V_1T_2}{P_2T_1} = \frac{(0.974 \text{ atm})(250 \text{ L})(221 \text{ K})}{(0.750 \text{ atm})(295 \text{ K})} = 243 \text{ L}$$

---

37. Given: $V_1 = 250$ L
$T_1 = 295$ K
$P_1 = 0.974$ atm
$V_2 = 400$ L
$P_2 = 0.475$ atm

Unknown: T_2

$$\frac{P_1V_1}{T_1} = \frac{P_2V_2}{T_2}$$

$$T_2 = \frac{P_2V_2T_1}{P_1V_1} = \frac{(0.475 \text{ atm})(400 \text{ L})(295 \text{ K})}{(0.974 \text{ atm})(250 \text{ L})} = 230 \text{ K} = -43°C$$

38. Given: $V_1 = 5.05$ m^3
$T_1 = 20.°C$
$= 293$ K
$P_1 = 9.95 \times 10^4$ Pa
$T_2 = 0°C = 273$ K
$P_2 = 1.01\,325 \times 10^5$ Pa

Unknown: V_2 in m^3 per day

$$\frac{P_1 V_1}{T_1} = \frac{P_2 V_2}{T_2}$$

$$V_2 = \frac{P_1 V_1 T_2}{P_2 T_1} = \frac{(9.95 \times 10^4 \text{ Pa})(5.05 \text{ m}^3)(273 \text{ K})}{(1.01\,325 \times 10^5 \text{ Pa})(293 \text{ K})}$$

$= 4.62 \times 10^{-4}$ m^3

$\left(\dfrac{15 \text{ breaths}}{\text{min}}\right)\left(\dfrac{60 \text{ min}}{\text{hour}}\right)\left(\dfrac{24 \text{ hours}}{\text{day}}\right) = 21\,600$ breaths/day

$\left(\dfrac{21\,600 \text{ breaths}}{\text{day}}\right)\left(\dfrac{4.62 \times 10^{-4} \text{ m}^3}{\text{breath}}\right) = 9.98$ m^3/day

---

39. Given: $P_{CO_2} = 0.285$ torr
$P_{N_2} = 593.525$ torr
$P_T = 1$ atm $= 760$ torr

Unknown: P_{O_2}

$P_T = P_{CO_2} + P_{N_2} + P_{O_2}$

$P_{O_2} = P_T - (P_{CO_2} + P_{N_2})$

$= 760$ torr $- (593.525 + 0.285) = 166.190$ torr

---

40. Given: $T = 20.0°C$
$P_T = 730.0$ torr
P_{H_2O} at $20.0°C = 17.5$ torr

Unknown: P_{O_2}

$P_T = P_{O_2} + P_{H_2O}$

$P_{O_2} = P_T - P_{H_2O} = 730.0 - 17.5 = 712.5$ torr

---

41. Given: $T = 35.0°C$
$P_T = 742.0$ torr

Unknown: P_{gas}

P_{H_2O} at $35°C = 42.2$ mm

$P_T = P_{H_2O} + P_{gas}$

$P_{gas} = P_T - P_{H_2O} = 742.0 - 42.2 = 699.8$ torr

---

42. Given: $V_1 = 175$ mL
$T_1 = 15°C = 288$ K
$P_T = 752.0$ torr
$P_2 = 770.0$ torr
$T_2 = 15°C = 288$ K

Unknown: V_2

P_{H_2O} at $15°C = 12.8$ mm Hg $= 12.8$ torr

$P_T = P_{O_2} + P_{H_2O}$

$P_{O_2} = P_T - P_{H_2O} = 752.0 - 12.8$ torr $= 739.2$ torr $= P_1$

$$\frac{P_1 V_1}{T_1} = \frac{P_2 V_2}{T_2}$$

$$V_2 = \frac{P_1 V_1 T_2}{P_2 T_1} = \frac{(739.2 \text{ torr})(175 \text{ mL})(288 \text{ K})}{(770 \text{ torr})(288 \text{ K})} = 168 \text{ mL}$$

---

43. Given: $V_1 = 120.$ mL
$T_1 = 25°C = 298$ K
$P_T = 780.0$ torr
$P_2 = 760.0$ torr
$T_2 = 0°C = 273$ K

Unknown: V_2

P_{H_2O} at $25°C = 23.8$ torr

$P_T = P_{Ar} + P_{H_2O}$

$P_{Ar} = P_T - P_{H_2O} = 780 - 23.8 = 756.2$ torr $= P_1$

$$\frac{P_1 V_1}{T_1} = \frac{P_2 V_2}{T_2}$$

$$V_2 = \frac{P_1 V_1 T_2}{P_2 T_1} = \frac{(756.2 \text{ torr})(120 \text{ mL})(273 \text{ K})}{(760 \text{ torr})(298 \text{ K})} = 109 \text{ mL}$$

Mixed Review

44. Given: $P_T = 6.11$ atm
$P_A = 1.68$ atm
$P_B = 3.89$ atm

Unknown: P_C

$P_T = P_A + P_B + P_C$
$P_C = P_T - (P_A + P_B) = 6.11 - (1.68 + 3.89) = 0.54$ atm

45. Given: $V_1 = 2.30$ L
$T_1 = 311$ K
$T_2 = 295$ K

Unknown: V_2

$\dfrac{V_1}{T_1} = \dfrac{V_1}{T_2}$

$V_2 = \dfrac{V_1 T_2}{T_1} = \dfrac{(2.30\text{ L})(295\text{ K})}{(311\text{ K})} = 2.18$ L

46. Given: $V_1 = 295$ mL
$T_1 = 36°C = 309$ K
$T_2 = 55°C = 328$ K

Unknown: V_2

$\dfrac{V_1}{T_1} = \dfrac{V_2}{T_2}$

$V_2 = \dfrac{V_1}{T_1} = \dfrac{(295\text{ mL})(328\text{ K})}{(309\text{ K})} = 313$ mL

47. Given: $V_1 = 638$ mL
$P_1 = 0.893$ atm
$T_1 = 12°C = 285$ K
$V_2 = 881$ mL
$T_2 = 18°C = 291$ K

Unknown: P_2

$\dfrac{P_1 V_1}{T_1} = \dfrac{P_2 V_2}{T_2}$

$P_2 = \dfrac{P_1 V_1 T_2}{T_1 V_2} = \dfrac{(0.893\text{ atm})(638\text{ mL})(291\text{ K})}{(285\text{ K})(881\text{ mL})} = 0.660$ atm

48. Given: $T_1 = 84°C = 357$ K
$P_1 = 0.503$ atm
$P_2 = 1.20$ atm

Unknown: T_2 in °C

$\dfrac{P_1}{T_1} = \dfrac{P_2}{T_2}$

$T_2 = \dfrac{P_2 T_1}{P_1} = \dfrac{(1.20\text{ atm})(357\text{ K})}{(0.503\text{ atm})} = 852$ K $= 579°C$

49. Given: $V_1 = 4.00$ L
$T_1 = 304$ K
$P_1 = 755$ mm
$V_2 = 4.08$ L
$P_2 = 728$ mm

Unknown: T_2

$\dfrac{P_1 V_1}{T_1} = \dfrac{P_2 V_2}{T_2}$

$T_2 = \dfrac{P_2 V_2 T_1}{P_1 V_1} = \dfrac{(728\text{ mm})(4.08\text{ L})(304\text{ K})}{(755\text{ mm})(4.00\text{ L})} = 299$ K

50. Given: $P_1 = 4.62$ atm
$V_1 = 2.33$ L
$V_2 = 1.03$ L

Unknown: P_2 (in torr)

$P_1 V_1 = P_2 V_2$

$P_2 = \dfrac{P_1 V_1}{V_2} = \dfrac{(4.62\text{ atm})(2.33\text{ L})}{1.03\text{ L}} = 10.45$ atm

$(10.45\text{ atm})\left(\dfrac{760\text{ torr}}{\text{atm}}\right) = 7940$ torr

51. Given: $V_2 = 2.00 \times 10^7$ L
$P_2 = 20.0$ atm
$P_1 = 1.0$ atm

Unknown: V_1

$P_1 V_1 = P_2 V_2$

$V_1 = \dfrac{P_2 V_2}{P_1} = \dfrac{(20.0\text{ atm})(2.00 \times 10^7\text{ L})}{1.0\text{ atm}} = 4.00 \times 10^8$ L

CHAPTER 11
Molecular Composition of Gases

Practice, pp. 336–337

1. Given: $n = 7.08$ mol N_2 at STP
Unknown: V of N_2 at STP

$V = 7.08 \text{ mol } N_2 \times \dfrac{22.4 \text{ L}}{\text{mol}} = 159 \text{ L } N_2$

2. Given: $V = 14.1$ L H_2 at STP
Unknown: n: number of moles of H_2 at STP

$n = \dfrac{V\ H_2}{22.4 \text{ L/mol}} = \dfrac{14.1 \text{ L}}{22.4 \text{ L/mol}} = 0.629 \text{ mol } H_2$

3. Given: $V = 550.$ cm^3 Ne at STP
Unknown: n: number of moles of Ne at STP

$n = \dfrac{V\ Ne}{22.4 \text{ L/mol}} = \dfrac{550 \text{ cm}^3}{22.4 \text{ L/mol}} \times \dfrac{1 \text{ L}}{1000 \text{ cm}^3} = 0.0246 \text{ mol Ne}$

ATE, Additional Sample Problem 11–1, p. 336

Given: $n = 0.0580$ mol NO at STP
Unknown: V of NO at STP

$V \text{ of NO} = 0.0580 \text{ mol NO} \times \dfrac{22.4 \text{ L}}{\text{mol}} = 1.30 \text{ L}$

Practice, p. 337

1. Given: $V = 1.33 \times 10^4$ mL O_2 at STP
Unknown: m of O_2 at STP

$m = (1.33 \times 10^4 \text{ mL})\left(\dfrac{1 \text{ L}}{1000 \text{ mL}}\right)\left(\dfrac{1 \text{ mol } O_2}{22.4 \text{ L}}\right)\left(\dfrac{32 \text{ g } O_2}{\text{mol } O_2}\right) = 19.0 \text{ g } O_2$

2. Given: $m = 77.0$ g NO_2 at STP
Unknown: V of NO_2 at STP

$V = (77.0 \text{ g } NO_2)\left(\dfrac{22.4 \text{ L}}{\text{mol } NO_2}\right)\left(\dfrac{\text{mol } NO_2}{46.0 \text{ g}}\right) = 37.4 \text{ L } NO_2$

3. Given: $V = 3$ L Cl_2
Unknown: m of Cl_2

$m = (3 \text{ L})\left(\dfrac{1 \text{ mol } Cl_2}{22.4 \text{ L}}\right)\left(\dfrac{70 \text{ g } Cl_2}{1 \text{ mol } Cl_2}\right) = 9 \text{ g } Cl_2$

ATE, Additional Sample Problem 11–2, p. 337

Given: $m = 4.22$ g Cl_2 at STP
Unknown: V of Cl_2 at STP

$V = (4.22 \text{ g } Cl_2)\left(\dfrac{22.4 \text{ L}}{\text{mol } Cl_2}\right)\left(\dfrac{\text{mol } Cl_2}{71 \text{ g } Cl_2}\right) = 1.33 \text{ L } Cl_2$

Section Review, p. 337

4. Given: $V\ O_2 = 135\ L\ O_2$ at STP

Unknown: n: number of moles of O_2 at STP

$$n = \frac{V}{22.4\ L/mol} = \frac{135\ L\ O_2}{22.4\ L/mol} = 6.03\ mol\ O_2$$

5. Given: $n = 0.0035$ mol CH_4 at STP

Unknown: V of CH_4 in mL at STP

$$V = (0.0035\ mol)\left(\frac{22.4\ L}{mol}\right)\left(\frac{1000\ mL}{L}\right) = 78\ ml\ CH_4$$

Practice, p. 343

1. Given: $n = 0.325$ mol H_2
 $V = 4.08$ L
 $T = 35°C = 308$ K

Unknown: P in atm

$$P = nRT/V = \frac{(0.325\ mol)\left(\frac{0.0821\ L\cdot atm}{mol\cdot K}\right)(308\ K)}{4.08\ L} = 2.01\ atm$$

2. Given: $V = 8.77$ L
 $n = 1.45$ mol
 $T = 20°C = 293\ K$

Unknown: P in atm

$$P = nRT/V = \frac{(1.45\ mol)\left(\frac{0.0821\ L\cdot atm}{mol\cdot K}\right)(293\ K)}{8.77\ L} = 3.98\ atm$$

ATE, Additional Sample Problems 11–3, p. 343

a. Given: $V = 2.07$ L He
 $n = 2.88$ mol
 $T = 22°C = 295\ K$

Unknown: P of He in atm

$$P = \frac{nRT}{V} = \frac{(2.88\ mol)\left(\frac{0.0821\ L\cdot atm}{mol\cdot K}\right)(295\ K)}{2.07\ L} = 33.7\ atm\ He$$

b. Given: $V = 22.9$ L H_2
 $n = 14.0$ mol
 $T = 12°C = 285$ K

Unknown: P in atm

$$P = \frac{nRT}{V} = \frac{(14.0\ mol)\left(\frac{0.0821\ L\cdot atm}{mol\cdot K}\right)(285\ K)}{22.9\ L} = 14.3\ atm$$

Practice, p. 344

1. Given: $n = 4.38$ mol
 $T = 250$ K
 $P = 0.857$ atm

Unknown: V

$$V = nRT/P = \frac{(4.38\ mol)\left(\frac{0.0821\ L\cdot atm}{mol\cdot K}\right)(250\ K)}{0.857\ atm} = 105\ L$$

2. Given: $n = 0.909$ mol N_2
$T = 125°C = 398$ K
$P = 0.901$ atm

Unknown: V of N_2 in L

$$V = nRT/P = \frac{(0.909 \text{ mol})\left(\frac{0.0821 \text{ L} \cdot \text{atm}}{\text{mol} \cdot \text{K}}\right)(398 \text{ K})}{0.901 \text{ atm}} = 33.0 \text{ L } N_2$$

ATE, Additional Sample Problems 11–4, p. 344

a. Given: $n = 0.00856$ mol O_2
$T = 43°C = 316$ K
$P = 0.926$ atm

Unknown: V of O_2 in mL

$$V = nRT/P = \frac{(0.00856 \text{ mol})\left(\frac{0.0821 \text{ L} \cdot \text{atm}}{\text{mol} \cdot \text{K}}\right)(316 \text{ K})}{0.926 \text{ atm}} = 240. \text{ mL } O_2$$

b. Given: $n = 9.09 \times 10^{-3}$ mol
$T = 16°C = 289$ K
$P = 0.873$ atm

Unknown: V of gas in mL

$$V = nRT/P = \frac{(9.09 \times 10^{-3} \text{ mol})\left(\frac{0.0821 \text{ L} \cdot \text{atm}}{\text{mol} \cdot \text{K}}\right)(289 \text{ K})}{0.873 \text{ atm}} = 247 \text{ mL}$$

Practice, p. 345

1. Given: $V = 45.1$ L CO_2
$T = 34°C = 307$ K
$P = 1.04$ atm

Unknown: m of CO_2 in g

$$n = \frac{PV}{RT} = \frac{(1.04 \text{ atm})(45.1 \text{ L})}{\left(\frac{0.0821 \text{ L} \cdot \text{atm}}{\text{mol} \cdot \text{K}}\right)(307 \text{ K})} = 1.86 \text{ mol } CO_2$$

$$m = (1.86 \text{ mol})\left(\frac{44 \text{ g}}{\text{mol}}\right) = 81.9 \text{ g } CO_2$$

2. Given: $V = 12.5$ L O_2
$T = 45°C = 318$ K
$P = 7.22$ atm

Unknown: m of O_2 in g

$$n = \frac{PV}{RT} = \frac{(7.22 \text{ atm})(12.5 \text{ L})}{\left(\frac{0.0821 \text{ L} \cdot \text{atm}}{\text{mol} \cdot \text{K}}\right)(318 \text{ K})} = 3.46 \text{ mol } O_2$$

$$m = (3.46 \text{ mol})\left(\frac{32 \text{ g}}{\text{mol}}\right) = 111 \text{ g } O_2$$

3. Given: $m = 0.30$ g CO_2
$V = 0.25$ L CO_2
$T = 400$ K

Unknown: P of CO_2

$$n = (0.30 \text{ g})\left(\frac{\text{mol}}{44 \text{ g}}\right) = 0.0068 \text{ mol}$$

$$P = nRT/V = (0.0068 \text{ mol})\left(\frac{0.0821 \text{ L} \cdot \text{atm}}{\text{mol} \cdot \text{K}}\right)(400 \text{ K})/0.25 \text{ L} = 0.90 \text{ atm } CO_2$$

ATE, Additional Sample Problems 11–5, p. 345

a. Given: $V = 15.0$ L C_2H_4
$P = 4.40$ atm
$T = 305$ K

Unknown: m of ethene gas (C_2H_4)

$n = \dfrac{PV}{RT} = \dfrac{(4.40 \text{ atm})(15.0 \text{ L})}{\left(\dfrac{0.0821 \text{ L} \cdot \text{atm}}{\text{mol} \cdot \text{K}}\right)(305 \text{ K})} = 2.64$ mol C_2H_4

$m = 2.64 \text{ mol} \times \dfrac{28.0 \text{ g } C_2H_4}{\text{mol}} = 74.0$ g C_2H_4

b. Given: $V = 19.4$ L NH_3
$P = 4.45$ atm
$T = 24°C = 297$ K

Unknown: m of NH_3 in kg

$n = \dfrac{PV}{RT} = \dfrac{(4.45 \text{ atm})(19.4 \text{ L})}{\left(\dfrac{0.0821 \text{ L} \cdot \text{atm}}{\text{mol} \cdot \text{K}}\right)(297 \text{ K})} = 3.54$ mol NH_3

$m = (3.54 \text{ mol})(17.03 \text{ g/mol})\left(\dfrac{1 \text{ kg}}{1000 \text{ g}}\right) = 6.03 \times 10^{-2}$ kg NH_3

Practice, p. 346

1. Given: $m = 0.427$ g
$V = 0.125$ L
$T = 20°C = 293$ K
$P = 0.980$ atm

Unknown: M

$M = \dfrac{mRT}{PV} = \dfrac{(0.427 \text{ g})\left(\dfrac{0.0821 \text{ L} \cdot \text{atm}}{\text{mol} \cdot \text{K}}\right)(293 \text{ K})}{(0.980 \text{ atm})(0.125 \text{ L})} = 83.8$ g/mol

2. Given: $P = 0.928$ atm NH_3
$T = 63.0°C = 336$ K

Unknown: D of NH_3

$D = \dfrac{MP}{RT} = \dfrac{(17 \text{ g/mol})(0.928 \text{ atm})}{\left(\dfrac{0.0821 \text{ L} \cdot \text{atm}}{\text{mol} \cdot \text{K}}\right)(336 \text{ K})} = 0.572$ g/L NH_3

3. Given: $D = 2.0$ g/L
$P = 1.50$ atm
$T = 27°C = 300$ K

Unknown: M

$M = \dfrac{DRT}{P} = \dfrac{(2.0 \text{ g/L})\left(\dfrac{0.0821 \text{ L} \cdot \text{atm}}{\text{mol} \cdot \text{K}}\right)(300 \text{ K})}{1.50 \text{ atm}} = 33$ g/mol

4. Given: $P = (551 \text{ torr})\left(\dfrac{1 \text{ atm}}{760 \text{ torr}}\right) = 0.725$ atm
$T = 25°C = 298$ K

Unknown: D of Ar

$D = \dfrac{MP}{RT} = \dfrac{(40 \text{ g/mol})(0.725 \text{ atm})}{\left(\dfrac{0.0821 \text{ L} \cdot \text{atm}}{\text{mol} \cdot \text{K}}\right)(298 \text{ K})} = 1.18$ g/L Ar

ATE, Additional Sample Problems 11-6, p. 346

a. Given: $m = 3.17$ g
$V = 0.942$ L
$T = 14°C = 287$ K
$P = 1.09$ atm

Unknown: M of gas in g/mol

$$M = \frac{mRT}{PV} = \frac{(3.17 \text{ g})\left(\frac{0.0821 \text{ L} \cdot \text{atm}}{\text{mol} \cdot \text{K}}\right)(287 \text{ K})}{(1.09 \text{ atm})(0.942 \text{ L})} = 72.7 \text{ g/mol}$$

b. Given: density = 1.225 g/L
$T = 15°C = 288$ K
$P = 1$ atm

Unknown: M

$$M = \frac{mRT}{PV}$$

$$D = \frac{m}{V}$$

$$M = \frac{DRT}{P} = \frac{(1.225 \text{ g/L})\left(\frac{0.0821 \text{ L} \cdot \text{atm}}{\text{mol} \cdot \text{K}}\right)(288 \text{ K})}{1 \text{ atm}} = 29.0 \text{ g/mol}$$

Section Review, p. 346

2. Given: $m = 0.100$ g
$P = 0.0928$ atm
$T = 22.3°C = 295.3$ K

Unknown: V of $C_2H_2F_4$ in L

$$V = nRT/P$$

$$n = (0.100 \text{ g})\left(\frac{1 \text{ mol } C_2H_2F_4}{102 \text{ g}}\right) = 0.00098 \text{ mol } C_2H_2F_4$$

$$V = \frac{(0.00098 \text{ mol})\left(\frac{0.0821/\text{L} \cdot \text{atm}}{\text{mol} \cdot \text{K}}\right)(295.3 \text{ K})}{0.0928 \text{ atm}} = 0.0256 \text{ L } C_2H_2F_4$$

4. Given: $m = 1.25$ g
$V = 1.00$ L
$P = 0.961$ atm
$T = 27.0°C = 300$ K

Unknown: M

$$M = \frac{mRT}{PV} = \frac{(1.25 \text{ g})\left(\frac{0.0821 \text{ L} \cdot \text{atm}}{\text{mol} \cdot \text{K}}\right)(300 \text{ K})}{(0.961 \text{ atm})(1.00 \text{ L})} = 32.0 \text{ g/mol}$$

Practice, p. 348

1. Given: $V = 4.55$ L O_2
Unknown: V of H_2 gas

$2H_2(g) + O_2(g) \rightarrow 2H_2O(g)$

$$V = (4.55 \text{ L } O_2)\left(\frac{2 \text{ L } H_2}{1 \text{ L } O_2}\right) = 9.10 \text{ L } H_2$$

2. Given: V of CO = 0.626 L $2O_2 + 4CO \rightarrow 4CO_2$

Unknown: V of O_2 gas

$$V = (0.626 \text{ L CO})\left(\frac{2 \text{ L } O_2}{4 \text{ L CO}}\right) = 0.313 \text{ L } O_2$$

ATE, Additional Sample Problems 11–7, p. 348

a. Given: V = 3.14 L XeF_6 $Xe(g) + 3F_2(g) \rightarrow XeF_6(g)$

Unknown: V of Xe
V of F

$$V \text{ of Xe} = (3.14 \text{ L } XeF_6)\left(\frac{1 \text{ L Xe}}{1 \text{ L } XeF_6}\right) = 3.14 \text{ L Xe}$$

$$V \text{ of F} = (3.14 \text{ L } XeF_6)\frac{(3 \text{ L } F_2)}{(1 \text{ L } XeF_6)} = 9.42 \text{ L } F_2$$

b. Given: V = 708 L NO_2 $3NO_2(g) + H_2O(l) \rightarrow 2HNO_3(l) + NO(g)$

Unknown: V of NO gas produced

$$V = (708 \text{ L } NO_2)\left(\frac{1 \text{ L NO}}{3 \text{ L } NO_2}\right) = 236 \text{ L NO}$$

Practice, p. 349

1. Given: $S_8(s) + 8O_2(g) \rightarrow 8SO_2(g)$

V = 12.61 L SO_2 at STP

Unknown: m of S_8

$$n = \frac{PV}{RT} = \frac{(1 \text{ atm})(12.61 \text{ L } SO_2)}{\left(\frac{0.0821 \text{ L} \cdot \text{atm}}{\text{mol} \cdot \text{K}}\right)(273 \text{ K})} = 0.5629 \text{ mol } SO_2$$

$$m = (0.5629 \text{ mol } SO_2)\left(\frac{1 \text{ mol } S_8}{8 \text{ mol } SO_2}\right)\left(\frac{256 \text{ g } S_8}{\text{mol } S_8}\right) = 18.0 \text{ g } S_8$$

2. Given: V = 3.44 L O_2 at STP

Unknown: m of H_2O

$2H_2(g) + O_2(g) \rightarrow 2H_2O(l)$

$$n = \frac{PV}{RT} = \frac{(1 \text{ atm})(3.44 \text{ L } O_2)}{\left(\frac{0.0821 \text{ L} \cdot \text{atm}}{\text{mol} \cdot \text{K}}\right)(273 \text{ K})} = 0.154 \text{ mol } O_2$$

$$m = (0.154 \text{ mol } O_2)\left(\frac{2 \text{ mol } H_2O}{1 \text{ mol } O_2}\right)\left(\frac{18 \text{ g } H_2O}{\text{mol } H_2O}\right) = 5.54 \text{ g } H_2O$$

ATE, Additional Sample Problem 11–8, p. 349

Given: V = 4.00 L H_2 at STP

Unknown: m of Al

$2NaOH(aq) + 2Al(s) + 6H_2O(l) \rightarrow 2NaAl(OH)_4(aq) + 3H_2(g)$

$$n = \frac{PV}{RT} = \frac{(1 \text{ atm})(4.00 \text{ L } H_2)}{\left(\frac{0.0821 \text{ L} \cdot \text{atm}}{\text{mol} \cdot \text{K}}\right)(273 \text{ K})} = 0.1786 \text{ mol } H_2$$

$$m = (0.1786 \text{ mol } H_2)\left(\frac{2 \text{ mol Al}}{3 \text{ mol } H_2}\right)\left(\frac{27 \text{ g Al}}{\text{mol Al}}\right) = 3.21 \text{ g Al}$$

Practice, p. 350

1. Given: $T = 38°C = 311$ K $2Na + Cl_2 \rightarrow 2NaCl$

  $P = 1.63$ atm

  reactant mass = 10.4 g Na

  $n = (10.4 \text{ g Na})\left(\dfrac{1 \text{ mol Na}}{23 \text{ g Na}}\right)\left(\dfrac{1 \text{ mol Cl}_2}{2 \text{ mol Na}}\right) = 0.226 \text{ mol Cl}_2$

  Unknown: V of Cl_2 needed

  $V = \dfrac{nRT}{P} = \dfrac{(0.226 \text{ mol Cl}_2)\left(\dfrac{0.0821 \text{ L} \cdot \text{atm}}{\text{mol} \cdot \text{K}}\right)(311 \text{ K})}{1.63 \text{ atm}} = 3.55 \text{ L Cl}_2$

2. Given: $2C(s) + O_2(g) \rightarrow 2CO(g)$

  $T = 27°C = 300$ K

  $P = 0.247$ atm

  reactant mass = 65.5 g C

  Unknown: V of CO produced

  $n = (65.5 \text{ g C})\left(\dfrac{1 \text{ mol C}}{12 \text{ g C}}\right)\left(\dfrac{2 \text{ mol CO}}{2 \text{ mol C}}\right) = 5.46 \text{ mol CO}$

  $V = \dfrac{nRT}{P} = \dfrac{(5.46 \text{ mol CO})\left(\dfrac{0.0821 \text{ L} \cdot \text{atm}}{\text{mol} \cdot \text{K}}\right)(300 \text{ K})}{0.247 \text{ atm}} = 543 \text{ L CO}$

ATE, Additional Sample Problem 11–9, p. 350

Given: $2NaN_3(s) \rightarrow 3N_2(g) + 2Na(s)$

$P = 1.30$ atm

$T = 87°C = 360$ K

reactant mass = 70.0 g NaN_3

Unknown: V of N_2 produced

$n = (70.0 \text{ g NaN}_3)\left(\dfrac{1 \text{ mol NaN}_3}{65 \text{ g NaN}_3}\right)\left(\dfrac{3 \text{ mol N}_2}{2 \text{ mol NaN}_3}\right) = 1.615 \text{ mol N}_2$

$V = \dfrac{nRT}{P} = \dfrac{(1.615 \text{ mol N}_2)\left(\dfrac{0.0821 \text{ L} \cdot \text{atm}}{\text{mol} \cdot \text{K}}\right)(360 \text{ K})}{1.30 \text{ atm}} = 36.7 \text{ L}$

Section Review, p. 350

1. Given: $V = 150.$ L H_2 $3H_2(g) + N_2(g) \rightarrow 2NH_3(g)$

  Unknown: V of NH_3

  $V = (150 \text{ L H}_2)\left(\dfrac{2 \text{ L NH}_3}{3 \text{ L H}_2}\right) = 100. \text{ L NH}_3$

2. Given: reactant mass = 4.60 g Na

  $2Na(s) + 2H_2O(l) \rightarrow H_2(g) + 2NaOH(aq)$

  Unknown: V of H_2 produced at STP

  $n = (4.60 \text{ g Na})\left(\dfrac{1 \text{ mol Na}}{23 \text{ g Na}}\right)\left(\dfrac{1 \text{ mol H}_2}{2 \text{ mol Na}}\right) = 0.1 \text{ mol H}_2$

  $V = \dfrac{nRT}{P} = \dfrac{(0.1 \text{ mol H}_2)\left(\dfrac{0.0821 \text{ L} \cdot \text{atm}}{\text{mol} \cdot \text{K}}\right)(273 \text{ K})}{1.00 \text{ atm}} = 2.24 \text{ L}$

3. Given: $V = 4.00 \times 10^2$ mL H_2 at STP

Unknown: m of Na in grams

$2Na + H_2O \rightarrow Na_2O + H_2$

$V = (4.00 \times 10^2 \text{ mL})\left(\dfrac{1 \text{ L}}{1000 \text{ mL}}\right) = 0.4 \text{ L } H_2$

$n = \dfrac{PV}{RT} = \dfrac{(1.00 \text{ atm})(0.4 \text{ L } H_2)}{\left(\dfrac{0.0821 \text{ L}\cdot\text{atm}}{\text{mol}\cdot\text{K}}\right)(273 \text{ K})} = 0.01785 \text{ mol } H_2$

$m = (0.01785 \text{ mol } H_2)\left(\dfrac{2 \text{ mol Na}}{1 \text{ mol } H_2}\right)\left(\dfrac{23 \text{ g Na}}{\text{mol Na}}\right) = 0.821 \text{ g Na}$

---

4. Given: $P = 0.987$ atm

$T = 25.0°C = 298$ K

reactant mass = 30.6 g

Unknown: V of O_2 in liters

$2KClO_3(s) \xrightarrow[MnO_2]{\Delta} 2KCl(s) + 3O_2(g)$

$n = (30.6 \text{ g } KClO_3)\left(\dfrac{1 \text{ mol } KClO_3}{122.5 \text{ g } KClO_3}\right)\left(\dfrac{3 \text{ mol } O_2}{2 \text{ mol } KClO_3}\right) = 0.375 \text{ mol } O_2$

$V = \dfrac{nRT}{P} = \dfrac{(0.375 \text{ mol } O_2)\left(\dfrac{0.0821 \text{ L}\cdot\text{atm}}{\text{mol}\cdot\text{K}}\right)(298 \text{ K})}{0.987 \text{ atm}} = 9.28 \text{ L}$

Practice, p. 355

1. Given: H_2 rate of effusion = 9 times that of unknown gas

Unknown: M of unknown gas (X)

$\dfrac{\text{rate of effusion of } H_2}{\text{rate of effusion of } X} = \dfrac{\sqrt{M_X}}{\sqrt{M_{H_2}}}$

$\sqrt{M_X} = \left(\dfrac{\text{rate of effusion of } H_2}{\text{rate of effusion of } X}\right)(\sqrt{M_{H_2}})$

$= \left(\dfrac{9}{1}\right)(\sqrt{2 \text{ g/mol}}) = 12.7$

$M_X \approx 160 \text{ g/mol}$

---

2. Given: identities of 2 gases, CO_2 and HCl

Unknown: relative rates of effusion

$\dfrac{\text{rate of effusion of } CO_2}{\text{rate of effusion of HCl}} = \dfrac{\sqrt{M_{HCl}}}{\sqrt{M_{CO_2}}} = \dfrac{\sqrt{36.5 \text{ g/mol}}}{\sqrt{44 \text{ g/mol}}} = 0.9$

---

3. Given: rate of effusion of Ne = 400 m/s

Unknown: rate of effusion of butane, C_4H_{10}, at same temperature

$\dfrac{\text{rate of effusion of neon}}{\text{rate of effusion of butane}} = \dfrac{\sqrt{M_{C_4H_{10}}}}{\sqrt{M_{Ne}}}$

rate of effusion of butane = (rate of effusion of neon)$\left(\dfrac{\sqrt{M_{Ne}}}{\sqrt{M_{C_4H_{10}}}}\right)$

$= (400 \text{ m/s})\left(\dfrac{\sqrt{20 \text{ g/mol}}}{\sqrt{58 \text{ g/mol}}}\right) = 235 \text{ m/s}$

ATE, Additional Sample Problems 11–10, p. 355

a. Given: N_2 rate of effusion = 1.7 times that of other gas

Unknown: (1) M of other gas (X)
(2) identity of other gas

$$\frac{\text{rate of effusion of } N_2}{\text{rate of effusion of X}} = \frac{\sqrt{M_X}}{\sqrt{M_{N_2}}}$$

$$\sqrt{M_X} = \left(\frac{\text{rate of effusion of } N_2}{\text{rate of effusion of X}}\right)(\sqrt{M_{N_2}})$$

$$= \left(\frac{1.7}{1}\right)(\sqrt{28 \text{ g/mol}}) = 8.995$$

(1) $M_X = 81$ g/mol

(2) Krypton (average atomic mass of Kr = 83.8)

b. Given:
$$\frac{\text{rate of diffusion of A}}{\text{rate of diffusion of B}} = \frac{16}{1}$$

Unknown: $\dfrac{M_B}{M_A}$

$$\frac{\sqrt{M_B}}{\sqrt{M_A}} = \frac{16}{1}$$

$$\frac{M_B}{M_A} = \frac{256}{1}$$

Section Review, p. 355

2. Given: rate of effusion of a gas = 1.6 times that of CO_2

Unknown: M of unknown gas (X)

$$\frac{\text{rate of effusion of X}}{\text{rate of effusion of } CO_2} = \frac{\sqrt{M_{CO_2}}}{\sqrt{M_X}}$$

$$\sqrt{M_X} = (\sqrt{M_{CO_2}})\left(\frac{\text{rate of effusion } CO_2}{\text{rate of effusion X}}\right)$$

$$= (\sqrt{44} \text{ g/mol})\left(\frac{1}{1.6}\right) = 4.14$$

$M_X = 4.14^2 = 17$ g/mol

3. Given: $T = 25°C = 298$ K

Unknown: molecular velocities of H_2O, He, HCl, BrF, NO_2

$$\frac{\text{rate of effusion of } H_2O}{\text{rate of effusion of He}} = \frac{\sqrt{M_{He}}}{\sqrt{M_{H_2O}}}$$

(Molecular velocities of 2 different gases are inversely proportional to the square root of their molar masses.)

$M_{H_2O} = 18$ g/mol

$M_{He} = 4$ g/mol

$M_{HCl} = 36.5$ g/mol

$M_{BrF} = 98.8$ g/mol

$M_{NO_2} = 46$ g/mol

Rates of effusion: BrF < NO_2 < HCl < H_2O < He

Review Problems

9. Given: $V = 5.00$ L O_2
$n = 1.08 \times 10^{23}$ molecules

  a. Unknown: number of molecules in 5.00 L H_2

$$\left(\frac{1.08 \times 10^{23} \text{ molecules}}{5.00 \text{ L O}_2}\right)\left(\frac{1 \text{ L O}_2}{\text{L H}_2}\right)(5.00 \text{ L H}_2) = 1.08 \times 10^{23} \text{ molecules H}_2$$

  b. Unknown: number of molecules in 5.00 L CO_2

$$\left(\frac{1.08 \times 10^{23} \text{ molecules}}{5.00 \text{ L O}_2}\right)\left(\frac{1 \text{ L O}_2}{\text{L CO}_2}\right)(5.00 \text{ L CO}_2) = 1.08 \times 10^{23} \text{ molecules CO}_2$$

  c. Unknown: number of molecules in 10.00 L NH_3

$$\left(\frac{1.08 \times 10^{23} \text{ molecules}}{5.00 \text{ L O}_2}\right)\left(\frac{1 \text{ L O}_2}{\text{L NH}_3}\right)(10.00 \text{ L NH}_3) = 2.16 \times 10^{23} \text{ molecules NH}_3$$

10. a. Unknown: number of molecules in 1.00 mol O_2

$$(1.00 \text{ mol O}_2)\left(\frac{6.022 \times 10^{23} \text{ molecules}}{\text{mol}}\right) = 6.022 \times 10^{23} \text{ molecules O}_2$$

  b. Unknown: number of molecules in 2.50 mol He

$$(2.5 \text{ mol He})\left(\frac{6.022 \times 10^{23} \text{ molecules}}{\text{mol}}\right) = 1.51 \times 10^{24} \text{ molecules He}$$

  c. Unknown: number of molecules in 0.0650 mol NH_3

$$(0.0650 \text{ mol NH}_3)\left(\frac{6.022 \times 10^{23} \text{ molecules}}{\text{mol}}\right) = 3.91 \times 10^{22} \text{ molecules NH}_3$$

  d. Unknown: number of molecules in 11.5 g NO_2

$$(11.5 \text{ g NO}_2)\left(\frac{1 \text{ mol NO}_2}{46 \text{ g NO}_2}\right)\left(\frac{6.022 \times 10^{23} \text{ molecules}}{\text{mol}}\right) = 1.51 \times 10^{23} \text{ molecules NO}_2$$

11. a. Unknown: m of 2.25 mol Cl_2

$$(2.25 \text{ mol Cl}_2)\left(\frac{71 \text{ g Cl}_2}{\text{mol Cl}_2}\right) = 160. \text{ g Cl}_2$$

  b. Unknown: m of 3.01×10^{23} molecules H_2S

$$(3.01 \times 10^{23} \text{ molecules})\left(\frac{\text{mol}}{6.022 \times 10^{23} \text{ molecules}}\right)\left(\frac{34 \text{ g H}_2\text{S}}{\text{mol}}\right) = 17.0 \text{ g H}_2\text{S}$$

  c. Unknown: mass of 25.0 molecules SO_2

$$(25 \text{ molecules})\left(\frac{\text{mol}}{6.022 \times 10^{23} \text{ molecules}}\right)\left(\frac{64 \text{ g SO}_2}{\text{mol}}\right) = 2.66 \times 10^{-21} \text{ g SO}_2$$

12. a. Unknown: V in L of 1.00 mol O_2 at STP

$V = (1.00 \text{ mol})\left(\dfrac{22.4 \text{ L}}{\text{mol}}\right) = 22.4 \text{ L } O_2$

b. Unknown: V in L of 3.50 mol F_2 at STP

$V = (3.50 \text{ mol})\left(\dfrac{22.4 \text{ L}}{\text{mol}}\right) = 78.4 \text{ L } F_2$

c. Unknown: V in L of 0.0400 mol CO_2 at STP

$V = (0.0400 \text{ mol})\left(\dfrac{22.4 \text{ L}}{\text{mol}}\right) = 0.896 \text{ L } CO_2$

d. Unknown: V in L of 1.20×10^{-6} mol He at STP

$V = (1.20 \times 10^{-6} \text{ mol})\left(\dfrac{22.4 \text{ L}}{\text{mol}}\right) = 2.69 \times 10^{-5} \text{ L He}$

---

13. a. Unknown: number of moles in 22.4 L N_2 at STP

$V = M \times \dfrac{22.4 \text{ L}}{\text{mol}}$

$n = \dfrac{V}{22.4 \text{ L/mol}} = \dfrac{22.4 \text{ L}}{22.4 \text{ L/mol}} = 1.00 \text{ mol } N_2$

b. Unknown: number of moles in 5.60 L Cl_2 at STP

$n = \dfrac{V}{22.4 \text{ L/mol}} = \dfrac{5.60 \text{ L}}{22.4 \text{ L/mol}} = 0.250 \text{ mol } Cl_2$

c. Unknown: number of moles in 0.125 L Ne at STP

$n = \dfrac{V}{22.4 \text{ L/mol}} = \dfrac{0.125 \text{ L}}{22.4 \text{ L/mol}} = 5.58 \times 10^{-3} \text{ mol Ne}$

d. Unknown: number of moles in 70.0 mL NH_3 at STP

$n = \dfrac{V}{22.4 \text{ L/mol}} = \left(\dfrac{70.0 \text{ mL}}{22.4 \text{ L/mol}}\right)\left(\dfrac{\text{L}}{1000 \text{ mL}}\right) = 3.13 \times 10^{-3} \text{ mol } NH_3$

---

14. a. Unknown: m in g of 11.2 L H_2 at STP

$m = (11.2 \text{ L})\left(\dfrac{1 \text{ mol}}{22.4 \text{ L}}\right)\left(\dfrac{2.014 \text{ g } H_2}{\text{mol}}\right) = 1.01 \text{ g } H_2$

b. Unknown: m in g of 2.80 L CO_2 at STP

$m = (2.80 \text{ L})\left(\dfrac{1 \text{ mol}}{22.4 \text{ L}}\right)\left(\dfrac{44 \text{ g } CO_2}{\text{mol}}\right) = 5.50 \text{ g } CO_2$

c. Unknown: m in g of 15.0 mL SO_2 at STP

$m = (15.0 \text{ mL})\left(\dfrac{\text{L}}{1000 \text{ mL}}\right)\left(\dfrac{1 \text{ mol}}{22.4 \text{ L}}\right)\left(\dfrac{64 \text{ g } SO_2}{\text{mol}}\right) = 0.0429 \text{ g } SO_2$

d. Unknown: m in g of 3.40 cm^3 F_2 at STP

$m = (3.40 \text{ cm}^3)\left(\dfrac{\text{mL}}{\text{cm}^3}\right)\left(\dfrac{\text{L}}{1000 \text{ mL}}\right)\left(\dfrac{1 \text{ mol}}{22.4 \text{ L}}\right)\left(\dfrac{38 \text{ g } F_2}{\text{mol}}\right) = 5.78 \times 10^{-3} \text{ g } F_2$

15. a. Unknown: V in L of 8.00 g O_2 at STP

$$V = (8.00 \text{ g } O_2)\left(\frac{22.4 \text{ L}}{\text{mol}}\right)\left(\frac{\text{mol}}{32 \text{ g } O_2}\right) = 5.60 \text{ L } O_2$$

b. Unknown: V in L of 3.50 g CO at STP

$$V = (3.50 \text{ g CO})\left(\frac{22.4 \text{ L}}{\text{mol}}\right)\left(\frac{\text{mol}}{28 \text{ g CO}}\right) = 2.80 \text{ L CO}$$

c. Unknown: V in L of 0.017 g H_2S at STP

$$V = (0.0170 \text{ g } H_2S)\left(\frac{22.4 \text{ L}}{\text{mol}}\right)\left(\frac{\text{mol}}{34 \text{ g } H_2S}\right) = 0.0112 \text{ L } H_2S$$

d. Unknown: V in L of 2.25×10^5 kg NH_3 at STP

$$V = (2.25 \times 10^5 \text{ kg})\left(\frac{1000 \text{ g}}{\text{kg}}\right)\left(\frac{22.4 \text{ L}}{\text{mol}}\right)\left(\frac{\text{mol}}{17 \text{ g } NH_3}\right) = 2.96 \times 10^8 \text{ L } NH_3$$

---

16. a. Given: $V = 2.50$ L HF
$n = 1.35$ mol
$T = 320.$ K

Unknown: P in atm

$$P = \frac{nRT}{V} = \frac{(1.35 \text{ mol})\left(\frac{0.0821 \text{ L} \cdot \text{atm}}{\text{mol} \cdot \text{K}}\right)(320.\text{ K})}{2.50 \text{ L}} = 14.2 \text{ atm}$$

b. Given: $V = 4.75$ L NO_2
$n = 0.86$ mol
$T = 300.$ K

Unknown: P

$$P = \frac{nRT}{V} = \frac{(0.86 \text{ mol})\left(\frac{0.0821 \text{ L} \cdot \text{atm}}{\text{mol} \cdot \text{K}}\right)(300.\text{ K})}{4.75 \text{ L}} = 4.4 \text{ atm}$$

c. Given: $V = 7.50 \times 10^2$ mL CO_2
$n = 2.15$ mol
$T = 57°C = 330$ K

Unknown: P

$$(7.50 \times 10^2 \text{ mL})\left(\frac{\text{L}}{1000 \text{ mL}}\right) = 0.75 \text{ L}$$

$$P = \frac{nRT}{V} = \frac{(2.15 \text{ mol})\left(\frac{0.0821 \text{ L} \cdot \text{atm}}{\text{mol} \cdot \text{K}}\right)(330.\text{ K})}{0.75 \text{ L}} = 77.7 \text{ atm}$$

---

17. a. Given: $n = 2.00$ mol H_2
$T = 300.$ K
$P = 1.25$ atm

Unknown: V in L

$$V = \frac{nRT}{P} = \frac{(2.00 \text{ mol})\left(\frac{0.0821 \text{ L} \cdot \text{atm}}{\text{mol} \cdot \text{K}}\right)(300 \text{ K})}{1.25 \text{ atm}} = 39.4 \text{ L } H_2$$

b. Given: $n = 0.425$ mol NH_3
$T = 37°C = 310$ K
$P = 0.724$ atm

Unknown: V in L

$$V = \frac{nRT}{P} = \frac{(0.425 \text{ mol})\left(\frac{0.0821 \text{ L} \cdot \text{atm}}{\text{mol} \cdot \text{K}}\right)(310 \text{ K})}{0.724 \text{ atm}} = 14.9 \text{ L } NH_3$$

c. Given: $m = 4.00$ g O_2

$T = 57°C = 330$ K

$P = 0.888$ atm

Unknown: V in L

$$V = \frac{nRT}{P} = \frac{(4.00 \text{ g})\left(\frac{\text{mol } O_2}{32 \text{ g}}\right)\left(\frac{0.0821 \text{ L} \cdot \text{atm}}{\text{mol} \cdot \text{K}}\right)(330 \text{ K})}{0.888 \text{ atm}} = 3.82 \text{ L } O_2$$

18. a. Given: $V = 1.25$ L

$T = 250.$ K

$P = 1.06$ atm

Unknown: n

$$n = \frac{PV}{RT} = \frac{(1.06 \text{ atm})(1.25 \text{ L})}{\left(\frac{0.0821 \text{ L} \cdot \text{atm}}{\text{mol} \cdot \text{K}}\right)(250 \text{ K})} = 0.0649 \text{ mol}$$

b. Given: $V = 0.80$ L

$T = 27°C = 300$ K

$P = 0.925$ atm

Unknown: n

$$n = \frac{PV}{RT} = \frac{(0.925 \text{ atm})(0.80 \text{ L})}{\left(\frac{0.0821 \text{ L} \cdot \text{atm}}{\text{mol} \cdot \text{K}}\right)(300 \text{ K})} = 0.030 \text{ mol}$$

c. Given: $V = 0.750$ L

$T = 50°C = 223$ K

$P = 0.921$ atm

Unknown: n

$$n = \frac{PV}{RT} = \frac{(0.921 \text{ atm})(0.75 \text{ L})}{\left(\frac{0.0821 \text{ L} \cdot \text{atm}}{\text{mol} \cdot \text{K}}\right)(223 \text{ K})} = 0.377 \text{ mol}$$

19. a. Given: $V = 5.60$ L O_2

$P = 1.75$ atm

$T = 250.$ K

Unknown: m in g

$$n = \frac{PV}{RT}$$

$$n = \frac{(1.75 \text{ atm})(5.60 \text{ L})}{\left(\frac{0.0821 \text{ L} \cdot \text{atm}}{\text{mol} \cdot \text{K}}\right)(250 \text{ K})} = 0.478 \text{ mol } O_2$$

$$m = (0.478 \text{ mol})\left(\frac{32 \text{ g } O_2}{\text{mol}}\right) = 15.3 \text{ g } O_2$$

b. Given: $V = 3.50$ L NH_3

$P = 0.921$ atm

$T = 27°C = 300$ K

Unknown: m

$$n = \frac{PV}{RT} = \frac{(0.921 \text{ atm})(3.50 \text{ L})}{\left(\frac{0.0821 \text{ L} \cdot \text{atm}}{\text{mol} \cdot \text{K}}\right)(300 \text{ K})} = 0.131 \text{ mol } NH_3$$

$$m = (0.131 \text{ mol})\left(\frac{17 \text{ g } NH_3}{\text{mol}}\right) = 2.23 \text{ g } NH_3$$

c. Given: $V = 0.125$ L SO_2
$P = 0.822$ atm
$T = -53°C = 220$ K

Unknown: m

$n = \dfrac{PV}{RT} = \dfrac{(0.822 \text{ atm})(0.125 \text{ L})}{\left(\dfrac{0.0821 \text{ L} \cdot \text{atm}}{\text{mol} \cdot \text{K}}\right)(220 \text{ K})} = 5.7 \times 10^{-3}$ mol SO_2

$m = (0.0057 \text{ mol})\left(\dfrac{64 \text{ g } SO_2}{\text{mol}}\right) = 0.364$ g SO_2

20. a. Given: $m = 0.650$ g
$V = 1.12$ L
$T = 280.$ K
$P = 1.14$ atm

Unknown: M in g/mol

$M = \dfrac{mRT}{PV} = \dfrac{(0.650 \text{ g})\left(\dfrac{0.0821 \text{ L} \cdot \text{atm}}{\text{mol} \cdot \text{K}}\right)(280 \text{ K})}{(1.14 \text{ atm})(1.12 \text{ L})} = 11.7$ g/mol

b. Given: $m = 1.05$ g
$V = 2.35$ L
$T = 37°C = 310$ K
$P = 0.840$ atm

Unknown: M

$M = \dfrac{mRT}{PV} = \dfrac{(1.05 \text{ g})\left(\dfrac{0.0821 \text{ L} \cdot \text{atm}}{\text{mol} \cdot \text{K}}\right)(310 \text{ K})}{(0.840 \text{ atm})(2.35 \text{ L})} = 13.6$ g/mol

c. Given: $m = 0.432$ g
$V = 0.75$ L
$T = -23°C = 250$ K
$P = 1.03$ atm

Unknown: M

$M = \dfrac{mRT}{PV} = \dfrac{(0.432 \text{ g})\left(\dfrac{0.0821 \text{ L} \cdot \text{atm}}{\text{mol} \cdot \text{K}}\right)(250 \text{ K})}{(1.03 \text{ atm})(0.75 \text{ L})} = 11.5$ g/mol

21. Given: $D = 3.20$ g/L
$T = -18°C = 255$ K
$P = 2.17$ atm

Unknown: M

$M = \dfrac{mRT}{PV}; D = \dfrac{m}{V}$

$M = \dfrac{DRT}{P} = \dfrac{(3.20 \text{ g})\left(\dfrac{0.0821 \text{ L} \cdot \text{atm}}{\text{mol} \cdot \text{K}}\right)(255 \text{ K})}{2.17 \text{ atm}} = 30.9$ g/mol

22. Given: $D = (1.40 \text{ g/cm}^3)\left(\dfrac{\text{cm}^3}{\text{mL}}\right)\left(\dfrac{1000 \text{ mL}}{\text{L}}\right) = 1400.$ g/L
$P = 1.30 \times 10^9$ atm
$M = 2.00$ g/mol

Unknown: T

$M = \dfrac{DRT}{P}; T = \dfrac{MP}{DR}$

$T = \dfrac{(2.00 \text{ g/mol})(1.30 \times 10^9 \text{ atm})}{(1400 \text{ g/L})\left(\dfrac{0.0821 \text{ L} \cdot \text{atm}}{\text{mol} \cdot \text{K}}\right)} = 2.26 \times 10^7$ K $= 2.26 \times 10^7 °$C

23. Given: $V = 1.0$ L CO $2CO + O_2 \rightarrow 2CO_2$

  a. Unknown: number of liters O_2 required

  $(1.0 \text{ L CO})\left(\dfrac{1 \text{ L } O_2}{2 \text{ L CO}}\right) = 0.50 \text{ L } O_2$

  b. Unknown: number of L CO_2 produced

  $(1.0 \text{ L CO})\left(\dfrac{2 \text{ L } CO_2}{2 \text{ L CO}}\right) = 1.0 \text{ L } CO_2$

---

24. Given: $V = 75.0$ L CO_2 $2C_2H_2 + 5O_2 \rightarrow 4CO_2 + 2H_2O$

  a. Unknown: number of L C_2H_2 required

  $(75.0 \text{ L } CO_2)\left(\dfrac{2 \text{ L } C_2H_2}{4 \text{ L } CO_2}\right) = 37.5 \text{ L } C_2H_2$

  b. Unknown: number of L H_2O produced

  $(75.0 \text{ L } CO_2)\left(\dfrac{2 \text{ L } H_2O}{4 \text{ L } CO_2}\right) = 37.5 \text{ L } H_2O$

  c. Unknown: number of L O_2 required

  $(75.0 \text{ L } CO_2)\left(\dfrac{5 \text{ L } O_2}{4 \text{ L } CO_2}\right) = 93.8 \text{ L } O_2$

---

25. Given: $V = 0.45$ L O_2 $CS_2(l) + 3O_2(g) \rightarrow CO_2(g) + 2SO_2(g)$

  Unknown: (a) V of CO_2
  (b) V of SO_2

  a. $(0.45 \text{ L } O_2)\left(\dfrac{1 \text{ L } CO_2}{3 \text{ L } O_2}\right) = 0.150 \text{ L } CO_2$

  b. $(0.45 \text{ L } O_2)\left(\dfrac{2 \text{ L } SO_2}{3 \text{ L } O_2}\right) = 0.300 \text{ L } SO_2$

---

26. Given: $CuO(s) + H_2(g) \rightarrow Cu(s) + H_2O(g)$

  $V = 5.60$ L H_2 at STP

  a. Unknown: n of H_2

  $n = \dfrac{PV}{RT} = \dfrac{(1.0 \text{ atm})(5.60 \text{ L})}{\left(\dfrac{0.0821 \text{ L} \cdot \text{atm}}{\text{mol} \cdot \text{K}}\right)(273 \text{ K})} = 0.250 \text{ mol } H_2$

  b. Unknown: n of Cu produced

  $n \text{ of Cu} = (0.250 \text{ mol } H_2)\left(\dfrac{1 \text{ mol Cu}}{1 \text{ mol } H_2}\right) = 0.250 \text{ mol Cu}$

  c. Unknown: number of grams Cu produced

  $(0.250 \text{ mol})\left(\dfrac{63.5 \text{ g Cu}}{\text{mol}}\right) = 15.9 \text{ g Cu}$

27. Given: $V = 0.75$ L $H_2O(g)$ at STP $2Fe(OH)_3(s) \rightarrow Fe_2O_3(s) + 3H_2O(g)$

   a. Unknown: number of grams $Fe(OH)_3$ used

$$n = (0.75 \text{ L } H_2O)\left(\frac{1 \text{ mol } H_2O}{22.4 \text{ L}}\right)\left(\frac{2 \text{ mol } Fe(OH)_3}{3 \text{ mol } H_2O}\right) = 0.022 \text{ mol } Fe(OH)_3$$

$$m = (0.022 \text{ mol})\left(\frac{106.8 \text{ g } Fe(OH)_3}{\text{mol}}\right) = 2.4 \text{ g } Fe(OH)_3$$

   b. Unknown: number of grams Fe_2O_3 produced

$$n = (0.75 \text{ L } H_2O)\left(\frac{1 \text{ mol } H_2O}{22.4 \text{ L}}\right)\left(\frac{1 \text{ mol } Fe_2O_3}{3 \text{ mol } H_2O}\right) = 0.011 \text{ mol } Fe_2O_3$$

$$m = (0.011 \text{ mol})\left(\frac{159.6 \text{ g } Fe_2O_3}{\text{mol}}\right) = 1.8 \text{ g } Fe_2O_3$$

28. Given: $P = 0.961$ atm
$V = 29.0$ L CH_4
$T = 20°C = 293$ K

Unknown: (a) V of CO_2
(b) V of H_2O

$CH_4 + 2O_2 \rightarrow CO_2 + 2H_2O$

   a. $V = (29.0 \text{ L } CH_4)\left(\dfrac{1 \text{ L } CO_2}{1 \text{ L } CH_4}\right) = 29.0 \text{ L } CO_2$

   b. $V = (29.0 \text{ L } CH_4)\left(\dfrac{2 \text{ L } H_2O}{1 \text{ L } CH_4}\right) = 58.0 \text{ L } H_2O$ vapor

29. Given: air = 20.9% O_2 by volume. $2C_8H_{18} + 25O_2 \rightarrow 16CO_2 + 18H_2O$

   a. $V = 25.0$ L C_8H_{18}

Unknown: V of air needed for combustion

$$V = (25.0 \text{ L } C_8H_{18})\left(\frac{25 \text{ L } O_2}{2 \text{ L } C_8H_{18}}\right) = 312.5 \text{ L } O_2$$

$$V \text{ air needed} = \left(\frac{100 \text{ L air}}{20.9 \text{ L } O_2}\right)(312.5 \text{ L } O_2) = 1.50 \times 10^3 \text{ L air}$$

   b. Unknown: (1) V CO_2 produced
(2) V H_2O produced

(1) $V = (25.0 \text{ L } C_8H_{18})\left(\dfrac{16 \text{ L } CO_2}{2 \text{ L } C_8H_{18}}\right) = 200. \text{ L } CO_2$

(2) $V = (25.0 \text{ L } C_8H_{18})\left(\dfrac{18 \text{ L } H_2O}{2 \text{ L } C_8H_{18}}\right) = 225 \text{ L } H_2O$ vapor

30. Given: $T = 550.°C = 823$ K
$P = 2.50 \times 10^2$ atm
Reactant mass = 10 000 g N_2
Unknown: V of NH_3 produced

$N_2 + 3H_2 \rightarrow 2NH_3$

$n = (10\ 000\ \text{g}\ N_2)\left(\dfrac{1\ \text{mol}\ N_2}{28\ \text{g}}\right)\left(\dfrac{2\ \text{mol}\ NH_3}{1\ \text{mol}\ N_2}\right) = 714\ \text{mol}\ NH_3$

$V = \dfrac{nRT}{P} = \dfrac{(714\ \text{mol}\ NH_3)\left(\dfrac{0.0821\ \text{L} \cdot \text{atm}}{\text{mol} \cdot \text{K}}\right)(823\ \text{K})}{250\ \text{atm}} = 193\ \text{L}\ NH_3$

31. Given: 5.00×10^2 g = reactant mass $C_3H_5(NO_3)_3$ at STP

Unknown: (a) V of CO_2 produced
(b) V of N_2 produced
(c) V of O_2 produced
(d) V of H_2O produced

$4C_3H_5(NO_3)_3 \rightarrow 12CO_2 + 6N_2 + O_2 + 10H_2O$

a. $(500\ \text{g}\ C_3H_5(NO_3)_3)\left(\dfrac{1\ \text{mol}\ C_3H_5(NO_3)_3}{227\ \text{g}}\right) = 2.20\ \text{mol}\ C_3H_5(NO_3)_3$

$n = (2.20\ \text{mol}\ C_3H_5(NO_3)_3)\left(\dfrac{12\ \text{mol}\ CO_2}{4\ \text{mol}\ C_3H_5(NO_3)_3}\right) = 6.61\ \text{mol}\ CO_2$

$V = (6.61\ \text{mol}\ CO_2)\left(\dfrac{22.4\ \text{L}}{\text{mol}}\right) = 148\ \text{L}\ CO_2$

b. $n = (2.20\ \text{mol}\ C_3H_5(NO_3)_3)\left(\dfrac{6\ \text{mol}\ N_2}{4\ \text{mol}\ C_3H_5(NO_3)_3}\right) = 3.30\ \text{mol}\ N_2$

$V = (3.30\ \text{mol}\ N_2)\left(\dfrac{22.4\ \text{L}}{\text{mol}}\right) = 73.9\ \text{L}\ N_2$

c. $n = (2.20\ \text{mol}\ C_3H_5(NO_3)_3)\left(\dfrac{1\ \text{mol}\ O_2}{4\ \text{mol}\ C_3H_5(NO_3)_3}\right) = 0.550\ \text{mol}\ O_2$

$V = (0.550\ \text{mol}\ O_2)\left(\dfrac{22.4\ \text{L}}{\text{mol}}\right) = 12.3\ \text{L}\ O_2$

d. $n = (2.20\ \text{mol}\ C_3H_5(NO_3)_3)\left(\dfrac{10\ \text{mol}\ H_2O}{4\ \text{mol}\ C_3H_5(NO_3)_3}\right) = 5.50\ \text{mol}\ H_2O$

$V = (5.50\ \text{mol}\ H_2O)\left(\dfrac{22.4\ \text{L}}{\text{mol}}\right) = 123\ \text{L}\ H_2O$

Total $V = 148 + 73.9 + 12.3 + 123 = 357$ L

32. $8SO_2(g) + 16 H_2S(g) \rightarrow 16 H_2O(l) + 3S_8$

Given: $P = 0.961$ atm
$T = 22°C = 295$ K
$m = 4.50 \times 10^8$ g S_8

Unknown: (a) V of SO_2 needed
(b) V of H_2S needed

$n = (4.50 \times 10^8 \text{ g}) \left(\dfrac{1 \text{ mol } S_8}{257 \text{ g}} \right) = 1.75 \times 10^6$ mol S_8

a. $n = (1.75 \times 10^6 \text{ mol } S_8) \left(\dfrac{8 \text{ mol } SO_2}{3 \text{ mol } S_8} \right) = 4.67 \times 10^6$ mol SO_2

$V = \dfrac{nRT}{P} = \dfrac{(4.67 \times 10^6 \text{ mol } SO_2) \left(\dfrac{0.0821 \text{ L} \cdot \text{atm}}{\text{mol} \cdot \text{K}} \right)(295 \text{ K})}{0.961 \text{ atm}}$

$= 1.18 \times 10^8$ L SO_2

b. $n = (1.75 \times 10^6 \text{ mol } S_8) \left(\dfrac{16 \text{ mol } H_2S}{3 \text{ mol } S_8} \right) = 9.33 \times 10^6$ mol H_2S

$V = \dfrac{nRT}{P} = \dfrac{(9.33 \times 10^6 \text{ mol } H_2S) \left(\dfrac{0.0821 \text{ L} \cdot \text{atm}}{\text{mol} \cdot \text{K}} \right)(295 \text{ K})}{0.961 \text{ atm}}$

$= 2.35 \times 10^8$ L H_2S

33. $CaC_2(s) + 2H_2O(l) \rightarrow C_2H_2(g) + Ca(OH)_2(aq)$

Given: $T = 17°C = 290$ K
$P = 0.974$ atm
$m = 3.25$ g CaC_2

Unknown: V of C_2H_2 produced in mL

$n = (3.25 \text{ g } CaC_2) \left(\dfrac{1 \text{ mol } CaC_2}{64.1 \text{ g}} \right) = 0.0507$ mol CaC_2

$(0.0507 \text{ mol } CaC_2) \left(\dfrac{1 \text{ mol } C_2H_2}{1 \text{ mol } CaC_2} \right) = 0.0507$ mol C_2H_2

$V = \dfrac{nRT}{P} = \dfrac{(0.0507 \text{ mol } C_2H_2) \left(\dfrac{0.0821 \text{ L} \cdot \text{atm}}{\text{mol} \cdot \text{K}} \right)(290 \text{ K})}{0.974 \text{ atm}}$

$= (1.24 \text{ L } C_2H_2) \left(\dfrac{1000 \text{ mL}}{1 \text{ L}} \right) = 1.24 \times 10^3$ mL C_2H_2

34. $2Mg(s) + O_2(g) \rightarrow 2MgO(s)$

Given: Conditions are at STP.

Unknown: amount of O_2 and MgO in mol

a. $(22.4 \text{ L } O_2) \left(\dfrac{1.00 \text{ mol}}{22.4 \text{ L}} \right) = 1.00$ mol O_2

$(1.00 \text{ mol } O_2) \left(\dfrac{2 \text{ mol MgO}}{1 \text{ mol } O_2} \right) = 2.00$ mol MgO

b. $(11.2 \text{ L } O_2) \left(\dfrac{1.00 \text{ mol}}{22.4 \text{ L}} \right) = 0.500$ mol O_2

$(0.500 \text{ mol } O_2) \left(\dfrac{2 \text{ mol MgO}}{1 \text{ mol } O_2} \right) = 1.00$ mol MgO

c. $(1.40 \text{ L } O_2) \left(\dfrac{1.00 \text{ mol}}{22.4 \text{ L}} \right) = 0.0625$ mol O_2

$(0.0625 \text{ mol } O_2) \left(\dfrac{2 \text{ mol MgO}}{1 \text{ mol } O_2} \right) = 0.125$ mol MgO

35. Given: $V = 8.50$ L I_2 at STP

$2KI(aq) + Cl_2(g) \rightarrow 2KCl(aq) + I_2(g)$

a. Unknown: number of moles I_2 produced

$n = (8.50 \text{ L } I_2)\left(\dfrac{\text{mol } I_2}{22.4 \text{ L } I_2}\right) = 0.379 \text{ mol } I_2$

b. Unknown: number of moles KI used

$n = (0.379 \text{ mol } I_2)\left(\dfrac{2 \text{ mol KI}}{1 \text{ mol } I_2}\right) = 0.758 \text{ mol KI}$

c. Unknown: number of grams KI used

$m = (0.758 \text{ mol KI})\left(\dfrac{166 \text{ g KI}}{\text{mol}}\right) = 126 \text{ g KI}$

36. Given: $V = 6.50 \times 10^2$ mL H_2 produced at STP.

Unknown: number of g of $FeSO_4$ produced

$Fe(s) + H_2SO_4 \rightarrow FeSO_4 + H_2$

$(6.50 \times 10^2 \text{ mL})\left(\dfrac{\text{L}}{1000 \text{ mL}}\right) = 0.65 \text{ L}$

$(0.65 \text{ L } H_2)\left(\dfrac{\text{mol } H_2}{22.4 \text{ L}}\right) = 0.029 \text{ mol } H_2$

$(0.029 \text{ mol } H_2)\left(\dfrac{1 \text{ mol } FeSO_4}{1 \text{ mol } H_2}\right) = 0.029 \text{ mol } FeSO_4$

$(0.029 \text{ mol } FeSO_4)\left(\dfrac{151.9 \text{ g } FeSO_4}{1 \text{ mol } FeSO_4}\right) = 4.41 \text{ g } FeSO_4$

37. Given: $V = 4.50 \times 10^2$ mL CO

$V = 825$ mL H_2

$CO(g) + 2H_2(g) \rightarrow CH_3OH(g)$

a. Unknown: reactant present in excess

$(4.50 \times 10^2 \text{ mL CO})\left(\dfrac{2 \text{ mL } H_2}{1 \text{ mL CO}}\right) = 900 \text{ mL } H_2 \text{ needed}$

$(825 \text{ mL } H_2)\left(\dfrac{1 \text{ mL CO}}{2 \text{ mL } H_2}\right) = 413 \text{ mL CO needed}$

CO is present in excess. $(4.50 \times 10^2 \text{ mL} > 413 \text{ mL})$

b. Unknown: amount of CO remaining after reaction

$450 \text{ mL} - 413 \text{ mL} = 37 \text{ mL CO}$

c. Unknown: volume of CH_3OH produced

$(825 \text{ mL } H_2)\left(\dfrac{1 \text{ mL } CH_3OH}{2 \text{ mL } H_2}\right) = 413 \text{ mL } CH_3OH$

38. Given: reactant mass = 13.5 g Al

$Al(s) + 6HCl(aq) \rightarrow 2AlCl_3(aq) + 3H_2(g)$

a. Unknown: n of Al

$(13.5 \text{ g Al})\left(\dfrac{\text{mol}}{27 \text{ g Al}}\right) = 0.500 \text{ mol Al}$

b. Unknown: n of H_2 produced

$(0.500 \text{ mol Al})\left(\dfrac{3 \text{ mol } H_2}{2 \text{ mol Al}}\right) = 0.750 \text{ mol } H_2$

c. Unknown: V of H_2 produced at STP

$V = \dfrac{nRT}{P} = \dfrac{(0.750 \text{ mol } H_2)\left(\dfrac{0.0821 \text{ L}\cdot\text{atm}}{\text{mol}\cdot\text{K}}\right)(273 \text{ K})}{1.00 \text{ atm}} = 16.8 \text{ L } H_2$

---

39. a. Unknown: relative rate of effusion of H_2 and N_2

$\dfrac{\text{rate of effusion of } H_2}{\text{rate of effusion of } N_2} = \dfrac{\sqrt{M_{N_2}}}{\sqrt{M_{H_2}}} = \dfrac{\sqrt{28.00 \text{ g/mol}}}{\sqrt{2.02 \text{ g/mol}}} = 3.72$

b. Unknown: relative rate of effusion of F_2 and Cl_2

$\dfrac{\text{rate of effusion of } F_2}{\text{rate of effusion of } Cl_2} = \dfrac{\sqrt{M_{Cl_2}}}{\sqrt{M_{F_2}}} = \dfrac{\sqrt{71 \text{ g/mol}}}{\sqrt{38 \text{ g/mol}}} = 1.37$

---

40. Unknown: relative average velocity of H_2 and Ne

$\dfrac{\text{velocity of } H_2}{\text{velocity of Ne}} = \dfrac{\sqrt{M_{Ne}}}{\sqrt{M_{H_2}}} = \dfrac{\sqrt{20.179 \text{ g/mol}}}{\sqrt{2.02 \text{ g/mol}}} = 3.16$

---

41. Given: velocity of Cl_2 molecules = 0.038 m/s

Unknown: velocity of SO_2 molecules

$\dfrac{\text{velocity of } Cl_2}{\text{velocity of } SO_2} = \dfrac{\sqrt{M_{SO_2}}}{\sqrt{M_{Cl_2}}}$

$\text{velocity of } SO_2 = (\text{velocity of } Cl_2)\left(\dfrac{\sqrt{M_{Cl_2}}}{\sqrt{M_{SO_2}}}\right)$

$= (0.0380 \text{ m/s})\left(\dfrac{\sqrt{71 \text{ g/mol}}}{\sqrt{64 \text{ g/mol}}}\right) = 0.0400 \text{ m/s}$

---

42. Given: effusion rate of He = 6.50 times effusion rate of gas X

Unknown: M_X

$\dfrac{\text{effusion rate of He}}{\text{effusion rate of X}} = \dfrac{\sqrt{M_X}}{\sqrt{M_{He}}}$

$\sqrt{M_X} = \left(\dfrac{\text{effusion rate of He}}{\text{effusion rate of X}}\right)(\sqrt{M_{He}})$

$= (6.5)(\sqrt{4.00 \text{ g/mol}}) = 13$

$M_X = 13^2 = 169 \text{ g/mol}$

Mixed Review

43. Given: effusion rate of gas X = 0.850 times effusion rate of NO_2

Unknown: M_X

$$\frac{\text{effusion rate of X}}{\text{effusion rate of } NO_2} = \frac{\sqrt{M_{NO_2}}}{\sqrt{M_X}}$$

$$\sqrt{M_X} = \left(\frac{\text{effusion rate of } NO_2}{\text{effusion rate of X}}\right)(\sqrt{M_{NO_2}})$$

$$= \left(\frac{1.00}{0.850}\right)(\sqrt{46 \text{ g/mol}}) = 8.0 \text{ g/mol}$$

$$M_X = (8.0)^2 = 64 \text{ g/mol}$$

45. Given: V = 265 mL = 0.265 L Cl_2 at STP

Unknown: m of Cl_2

$$n = (0.265 \text{ L } Cl_2)\left(\frac{1 \text{ mol}}{22.4 \text{ L}}\right) = 0.0118 \text{ mol } Cl_2$$

$$m = (0.0118 \text{ mol } Cl_2)\left(\frac{71.0 \text{ g } Cl_2}{\text{mol}}\right) = 0.838 \text{ g } Cl_2$$

46. Given: n = 3.11 mol CO_2

P = 0.820 atm

T = 39°C = 312 K

Unknown: V in L

$$PV = nRT$$

$$V = \frac{nRT}{P} = \frac{(3.11 \text{ mol } CO_2)\left(\frac{0.0821 \text{ L} \cdot \text{atm}}{\text{mol} \cdot \text{K}}\right)(312 \text{ K})}{0.820 \text{ atm}} = 97.2 \text{ L}$$

47. Unknown: $\dfrac{\text{effusion rate of CO}}{\text{effusion rate of } SO_3}$

$$\frac{\text{effusion rate of CO}}{\text{effusion rate of } SO_3} = \frac{\sqrt{M_{SO_3}}}{\sqrt{M_{CO}}} = \frac{\sqrt{80 \text{ g/mol}}}{\sqrt{28 \text{ g/mol}}} = 1.7$$

48. Given: M = 0.993 g

V = 0.570 L

T = 281 K

P = 1.44 atm

Unknown: M

$$M = \frac{mRT}{PV} = \frac{(0.993 \text{ g})\left(\frac{0.0821 \text{ L} \cdot \text{atm}}{\text{mol} \cdot \text{K}}\right)(281 \text{ K})}{(1.44 \text{ atm})(0.570 \text{ L})} = 27.9 \text{ g/mol}$$

49. Given: D = 3.07 g/L at STP

Unknown: M

$$D = \frac{M}{V}$$

$$M = \frac{mRT}{PV} = \frac{DRT}{P}$$

$$= \frac{(3.07 \text{ g/L})\left(\frac{0.0821 \text{ L} \cdot \text{atm}}{\text{mol} \cdot \text{K}}\right)(273 \text{ K})}{1.00 \text{ atm}} = 68.8 \text{ g/mol}$$

50. Given: $V = 1000.\text{ cm}^3 = 1000.\text{ mL} = 1\text{ L}$

$T = 32°\text{C} = 305\text{ K}$

$P = (752\text{ mm Hg})\left(\dfrac{\text{atm}}{760\text{ mm Hg}}\right) = 0.99\text{ atm}$

Unknown: n

$PV = nRT$

$n = \dfrac{PV}{RT} = \dfrac{(0.99\text{ atm})(1\text{ L})}{\left(\dfrac{0.0821\text{ L}\cdot\text{atm}}{\text{mol}\cdot\text{K}}\right)(305\text{ K})} = 0.0396\text{ mol He}$

---

51. Given: $T = 16°\text{C} = 289\text{ K}$

$P = 0.982\text{ atm}$

$M = 7.40\text{ g}$

$V = 3.96\text{ L}$

Unknown: V at STP; M

$\dfrac{P_1V_1}{T_1} = \dfrac{P_2V_2}{T_2}$

$V_2 = \dfrac{P_1V_1T_2}{P_2T_1}$

$= \dfrac{(0.982\text{ atm})(3.96\text{ L})(273\text{ K})}{(1.00\text{ atm})(289\text{ K})} = 3.67\text{ L}$

$M = \dfrac{mRT}{PV} = \dfrac{(7.40\text{ g})\left(\dfrac{0.0821\text{ L}\cdot\text{atm}}{\text{mol}\cdot\text{K}}\right)(289\text{ K})}{(0.982\text{ atm})(3.96\text{ L})} = 45.2\text{ g/mol}$

CHAPTER 12

Liquids and Solids

Practice, p. 386

1. Given: mass of H_2O (l) = 506 g
   molar heat of fusion of ice = 6.008 kJ/mol
   molar mass H_2O = 18.02 g

   Unknown: energy released when water freezes

   $(506 \text{ g } H_2O)\left(\dfrac{1 \text{ mol } H_2O}{18.02 \text{ g } H_2O}\right) = 28.1 \text{ mol } H_2O$

   $(28.1 \text{ mol } H_2O)(6.008 \text{ kJ/mol}) = 169 \text{ kJ}$

2. Given: energy released on condensation of steam = 4.97×10^5 kJ
   molar mass of H_2O = 18.02 g
   molar heat of vaporization = 40.79 kJ/mol

   Unknown: mass of steam required

   $m = (4.97 \times 10^5 \text{ kJ})\left(\dfrac{1 \text{ mol } H_2O}{40.79 \text{ kJ}}\right)\left(\dfrac{18.02 \text{ g}}{\text{mol } H_2O}\right)$

   $= 2.19 \times 10^5 \text{ g}$

ATE, Additional Sample Problems 12–1, p. 386

1. Given: energy absorbed on boiling = 5.23×10^4 kJ
   molar mass of H_2O = 18.02 g
   molar heat of vaporization = 40.79 kJ/mol

   Unknown: mass of liquid water required

   $m = (5.23 \times 10^4 \text{ kJ})\left(\dfrac{1 \text{ mol } H_2O}{40.79 \text{ kJ}}\right)\left(\dfrac{18.02 \text{ g}}{\text{mol } H_2O}\right)$

   $= 2.31 \times 10^4 \text{ g}$

2. Given: mass of $H_2O(s)$ = 16.3 g
   heat of fusion of ice = 6.009 kJ/mol
   molar mass of H_2O = 18.02 g

   Unknown: heat energy absorbed when ice melts

   $m = (16.3 \text{ g } H_2O)\left(\dfrac{1 \text{ mol } H_2O}{18.02 \text{ g } H_2O}\right)\left(\dfrac{6.009 \text{ kJ}}{\text{mol}}\right)$

   $= 5.44 \text{ kJ}$

3. Given: mass of H_2O (g) = 783 g
heat of vaporization = 40.79 kJ/mol
molar mass H_2O = 18.02 g

Unknown: heat energy released when steam condenses

$$m = (783 \text{ g } H_2O)\left(\frac{1 \text{ mol } H_2O}{18.02 \text{ g } H_2O}\right)(40.79 \text{ kJ/mol})$$

$$= 1.77 \times 10^3 \text{ kJ}$$

Review Problems

18. a. Given: molar heat of vaporization for H_2O = 40.79 kJ/mol
molar mass of H_2O = 18.02 g

Unknown: heat of vaporization in joules per gram

$$(40.79 \text{ kJ/mol})\left(\frac{\text{mol } H_2O}{18.02 \text{ g}}\right)\left(\frac{1000 \text{ J}}{\text{kJ}}\right) = 2264 \text{ J/g}$$

b. Given: heat of fusion of H_2O = 6.009 kJ/mol

Unknown: heat of fusion in joules per gram

$$(6.009 \text{ kJ/mol})\left(\frac{\text{mol } H_2O}{18.02 \text{ g}}\right)\left(\frac{1000 \text{ J}}{\text{kJ}}\right) = 333.5 \text{ J/g}$$

19. Given: heat of vaporization for H_2O = 40.79 kJ/mol
molar mass of H_2O = 18.02 g

a. Unknown: energy required to vaporize 5.00 mol H_2O

$(40.79 \text{ kJ/mol})(5.00 \text{ mol}) = 204 \text{ kJ}$

b. Unknown: energy required to vaporize 45.0 g H_2O

$$(45.0 \text{ g } H_2O)\left(\frac{\text{mol } H_2O}{18.02 \text{ g } H_2O}\right)\left(\frac{40.79 \text{ kJ}}{\text{mol}}\right)$$

$= 102 \text{ kJ}$

c. Unknown: energy required to vaporize 8.45×10^{10} molecules H_2O

$$(8.45 \times 10^{10} \text{ molecules } H_2O)\left(\frac{\text{mol } H_2O}{6.022 \times 10^{23} \text{ molecules } H_2O}\right)(40.79 \text{ kJ/mol})$$

$= 5.72 \times 10^{-12} \text{ kJ}$

20. Given: heat of fusion of water = 6.009 kJ/mol
molar mass of H_2O = 18.02 g

  a. Unknown: energy required to melt 12.75 mol ice

  $(12.75 \text{ mol})\left(\dfrac{6.009 \text{ kJ}}{\text{mol}}\right) = 76.61 \text{ kJ}$

  b. Unknown: energy required to melt 6.48×10^5 kg ice

  $(6.48 \times 10^5 \text{ kg})\left(\dfrac{\text{g}}{1000 \text{ kg}}\right)\left(\dfrac{\text{mol}}{18.02 \text{ g}}\right)\left(\dfrac{6.009 \text{ kJ}}{\text{mol}}\right)$
  $= 2.16 \times 10^8 \text{ kJ}$

21. Given: mass of substance = 0.433 mol
energy absorbed when substance is vaporized = 36.5 kJ

Unknown: molar heat of vaporization

$\dfrac{36.5 \text{ kJ}}{0.433 \text{ mol}} = 54.3 \text{ kJ/mol}$

22. Given: molar mass of substance = 259.0 g/mol
mass of substance = 71.8 g
energy absorbed by substance when it melts = 4.307 kJ

  a. Unknown: moles in the sample

  $(71.8 \text{ g})\left(\dfrac{\text{mol}}{259.0 \text{ g}}\right) = 0.277 \text{ mol}$

  b. Unknown: molar heat of fusion

  $\dfrac{4.307 \text{ kJ}}{0.277 \text{ mol}} = 15.5 \text{ kJ/mol}$

23. a. Given: molar heat of fusion of substance = 3.811 kJ/mol
energy released when substance freezes = 83.2 kJ

Unknown: moles of substance

$(83.2 \text{ kJ})\left(\dfrac{\text{mol}}{3.811 \text{ kJ}}\right) = 21.8 \text{ mol}$

  b. Given: mass of sample = 5519 g

  Unknown: molar mass of substance

  $\dfrac{5519 \text{ g}}{21.8 \text{ mol}} = 253 \text{ g/mol}$

24. Given: volume of ice = 5.00 cm³ at 0°C
volume of liquid water = 5.00 cm³ at 0°C

Unknown: **a.** which substance contains more molecules

b. number of molecules more

c. ratio of molecules in both samples

a. $D = \dfrac{m}{V}$

$m = DV$

Density of ice at 0°C = 0.917 g/cm³

Density of water at 0°C = 1.0 g/cm³

mass of ice = (0.917 g/cm³)(5.00 cm³) = 4.6 g

moles of ice = $(4.6 \text{ g})\left(\dfrac{\text{mol}}{18.02 \text{ g}}\right) = 0.255$ mol

$(0.255 \text{ mol})\left(\dfrac{6.022 \times 10^{23} \text{ molecules}}{\text{mol}}\right) = 1.53 \times 10^{23}$ molecules of ice

mass of liquid water = (1.0 g/cm³)(5.00 cm³)
= 5.0 g

moles of water = $(5.0 \text{ g})\left(\dfrac{\text{mol}}{18.02 \text{ g}}\right) = 0.277$ mol

$(0.277 \text{ mol})\left(\dfrac{6.022 \times 10^{23} \text{ molecules}}{\text{mol}}\right) = 1.67 \times 10^{23}$ molecules

Liquid water contains more molecules.

b. $1.617 \times 10^{23} - 1.53 \times 10^{23} = 1.4 \times 10^{22}$ more molecules

c. Ratio = $\dfrac{1.67 \times 10^{23}}{1.53 \times 10^{23}} = \dfrac{1.09}{1.00}$

---

25. Given: volume of liquid H₂O at 0°C = 5.00 cm³
volume of water vapor at STP = 5.00 cm³
Density of water vapor at STP = 4.87×10^{-6} g/cm³

Unknown: **a.** which substance contains more molecules

b. number of molecules more

c. ratio of molecules in both samples

a. *Liquid water:*

From Problem 24, liquid water contains 1.67×10^{23} molecules

Water vapor:

$m = DV = (4.87 \times 10^{-6} \text{ g/cm}^3)(5.00 \text{ cm}^3) = 2.44 \times 10^{-5}$ g

number of molecules = $(2.44 \times 10^{-5} \text{ g})\left(\dfrac{1 \text{ mol}}{18.02 \text{ g}}\right)\left(\dfrac{6.022 \times 10^{23} \text{ molecules}}{\text{mol}}\right)$
$= 8.15 \times 10^{17}$ molecules

Liquid water contains more molecules.

b. $1.67 \times 10^{23} - 8.15 \times 10^{17} = 1.67 \times 10^{23}$ more molecules

c. Ratio = $\dfrac{1.67 \times 10^{23}}{8.15 \times 10^{17}} = \dfrac{205\,000}{1}$

26. a. Given: T of steam = 100.°C = 373 K
P of steam = 1.00 atm
heat of fusion of ice = 6.009 kJ/mol

Unknown: volume and mass of steam that would release the same amount of heat as the liquid water during freezing

Liquid water:

$(100.\text{ cm}^3)\left(\dfrac{1.00\text{ g}}{\text{cm}^3}\right)\left(\dfrac{1\text{ mol}}{18.02\text{ g}}\right)(6.009\text{ kJ/mol})$

= 33.3 kJ = heat released when water is frozen

Steam:

molar heat of vaporization = 40.79 kJ/mol

$(33.3\text{ kJ})\left(\dfrac{1\text{ mol}}{40.79\text{ kJ}}\right) = 0.816\text{ mol}$

$V = \dfrac{nRT}{P} = \dfrac{(0.816\text{ mol})\left(0.0821\dfrac{\text{L}\cdot\text{atm}}{\text{mol}\cdot\text{K}}\right)(373\text{ K})}{1.00\text{ atm}}$

= 25 L = volume of steam

$(0.816\text{ mol})\left(\dfrac{18.02\text{ g}}{\text{mol}}\right) = 14.7\text{ g}$ = mass of steam

b. Unknown: relative volumes and masses of steam and liquid water required to release same amount of heat

Volumes:

Steam = 25 L

Liquid water = $(100\text{ cm}^3)\left(\dfrac{\text{mL}}{\text{cm}^3}\right)\left(\dfrac{\text{L}}{1000\text{ mL}}\right)$

= 0.1 L

Masses:

Steam: 14.7 g

Liquid water: $(100.\text{ cm}^3)\left(\dfrac{1.00\text{ g}}{\text{cm}^3}\right) = 100$ g

A larger volume of steam is required than water; a smaller mass of steam is required than water.

27. Given: mass of substance = 3.21 mol
energy absorbed on vaporization = 28.4 kJ

Unknown: molar heat of vaporization

$\dfrac{28.4\text{ kJ}}{3.21\text{ mol}} = 8.85\text{ kJ/mol}$

28. Given: molar heat of fusion of water = 6.009 kJ/mol

Unknown: energy required to melt 7.95×10^5 g ice

$(7.95 \times 10^5\text{ g})\left(\dfrac{1\text{ mol}}{18.02\text{ g}}\right)\left(\dfrac{6.009\text{ kJ}}{\text{mol}}\right) = 2.65 \times 10^5\text{ kJ}$

29. Given: molar heat of vaporization of substance = 31.6 kJ/mol

Unknown: amount of substance requiring 57 kJ to vaporize

$\dfrac{57\text{ kJ}}{31.6\text{ kJ/mol}} = 1.80\text{ mol}$

30. Given: heat of vaporization of water = 40.79 kJ/mol
Unknown: grams of water vaporized by 0.545 kJ

$$\left(\frac{0.545 \text{ kJ}}{40.79 \text{ kJ/mol}}\right)\left(\frac{18.02 \text{ g}}{\text{mol}}\right) = 0.241 \text{ g}$$

31. Given: mass of liquid = 13.3 g
molar mass = 82.9 g/mol
heat of fusion = 4.60 kJ/mol
Unknown: energy released by freezing 13.3 g

$$(13.3 \text{ g})\left(\frac{\text{mol}}{82.9 \text{ g}}\right)\left(\frac{4.60 \text{ kJ}}{\text{mol}}\right) = 0.738 \text{ kJ}$$

32. Given: $T = 100.°C = 373$ K
$P = 760.$ torr
heat of fusion of ice = 6.009 kJ/mol
Unknown: volume and mass of steam that would release same amount of heat during condensation as 65.5 cm³ liquid water would release during freezing

Liquid water:

$$(65.5 \text{ cm}^3)\left(\frac{1.00 \text{ g}}{\text{cm}^3}\right)\left(\frac{1 \text{ mol}}{18.02 \text{ g}}\right)\left(\frac{6.009 \text{ kJ}}{\text{mol}}\right)$$

= 21.8 kJ = heat released during freezing

Steam:

molar heat of vaporization = 40.79 kJ/mol

$$(21.8 \text{ kJ})\left(\frac{1 \text{ mol}}{40.79 \text{ kJ}}\right) = 0.5344 \text{ mol}$$

$$V = \frac{nRT}{P} = \frac{(0.5344 \text{ mol})\left(\frac{0.0821 \text{ L·atm}}{\text{mol·K}}\right)(373 \text{ K})}{1.00 \text{ atm}}$$

= 16.4 L = volume of steam

$$(0.5344 \text{ mol})\left(\frac{18.02 \text{ g}}{\text{mol}}\right) = 9.63 \text{ g} = \text{mass of steam}$$

33. Given: heat of fusion = 3.43 kJ/mol
mass = 64.2 g
energy absorbed on melting = 2.77 kJ
Unknown: molar mass of substance

$$(2.77 \text{ kJ})\left(\frac{\text{mol}}{3.43 \text{ kJ}}\right) = 0.807 \text{ mol}$$

$$\frac{64.2 \text{ g}}{0.807 \text{ mol}} = 79.5 \text{ g/mol}$$

CHAPTER 13

Solutions

ATE, Additional Sample Problem 13-1, p. 414

a. Given: $V = 2.0$ L
  mass of solute = 14.6 g NaCl
  Unknown: molarity (M)

$(14.69 \text{ NaCl})\left(\dfrac{\text{mol NaCl}}{58.44 \text{ g NaCl}}\right) = 0.249$ mol NaCl

$M = \dfrac{0.249 \text{ mol}}{2.0 \text{ L}} = 0.12$ M NaCl

b. Given: mass of solute = 10.0 g HCl
  $V = 0.250$ L
  Unknown: molarity (M)

$(10.0 \text{ g HCl})\left(\dfrac{\text{mol HCl}}{36.46 \text{ g HCl}}\right) = 0.274$ mol

$M = \dfrac{0.274 \text{ mol}}{0.250 \text{ L}} = 1.1$ M HCl

c. Given: molarity = 0.330 M NaCl
  $V = 1.25$ L
  Unknown: moles NaCl

mol solute = MV
= (0.330 mol/L)(1.25 L)
= 0.413 mol

d. Given: $V = 0.50$ L
  molarity = 0.50 M HCl
  Unknown: moles HCl

mol solute = MV
= (0.50 mol/L)(0.50 L)
= 0.25 mol

Practice, p. 415

1. Given: mass of solute = 5.85 g KI
  $V = 0.125$ L
  Unknown: molarity (M)

$(5.85 \text{ g KI})\left(\dfrac{1 \text{ mol}}{166 \text{ g}}\right) = 0.0352$ mol KI

$M = \dfrac{0.0352 \text{ mol KI}}{0.125 \text{ L}} = 0.282$ M KI

2. Given: $V = 0.500$ L
  molarity = 0.150 M H_2SO_4
  Unknown: moles H_2SO_4

mol solute = MV
= (0.150 mol/L H_2SO_4)(0.500 L)
= 0.075 mol

3. Given: molarity = 3.00 M NaCl
  mass of NaCl = 146.3 g
  Unknown: V

$(146.3 \text{ g NaCl})\left(\dfrac{\text{mol NaCl}}{58.44 \text{ g}}\right) = 2.503$ mol NaCl

$V = \dfrac{2.503 \text{ mol}}{3.00 \text{ mol/L}} = 0.834$ L

ATE, Additional Sample Problem 13–2, p. 415

a. Given: mass of solute = 6.25 g HCl
$V = 0.300$ L
Unknown: molarity (M)

$(6.25 \text{ g HCl})\left(\dfrac{\text{mol HCl}}{36.46 \text{ g HCl}}\right) = 0.171 \text{ mol HCl}$

$M = \dfrac{0.171 \text{ mol}}{0.300 \text{ L}} = 0.570 \text{ M HCl}$

b. Given: $V = 0.250$ L
molarity = 2.30 M KI
Unknown: moles KI

mol solute = MV
$= (2.30 \text{ mol/L})(0.250 \text{ L})$
$= 0.575 \text{ mol}$

c. Given: molarity = 0.500 M HBr
mass of HBr = 32.5 g
Unknown: V

$(32.5 \text{ g HBr})\left(\dfrac{\text{mol HBr}}{80.91 \text{ g HBr}}\right) = 0.402 \text{ mol}$

$V = \dfrac{0.402 \text{ mol}}{0.500 \text{ mol/L}} = 0.804 \text{ L}$

Practice, p. 418

1. Given: mass of solute = 255 g $(CH_3)_2CO$
mass of solvent = 200. g H_2O
Unknown: molality (m)

$(255 \text{ g } (CH_3)_2CO)\left(\dfrac{\text{mol }(CH_3)_2CO}{58.08 \text{ g } (CH_3)_2CO}\right) = 4.39 \text{ mol}$

$(200. \text{ g } H_2O)\left(\dfrac{\text{kg}}{1000 \text{ g}}\right) = 0.20 \text{ kg}$

$m = \dfrac{4.39 \text{ mol}}{0.20 \text{ kg}} = 22.0 \; m$ acetone

2. Given: molality = 0.244 m CH_3OH
mass of solvent = 400. g H_2O
Unknown: mass of solute, CH_3OH

moles solute = $(m)(\text{kg solvent})$

moles solute = $\left(0.244 \dfrac{\text{mol}}{\text{kg}}\right)(0.400 \text{ kg}) = 0.0976 \text{ mol } CH_3OH$

$(0.0976 \text{ mol } CH_3OH)\left(\dfrac{32 \text{ g } CH_3OH}{\text{mol } CH_3OH}\right) = 3.12 \text{ g } CH_3OH$

3. Given: molality = 0.125 m $AgNO_3$
volume of solvent = 250. mL H_2O
Unknown: mass of solute, $AgNO_3$

$m = \dfrac{\text{moles solute}}{\text{mass of solvent (kg)}}$

moles solute = $(m)(\text{kg solvent})$

$(250. \text{ mL } H_2O)\left(\dfrac{L}{1000 \text{ mL}}\right) = 0.250 \text{ L} = 0.250 \text{ kg } H_2O$

\# mol $AgNO_3 = (0.125)(0.250) = 0.0313$ mol

$(0.0313 \text{ mol } AgNO_3)\left(\dfrac{169.88 \text{ g } AgNO_3}{\text{mol } AgNO_3}\right) = 5.32 \text{ g } AgNO_3$

4. Given: mass of solute = 18.2 g HCl
mass of solvent = 250. g H_2O
Unknown: molality (m)

$(18.2 \text{ g HCl})\left(\dfrac{\text{mol HCl}}{36.46 \text{ g HCl}}\right) = 0.499 \text{ mol HCl}$

$(250. \text{ g})\left(\dfrac{\text{kg}}{1000 \text{ g}}\right) = 0.250 \text{ kg } H_2O$

$m = \dfrac{0.499 \text{ mol}}{0.250 \text{ kg}} = 2.00 \; m$

Section Review, p. 418

2. Given: mass of solute = 5.0 g $C_{12}H_{22}O_{11}$
$V = 1$ L
Unknown: molarity (M)

$(5.0 \text{ g } C_{12}H_{22}O_{11})\left(\dfrac{\text{mol } C_{12}H_{22}O_{11}}{342.23 \text{ g } C_{12}H_{22}O_{11}}\right) = 0.01461 \text{ mol}$

$M = \dfrac{0.01461 \text{ mol}}{1 \text{ L}} = 0.01 \text{ M}$

ATE, Additional Sample Problem 13–3, p. 418

a. Given: mass of solvent = 13.0 g NaCl
mass of solute = 500. g H_2O
Unknown: molality (m)

$(13.0 \text{ g NaCl})\left(\dfrac{\text{mol NaCl}}{58.44 \text{ g}}\right) = 0.222 \text{ mol NaCl}$

$(500 \text{ g } H_2O)\left(\dfrac{1 \text{ kg}}{1000 \text{ g}}\right) = 0.500 \text{ kg}$

$m = \dfrac{0.222 \text{ mol NaCl}}{0.500 \text{ kg } H_2O} = 0.444 \; m$

b. Given: $m = 1.0$ NaCl
mass of solvent = 250 g
Unknown: mass of solute

mol NaCl = (m)(kg solvent) = (1.0)(0.25) = 0.25 mol

$(0.25 \text{ mol})\left(\dfrac{58.44 \text{ g NaCl}}{\text{mol}}\right) = 15 \text{ g NaCl}$

c. Given: molality = 0.245 m NaCl
moles solute = 1.0 mol
Unknown: volume (mass of solvent, water, in kg)

$m = \dfrac{\text{moles solute}}{\text{mass of solvent (kg)}}$

mass of solvent = $\dfrac{\text{moles solute}}{m} = \dfrac{1.0}{0.245} = 4.1 \text{ kg} = 4.1 \text{ L}$

Review Problems

14. Solubility at 60°C = 106
Solubility at 20°C = 31.6
106 − 31.6 = 74.4 g

15. a. Given: mass of solute
= 40.0 g NaOH
$V = 6.00$ L

Unknown #2: molar mass of NaOH

Unknown #3: molarity (M)

Atomic mass Na = 22.99
Atomic mass O = 15.99
Atomic mass H = <u>1.01</u>
Total: 39.99 = 40 g = molar mass NaOH

$(40.0 \text{ g NaOH})\left(\dfrac{\text{mol NaOH}}{40 \text{ g NaOH}}\right) = 1 \text{ mol NaOH}$

$M = \dfrac{1.00 \text{ mol}}{6.00 \text{ L}} = 0.167 \text{ M NaOH}$

b. Given: mass of NH_4Br = 14.0 g
$V = 0.150$ L

Unknown: molarity (M)

$(14.0 \text{ g NH}_4\text{Br})\left(\dfrac{\text{mol NH}_4\text{Br}}{97.95 \text{ g NH}_4\text{Br}}\right) = 0.143 \text{ mol NH}_4\text{Br}$

$M = \dfrac{0.143 \text{ mol}}{0.150 \text{ L}} = 0.953 \text{ M NH}_4\text{Br}$

16. a. Given: $V = 1.00$ L
molarity = 3.50 M H_2SO_4

Unknown #3: mass of solute

amount of solute (mol) = (M)(volume)
= (3.50 mol/L)(1.00 L)
= 3.50 mol

mass of solute = $(3.50 \text{ mol H}_2\text{SO}_4)\left(\dfrac{98.05 \text{ g H}_2\text{SO}_4}{\text{mol H}_2\text{SO}_4}\right) = 343 \text{ g H}_2\text{SO}_4$

b. Given: $V = 2.50$ L
molarity = 1.75 M $Ba(NO_3)_2$

Unknown: mass of $Ba(NO_3)_2$

moles solute = (M)(volume)
= (1.75 mol/L)(2.50 L)
= 4.38 mol

mass of solute = $(4.38 \text{ mol Ba(NO}_3)_2)\left(\dfrac{261.3 \text{ g Ba(NO}_3)_2}{\text{mol Ba(NO}_3)_2}\right)$

= 1140 g $Ba(NO_3)_2$

17. Given: $V = 0.065$ L
molarity = 2.20 M NaOH

Unknown: moles NaOH

mol solute = (M)(L solution)
= (2.20 mol/L NaOH)(0.065 L)
= 0.143 mol NaOH

18. Given: mass of solute = 26.42 g $(NH_4)_2SO_4$
$V = 0.050$ L

a. Unknown: molar mass $(NH_4)_2SO_4$

Atomic mass N = 14.01 × 2 = 28.02
Atomic mass H = 1.01 × 8 = 8.08
Atomic mass S = 32.07
Atomic mass O = 15.99 × 4 = <u>63.96</u>
Total: 132.13 g

= 132.1 g = molar mass $(NH_4)_2SO_4$

b. Unknown: Products of the solution

$(NH_4)_2SO_4 + H_2O \rightarrow NH_4OH + H_2SO_4$

c. Unknown: molarity (M)

$$\frac{26.42 \text{ g (NH}_4)_2\text{SO}_4}{132.1 \text{ g/mol (NH}_4)_2\text{SO}_4} = 0.2000 \text{ mol (NH}_4)_2\text{SO}_4$$

$$M = \frac{0.2000 \text{ mol}}{0.05000 \text{ L}} = 4.000 \text{ M}$$

---

19. Given: molarity = 1.0 M AgNO$_3$
mass of solute = 166.88 g AgNO$_3$

b. Unknown: molar mass AgNO$_3$
Atomic mass Ag = 107.87
Atomic mass N = 14.007
Atomic mass O = $15.999 \times 3 = 47.997$
Total: 169.87 g = molar mass AgNO$_3$

c. Unknown: mL of solution

$$(169.87 \text{ g AgNO}_3)\left(\frac{\text{mol AgNO}_3}{169.87 \text{ g AgNO}_3}\right) = 1.0 \text{ mol AgNO}_3$$

$$V = \frac{\text{moles}}{M} = \frac{1.0 \text{ mol}}{1.0 \text{ mol/L}} = 1.0 \text{ L} = 1000 \text{ mL}$$

---

20. b. Given: V = 0.750 L H$_3$PO$_4$
molarity = 6.00 M H$_3$PO$_4$

Unknown: mass of products

moles of solute = (M)(volume of solution)
= (6.00 M)(0.750 L)
= 4.50 mol H$_3$PO$_4$

$$(4.50 \text{ mol H}_3\text{PO}_4)\left(\frac{1 \text{ mol Ca}_3(\text{PO}_4)_2}{2 \text{ mol H}_3\text{PO}_4}\right)\left(\frac{310.1 \text{ g Ca}_3(\text{PO}_4)_2}{\text{mol Ca}_3(\text{PO}_4)_2}\right)$$

= 698 g Ca$_3$(PO$_4$)$_2$

$$(4.50 \text{ mol H}_3\text{PO}_4)\left(\frac{6 \text{ mol H}_2\text{O}}{2 \text{ mol H}_3\text{PO}_4}\right)\left(\frac{18.01 \text{ g H}_2\text{O}}{\text{mol H}_2\text{O}}\right) = 243 \text{ g H}_2\text{O}$$

---

21. Given: molarity H$_2$SO$_4$ = 18.0 M
V Al(OH)$_3$ solution = 0.250 L
molarity Al(OH)$_3$ = 2.50 M

Unknown: volume of H$_2$SO$_4$ solution required to react

3H$_2$SO$_4$ + 2Al(OH)$_3$ → Al$_2$(SO$_4$)$_3$ + 6H$_2$O

moles solute = (M)(volume of solution)
= (2.50 mol/L)(0.250 L) = 0.625 mol Al(OH)$_3$

$$(0.625 \text{ mol Al(OH)}_3)\left(\frac{3 \text{ mol H}_2\text{SO}_4}{2 \text{ mol Al(OH)}_3}\right)\left(\frac{\text{L}}{18.0 \text{ mol H}_2\text{SO}_4}\right) = 0.0521 \text{ L}$$

$$(0.0521 \text{ L H}_2\text{SO}_4)\left(\frac{1000 \text{ ml}}{\text{L}}\right) = 52.1 \text{ mL H}_2\text{SO}_4$$

---

22. Given: V AgNO$_3$ solution = 0.0750 L
mass of solute = 0.250 g Ag

Unknown: molarity (M) of AgNO$_3$ if other product is Cu(NO$_3$)$_2$

$$(0.250 \text{ g Ag})\left(\frac{\text{mol Ag}}{108 \text{ g Ag}}\right) = 2.30 \times 10^{-3} \text{ mol Ag}$$

$$M = \frac{2.3 \times 10^{-3} \text{ mol Ag}}{0.0750 \text{ L AgNO}_3} = 0.0309 \text{ M AgNO}_3$$

23. a. Given: mass of H_2SO_4 = 294.3 g
mass of H_2O = 1.000 kg

Unknown 3: molality (m)

$(294.3 \text{ g } H_2SO_4)\left(\dfrac{\text{mol } H_2SO_4}{98.05 \text{ g } H_2SO_4}\right) = 3.000 \text{ mol } H_2SO_4$

$m = \dfrac{3.000 \text{ mol } H_2SO_4}{1.000 \text{ kg } H_2O} = 3.000 \; m \; H_2SO_4$

b. Given: mass of solute = 63.0 g HNO_3
mass of solvent = 0.250 kg H_2O

Unknown: molality (m)

$(63.0 \text{ g } HNO_3)\left(\dfrac{\text{mol } HNO_3}{62.99 \text{ g } HNO_3}\right) = 1.00 \text{ mol } HNO_3$

$m = \dfrac{1.00 \text{ mol } HNO_3}{0.250 \text{ kg } H_2O} = 4.00 \; m$

24. a. Given: molality = 4.50 m H_2SO_4
mass of solvent = 1.00 kg H_2O

Unknown: mass of solute

moles solute = (m)(kg solvent)
= (4.50)(1.00) = 4.50 mol H_2SO_4

$(4.50 \text{ mol } H_2SO_4)\left(\dfrac{98.05 \text{ g } H_2SO_4}{\text{mol } H_2SO_4}\right) = 441 \text{ g } H_2SO_4$

b. Given: molality = 1.00 m HNO_3
mass of solvent = 2.00 kg H_2O

Unknown: mass of solute

moles solute = (m)(kg solvent)
= (1.00)(2.00) = 2.00 mol HNO_3

$(2.00 \text{ mol } HNO_3)\left(\dfrac{62.99 \text{ g } HNO_3}{\text{mol } HNO_3}\right) = 126 \text{ g } HNO_3$

25. Given: mass of solute = 17.1 g $C_{12}H_{22}O_{11}$
mass of solvent = 0.275 kg H_2O

a. Unknown: molar mass of sucrose

Atomic mass C = 12.01 × 12 = 144.12
Atomic mass H = 1.01 × 22 = 22.22
Atomic mass O = 15.99 × 11 = 175.89

Total: 342.23 g = 342 g = molar mass sucrose

b. Unknown: molality (m)

$(17.1 \text{ g } C_{12}H_{22}O_{11})\left(\dfrac{\text{mol } C_{12}H_{22}O_{11}}{342 \text{ g } C_{12}H_{22}O_{11}}\right) = 0.05 \text{ mol } C_{12}H_{22}O_{11}$

$m = \dfrac{0.05 \text{ mol}}{0.275 \text{ kg}} = 0.182 \; m$

26. Given: mass of solute = 75.5 g $Ca(NO_3)_2$
m = 0.500

Unknown: mass of solvent (H_2O) in kg

$(75.5 \text{ g } Ca(NO_3)_2)\left(\dfrac{\text{mol } Ca(NO_3)_2}{164.04 \text{ g } Ca(NO_3)_2}\right) = 0.46 \text{ mol } Ca(NO_3)_2$

kg solvent = $\dfrac{0.46 \text{ mol } Ca(NO_3)_2}{0.500 \; m} = 0.920 \text{ kg } H_2O$

27. Given: $m = 1.75$
mass of solvent = 0.250 kg

Unknown: mass of solute (C_2H_5OH) in g

moles solute = (m)(kg solvent)
$= (1.75)(0.250) = 0.438$ mol C_2H_5OH

$(0.438 \text{ mol } C_2H_5OH)\left(\dfrac{46.07 \text{ g } C_2H_5OH}{\text{mol } C_2H_5OH}\right) = 20.2$ g C_2H_5OH

Mixed Review

28. Given: $V = 0.450$ L
molarity = 0.250 M Na_2SO_4

a. Unknown: molar mass Na_2SO_4

Atomic mass Na = $22.99 \times 2 = 45.98$
Atomic mass S = 32.07
Atomic mass O = $15.99 \times 4 = 63.96$
Total: $\overline{142.01}$ = molar mass Na_2SO_4

b. Unknown: moles Na_2SO_4 needed

mol Na_2SO_4 = (M)(volume of solution)
$= (0.250 \text{ mol/L})(0.450 \text{ L}) = 0.113$ mol Na_2SO_4

29. Given: $V = 2$ L
mass of solute = 0.150 g $C_6H_8O_7$

a. Unknown: molar mass citric acid, $C_6H_8O_7$

Atomic mass C = $12.01 \times 6 = 72.06$
Atomic mass H = $1.01 \times 8 = 8.08$
Atomic mass O = $15.99 \times 7 = \overline{111.93}$
Total: $\overline{192.07}$ = 192 g = molar mass $C_6H_8O_7$

d. Unknown: M (molarity of $C_6H_8O_7$)

$(0.150 \text{ g } C_6H_8O_7)\left(\dfrac{\text{mol } C_6H_8O_7}{192 \text{ g } C_6H_8O_7}\right) = 7.8 \times 10^{-4}$ mol $C_6H_8O_7$

$M = \dfrac{7.8 \times 10^{-4} \text{ mol}}{2 \text{ L}} = 3.9 \times 10^{-4}$ M $C_6H_8O_7$

30. Given: $V = 0.350$ L
molarity = 6.0 M KCl

a. Unknown: molar mass of KCl

Atomic mass K = 39.1
Atomic mass Cl = $\underline{35.5}$
Total $\overline{74.6}$ g = molar mass KCl

c. Unknown: mass of solute (KCl)

moles solute = (M)(volume of solution)
$= (6.0 \text{ mol/L})(0.350 \text{ L}) = 2.1$ mol KCl

$(2.1 \text{ mol KCl})\left(\dfrac{74.6 \text{ g KCl}}{\text{mol KCl}}\right) = 160$ g KCl

31. Given:
mass of Na = 10.0 g
volume of H_2O = 1.00 L
final volume of system = 1 L

a. Unknown: molar mass NaOH

Atomic mass Na = 22.99
Atomic mass O = 15.99
Atomic mass H = $\underline{1.01}$
Total: $\overline{39.99}$ = 40 g = molar mass NaOH

c. Unknown: molarity of NaOH

$(10.0 \text{ g Na})\left(\dfrac{\text{mol Na}}{22.99 \text{ g Na}}\right) = 0.435 \text{ mol Na}$

$(0.435 \text{ mol Na})\left(\dfrac{1 \text{ mol NaOH}}{1 \text{ mol Na}}\right) = 0.435 \text{ mol NaOH}$

$M = \dfrac{0.435 \text{ mol}}{1.00 \text{ L}} = 0.435 \text{ M}$

---

32. Given: mass of solvent = 6.5 kg $C_2H_6O_2$
mass of solute = 1500 g H_2O

a. Unknown: molar mass $C_2H_6O_2$

Atomic mass C = 12.01 × 2 = 24.02
Atomic mass H = 1.01 × 6 = 6.06
Atomic mass O = 15.99 × 2 = 31.98
Total: $\overline{62.06}$ = 62 g = molar mass $C_2H_6O_2$

b. Unknown: molality (m) of H_2O

$(1500 \text{ g } H_2O)\left(\dfrac{\text{mol } H_2O}{18 \text{ g } H_2O}\right) = 83.3 \text{ mol } H_2O$

$m = \dfrac{83.3 \text{ mol}}{6.5 \text{ kg}} = 13 \; m \; H_2O$

CHAPTER 14

Ions in Aqueous Solutions and Colligative Properties

ATE, Additional Sample Problems 14–1, p. 426

1. Given: amount of solute = 1 mol Mg(ClO$_3$)$_2$
 solvent = H$_2$O

Unknown: number of moles of ions produced

$$Mg(ClO_3)_2(s) \xrightarrow{H_2O} Mg^{2+}(aq) + 2\ ClO_3^-(aq)$$

1 mol Mg^{2+} + 2 mol ClO$_3^-$ = 3 mol ions produced

2. Given: amount of solute = 3.5 mol NH$_4$NO$_3$
 solvent = H$_2$O

Unknown: number of moles of ions produced

$$NH_4NO_3(s) \xrightarrow{H_2O} NH_4^-(aq) + NO_3^-(aq)$$

3.5 mol NH$_4^+$ + 3.5 mol NO$_3^-$ = 7.0 mol ions produced

3. Given: ions produced = K$^+$, SO$_3^{2-}$
 amount of K$^+$ produced = 0.5 mol
 solvent = H$_2$O

Unknowns: salt that produced the solution, number of moles of SO$_3^{2-}$ produced

$$K_2SO_3(s) \xrightarrow{H_2O} 2K^+(aq) + SO_3^{2-}(aq)$$

K$_2$SO$_3$ is the salt; $0.5 \text{ mol K}^+ \times \dfrac{1 \text{ mol SO}_3^{2-}}{2 \text{ mol K}^+} = 0.25$ mol SO$_3^{2-}$ ion produced

ATE, Additional Sample Problems 14-3, p. 439

1. Given: mass of solute = 58.0 g C$_6$H$_{12}$O$_6$
 mass of solvent = 0.185 kg H$_2$O

Unknown: freezing point of solution

$\Delta t_f = K_f m$

$$(58.0 \text{ g C}_6\text{H}_{12}\text{O}_6)\left(\dfrac{\text{mol C}_6\text{H}_{12}\text{O}_6}{180.\text{ g C}_6\text{H}_{12}\text{O}_6}\right) = 0.322 \text{ mol C}_6\text{H}_{12}\text{O}_6$$

$$m = \dfrac{\text{moles of solute}}{\text{mass of solvent (kg)}}$$

$$m = \dfrac{0.322 \text{ mol}}{0.185 \text{ kg}} = 1.74\ m$$

$\Delta t_f = (-1.86°C/m)(1.74\ m) = -3.24°C$

f.p. solution = f.p. solvent + Δt_f
 = 0.000°C + (−3.24°C) = −3.24°C

2. Given: mass of solute =
39.2 g H_2NCONH_2
mass of solvent =
0.485 kg acetic acid

Unknowns: molality (m) of solution, freezing point of solution

$$(39.2 \text{ g } H_2NCONH_2)\left(\frac{\text{mol } H_2NCONH_2}{60.0 \text{ g } H_2NCONH_2}\right) = 0.653 \text{ mol } H_2NCONH_2$$

$$m = \frac{\text{moles of solute}}{\text{mass of solvent (kg)}}$$

$$m = \frac{0.653 \text{ mol}}{0.485 \text{ kg}} = 1.35 \, m$$

$$\Delta t_f = K_f m$$
$$= (-3.90°C/m)(1.35 \, m) = -5.27°C$$

f.p. solution = f.p. solvent + Δt_f

f.p. = 16.6°C + (−5.27°C) = 11.3°C

3. Given: f.p. of solution = −6.40°C
mass of solvent = (500. g H_2O)

Unknowns: molality (m) of solution of $HOCH_2CH_2OH$, mass of solute needed

f.p. solution = f.p. solvent + Δt_f

Δt_f = f.p. solution − f.p. solvent
= −6.40° C − 0.000° C = −6.40° C

$\Delta t_f = K_f m$

$m = \Delta t_f / K_f$
= (−6.40°C)/(−1.86°C/m) = 3.44 m

$$m = \frac{\text{moles of solute}}{\text{mass of solvent (kg)}}$$

moles of solute = (m)(mass of solvent)
= (3.44)(0.500) = 1.72 mol

$$(1.72 \text{ mol } HOCH_2CH_2OH)\left(\frac{62.0 \text{ g } HOCH_2CH_2OH}{\text{mol } HOCH_2CH_2OH}\right) = 107 \text{ g } HOCH_2CH_2OH$$

Practice, p. 440

1. Given: mass of solute =
10.3 g $C_6H_{12}O_6$
mass of solvent =
0.250 kg H_2O

Unknown: freezing-point depression of solution

$$(10.3 \text{ g } C_6H_{12}O_6)\left(\frac{\text{mol } C_6H_{12}O_6}{180. \text{ g } C_6H_{12}O_6}\right) = 0.0572 \text{ mol } C_6H_{12}O_6$$

$$m = \frac{\text{moles of solute}}{\text{mass of solvent (kg)}}$$

$$m = \frac{0.0572 \text{ mol}}{0.250 \text{ kg}} = 0.22 \, m$$

$\Delta t_f = K_f m$

$\Delta t_f = (-1.86°C/m)(0.2288 \, m) = -0.426°C$

2. Given: f.p. of solution = −0.325°C
solute = $C_6H_{12}O_6$
solvent = H_2O

Unknown: molality (m) of solution

f.p. solution = f.p. solvent + Δt_f

Δt_f = f.p. solution − f.p. solvent
= −0.325°C − 0.000°C = −0.325°C

$\Delta t_f = K_f m$

$m = \Delta t_f / K_f$

$$= \frac{(-0.325°C)}{(-1.86°C/m)} = 0.175 \, m$$

3. Given: amount of solute
   = 0.500 mol
   amount of solvent
   0.5000 kg ether
   f.p. ether =
   −116.3°C

   Unknown: freezing point of solution

   $m = \dfrac{\text{moles of solute}}{\text{mass of solvent (kg)}}$

   $m = \dfrac{0.500 \text{ mol}}{0.500 \text{ kg}} = 1.00 \; m$

   $\Delta t_f = K_f m$

   $\Delta t_f = (-1.79°\text{C}/m)(1.00 \; m) = -1.79°\text{C}$

   f.p. solution = f.p. ether + Δt_f
   = −116.3°C + (−1.79°C) = −118.1°C

4. Given: f.p. of solution =
   −9.0°C
   solvent = H_2O
   f.p. H_2O = 0.0°C

   a. Unknown: freezing-point depression of solution

   f.p. solution = f.p. solvent + Δt_p

   Δt_p = f.p. solution − f.p. solvent
   = −9.0°C − 0.0°C = −9.0°C

   b. Unknown: molality (m) of solution

   $\Delta t_f = K_f m$

   $m = \Delta t_f / K_f$

   $= \dfrac{(-9.0°\text{C})}{(-1.86°\text{C}/m)} = 4.8 \; m$

Practice, p. 441

1. Given: mass of solute =
   50.0 g $C_{12}H_{22}O_{11}$
   (sucrose)
   mass of solvent =
   0.500 kg H_2O

   Unknown: boiling-point elevation

   $(50.0 \text{ g } C_{12}H_{22}O_{11})\left(\dfrac{\text{mol } C_{12}H_{22}O_{11}}{342 \text{ g } C_{12}H_{22}O_{11}}\right) = 0.146 \text{ mol } C_{12}H_{22}O_{11}$

   $m = \dfrac{\text{moles of solute}}{\text{mass of solvent (kg)}}$

   $m = \dfrac{0.146 \text{ mol}}{0.500 \text{ kg}} = 0.292 \; m$

   $\Delta t_b = K_b m$

   $\Delta t_b = (0.51°\text{C}/m)(0.292 \; m) = 0.15°\text{C}$

2. Given: mass of solute =
   450.0 g $C_{12}H_{22}O_{11}$
   (sucrose)
   mass of solvent =
   0.250 kg H_2O
   b.p. H_2O = 100.0°C

   Unknown: boiling point of solution

   $(450.0 \text{ g } C_{12}H_{22}O_{11})\left(\dfrac{\text{mol } C_{12}H_{22}O_{11}}{342 \text{ g } C_{12}H_{22}O_{11}}\right) = 1.32 \text{ mol } C_{12}H_{22}O_{11}$

   $m = \dfrac{\text{moles of solute}}{\text{mass of solvent (kg)}}$

   $m = \dfrac{1.32 \text{ mol}}{0.250 \text{ kg}} = 5.28 \; m$

   $\Delta t_b = K_b m$

   $\Delta t_b = (0.51°\text{C}/m)(5.28 \; m) = 2.7°\text{C}$

   b.p. solution = b.p. solvent + Δt_b
   = 100.0°C + 2.7°C = 102.7°C

3. Given: boiling-point elevation = 1.02°C
solvent = H$_2$O

Unknown: molality (m) of solution

$\Delta t_b = K_b m$

$m = \Delta t_b / K_b$
$= 1.02°C/(0.51°C/m) = 2.0\ m$

---

4. Given: boiling point of solution = 100.7°C
solvent = H$_2$O

a. Unknown: boiling-point elevation

b.p. solution = b.p. solvent + Δt_b

Δt_b = b.p. solution − b.p. solvent
$= 100.75°C − 100.0°C = 0.75°C$

b. Unknown: molality (m) of solution

$\Delta t_b = K_b m$

$m = \Delta t_b / K_b$

$= \dfrac{0.75°C}{(0.51°C/m)} = 1.5\ m$

ATE, Additional Sample Problems 14–5, p. 441

1. Given: mass of solute = 25.0 g HOCH$_2$CH$_2$OC$_4$H$_9$ (butyl cellosolve)
mass of solvent = 0.0687 kg ether

Unknown: boiling point of solution

$(25.0\ g\ HOCH_2CH_2OC_4H_9)\left(\dfrac{mol\ HOCH_2CH_2OC_4H_9}{118\ g\ HOCH_2CH_2OC_4H_9}\right)$

$= 0.212\ mol\ HOCH_2CH_2OC_4H_9$

$m = \dfrac{\text{moles of solute}}{\text{mass of solvent (kg)}}$

$m = \dfrac{0.212\ mol}{0.0687\ kg} = 3.09\ m$

$\Delta t_b = K_b m$
$= (2.02°C/m)(4.37\ m) = 6.24°C$

b.p. solution = b.p. solvent + Δt_b
$= 34.6°C + 6.24°C = 40.8°C$

---

2. Given: mass of solvent = 1.00 kg H$_2$O
boiling point of solution = 104.5°C
solute = CH$_2$OH-CHOHCH$_2$OH (glycerol)

Unknown: mass of solute

b.p. solution = b.p. solvent + Δt_b

Δt_b = b.p. solution − b.p. solvent
$= 104.5°C − 100.0°C = 4.5°C$

$\Delta t_b = K_b m$

$m = \Delta t_b / K_b$
$= (4.5°C/m)/0.51°C = 8.8\ m$

$m = \dfrac{\text{moles of solute}}{\text{mass of solvent (kg)}}$

moles of solute = (m)(mass of solvent)
$= (8.8\ mol/kg)(1.00\ kg) = 8.8\ mol$

$(8.8\ mol\ glycerol)\left(\dfrac{92.08\ g\ glycerol}{mol\ glycerol}\right) = 810\ g\ glycerol$

Practice, p. 445

1. Given: solute = 2.0 mol MgSO$_4$
mass of solvent = 1.0 kg H$_2$O

Unknown: expected freezing-point depression of solution

$m = \dfrac{\text{moles of solute}}{\text{mass of solvent (kg)}} = \dfrac{2.0 \text{ mol}}{1.0 \text{ kg}}$

MgSO$_4$(s) → Mg^{2+}(aq) + SO$_4^{2-}$(aq)

Each formula unit of MgSO$_4$ yields two ions in solution.

$\Delta t_f = K_f m$

$= \left(\dfrac{-1.86°\text{C} \cdot \text{kg H}_2\text{O}}{\text{mol ions}}\right)\left(\dfrac{2.0 \text{ mol MgSO}_4}{\text{kg H}_2\text{O}}\right)\left(\dfrac{2 \text{ mol ions}}{\text{mol MgSO}_4}\right)$

$= -7.4°\text{C}$

2. Given: mass of solute = 150 g NaCl
mass of solvent = 1.0 kg H$_2$O

Unknown: expected boiling-point elevation of solution

$(150 \text{ g NaCl})\left(\dfrac{\text{mol NaCl}}{58.44 \text{ g NaCl}}\right) = 2.6 \text{ mol NaCl}$

$m = \dfrac{\text{moles of solute}}{\text{mass of solvent (kg)}} = \dfrac{2.6 \text{ mol}}{1.0 \text{ kg}}$

NaCl(s) → Na$^+$(aq) + Cl$^-$(aq)

Each formula unit of NaCl yields two ions in solution.

$\Delta t_b = K_b m$

$= \left(\dfrac{0.51°\text{C} \cdot \text{kg H}_2\text{O}}{\text{mol ions}}\right)\left(\dfrac{2.6 \text{ mol NaCl}}{\text{kg H}_2\text{O}}\right)\left(\dfrac{2 \text{ mol ions}}{\text{mol NaCl}}\right)$

$= 2.7°\text{C}$

3. Given: solute = NaCl
solvent = H$_2$O
freezing point of solution = −20.0°C
freezing point of H$_2$O = 0.0°C

Unknown: molality (m)

f.p. solution = f.p. solvent + Δt_f

Δt_f = f.p. solution − f.p. solvent
$= -20.0°\text{C} - 0.0°\text{C} = -20.0°\text{C}$

NaCl(s) → Na$^+$(aq) + Cl$^-$(aq)

Each formula unit of NaCl yields two ions in solution.

$\Delta t_f = K_f m$

$m = \Delta t_f / K_f$

$= \left(\dfrac{-20.0°\text{C} \cdot \text{mol ions}}{-1.86°\text{C}}\right)\left(\dfrac{1 \text{ mol NaCl}}{2 \text{ mol ions}}\right) = 5.4 \, m \text{ NaCl}$

Section Review, p. 446

2. Given: amount of solute = 2 mol
amount of solvent = 1 kg
freezing point of solution = 7.8°C below its normal freezing point

Unknowns: molal freezing-point constant of unknown solvent (K_f), identity of solvent

$m = \dfrac{\text{moles of solute}}{\text{mass of solvent (kg)}}$

$= \dfrac{2 \text{ mol}}{1 \text{ kg}} = 2 \, m$

$\Delta t_f = K_f m$

$K_f = \Delta t_f / m$

$= -7.8°\text{C}/2 \, m = -3.9°\text{C}/m$

Identity of solvent: acetic acid

4. a. Given: molality (m) of solution = 0.2 m
solute = KNO_3

Unknown: expected freezing-point depression

$KNO_3(s) \rightarrow K^+(aq) + NO_3^-(aq)$

Each formula unit of KNO_3 yields two ions in solution.

$\Delta t_f = K_f m$

$= \left(\dfrac{-1.86°C \cdot kg\ H_2O}{mol/ions}\right)\left(\dfrac{0.2\ mol\ KNO_3}{kg\ H_2O}\right)\left(\dfrac{2\ mol/ions}{mol/KNO_3}\right)$

$= -0.744°C$

Chapter 14 Review, p. 447

1. a. Given: solute = KCl

Unknown: number of moles of ions present

$KCl(s) \xrightarrow{H_2O} K^+(aq) + Cl^-(aq)$

Ions produced = 1 mol K^+ + 1 mol Cl^- = 2 mol

b. Given: amount of solvent = 1 L
M of solution = 1 M
solute = $Mg(NO_3)_2$

Unknown: number of moles of ions present

$Mg(NO_3)_2 \xrightarrow{H_2O} Mg^{2+}(aq) + 2NO_3^-(aq)$

Ions produced = 1 mol Mg^{2+} + 2 mol NO_3^- = 3 mol

Review Problems

15. a. Given: amount of solute = 0.50 mol $Sr(NO_3)_2$
solvent = H_2O

Unknown: total number of moles of solute ions formed

$Sr(NO_3)_2(s) \xrightarrow{H_2O} Sr^{2+}(aq) + 2NO_3^-(aq)$

Ions produced: 0.5 mol $Sr(NO_3)_2 \times \dfrac{1\ mol\ Sr^{2+}}{1\ mol\ Sr(NO_3)_2} = 0.5$ mol Sr^{2+}

0.5 mol $Sr(NO_3)_2 \times \dfrac{2\ mol\ NO_3^-}{1\ mol\ Sr(NO_3)_2} = 1.00$ mol NO_3^-

0.50 mol Sr^{2+} + 1.00 mol NO_3^- = 1.50 mol ions

b. Given: amount of solute = 0.50 mol Na_3PO_4

Unknown: total number of moles of solute ions formed

$Na_3PO_4(s) \xrightarrow{H_2O} 3Na^+(aq) + PO_4^{3-}(aq)$

Ions produced: 0.5 mol $Na_3PO_4 \times \dfrac{3\ mol\ Na^+}{1\ mol\ Na_3PO_4} = 1.5$ mol Na^+

0.5 mol $Na_3PO_4 \times \dfrac{1\ mol\ PO_4^{3-}}{1\ mol\ Na_3PO_4} = 0.5$ mol PO_4^{3-}

1.50 mol Na^+ + 0.50 mol PO_4^{3-} = 2.00 mol ions

18. Given: mass of $CuCl_2$ = 13.45 g

Unknown: maximum amount of precipitate formed

$(13.45\ g\ CuCl_2)\left(\dfrac{mol\ CuCl_2}{134.5\ g\ CuCl_2}\right) = 0.1000$ mol $CuCl_2$

0.1000 mol $CuCl_2 \times \dfrac{1\ mol\ PbCl_2}{1\ mol\ CuCl_2} = 0.1000$ mol $PbCl_2$

$(0.1000\ mol\ PbCl_2)\left(\dfrac{278.1\ g\ PbCl_2}{mol\ PbCl_2}\right) = 27.81$ g $PbCl_2$

19. a. Given: molality = 1.50 m $C_{12}H_{22}O_{11}$ (sucrose)
solvent = H_2O

Unknown: freezing-point depression

$\Delta t_f = K_f m$
$= (-1.86°C/m)(1.50\ m) = -2.79°C$

b. Given: mass of solute = 171 g $C_{12}H_{22}O_{11}$ (sucrose)
mass of solvent = 1.00 kg H_2O

Unknown: freezing-point depression

$(171\ g\ sucrose)\left(\dfrac{mol\ sucrose}{342\ g\ sucrose}\right) = 0.500$ mol sucrose

$m = \dfrac{moles\ of\ solute}{mass\ of\ solvent\ (kg)}$

$m = \dfrac{0.500\ mol}{1.00\ kg} = 0.5\ m$

$\Delta t_f = K_f m$

$\Delta t_f = (-1.86°C/m)(0.500\ m) = -0.930°C$

c. Given: mass of solute = 77.0 g $C_{12}H_{22}O_{11}$ (sucrose)
mass of solvent = 0.400 kg H_2O

Unknown: freezing-point depression

$(77.0\ g\ sucrose)\left(\dfrac{mol\ sucrose}{342\ g\ sucrose}\right) = 0.225$ mol sucrose

$m = \dfrac{moles\ of\ solute}{mass\ of\ solvent\ (kg)}$

$m = \dfrac{0.225\ mol}{0.400\ kg} = 0.563\ m$

$\Delta t_f = K_f m$

$\Delta t_f = (-1.86°C/m)(0.563\ m) = -1.05°C$

20. Given: solvent = H_2O; solute = nonelectrolyte

a. freezing-point depression = -0.93°C

b. freezing-point depression = -3.72°C

c. freezing-point depression = -8.37°C

Unknown: molality (m) of each solution

$\Delta t_f = K_f m$

$m = \Delta t_f / K_f$

a. $m = (-0.930°C)/(-1.86°C/m) = 0.500\ m$

b. $m = (-3.72°C)/(-1.86°C/m) = 2.00\ m$

c. $m = (-8.37°C)/(-1.86°C/m) = 4.50\ m$

21. Given: mass of solute = 20.0 g $C_6H_{12}O_6$ (glucose)
mass of solvent = 0.250 kg H_2O

    a. Unknown: freezing-point depression of solvent

$\Delta t_f = K_f m$

$(20.0 \text{ g glucose})\left(\dfrac{\text{mol glucose}}{180 \text{ g glucose}}\right) = 0.111 \text{ mol glucose}$

$m = \dfrac{\text{moles of solute}}{\text{mass of solvent (kg)}}$

$m = \dfrac{0.111 \text{ mol}}{0.250 \text{ kg}} = 0.444 \text{ m}$

$\Delta t_f = K_f m$

$\Delta t_f = (-1.86°C/m)(0.444 \text{ m}) = -0.826°C$

    b. Unknown: freezing point of solution

f.p. solution = f.p. solvent + Δt_f
= 0.000°C + (−0.826°C) = −0.826°C

22. Given: solute = $C_2H_4(OH)_2$ (antifreeze)
mass of solvent = 0.500 kg H_2O
freezing point of solution = −20.0°C

Unknown: mass of solute in g

f.p. solution = f.p. solvent + Δt_f

Δt_f = f.p. solution − f.p. solvent
= −20.0°C − 0.000°C = −20.0°C

$\Delta t_f = K_f m$

$m = \Delta t_f / K_f$
= (−20.0°C)/(−1.86°C/m) = 10.8 m

$m = \dfrac{\text{moles of solute}}{\text{mass of solvent (kg)}}$

moles of solute = (m)(mass of solvent)
= (10.8 mol/kg)(0.500 kg) = 5.40 mol

$(5.40 \text{ mol } C_2H_4(OH)_2)\left(\dfrac{62.06 \text{ g } C_2H_4(OH)_2}{\text{mol } C_2H_4(OH)_2}\right) = 335 \text{ g } C_2H_4(OH)_2$

23. Given: freezing point C_6H_6 (benzene) = 5.45°C
mass of solute = 7.24 g $C_2Cl_4H_2$
mass of solvent = 0.115 kg benzene
specific gravity benzene = 0.879
freezing point of solution = 3.55°C

Unknown: molal freezing-point constant (K_f) for benzene

$(7.24 \text{ g } C_2Cl_4H_2)\left(\dfrac{\text{mol } C_2Cl_4H_2}{168 \text{ g } C_2Cl_4H_2}\right) = 0.0431 \text{ mol } C_2Cl_4H_2$

$m = \dfrac{\text{moles of solute}}{\text{mass of solvent (kg)}}$

$m = \dfrac{0.0431 \text{ mol}}{0.115 \text{ kg}} = 0.375 \text{ m}$

f.p. solution = f.p. solvent + Δt_f

Δt_f = f.p. solution − f.p. solvent
= 3.55°C − 5.45°C = −1.90°C

$\Delta t_f = K_f m$

$K_f = \Delta t_f / m$
= (−1.90°C)/(0.375 m)
= −5.07°C/m

24. Given: mass of solute = 1.500 g
molar mass of solute = 125.0 g
mass of solvent = 0.03500 kg camphor

Unknown: freezing point of solution

$$(1.500 \text{ g solute})\left(\frac{\text{mol solute}}{125.0 \text{ g solute}}\right) = 0.01200 \text{ mol solute}$$

$$m = \frac{\text{moles of solute}}{\text{mass of solvent (kg)}}$$

$$m = \frac{0.01200 \text{ mol}}{0.03500 \text{ kg}} = 0.3429 \, m$$

$$\Delta t_f = K_f m$$

$$\Delta t_f = (-39.7°C/m)(0.3429 \, m)$$

$$= -13.6°C$$

f.p. solution = f.p. solvent + Δt_f
$= 178.8°C + (-13.6°C)$
$= 165.2°C$

---

25. a. Given: molality = 2.5 m $C_6H_{12}O_6$ (glucose)
solvent = H_2O

Unknown: boiling-point elevation of H_2O

$\Delta t_b = K_b m$
$= (0.51°C/m)(2.5 \, m) = 1.3°C$

b. Given: mass of solute = 3.20 g $C_6H_{12}O_6$ (glucose)
mass of solvent = 1.00 kg H_2O

Unknown: boiling-point elevation of H_2O

$$(3.20 \text{ g glucose})\left(\frac{\text{mol glucose}}{180. \text{ g glucose}}\right) = 0.0178 \text{ mol glucose}$$

$$m = \frac{\text{moles of solute}}{\text{mass of solvent (kg)}}$$

$$m = \frac{0.0178 \text{ mol}}{1.00 \text{ kg}} = 0.0178 \, m$$

$\Delta t_b = K_b m$

$\Delta t_b = (0.51°C/m)(0.0178 \, m) = 0.0091°C$

c. Given: mass of solute = 20.0 g $C_{12}H_{22}O_{11}$ (sucrose)
mass of solvent 0.500 kg H_2O

Unknown: boiling-point elevation of H_2O

$$(20.0 \text{ g sucrose})\left(\frac{\text{mol sucrose}}{342 \text{ g sucrose}}\right) = 0.0585 \text{ mol sucrose}$$

$$m = \frac{\text{moles of solute}}{\text{mass of solvent (kg)}}$$

$$m = \frac{0.585 \text{ mol}}{0.500 \text{ kg}} = 0.117 \, m$$

$\Delta t_b = K_b m$

$\Delta t_b = (0.51°C/m)(0.117 \, m) = 0.060°C$

26. Given: solvent = H_2O
    a. boiling point of solution = 100.25°C
    b. boiling point of solution = 101.53°C
    c. boiling point of solution = 102.805°C

Unknown: molality (m) of each solution

$\Delta t_b = K_b m$

$m = \Delta t_b / K_b$

b.p. solution = b.p. solvent + Δt_b

Δt_b = b.p. solution − b.p. solvent

a. $\Delta t_b = 100.25°C - 100.00°C = 0.25°C$

$$m = \frac{0.25°C}{0.51°C/m} = 0.49\ m$$

b. $\Delta t_b = 101.53°C - 100.00°C = 1.53°C$

$$m = \frac{1.53°C}{0.51°C/m} = 3.0\ m$$

c. $\Delta t_b = 102.805°C - 100.000°C = 2.805°C$

$$m = \frac{2.805°C}{0.51°C/m} = 5.5\ m$$

---

27. Given: molality = 1.00 m
    solvent = H_2O
    a. solute = KI
    b. solute = $CaCl_2$
    c. solute = $Ba(NO_3)_2$

Unknown: expected freezing-point depression of solutions

$\Delta t_f = K_f m$

a. $KI(s) \rightarrow K^+(aq) + I^-(aq)$

Each formula unit of KI yields two ions in solution.

$$\left(\frac{-1.86°C \cdot kg\ H_2O}{mol\ ions}\right)\left(\frac{1.00\ mol\ KI}{kg\ H_2O}\right)\left(\frac{2\ mol\ ions}{mol\ KI}\right) = -3.72°C$$

b. $CaCl_2(s) \rightarrow Ca^{2+}(aq) + 2Cl^-(aq)$

Each formula unit of KI yields three ions in solution.

$$\left(\frac{-1.86°C \cdot kg\ H_2O}{mol\ ions}\right)\left(\frac{1.00\ mol\ CaCl_2}{kg\ H_2O}\right)\left(\frac{3\ mol\ ions}{mol\ CaCl_2}\right) = -5.58°C$$

c. $Ba(NO_3)_2(s) \rightarrow Ba^{2+}(aq) + 2NO_3^-(aq)$

Each formula unit $Ba(NO_3)_2$ yields three ions in solution.

$$\left(\frac{-1.86°C \cdot kg\ H_2O}{mol\ ions}\right)\left(\frac{1.00\ mol\ Ba(NO_3)_2}{kg\ H_2O}\right)\left(\frac{3\ mol\ ions}{mol\ Ba(NO_3)_2}\right) = -5.58°C$$

---

28. Given: molality = 0.015 m
    $AlCl_3$
    solvent = H_2O

Unknown: expected freezing-point depression of solution

$\Delta t_f = K_f m$

$AlCl_3(s) \rightarrow Al^{3+}(aq) + 3Cl^-(aq)$

Each formula unit of $AlCl_3$ yields four ions in solution.

$$\left(\frac{-1.86°C \cdot kg\ H_2O}{mol\ ions}\right)\left(\frac{0.015\ mol\ AlCl_3}{kg\ H_2O}\right)\left(\frac{4\ mol\ ions}{mol\ AlCl_3}\right) = -0.11°C$$

29. Given: mass of solute = 85.0 g NaCl
mass of solvent = 0.450 kg H₂O

Unknown: expected freezing point of solution

f.p. solution = f.p. solvent + Δt_f

$\Delta t_f = K_f m$

$m = \dfrac{\text{moles of solute}}{\text{mass of solvent (kg)}}$

$(85.0 \text{ g NaCl})\left(\dfrac{\text{mol NaCl}}{58.44 \text{ g NaCl}}\right) = 1.45 \text{ mol NaCl}$

NaCl(s) → Na⁺(aq) + Cl⁻(aq)

Each formula unit of NaCl yields two ions in solution.

$\left(\dfrac{-1.86°\text{C} \cdot \text{kg H}_2\text{O}}{\text{mol ions}}\right)\left(\dfrac{1.45 \text{ mol NaCl}}{0.450 \text{ kg H}_2\text{O}}\right)\left(\dfrac{2 \text{ mol ions}}{\text{mol NaCl}_3}\right) = -12.0°\text{C} = \Delta t_f$

f.p. solution = 0.0°C + (−12.0°C) = −12.0°C

---

30. Given: mass of solute = 25.0 g BaCl₂
mass of solvent = 0.150 kg H₂O

Unknown: expected boiling point of solution

b.p. solution = b.p. solvent + Δt_b

$\Delta t_b = K_b m$

$m = \dfrac{\text{moles solute}}{\text{mass of solvent (kg)}}$

$(25.0 \text{ g BaCl}_2)\left(\dfrac{\text{mol BaCl}_2}{208.23 \text{ g BaCl}_2}\right) = 0.120 \text{ mol BaCl}_2$

BaCl₂(s) → Ba²⁺(aq) + 2Cl⁻(aq)

Each formula unit of BaCl₂ yields three ions in solution.

$\left(\dfrac{0.51°\text{C} \cdot \text{kg H}_2\text{O}}{\text{mol ions}}\right)\left(\dfrac{0.120 \text{ mol BaCl}_2}{0.150 \text{ kg H}_2\text{O}}\right)\left(\dfrac{3 \text{ mol ions}}{\text{mol BaCl}_2}\right) = 1.2°\text{C} = \Delta t_b$

b.p. solution = 100.0°C + 1.2°C = 101.2°C

---

31. Given: boiling-point elevation of solution = 0.65°C
solvent = H₂O
solute = KI

Unknown: molality (m)

$\Delta t_b = K_b m$

KI(s) → K⁺(aq) + I⁻(aq)

Each formula unit of KI yields two ions in solution.

$m = \Delta t_b / K_b$

$= \left(\dfrac{0.65°\text{C} \cdot \text{mol ions}}{0.51°\text{C}}\right)\left(\dfrac{1 \text{ mol KI}}{2 \text{ mol ions}}\right) = 0.64 \ m \text{ KI}$

---

34. Given: molality = 1.00 m MgI₂
solvent = H₂O
freezing-point depression = −4.78°C

Unknowns: expected freezing-point depression of H₂O; reason for discrepancy between experimental and expected values

$\Delta t_f = K_f m$

MgI₂(s) → Mg²⁺(aq) + 2I⁻(aq)

Each formula unit of MgI₂ yields three ions in solution.

$\left(\dfrac{-1.86°\text{C} \cdot \text{kg H}_2\text{O}}{\text{mol ions}}\right)\left(\dfrac{1.00 \text{ mol MgI}_2}{\text{kg H}_2\text{O}}\right)\left(\dfrac{3 \text{ mol ions}}{\text{mol MgI}_2}\right) = -5.58°\text{C}$

Reason for discrepancy: The ions do not move freely because of forces of attraction between them.

Mixed Review

35. Given: molality = 0.01 m
solvent = H_2O

Unknown: freezing-point depression of each solution

$\Delta t_f = K_f m$

a. Given: solute = NaI

$NaI(s) \rightarrow Na^+(aq) + I^-(aq)$

Each formula unit of NaI yields two ions in solution.

$\left(\dfrac{-1.86°C \cdot kg\ H_2O}{mol\ ions}\right)\left(\dfrac{0.01\ mol\ NaI}{kg\ H_2O}\right)\left(\dfrac{2\ mol\ ions}{mol\ NaI}\right) = -0.04°C$

b. Given: solute = $CaCl_2$

$CaCl_2(s) \rightarrow Ca^{2+}(aq) + 2Cl^-(aq)$

Each formula unit of $CaCl_2$ yields three ions in solution.

$\left(\dfrac{-1.86°C \cdot kg\ H_2O}{mol\ ions}\right)\left(\dfrac{0.01\ mol\ CaCl_2}{kg\ H_2O}\right)\left(\dfrac{3\ mol\ ions}{mol\ CaCl_2}\right) = -0.06°C$

c. Given: solute = K_3PO_4

$K_3PO_4(s) \rightarrow 3K^+(aq) + PO_4^{3-}(aq)$

Each formula unit of K_3PO_4 yields four ions in solution.

$\left(\dfrac{-1.86°C \cdot kg\ H_2O}{mol\ ions}\right)\left(\dfrac{0.01\ mol\ K_3PO_4}{kg\ H_2O}\right)\left(\dfrac{4\ mol\ ions}{mol\ K_3PO_4}\right) = -0.07°C$

d. Given: solute = $C_6H_{12}O_6$ (glucose)

$\Delta t_f = (-1.86°C/m)(0.01\ m) = -0.02°C$

order of increasing Δt_f = d(-0.02°C), a(-0.04°C), b(-0.06°C), c(-0.07°C)

36. Given: solute = $CaCl_2$
solvent = H_2O
freezing point of solution = -2.43°C

Unknown: molality (m)

$\Delta t_f = K_f m$

$CaCl_2(s) \rightarrow Ca^{2+}(aq) + 2Cl^-(aq)$

Each formula unit of $CaCl_2$ yields three ions in solution.

$m = \Delta t_f / K_f$

$\left(\dfrac{-2.43°C \cdot mol\ ions}{-1.86°C}\right)\left(\dfrac{1\ mol\ CaCl_2}{3\ mol/ions}\right) = 0.435\ m\ CaCl_2$

39. a. Given: solute = 0.275 mol K_2S
solvent = H_2O

Unknowns: balanced equation, total number of moles of solute ions formed

$K_2S(s) \xrightarrow{H_2O} 2K^+(aq) + S^{2-}(aq)$

Ions produced: 2(0.275 mol K^+) = 0.550 mol K^+; 0.275 mol S^{2-}

0.550 + 0.275 = 0.825 mol solute ions formed

b. Given: solute = 0.15 mol Al$_2$(SO$_4$)$_3$

Unknowns: balanced equation, total number of moles of solute ions formed

Al$_2$(SO$_4$)$_3$(s) $\xrightarrow{H_2O}$ 2Al^{3+}(aq) + 3SO$_4^{2-}$(aq)

Ions produced: 2(0.15 mol Al^{3+}) = 0.30 mol Al^{3+} + 3(0.15 mol SO$_4^{2-}$)
= 0.45 mol SO$_4^{2-}$

0.30 + 0.45 = 0.75 mol solute ions formed

---

40. Given: mass of solute = 131.2 g AgNO$_3$
mass of solvent = 2.00 kg H$_2$O

Unknown: expected boiling-point elevation of solution

$(131.2 \text{ g AgNO}_3)\left(\dfrac{\text{mol AgNO}_3}{169.85 \text{ g AgNO}_3}\right) = 0.7724 \text{ mol AgNO}_3$

$m = \dfrac{\text{moles of solute}}{\text{mass of solvent (kg)}} = \dfrac{0.7724 \text{ mol}}{2.00 \text{ kg}} = 0.386\ m$

$\Delta t_b = K_b m$

AgNO$_3$(s) → Ag$^+$(aq) + NO$_3^-$(aq)

Each formula unit AgNO$_3$ yields two ions in solution.

$\left(\dfrac{0.51°C \cdot \text{kg H}_2\text{O}}{\text{mol ions}}\right)\left(\dfrac{0.385 \text{ mol AgNO}_3}{\text{kg H}_2\text{O}}\right)\left(\dfrac{2 \text{ mol ions}}{\text{mol AgNO}_3}\right) = 0.39°C$

---

42. Given: solvent = H$_2$O
solute = nonelectrolyte
freezing point = −6.51°C

Unknown: boiling point

f.p. solution = f.p. solvent + Δt_f

Δt_f = f.p. solution − f.p. solvent
 = −6.51°C − 0°C = −6.51°C

$\Delta t_f = K_f m$

$m = \Delta t_f / K_f$

$= \dfrac{-6.51°C}{-1.86°C \cdot \text{kg H}_2\text{O/mol solute}} = 3.50\ m$

$\Delta t_b = K_b m$
 $= (0.51°C/m)(3.50\ m) = 1.8°C$

Δt_b = b.p. solution − b.p. solvent

b.p. solution = Δt_b + b.p. solvent
 = 1.8°C + 100.0°C = 101.8°C

---

43. Given: solute = Na$_2$CO$_3$
solvent = H$_2$O
amount of solute = 0.20 mol Na$_2$CO$_3$

Unknowns: **a.** balanced equation
b. number of moles of each ion produced
c. total number of moles of ions

a. Na$_2$CO$_3$(s) $\xrightarrow{H_2O}$ 2Na$^+$(aq) + CO$_3^{2-}$(aq)

b. 2(0.20 mol Na$^+$) = 0.40 mol Na$^+$

1(0.20 mol CO$_3^{2-}$) = 0.20 mol CO$_3^{2-}$

c. 0.40 + 0.20 = 0.60 moles of solute ions produced

44. Given: $K_3PO_4(aq)$ + $Pb(NO_3)_2(aq)$

Unknown: net ionic equation

Ionic equation:

$6K^+(aq) + 2PO_4^{3-}(aq) + 3Pb^{2+}(aq) + 6NO_3^-(aq) \rightarrow 6K^+(aq) + 6NO_3^-(aq) + Pb_3(PO_4)_2(s)$

Net ionic equation:

$3Pb^{2+}(aq) + 2PO_4^{3-}(aq) \rightarrow Pb_3(PO_4)_2(s)$

---

45. Given: solvent = H_2O
mass of solute = 268 g $Al(NO_3)_3$
mass of solvent = 8.50 kg H_2O

Unknown: expected freezing point of solution

$m = \dfrac{\text{moles of solute}}{\text{mass of solvent (kg)}}$

$(268 \text{ g } Al(NO_3)_3)\left(\dfrac{\text{mol } Al(NO_3)_3}{212.9 \text{ g } Al(NO_3)_3}\right) = 1.26 \text{ mol } Al(NO_3)_3$

$Al(NO_3)_3(s) \rightarrow Al^{3+}(aq) + 3NO_3^-(aq)$

Each formula unit $Al(NO_3)_3$ yields four ions in solution.

$\Delta t_f = K_f m$

$\left(\dfrac{-1.86°C \cdot \text{kg } H_2O}{\text{mol ions}}\right)\left(\dfrac{1.26 \text{ mol } Al(NO_3)_3}{8.50 \text{ kg } H_2O}\right)\left(\dfrac{4 \text{ mol ions}}{\text{mol } Al(NO_3)_3}\right) = -1.10°C$

f.p. solution = f.p. solvent + Δt_f
= 0.00°C + (−1.10°C) = −1.10°C

---

46. a. Given: solvent = H_2O
solute = KNO_3
observed freezing point of solution = −1.15°C
observed freezing point of pure H_2O = 0.25°C

Unknown: molality (m) of KNO_3

f.p. solution = f.p. solvent + Δt_f
Δt_f = f.p. solution − f.p. solvent
= −1.15°C − 0.25°C = −1.40°C

$\Delta t_f = K_f m$

$KNO_3(s) \rightarrow K^+(aq) + NO_3^-(aq)$

Each formula unit KNO_3 yields two ions in solution.

$m = \Delta t_f / K_f$

$= \left(\dfrac{-1.40°C \cdot \text{mol ions}}{-1.86°C}\right)\left(\dfrac{1 \text{ mol } KNO_3}{2 \text{ mol ions}}\right)$

$= 0.376 \, m$

b. Given: mass of solute = 0.415 g KNO_3
volume of solution = 0.010 L
density of solution = 1.00 g/mL

Unknowns: actual molality (m) of KNO_3; percentage difference between predicted concentration and actual concentration of KNO_3

$m = \dfrac{\text{moles of solute}}{\text{mass of solvent (kg)}}$

mass of solvent = $10.00 \text{ mL} \times \dfrac{1.00 \text{ g}}{\text{mL}} = 10.0 \text{ g} \times \dfrac{\text{kg}}{1000 \text{ g}} = 0.0100 \text{ kg}$

$(0.415 \text{ g } KNO_3)\left(\dfrac{\text{mol } KNO_3}{101 \text{ g } KNO_3}\right) = 0.00411 \text{ mol } KNO_3$

$\dfrac{0.00411 \text{ mol}}{0.0100 \text{ kg}} = 0.411 \, m \text{ } KNO_3$

Predicted $m = 0.376 \, m$
Actual $m = 0.411 \, m$

$\dfrac{0.376 \, m}{0.41 \, m} \times 100 = 91.7\% = \text{percentage difference}$

CHAPTER 15

Acids and Bases

Chapter 15 Review, p. 478
Stoichiometry

30. Given: reactants = Zn, H_2SO_4

volume of H_2SO_4 = 0.100 L

molarity = 6.00 M H_2SO_4

$Zn(s) + H_2SO_4(aq) \rightarrow ZnSO_4(aq) + H_2(g)$

$M = \dfrac{\text{amount of solute (mol)}}{\text{volume of solution (L)}}$

amount of solute = (M)(volume)

= (6.00)(0.100) = 0.600 mol H_2SO_4

a. Unknown: number of grams of $ZnSO_4$ produced

$(0.600 \text{ mol } H_2SO_4)\left(\dfrac{1 \text{ mol } ZnSO_4}{1 \text{ mol } H_2SO_4}\right)\left(\dfrac{161.42 \text{ g } ZnSO_4}{\text{mol } ZnSO_4}\right) = 98.9 \text{ g } ZnSO_4$

b. Unknown: number of liters of H_2 gas released at STP

$(0.600 \text{ mol } H_2SO_4)\left(\dfrac{1 \text{ mol } H_2}{1 \text{ mol } H_2SO_4}\right)\left(\dfrac{22.4 \text{ L } H_2}{\text{mol } H_2}\right) = 13.4 \text{ L}$

31. Given: mass of solute = 211 g $BaCO_3$

solvent = HNO_3

(Assume acid is present in excess.)

Unknown: mass and volume of dry CO_2 gas produced at STP

$BaCO_3(s) + 2HNO_3(aq) \rightarrow Ba(NO_3)_2(aq) + H_2O(l) + CO_2(g)$

$(211 \text{ g } BaCO_3)\left(\dfrac{\text{mol } BaCO_3}{197.3 \text{ g } BaCO_3}\right)\left(\dfrac{1 \text{ mol } CO_2}{1 \text{ mol } BaCO_3}\right) = 1.07 \text{ mol } CO_2$

$(1.07 \text{ mol } CO_2)\left(\dfrac{44.0 \text{ g } CO_2}{\text{mol } CO_2}\right) = 47.1 \text{ g } CO_2$

$(1.07 \text{ mol } CO_2)\left(\dfrac{22.4 \text{ L } CO_2}{\text{mol } CO_2}\right) = 24.0 \text{ L } CO_2$

32. Given: solute = $CaCO_3$
solvent = HCl
products = CO_2, $CaCl_2$, H_2O
volume CO_2 = 1.5 L

$CaCO_3(s) + 2HCl(aq) \rightarrow CaCl_2(aq) + H_2O(l) + CO_2(g)$

$(1.5 \text{ L } CO_2)\left(\dfrac{\text{mol } CO_2}{22.4 \text{ L } CO_2}\right)\left(\dfrac{1 \text{ mol } CaCO_3}{1 \text{ mol } CO_2}\right) = 0.067 \text{ mol } CaCO_3$

a. Unknown: mass of $CaCO_3$ in g

$(0.067 \text{ mol } CaCO_3)\left(\dfrac{100 \text{ g } CaCO_3}{\text{mol } CaCO_3}\right) = 6.7 \text{ g } CaCO_3$

b. Given: molarity of HCl (M) = 2.00 M HCl

Unknown: volume of HCl used in reaction

$(0.067 \text{ mol } CaCO_3)\left(\dfrac{2 \text{ mol HCl}}{\text{mol } CaCO_3}\right) = 0.13 \text{ mol HCl}$

volume HCl = $\dfrac{0.13 \text{ mol}}{2.00 \text{ mol/L}} = 0.65 \text{ L}$

---

33. Given: mass of SO_2 = 3.5×10^{11} g

$SO_2 + \dfrac{1}{2}O_2 \rightarrow SO_3$
$SO_3 + H_2O \rightarrow H_2SO_4$

Unknown: mass of H_2SO_4 produced in kg

$(3.50 \times 10^{11} \text{ g } SO_2)\left(\dfrac{\text{mol } SO_2}{64.1 \text{ g } SO_2}\right)\left(\dfrac{1 \text{ mol } SO_3}{1 \text{ mol } SO_2}\right) = 5.46 \times 10^9 \text{ mol } SO_3$

$(5.46 \times 10^9 \text{ mol } SO_3)\left(\dfrac{1 \text{ mol } H_2SO_4}{1 \text{ mol } SO_3}\right)\left(\dfrac{98.1 \text{ g } H_2SO_4}{\text{mol } H_2SO_4}\right)\left(\dfrac{\text{kg}}{1000 \text{ g}}\right) = 5.36 \times 10^8 \text{ kg } H_2SO_4$

CHAPTER 16

Acid-Base Titration and pH

Practice, p. 484

1. Given: [HCl] = 1×10^{-4} M
Unknown: [H_3O^+], [OH^-]

$HCl(l) + H_2O(l) \rightarrow H_3O^+(aq) + Cl^-(aq)$

$M = \dfrac{\text{amount of solute (mol)}}{\text{volume of solution (L)}}$

$\left(\dfrac{1 \times 10^{-4} \text{ mol HCl}}{\text{L solution}}\right)\left(\dfrac{1 \text{ mol } H_3O^+}{1 \text{ mol HCl}}\right) = 1 \times 10^{-4}$ mol H_3O^+/L

$= 1 \times 10^{-4}$ M H_3O^+

$[H_3O^+][OH^+] = 1.0 \times 10^{-14}$ M^2

$[OH^-] = \dfrac{1.0 \times 10^{-14} \text{ M}^2}{1 \times 10^{-4} \text{ M}} = 1 \times 10^{-10}$ M

---

2. Given: [HNO_3] = 1.0×10^{-3} M
Unknown: [H_3O^+], [OH^-]

$HNO_3(l) + H_2O(l) \rightarrow H_3O^+(aq) + NO_3^-(aq)$

$\left(\dfrac{1.0 \times 10^{-3} \text{ mol } HNO_3}{\text{L solution}}\right)\left(\dfrac{1 \text{ mol } H_3O^+}{1 \text{ mol } HNO_3}\right) = 1.0 \times 10^{-3}$ mol H_3O^+/L

$= 1.0 \times 10^{-3}$ M H_3O^+

$[H_3O^+][OH^-] = 1.0 \times 10^{-14}$ M^2

$[OH^-] = \dfrac{1.0 \times 10^{-14} \text{ M}^2}{1.0 \times 10^{-3} \text{ M}} = 1.0 \times 10^{-11}$ M

---

3. Given: [NaOH] = 3.0×10^{-2} M
Unknown: [H_3O^+], [OH^-]

$NaOH(s) \xrightarrow{H_2O} Na^+(aq) + OH^-(aq)$

$\left(\dfrac{3.0 \times 10^{-2} \text{ mol NaOH}}{\text{L solution}}\right)\left(\dfrac{1 \text{ mol } OH^-}{1 \text{ mol NaOH}}\right) = 3.0 \times 10^{-2}$ mol OH^-/L

$= 3.0 \times 10^{-2}$ M OH^-

$[H_3O^+][OH^-] = 1.0 \times 10^{-14}$ M^2

$[H_3O^+] = \dfrac{1.0 \times 10^{-14} \text{ M}^2}{3.0 \times 10^{-2} \text{ M}} = 3.3 \times 10^{-13}$ M

---

4. Given: [$Ca(OH)_2$] = 1.0×10^{-4} M
Unknown: [H_3O^+], [OH^-]

$Ca(OH)_2(s) \xrightarrow{H_2O} Ca^{2+}(aq) + 2OH^-(aq)$

$\left(\dfrac{1.0 \times 10^{-4} \text{ mol } Ca(OH)_2}{\text{L solution}}\right)\left(\dfrac{2 \text{ mol } OH^-}{1 \text{ mol } Ca(OH)_2}\right) = 2.0 \times 10^{-4}$ M OH^-

$[H_3O^+][OH^-] = 1.0 \times 10^{-14}$ M^2

$[H_3O^+] = \dfrac{1.0 \times 10^{-14} \text{ M}^2}{2.0 \times 10^{-4} \text{ M}} = 5.0 \times 10^{-4}$ M

ATE, Additional Sample Problems 16-1, p. 484

1. Given: $[HClO_4] = 0.01$ M
   Unknown: $[H_3O^+]$, $[OH^-]$

   $HClO_4(l) + H_2O(l) \rightarrow H_3O^+(aq) + ClO_4^-(aq)$

   $\left(\dfrac{0.01 \text{ mol } HClO_4}{\text{L solution}}\right)\left(\dfrac{1 \text{ mol } H_3O^+}{1 \text{ mol } HClO_4}\right) = \dfrac{0.01 \text{ mol } H_3O^+}{\text{L}}$

   $= 0.01 \text{ M } H_3O^+$

   $[H_3O^+][OH^-] = 1.0 \times 10^{-14} \text{ M}^2$

   $[OH^-] = \dfrac{1.0 \times 10^{-14} \text{ M}^2}{0.01 \text{ M}} = 1.0 \times 10^{-12} \text{ M}$

2. Given: aqueous solution of $Ba(OH)_2$
   $[H_3O^+] = 1 \times 10^{-11}$ M
   Unknowns: **a.** $[OH^-]$
   b. $[Ba(OH)_2]$

   $[H_3O^+][OH^-] = 1.0 \times 10^{-14} \text{ M}^2$

   $[OH^-] = \dfrac{1.0 \times 10^{-14} \text{ M}^2}{1 \times 10^{-11} \text{ M}} = 1 \times 10^{-3} \text{ M}$

   $Ba(OH)_2 \xrightarrow{H_2O} Ba^{2+} + 2OH^-$

   $\left(\dfrac{1 \times 10^{-3} \text{ mol } OH^-}{\text{L}}\right)\left(\dfrac{1 \text{ mol } Ba(OH)_2}{2 \text{ mol } OH^-}\right)$

   $= 5 \times 10^{-4} \text{ mol/L } Ba(OH)_2 = 5 \times 10^{-4} \text{ M } Ba(OH)_2$

Practice, p. 487

1. *Given:* Identity and concentrations of solutions:
   a. 1×10^{-3} M HCl
   b. 1×10^{-5} M HNO_3
   c. 1×10^{-4} M NaOH
   d. 1×10^{-2} M KOH
   Unknown: pH of solutions

   $[H_3O^+][OH^-] = 1.0 \times 10^{-14} \text{ M}^2$

   a. $HCl + H_2O \rightarrow H_3O^+ + Cl^-$

   $[H_3O^+] = 1 \times 10^{-3}$ M

   $pH = -\log [H_3O^+] = -\log (1 \times 10^{-3}) = 3.0$

   b. $HNO_3 + H_2O \rightarrow H_3O^+ + NO_3^-$

   $[H_3O^+] = 1 \times 10^{-5}$ M

   $pH = -\log (1 \times 10^{-5}) = 5.0$

   c. $NaOH + H_2O \rightarrow Na^+ + OH^-$

   $[OH^-] = 1 \times 10^{-4}$ M

   $[H_3O^+] = \dfrac{1.0 \times 10^{-14} \text{ M}^2}{1 \times 10^{-4} \text{ M}} = 1 \times 10^{-10} \text{ M}$

   $pH = -\log (1 \times 10^{-10}) = 10.0$

   d. $KOH + H_2O \rightarrow K^+ + OH^-$

   $[OH^-] = 1.0 \times 10^{-2}$ M

   $[H_3O^+] = \dfrac{1.0 \times 10^{-14} \text{ M}^2}{1.0 \times 10^{-2} \text{ M}} = 1.0 \times 10^{-12} \text{ M}$

   $pH = -\log (1 \times 10^{-12}) = 12.00$

ATE, Additional Sample Problems, p. 487

1. Given: [HBr] = 1×10^{-4} M
Unknown: pH

$HBr + H_2O \rightarrow H_3O^+ + Br^-$

$[H_3O^+] = 1 \times 10^{-4}$ M

$pH = -\log[H_3O^+]$

$pH = -\log(1 \times 10^{-4}) = 4.0$

2. Given: [Ca(OH)$_2$] = 5×10^{-4} M
Unknown: pH

$Ca(OH)_2 \xrightarrow{H_2O} Ca^{2+} + 2OH^-$

$\left(\dfrac{5 \times 10^{-4} \text{ mol Ca(OH)}_2}{\text{L solution}}\right)\left(\dfrac{2 \text{ mol OH}^-}{1 \text{ mol Ca(OH)}_2}\right) = 1 \times 10^{-3}$ M OH$^-$

$[H_3O^+][OH^-] = 1.0 \times 10^{-14}$ M^2

$[H_3O^+] = \dfrac{1.0 \times 10^{-14} \text{ M}^2}{1 \times 10^{-3} \text{ M}} = 1 \times 10^{-11}$ M

$pH = -\log(1 \times 10^{-11}) = 11.0$

Practice, p. 488

1. Given: [H$_3$O$^+$] = 6.7×10^{-4} M
Unknown: pH

$pH = -\log(6.7 \times 10^{-4}) = 3.17$

2. Given: [H$_3$O$^+$] = 2.5×10^{-2} M
Unknown: pH

$pH = -\log(2.5 \times 10^{-2}) = 1.60$

3. Given: [HNO$_3$] = 2.5×10^{-6} M
Unknown: pH

$HNO_3 + H_2O \rightarrow H_3O^+ + NO_3^-$

$[H_3O^+] = 2.5 \times 10^{-6}$ M

$pH = -\log[H_3O^+] = -\log(2.5 \times 10^{-6}) = 5.60$

4. Given: [Sr(OH)$_2$] = 2.0×10^{-2} M
Unknown: pH

$Sr(OH)_2 \xrightarrow{H_2O} Sr^{2+} + 2OH^-$

$\left(\dfrac{2.0 \times 10^{-2} \text{ mol Sr(OH)}_2}{\text{L solution}}\right)\left(\dfrac{2 \text{ mol OH}^-}{1 \text{ mol Sr(OH)}_2}\right) = 4.0 \times 10^{-2}$ M OH$^-$

$[H_3O^+][OH^-] = 1.0 \times 10^{-14}$ M^2

$[H_3O^+] = \dfrac{1.0 \times 10^{-14} \text{ M}^2}{4.0 \times 10^{-2} \text{ M}} = 2.5 \times 10^{-13}$ M

$pH = -\log[H_3O^+] = -\log(2.5 \times 10^{-13}) = 12.60$

ATE, Additional Sample Problems 16–3, p. 488

1. Given: $[H_3O^+] = 6.2 \times 10^{-9}$ M
Unknown: pH

$pH = -\log [H_3O^+] = -\log (6.2 \times 10^{-9}) = 8.21$

2. Given: [NaOH] = 0.00074 M
$= 7.4 \times 10^{-4}$ M
Unknown: pH

$NaOH + H_2O \rightarrow Na^+ + OH^-$

$[OH^-] = 7.4 \times 10^{-4}$ M

$[H_3O^+][OH^-] = 1.0 \times 10^{-14}$ M^2

$[H_3O^+] = \dfrac{1.0 \times 10^{-14} \text{ M}^2}{7.4 \times 10^{-4} \text{ M}} = 1.4 \times 10^{-11}$ M

$pH = -\log [H_3O^+] = -\log (1.4 \times 10^{-11}) = 10.87$

ATE, Additional Sample Problems 16–4, p 489

1. Given: pH = 9.0
Unknown: $[H_3O^+]$, $[OH^-]$

$pH = -\log [H_3O^+]$

$\log [H_3O^+] = -pH$

$[H_3O^+] = \text{antilog} (-pH)$

$[H_3O^+] = 1 \times 10^{-pH} = 1 \times 10^{-9}$ M

$[H_3O^+][OH^-] = 1.0 \times 10^{-14}$ M^2

$[OH^-] = \dfrac{1.0 \times 10^{-14} \text{ M}^2}{1 \times 10^{-9} \text{ M}} = 1 \times 10^{-5}$ M

2. Given: pH = 10.0
Solution = $Sr(OH)_2$
Unknowns: **a.** $[OH^-]$
b. $[Sr(OH)_2]$

$Sr(OH)_2 \xrightarrow{H_2O} Sr^{2+} + 2OH^-$

a. $pH = -\log [H_3O^+]$

$\log [H_3O^+] = -pH$

$[H_3O^+] = \text{antilog} (-pH)$

$[H_3O^+] = 1 \times 10^{-pH} = 1 \times 10^{-10}$ M

$[H_3O^+][OH^-] = 1.0 \times 10^{-14}$ M^2

$[OH^-] = \dfrac{1.0 \times 10^{-14} \text{ M}^2}{1 \times 10^{-10} \text{ M}} = 1 \times 10^{-4}$ M

b. Molarity (M) of $Sr(OH)_2 = \dfrac{\text{mol } Sr(OH)_2}{\text{L solution}}$

$\left(\dfrac{1 \times 10^{-4} \text{ mol } OH^-}{L}\right)\left(\dfrac{1 \text{ mol } Sr(OH)_2}{2 \text{ mol } OH^-}\right) = 5 \times 10^{-5}$ M $Sr(OH)_2$

Practice, p. 490

1. Given: pH = 5.0
Unknown: $[H_3O^+]$

$pH = -\log [H_3O^+]$

$\log [H_3O^+] = -pH$

$[H_3O^+] = \text{antilog}(-pH)$

$[H_3O^+] = 1 \times 10^{-pH} = 1 \times 10^{-5}$ M

2. Given: pH = 12.0
Unknown: $[H_3O^+]$

$pH = -\log [H_3O^+]$

$\log [H_3O^+] = -pH$

$[H_3O^+] = \text{antilog}(-pH)$

$[H_3O^+] = 1 \times 10^{-pH} = 1 \times 10^{-12}$ M

3. Given: pH = 1.50
Unknown: $[H_3O^+]$, $[OH^-]$

$pH = -\log [H_3O^+]$

$\log [H_3O^+] = -pH$

$[H_3O^+] = \text{antilog}(-pH)$

$[H_3O^+] = 1 \times 10^{-pH} = 1 \times 10^{-1.50} = 3.2 \times 10^{-2}$ M

$[H_3O^+][OH^-] = 1.0 \times 10^{-14}$ M^2

$[OH^-] = \dfrac{1.0 \times 10^{-14} \text{ M}^2}{3.2 \times 10^{-2} \text{ M}} = 3.2 \times 10^{-13}$ M

4. Given: pH = 3.67
Unknown: $[H_3O^+]$

$pH = -\log [H_3O^+]$

$\log [H_3O^+] = -pH$

$[H_3O^+] = \text{antilog}(-pH)$

$[H_3O^+] = 1 \times 10^{-pH} = 1 \times 10^{-3.67} = 2.1 \times 10^{-4}$ M

ATE, Additional Sample Problems 16–5, p. 490

1. Given: pH = 0.45
Unknown: $[H_3O^+]$

$pH = -\log [H_3O^+]$

$\log [H_3O^+] = -pH$

$[H_3O^+] = \text{antilog}(-pH)$

$[H_3O^+] = 1 \times 10^{-pH} = 1 \times 10^{-0.45} = 0.35$ M

2. Given: pH = 8.7
Unknown: $[H_3O^+]$, $[OH^-]$

$pH = -\log [H_3O^+]$

$\log [H_3O^+] = -pH$

$[H_3O^+] = \text{antilog}(-pH)$

$[H_3O^+] = 1 \times 10^{-pH} = 1 \times 10^{-8.7} = 2 \times 10^{-9}$ M

$[H_3O^+][OH^-] = 1.0 \times 10^{-14}$ M^2

$[OH^-] = \dfrac{1.0 \times 10^{-14} \text{ M}^2}{2 \times 10^{-9} \text{ M}} = 5 \times 10^{-6}$ M

Section Review, p. 491

4. Given:
  a. $[H_3O^+] = 1 \times 10^{-3}$ M
  b. $[OH^-] = 1 \times 10^{-4}$ M
  c. pH = 5.0
  d. pH = 8.0
  Unknown: if solutions are acidic or basic at 25°C

  a. pH = $-\log[H_3O^+]$
     = $-\log(1 \times 10^{-3})$
     = 3.0 = acidic

  b. pH = $-\log[H_3O^+]$
     $[H_3O^+][OH^-] = 1.0 \times 10^{-14}$ M^2
     $[H_3O^+] = \dfrac{1.0 \times 10^{-14} \text{ M}^2}{1 \times 10^{-4} \text{ M}} = 1 \times 10^{-10}$ M
     pH = $-\log(1.0 \times 10^{-10}) = 10.0$ = basic

  c. pH = 5.0 = acidic

  d. pH = 8.0 = basic

---

5. Given: $[HCl] = 4.5 \times 10^{-3}$ M
  Unknown: **a.** $[H_3O^+]$
  b. $[OH^-]$
  c. pH

  a. $HCl(l) + H_2O(l) \rightarrow H_3O^+(aq) + Cl^-(aq)$

  $\left(\dfrac{4.5 \times 10^{-3} \text{ mol HCl}}{\text{L solution}}\right)\left(\dfrac{1 \text{ mol } H_3O^+}{1 \text{ mol HCl}}\right)$

  = 4.5×10^{-3} mol H_3O^+/L
  = 4.5×10^{-3} M H_3O^+

  b. $[H_3O^+][OH^-] = 1.0 \times 10^{-14}$ M^2

  $[OH^-] = \dfrac{1.0 \times 10^{-14} \text{ M}^2}{4.5 \times 10^{-3} \text{ M}} = 2.2 \times 10^{-12}$ M OH^-

  c. pH = $-\log[H_3O^+] = -\log[4.5 \times 10^{-3}] = 2.35$

---

6. Given: pH = 8.0
  Unknown: **a.** $[H_3O^+]$
  b. $[OH^-]$
  c. $[Ca(OH)_2]$

  $Ca(OH)_2(s) \xrightarrow{H_2O} Ca^{2+}(aq) + 2OH^-(aq)$

  a. pH = $-\log[H_3O^+]$
  $\log[H_3O^+] = -\text{pH}$
  $[H_3O^+]$ = antilog$(-\text{pH}) = 1 \times 10^{-\text{pH}} = 1 \times 10^{-8}$ M

  b. $[H_3O^+][OH^-] = 1.0 \times 10^{-14}$ M^2

  $[OH^-] = \dfrac{1.0 \times 10^{-14} \text{ M}^2}{1 \times 10^{-8} \text{ M}} = 1 \times 10^{-6}$ M

  c. $\left(\dfrac{1 \times 10^{-6} \text{ mol } OH^-}{\text{L solution}}\right)\left(\dfrac{1 \text{ mol } Ca(OH)_2}{2 \text{ mol } OH^-}\right)$

  = 5×10^{-7} M $Ca(OH)_2$

ATE, Additional Sample Problems 16–6, p. 502

1. Given: $V_{RbOH} = 25.00$ mL
$V_{HBr} = 19.22$ mL
$[HBr] = 1.017$ M
Unknown: $[RbOH]$

$RbOH + HBr \rightarrow RbBr + H_2O$

$\left(\dfrac{1.017 \text{ mol HBr}}{L}\right)(19.22 \text{ mL})\left(\dfrac{L}{1000 \text{ mL}}\right) = 0.01955 \text{ mol HBr}$

$(0.01955 \text{ mol HBr})\left(\dfrac{1 \text{ mol RbOH}}{1 \text{ mol HBr}}\right)\left(\dfrac{1}{25.00 \text{ mL}}\right)\left(\dfrac{1000 \text{ mL}}{L}\right) = 0.7819 \dfrac{\text{mol}}{L} \text{ RbOH}$

$= 0.7819$ M RbOH

---

2. Given: $V_{Ba(OH)_2} = 29.96$ mL
$V_{HNO_3} = 16.08$ mL
$[HNO_3] = 2.303$ M
Unknown: $[Ba(OH)_2]$

$Ba(OH)_2 + 2HNO_3 \rightarrow Ba(NO_3)_2 + 2H_2O$

$\left(\dfrac{2.303 \text{ mol HNO}_3}{L}\right)(16.08 \text{ mL})\left(\dfrac{L}{1000 \text{ mL}}\right) = 0.03703 \text{ mol HNO}_3$

$(0.03703 \text{ mol HNO}_3)\left(\dfrac{1 \text{ mol Ba(OH)}_2}{2 \text{ mol HNO}_3}\right)\left(\dfrac{1}{29.96 \text{ mL}}\right)\left(\dfrac{1000 \text{ mL}}{L}\right)$

$= 0.6180$ M $Ba(OH)_2$

---

3. Given: $[CH_3COOH] = 0.83$ M
$V_{CH_3COOH} = 20.00$ mL
$[NaOH] = 0.519$ M
Unknown: V_{NaOH}

$CH_3COOH + NaOH \rightarrow NaC_2H_3O_2 + H_2O$

$\left(\dfrac{0.83 \text{ mol CH}_3COOH}{L}\right)(20.00 \text{ mL})\left(\dfrac{L}{1000 \text{ mL}}\right)$

$= 0.017$ mol CH_3COOH

$(0.017 \text{ mol CH}_3COOH)\left(\dfrac{1 \text{ mol NaOH}}{1 \text{ mol CH}_3COOH}\right)$

$= 0.017$ mol NaOH

$\dfrac{0.017 \text{ mol NaOH}}{\text{volume in L}} = 0.519$ M

volume in L $= \dfrac{0.017 \text{ mol NaOH}}{0.519 \text{ mol/L}}$

$= (0.033 \text{ L})\left(\dfrac{1000 \text{ mL}}{L}\right) = 33$ mL NaOH

Practice, p. 503

1. Given: $[KOH] = 0.215$ M
$V_{KOH} = 15.5$ mL
$V_{CH_3COOH} = 21.2$ mL
Unknown: $[CH_3COOH]$

$KOH + CH_3COOH \rightarrow KC_2H_3O_2 + H_2O$

$\left(\dfrac{0.215 \text{ mol KOH}}{4}\right)(15.5 \text{ mL})\left(\dfrac{L}{1000 \text{ mL}}\right)$

$= 0.00333$ mol KOH

$(0.00333 \text{ mol KOH})\left(\dfrac{1 \text{ mol CH}_3COOH}{1 \text{ mol KOH}}\right)\left(\dfrac{1}{21.2 \text{ mL}}\right)\left(\dfrac{1000 \text{ mL}}{L}\right)$

$= 0.157$ mol $CH_3COOH/L = 0.157$ M

2. Given: $V_{H_2SO_4} = 17.6$ mL
$V_{LiOH} = 27.4$ mL
[LiOH] = 0.0165 M
Unknown: [H_2SO_4]

$2LiOH + H_2SO_4 \rightarrow Li_2SO_4 + 2H_2O$

$$\left(\frac{0.0165 \text{ mol LiOH}}{L}\right)(27.4 \text{ mL})\left(\frac{L}{1000 \text{ mL}}\right)$$

$= 0.000452$ mol LiOH

$$(0.000452 \text{ mol LiOH})\left(\frac{1 \text{ mol } H_2SO_4}{2 \text{ mol LiOH}}\right)\left(\frac{1}{17.6 \text{ mL}}\right)\left(\frac{1000 \text{ mL}}{L}\right)$$

$= 0.0128$ M H_2SO_4

Section Review, p. 503

3. Given: [HCl] = 0.0100 M
$V_{HCl} = 20.0$ mL
$V_{NaOH} = 30.0$ mL
Unknown: [NaOH]

$HCl + NaOH \rightarrow NaCl + H_2O$

$$\left(\frac{0.0100 \text{ mol HCl}}{L}\right)(20.0 \text{ mL})\left(\frac{L}{1000 \text{ mL}}\right)$$

$= 0.000200$ mol HCl

$$(0.000200 \text{ mol HCl})\left(\frac{1 \text{ mol NaOH}}{1 \text{ mol HCl}}\right)\left(\frac{1}{30.0 \text{ mL}}\right)\left(\frac{1000 \text{ mL}}{L}\right)$$

$= 6.67 \times 10^{-3}$ M NaOH

4. Given: [Ca(OH)$_2$] = 0.10 M
$V_{Ca(OH)_2} = 20.0$ mL
$V_{HCl} = 12.0$ mL
Unknown: [HCl]

$Ca(OH)_2 + 2HCl \rightarrow CaCl_2 + 2H_2O$

$$\left(\frac{0.10 \text{ mol Ca(OH)}_2}{L}\right)(20.0 \text{ mL})\left(\frac{L}{1000 \text{ mL}}\right)$$

$= 0.0020$ mol Ca(OH)$_2$

$$(0.0020 \text{ mol Ca(OH)}_2)\left(\frac{2 \text{ mol HCl}}{1 \text{ mol Ca(OH)}_2}\right)\left(\frac{1}{12.0 \text{ mL}}\right)\left(\frac{1000 \text{ mL}}{L}\right) = 0.33 \text{ M HCl}$$

Reviewing Concepts

6. Given: $T = 25°C$
   a. [H_3O^+] = 1.0×10^{-7} M
   b. [H_3O^+] = 1.0×10^{-10} M
   c. [OH^-] = 1.0×10^{-7} M
   d. [OH^-] = 1.0×10^{-11} M
   e. [H_3O^+] = [OH^-]
   f. pH = 3.0
   g. pH = 13.0
Unknown: if solution is acidic, basic, or neutral

a. pH = $-\log$ [H_3O^+]
   = $-\log (1.0 \times 10^{-7})$ M
   = 7.0 = neutral

b. pH = $-\log[H_3O^+]$
   = $-\log (1.0 \times 10^{-10})$ M
   = 10 = basic

c. pH = $-\log[H_3O^+]$
   [H_3O^+][OH^-] = 1.0×10^{-14} M^2
   [H_3O^+] = $\dfrac{1.0 \times 10^{-14} \text{ M}^2}{1.0 \times 10^{-7} \text{ M}}$ = 1.0×10^{-7} M
   pH = $-\log (1.0 \times 10^{-7})$ M
   = 7.0 = neutral

d. pH = $-\log [H_3O^+]$

$[H_3O^+][OH^-] = 1.0 \times 10^{-14}$ M^2

$[H_3O^+] = \dfrac{1.0 \times 10^{-14} \text{ M}^2}{1.0 \times 10^{-11} \text{ M}} = 1.0 \times 10^{-3}$ M

pH = $-\log (1.0 \times 10^{-3})$ M

= 3.0 = acidic

e. $[H_3O^+][OH^-] = 1.0 \times 10^{-14}$ M^2

$[H_3O^+]^2 = 1.0 \times 10^{-14}$ M^2

$\sqrt{[H_3O^+]^2} = \sqrt{1.0 \times 10^{-14} \text{ M}^2}$

$[H_3O^+] = 1.0 \times 10^{-7}$ M

pH = $-\log (1.0 \times 10^{-7})$ M

= 7.0 = neutral

f. pH = 3.0 = acidic

g. pH = 13.0 = basic

Review Problems

15. Given: concentration of solution:
  a. 0.03 M HCl
  b. 1×10^{-4} M NaOH
  c. 5×10^{-3} M H$_2$SO$_4$
  d. 0.01 M Ca(OH)$_2$
  Unknown: [H$_3$O$^+$], [OH$^-$]

a. HCl(l) + H$_2$O(l) → H$_3$O$^+$(aq) + Cl$^-$(aq)

M = $\dfrac{\text{amount of solute (mol)}}{\text{volume of solution (L)}}$

$\left(\dfrac{0.03 \text{ mol HCl}}{\text{L solution}}\right)\left(\dfrac{1 \text{ mol H}_3\text{O}^+}{1 \text{ mol HCl}}\right)$ = 0.03 mol H$_3$O$^+$/L

$[H_3O^+] = 3 \times 10^{-2}$ M

$[H_3O^+][OH^-] = 1.0 \times 10^{-14}$ M^2

$[OH^-] = \dfrac{1.0 \times 10^{-14} \text{ M}^2}{3 \times 10^{-2} \text{ M}} = 3 \times 10^{-13}$ M

b. NaOH(s) $\xrightarrow{\text{H}_2\text{O}}$ Na$^+$(aq) + OH$^-$(aq)

$\left(\dfrac{1 \times 10^{-4} \text{ mol NaOH}}{\text{L solution}}\right)\left(\dfrac{1 \text{ mol OH}^-}{1 \text{ mol NaOH}}\right)$ = 1×10^{-4} mol OH$^-$/L

$[OH^-] = 1 \times 10^{-4}$ M

$[H_3O^+][OH^-] = 1.0 \times 10^{-14}$ M^2

$[H_3O^+] = \dfrac{1.0 \times 10^{-14} \text{ M}^2}{1 \times 10^{-4} \text{ M}} = 1 \times 10^{-10}$ M

c. H$_2$SO$_4$(l) + 2H$_2$O(l) → 2H$_3$O$^+$(aq) + H$_2$SO$_4^{2-}$(aq)

$\left(\dfrac{5 \times 10^{-3} \text{ mol H}_2\text{SO}_4}{\text{L}}\right)\left(\dfrac{2 \text{ mol H}_3\text{O}^+}{1 \text{ mol H}_2\text{SO}_4}\right)$ = 1×10^{-2} mol H$_3$O$^+$/L

$[H_3O^+] = 1 \times 10^{-2}$ M

$[H_3O^+][OH^-] = 1.0 \times 10^{-14}$ M^2

$[OH^-] = \dfrac{1.0 \times 10^{-14} \text{ M}^2}{1 \times 10^{-2} \text{ M}} = 1 \times 10^{-12}$ M

d. $Ca(OH)_2(s) \xrightarrow{H_2O} Ca^{2+}(aq) + 2OH^-(aq)$

$$\left(\frac{0.01 \text{ mol } Ca(OH)_2}{L}\right)\left(\frac{2 \text{ mol } OH^-}{1 \text{ mol } Ca(OH)_2}\right) = 0.02 \text{ mol } OH^-/L$$

$[OH^-] = 2 \times 10^{-2} \text{ M}$

$[H_3O^+][OH^-] = 1.0 \times 10^{-14} \text{ M}^2$

$[H_3O^+] = \dfrac{1.0 \times 10^{-14} \text{ M}^2}{2 \times 10^{-2} \text{ M}} = 5 \times 10^{-13} \text{ M}$

---

16. Given: Identity and concentration of solution:
a. 1.0×10^{-2} M HCl
b. 1.0×10^{-3} M HNO$_3$
c. 1.0×10^{-5} M HI
d. 1.0×10^{-4} M HBr
Unknown: pH

a. $HCl + H_2O \rightarrow H_3O^+ + Cl^-$

$[H_3O^+] = 1.0 \times 10^{-2} \text{ M}$

$\text{pH} = -\log [H_3O^+] = -\log (1.0 \times 10^{-2}) = 2.00$

b. $HNO_3 + H_2O \rightarrow H_3O^+ + NO_3^-$

$[H_3O^+] = 1.0 \times 10^{-3} \text{ M}$

$\text{pH} = -\log [H_3O^+] = -\log (1.0 \times 10^{-3}) = 3.00$

c. $HI + H_2O \rightarrow H_3O^+ + I^-$

$[H_3O^+] = 1.0 \times 10^{-5} \text{ M}$

$\text{pH} = -\log [H_3O^+] = -\log(1.0 \times 10^{-5}) = 5.00$

d. $HBr + H_2O \rightarrow H_3O^+ + Br^-$

$[H_3O^+] = 1.0 \times 10^{-4} \text{ M}$

$\text{pH} = -\log [H_3O^+] = -\log (1.0 \times 10^{-4}) = 4.00$

---

17. Given: $[OH^-] =$
a. 1.0×10^{-6} M
b. 1.0×10^{-9} M
c. 1.0×10^{-2} M
d. 1.0×10^{-7} M
Unknown: pH

$[H_3O^+][OH^-] = 1.0 \times 10^{-14} \text{ M}^2$

a. $[H_3O^+] = \dfrac{1.0 \times 10^{-14} \text{ M}^2}{1.0 \times 10^{-6} \text{ M}}$

$= 1.0 \times 10^{-8} \text{ M}$

$\text{pH} = -\log [H_3O^+] = -\log (1.0 \times 10^{-8}) = 8.00$

b. $[H_3O^+] = \dfrac{1.0 \times 10^{-14} \text{ M}^2}{1.0 \times 10^{-9} \text{ M}}$

$= 1.0 \times 10^{-5} \text{ M}$

$\text{pH} = -\log [H_3O^+] = -\log (1.0 \times 10^{-5}) = 5.00$

c. $[H_3O^+] = \dfrac{1.0 \times 10^{-14} \text{ M}^2}{1.0 \times 10^{-2} \text{ M}}$

$= 1.0 \times 10^{-12} \text{ M}$

$\text{pH} = -\log [H_3O^+] = -\log (1.0 \times 10^{-12}) = 12.00$

d. $[H_3O^+] = \dfrac{1.0 \times 10^{-14} \text{ M}^2}{1.0 \times 10^{-7} \text{ M}}$

$= 1.0 \times 10^{-7} \text{ M}$

$\text{pH} = -\log [H_3O^+] = -\log (1.0 \times 10^{-7}) = 7.00$

18. Given: Identity and concentration of solution:
 a. 1.0×10^{-2} M NaOH
 b. 1.0×10^{-3} M KOH
 c. 1.0×10^{-4} M LiOH
Unknown: pH

$[H_3O^+][OH^-] = 1.0 \times 10^{-14}$ M^2

a. NaOH $\xrightarrow{H_2O}$ Na$^+$ + OH$^-$

$[OH^-] = 1.0 \times 10^{-2}$ M

$[H_3O^+] = \dfrac{1.0 \times 10^{-14} \text{ M}^2}{1.0 \times 10^{-2} \text{ M}} = 1.0 \times 10^{-12}$ M

pH = $-\log[H_3O^+] = -\log(1.0 \times 10^{-12}) = 12.00$

b. KOH $\xrightarrow{H_2O}$ K$^+$ + OH$^-$

$[OH^-] = 1.0 \times 10^{-3}$ M

$[H_3O^+] = \dfrac{1.0 \times 10^{-14} \text{ M}^2}{1.0 \times 10^{-3} \text{ M}} = 1.0 \times 10^{-11}$ M

pH = $-\log[H_3O^+] = -\log(1.0 \times 10^{-11}) = 11.00$

c. LiOH $\xrightarrow{H_2O}$ Li$^+$ + OH$^-$

$[OH^-] = 1.0 \times 10^{-4}$ M

$[H_3O^+] = \dfrac{1.0 \times 10^{-14} \text{ M}^2}{1.0 \times 10^{-4} \text{ M}} = 1.0 \times 10^{-10}$ M

pH = $-\log[H_3O^-] = -\log(1.0 \times 10^{-10}) = 10.00$

19. Given: $[H_3O^+] =$
 a. 2.0×10^{-5} M
 b. 4.7×10^{-7} M
 c. 3.8×10^{-3} M
Unknown: pH

a. pH = $-\log[H_3O^+] = -\log(2.0 \times 10^{-5}) = 4.70$

b. pH = $-\log[H_3O^+] = -\log(4.7 \times 10^{-7}) = 6.33$

c. pH = $-\log[H_3O^+] = -\log(3.8 \times 10^{-3}) = 2.42$

20. Given: pH value:
 a. 3.0
 b. 7.00
 c. 11.0
 d. 5.0
Unknown: $[H_3O^+]$

pH = $-\log[H_3O^+]$

$\log[H_3O^+] = -$pH

$[H_3O^+] = $ antilog$(-$pH$) = 1 \times 10^{-\text{pH}}$

a. $[H_3O^+] = 1 \times 10^{-3}$

b. $[H_3O^+] = 1.0 \times 10^{-7}$

c. $[H_3O^+] = 1 \times 10^{-11}$

d. $[H_3O^+] = 1 \times 10^{-5}$

21. Given: pH value:
  a. 7.00
  b. 11.00
  c. 4.00
  d. 6.00
  Unknown: [OH$^-$]

$$pH = -\log[H_3O^+]$$
$$\log[H_3O^+] = -pH$$
$$[H_3O^+] = \text{antilog}(-pH) = 1 \times 10^{-pH}$$
$$[H_3O^+][OH^-] = 1 \times 10^{-14}\ M$$

a. $[H_3O^+] = 1.0 \times 10^{-7}\ M$

$$[OH^-] = \frac{1.0 \times 10^{-14}\ M^2}{1.0 \times 10^{-7}\ M} = 1.0 \times 10^{-7}\ M$$

b. $[H_3O^+] = 1.0 \times 10^{-11}\ M$

$$[OH^-] = \frac{1.0 \times 10^{-14}\ M^2}{1.0 \times 10^{-11}\ M} = 1.0 \times 10^{-3}\ M$$

c. $[H_3O^+] = 1.0 \times 10^{-4}\ M$

$$[OH^-] = \frac{1.0 \times 10^{-14}\ M^2}{1.0 \times 10^{-4}\ M} = 1.0 \times 10^{-10}\ M$$

d. $[H_3O^+] = 1.0 \times 10^{-6}\ M$

$$[OH^-] = \frac{1.0 \times 10^{-14}\ M^2}{1.0 \times 10^{-6}\ M} = 1.0 \times 10^{-8}\ M$$

22. Given: pH value:
  a. 4.23
  b. 7.65
  c. 9.48
  Unknown: [H$_3$O$^+$]

$$pH = -\log[H_3O^+]$$
$$\log[H_3O^+] = -pH$$
$$[H_3O^+] = \text{antilog}(-pH)$$

a. $[H_3O^+] = \text{antilog}(-4.23)$
$$= 1.0 \times 10^{-4.23} = 5.9 \times 10^{-5}\ M$$

b. $[H_3O^+] = \text{antilog}(-7.65)$
$$= 1.0 \times 10^{-7.65} = 2.2 \times 10^{-8}\ M$$

c. $[H_3O^+] = \text{antilog}(-9.48)$
$$= 1.0 \times 10^{-9.48} = 3.3 \times 10^{-10}\ M$$

23. Given: pH = 2.70
  a. Unknown: [H$_3$O$^+$]

$$pH = -\log[H_3O^+]$$
$$\log[H_3O^+] = -pH$$
$$[H_3O^+] = \text{antilog}(-pH)$$
$$[H_3O^+] = 1 \times 10^{-pH}$$
$$= 1 \times 10^{-2.70} = 2.0 \times 10^{-3}\ M$$

  b. Unknown: [OH$^-$]

$$[H_3O^+][OH^-] = 1.0 \times 10^{-14}\ M^2$$
$$[OH^-] = \frac{1.0 \times 10^{-14}\ M^2}{2.0 \times 10^{-3}\ M}$$
$$= 5.0 \times 10^{-12}\ M$$

c. Unknown: number of moles HNO_3 required to prepare 5.50 L of this solution

$HNO_3 + H_2O \rightarrow H_3O^+ + NO_3^-$

$(2.0 \times 10^{-3} \text{ mol } H_3O^+)\left(\dfrac{1 \text{ mol } HNO_3}{1 \text{ mol } H_3O^+}\right)$

$= 2.0 \times 10^{-3} \text{ mol } HNO_3$

$\left(\dfrac{2.0 \times 10^{-3} \text{ mol}}{L}\right)(5.5L) = 1.1 \times 10^{-2} \text{ mol } HNO_3$

d. Unknown: mass of 1.1×10^{-2} mol HNO_3

$(1.1 \times 10^{-2} \text{ mol } HNO_3)\left(\dfrac{63 \text{ g } HNO_3}{\text{mol } HNO_3}\right) = 0.69 \text{ g } HNO_3$

e. Given: conc. HNO_3 = 69.5% HNO_3; $D = 1.42$ g/mL
Unknown: V_{HNO_3} needed to prepare 2.0×10^{-3} M HNO_3

$(0.69 \text{ g } HNO_3)\left(\dfrac{100 \text{ g concentrated } HNO_3}{69.5 \text{ g } HNO_3}\right)\left(\dfrac{mL}{1.42 \text{ g}}\right)$

$= 0.70$ mL concentrated HNO_3

---

24. Given: acid-base titrations:
a. NaOH with 1.0 mol HCl
b. HNO_3 with 0.75 mol KOH
c. $Ba(OH)_2$ with 0.20 mol HF
d. H_2SO_4 with 0.60 mol Al$(OH)_3$
Unknown: chemical equivalent in moles

a. $NaOH + HCl \rightarrow NaCl + H_2O$

$(1.0 \text{ mol } HCl)\left(\dfrac{1 \text{ mol } NaOH}{1 \text{ mol } HCl}\right) = 1.0 \text{ mol } NaOH$

b. $HNO_3 + KOH \rightarrow KNO_3 + H_2O$

$(0.75 \text{ mol } KOH)\left(\dfrac{1 \text{ mol } HNO_3}{1 \text{ mol } KOH}\right) = 0.75 \text{ mol } HNO_3$

c. $Ba(OH)_2 + 2HF \rightarrow BaF_2 + 2H_2O$

$(0.20 \text{ mol } HF)\left(\dfrac{1 \text{ mol } Ba(OH)_2}{2 \text{ mol } HF}\right) = 0.10 \text{ mol } Ba(OH)_2$

d. $3H_2SO_4 + 2Al(OH)_3 \rightarrow Al_2(SO_4)_3 + 6H_2O$

$(0.60 \text{ mol } Al(OH)_3)\left(\dfrac{3 \text{ mol } H_2SO_4}{2 \text{ mol } Al(OH)_3}\right) = 0.90 \text{ mol } H_2SO_4$

---

25. Given: $[H_2SO_4] = 2.50 \times 10^{-2}$ M
$V_{H_2SO_4} = 15.0$ mL
$V_{KOH} = 10.0$ mL
Unknown: [KOH]

$H_2SO_4 + 2KOH \rightarrow K_2SO_4 + 2H_2O$

$\left(\dfrac{2.50 \times 10^{-2} \text{ mol } H_2SO_4}{L}\right)(15.0 \text{ mL})\left(\dfrac{L}{1000 \text{ mL}}\right)$

$= 3.75 \times 10^{-4} \text{ mol } H_2SO_4$

$(3.75 \times 10^{-4} \text{ mol } H_2SO_4)\left(\dfrac{2 \text{ mol } KOH}{1 \text{ mol } H_2SO_4}\right) = 7.50 \times 10^{-4} \text{ mol } KOH$

$\left(\dfrac{7.50 \times 10^{-4} \text{ mol } KOH}{10 \text{ mL}}\right)\left(\dfrac{1000 \text{ mL}}{L}\right) = 7.50 \times 10^{-2}$ M KOH

26. Given: $[Ba(OH)_2] = 1.75 \times 10^{-2}$ M
$V_{Ba(OH)_2} = 12.5$ mL
$V_{HNO_3} = 14.5$ mL
Unknown: $[HNO_3]$

$Ba(OH)_2 + 2HNO_3 \rightarrow Ba(NO_3)_2 + 2H_2O$

$\left(\dfrac{1.75 \times 10^{-2} \text{ mol Ba(OH)}_2}{L}\right)(12.5 \text{ mL})\left(\dfrac{1 \text{ L}}{1000 \text{ mL}}\right)$

$= 2.19 \times 10^{-4}$ mol $Ba(OH)_2$

$(2.19 \times 10^{-4} \text{ mol Ba(OH)}_2)\left(\dfrac{2 \text{ mol HNO}_3}{1 \text{ mol Ba(OH)}_2}\right)$

$= 4.38 \times 10^{-4}$ mol HNO_3

$\left(\dfrac{4.38 \times 10^{-4} \text{ mol HNO}_3}{14.5 \text{ mL}}\right)\left(\dfrac{1000 \text{ mL}}{L}\right)$

$= 3.02 \times 10^{-2}$ M HNO_3

Mixed Review

27. Given: $[Ca(OH)_2] = 4.0 \times 10^{-4}$ M

a. Unknown: $[OH^-]$

$Ca(OH)_2(s) \xrightarrow{H_2O} Ca^{2+}(aq) + 2OH^-(aq)$

$\left(\dfrac{4.0 \times 10^{-4} \text{ mol Ca(OH)}_2}{L}\right)\left(\dfrac{2 \text{ mol OH}^-}{1 \text{ mol Ca(OH)}_2}\right)$

$= 8.0 \times 10^{-4}$ M OH^-

b. Unknown: $[H_3O^+]$

$[H_3O^+][OH^-] = 1.0 \times 10^{-14}$ M^2

$[H_3O^+] = \dfrac{1.0 \times 10^{-14} \text{ M}^2}{8.0 \times 10^{-4} \text{ M}} = 1.3 \times 10^{-11}$ M

28. Given: $[H_3O^+]$:
a. 1.0×10^{-7} M
b. 1.0×10^{-3} M
c. 1.0×10^{-12} M
d. 1.0×10^{-5} M
Unknown: pH

pH = $-\log [H_3O^+]$

a. pH = $-\log (1.0 \times 10^{-7}) = 7.00$
b. pH = $-\log (1.0 \times 10^{-3}) = 3.00$
c. pH = $-\log (1.0 \times 10^{-12}) = 12.00$
d. pH = $-\log (1.0 \times 10^{-5}) = 5.00$

29. Given: pH = 6.0
Unknown: $[H_3O^+]$

pH = $-\log [H_3O^+]$

$\log [H_3O^+] = -$pH

$[H_3O^+] = $ antilog $(-$pH$) = 1 \times 10^{-pH} = 1 \times 10^{-6}$ M

30. Given: $[Ba(OH)_2] = 5.0 \times 10^{-5}$ M
Unknown: pH

$Ba(OH)_2(s) \xrightarrow{H_2O} Ba^{2+}(aq) + 2OH^-(aq)$

$\left(\dfrac{5.0 \times 10^{-5} \text{ mol Ba(OH)}_2}{L}\right)\left(\dfrac{2 \text{ mol OH}^-}{1 \text{ mol Ba(OH)}_2}\right)$

$= 1.0 \times 10^{-4}$ M OH^-

$[H_3O^+][OH^-] = 1.0 \times 10^{-14}$ M^2

$[H_3O^+] = \dfrac{1.0 \times 10^{-14} \text{ M}^2}{1.0 \times 10^{-4} \text{ M}} = 1.0 \times 10^{-10}$ M

pH = $-\log [H_3O^+] = -\log (1.0 \times 10^{-10}) = 10.00$

31. a. Given: $[H_3O^+] = 8.4 \times 10^{-11}$ M
Unknown: pH

$pH = -\log [H_3O^+] = -\log (8.4 \times 10^{-11})$
$= 10.08$

b. Given: pH = 2.50
Unknown: $[H_3O^+]$

$pH = -\log [H_3O^+]$
$\log [H_3O^+] = -pH$
$[H_3O^+] = \text{antilog } (-pH)$
$= 1 \times 10^{-pH}$
$= 1 \times 10^{-2.50}$
$= 3.2 \times 10^{-3}$ M

32. a. Given: $[Mg(OH)_2] = 5.4 \times 10^{-5}$ M
Unknown: $[OH^-]$

$Mg(OH)_2(s) \xrightarrow{H_2O} Mg^{2+}(aq) + 2OH^-(aq)$

$\left(\dfrac{5.4 \times 10^{-5} \text{ mol Mg (OH)}_2}{L}\right)\left(\dfrac{2 \text{ mol OH}^-}{1 \text{ mol Mg(OH)}_2}\right)$

$= 1.1 \times 10^{-4}$ M OH^-

b. Unknown: $[H_3O^+]$

$[H_3O^+][OH^-] = 1.0 \times 10^{-14}$ M^2

$[H_3O^+] = \dfrac{1.0 \times 10^{-14} \text{ M}^2}{1.1 \times 10^{-4} \text{ M}} = 9.1 \times 10^{-11}$ M

33. Given: pH = 8.90
a. Unknown: $[H_3O^+]$

$pH = -\log [H_3O^+]$
$\log [H_3O^+] = -pH$
$[H_3O^+] = \text{antilog } (-pH) = 1 \times 10^{-8.90} = 1.3 \times 10^{-9}$ M

b. Unknown: $[OH^-]$

$[H_3O^+][OH^-] = 1.0 \times 10^{-14}$ M^2

$[OH^-] = \dfrac{1.0 \times 10^{-14} \text{ M}^2}{1.3 \times 10^{-9} \text{ M}} = 7.7 \times 10^{-6}$ M

34. Given: $[OH^-] = 6.9 \times 10^{-10}$ M
Unknown: pH

$pH = -\log [H_3O^+]$
$[H_3O^+][OH^-] = 1.0 \times 10^{-14}$ M^2

$[H_3O^+] = \dfrac{1.0 \times 10^{-14} \text{ M}^2}{6.9 \times 10^{-10} \text{ M}} = 1.4 \times 10^{-5}$ M

$pH = -\log (1.4 \times 10^{-5}) = 4.85$

35. Given: $[Ba(OH)_2] = 3.4 \times 10^{-3}$ M
$V_{Ba(OH)_2} = 25.9$ mL
$V_{HCl} = 16.6$ mL
Unknown: [HCl]

$Ba(OH)_2 + 2HCl \rightarrow BaCl_2 + 2H_2O$

$\left(\dfrac{3.4 \times 10^{-3} \text{ mol Ba(OH)}_2}{L}\right)(25.9 \text{ mL})\left(\dfrac{L}{1000 \text{ mL}}\right)$

$= 8.8 \times 10^{-5}$ mol $Ba(OH)_2$

$(8.8 \times 10^{-5} \text{ mol Ba(OH)}_2)\left(\dfrac{2 \text{ mol HCl}}{1 \text{ mol Ba(OH)}_2}\right)$

$= 1.8 \times 10^{-4}$ mol HCl

$\left(\dfrac{1.8 \times 10^{-4} \text{ mol HCl}}{16.6 \text{ mL}}\right)\left(\dfrac{1000 \text{ mL}}{L}\right)$

$= 1.1 \times 10^{-2}$ M HCl

36. Given: $V_{Ca(OH)_2} = 428$ mL
$V_{HNO_3} = 115$ mL
$[HNO_3] = 6.7 \times 10^{-3}$ M
Unknown: $[Ca(OH)_2]$

$Ca(OH)_2 + 2HNO_3 \rightarrow Ca(NO_3)_2 + 2H_2O$

$\left(\dfrac{6.7 \times 10^{-3} \text{ mol HNO}_3}{L}\right)(115 \text{ mL})\left(\dfrac{L}{1000 \text{ mL}}\right)$

$= 7.7 \times 10^{-4}$ mol HNO_3

$(7.7 \times 10^{-4} \text{ mol HNO}_3)\left(\dfrac{1 \text{ mol Ca(OH)}_2}{2 \text{ mol HNO}_3}\right)$

$= 3.9 \times 10^{-4}$ mol $Ca(OH)_2$

$\left(\dfrac{3.9 \times 10^{-4} \text{ mol Ca(OH)}_2}{428 \text{ mL}}\right)\left(\dfrac{1000 \text{ mL}}{L}\right)$

$= 9.1 \times 10^{-4}$ M $Ca(OH)_2$

37. Given: $V_{HNO_3} = 10.1$ mL
$V_{KOH} = 71.4$ mL
$[KOH] = 4.2 \times 10^{-3}$ M
Unknown: $[HNO_3]$

$KOH + HNO_3 \rightarrow KNO_3 + H_2O$

$\left(\dfrac{4.2 \times 10^{-3} \text{ mol KOH}}{L}\right)(71.4 \text{ mL})\left(\dfrac{L}{1000 \text{ mL}}\right)$

$= 3.0 \times 10^{-4}$ mol KOH

$(3.0 \times 10^{-4} \text{ mol KOH})\left(\dfrac{1 \text{ mol HNO}_3}{1 \text{ mol KOH}}\right)$

$= 3.0 \times 10^{-4}$ mol HNO_3

$\left(\dfrac{3.0 \times 10^{-4} \text{ mol HNO}_3}{10.1 \text{ mL}}\right)\left(\dfrac{1000 \text{ mL}}{L}\right)$

$= 3.0 \times 10^{-2}$ M HNO_3

CHAPTER 17

Reaction Energy and Reaction Kinetics

Practice, p. 514

1. Given: $m = 35$ g
$\Delta T = 313$ K $- 293$ K
$= 20.$ K
$q = 48$ J
Unknown: c_p in J/(g·K)

$c_p = \dfrac{q}{m \times \Delta T}$

$= \dfrac{48 \text{ J}}{(35 \text{ g})(20. \text{ K})} = 0.069$ J/(g·K)

2. Given: $q = 9.8 \times 10^5$ J
volume of $H_2O = 6.2$ L
$= 6200$ g
$T_i = 291$ K
Unknown: T_f

c_p of $H_2O = 4.18$ J/(g·K)

$c_p = \dfrac{q}{m \times \Delta T}$

$\Delta T = \dfrac{q}{(m)(c_p)} = \dfrac{9.8 \times 10^5 \text{ J}}{(6.2 \times 10^3 \text{ g})(4.18 \text{ J/(g·K)})}$

$= 38$ K

$\Delta T = T_f - T_i$

$T_f = \Delta T + T_i$

$= 38$ K $+ 291$ K $= 329$ K

ATE, Additional Sample Problems 17–1, p. 514

1. Given: $m = 85.0$ g
$\Delta T = 45°C - 30°C$
$= 15°C = 15$ K
$q = 523$ J

$c_p = \dfrac{q}{m \times \Delta T}$

$c_p = \dfrac{523 \text{ J}}{(85.0 \text{ g})(15 \text{ K})} = 0.41$ J/(g·K)

a. Unknown: c_p

b. Unknown: amount of energy the sample will lose if cooled to 25°C

$\Delta T = 45°C - 25°C = 20.°C = 20.$ K

$q = c_p \times m \times \Delta T$

$= (0.41$ J/(g·K)$)(85.0$ g$)(20.$ K$)$

$= 7.0 \times 10^2$ J

2. Given: $m = 74$ g
$\Delta T = 45°C - 15°C$
$= 30.°C = 30.$ K
$q = 2000$ J

Unknown: c_p

$c_p = \dfrac{q}{m \times \Delta T}$

$= \dfrac{2000 \text{ J}}{(74 \text{ g})(30. \text{ K})}$

$= 0.90$ J/(g·K)

3. Given: $m = 5.0$ g
$\Delta T = 25°C = 25$ K

Unknown: q

c_p of gold $= 0.129$ J/(g·K)

$c_p = \dfrac{q}{m \times \Delta T}$

$q = c_p \times m \times \Delta T$

$= (0.129 \text{ J/(g·K)})(5.0 \text{ g})(25 \text{ K})$

$= 16$ J

4. Given: $q = 420$ J
$m = 35$ g
$T_i = 10.°C$

Unknown: T_f

c_p of $H_2O = 4.18$ J/(g·K)

$c_p = \dfrac{q}{m \times \Delta T}$

$\Delta T = \dfrac{q}{m \times c_p} = \dfrac{420 \text{ J}}{(35 \text{ g})(4.18 \text{ J/(g·K)})}$

$= 3$ K $= 3°C$

$\Delta T = T_f - T_i$

$T_f = \Delta T + T_i$

$= 3°C + 10°C = 13°C$

5. Given: $q = 24\,000$ J
$\Delta T = 100.0°C - 25°C = 75°C = 75$ K

Unknown: m

c_p of $H_2O = 4.18$ J/(g·K)

$c_p = \dfrac{q}{m \times \Delta T}$

$m = \dfrac{q}{(\Delta T)(c_p)}$

$= \dfrac{24\,000 \text{ J}}{(75 \text{ K})(4.18 \text{ J/(g·K)})}$

$= 77$ g

Practice, p. 522

1. Given: $CH_4(g) + 2O_2 \rightarrow CO_2(g) + H_2O(l)$

Unknown: $\Delta H°$ for combustion of CH_4

$CH_4(g) \rightarrow C(s) + 2H_2(g)$ $\Delta H_f^0 = 74.9$ kJ/mol

$C(s) + O_2(g) \rightarrow CO_2$ $\Delta H_f^0 = -393.5$ kJ/mol

$2H_2(g) + O_2(g) \rightarrow 2H_2O(l)$ $\Delta H_f^0 = 2(-285.8)$

$= -571.6$ kJ/mol

$CH_4(g) + 2O_2(g) \rightarrow CO_2(g) + 2H_2O(l)$

$\Delta H_e^0 = -890.2$ kJ/mol

2. Given: $C_{graphite}(s) \rightarrow C_{diamond}(s)$
Unknown: ΔH^0

$C_{graphite}(s) + O_2(g) \rightarrow CO_2(g)$	$\Delta H_c^0 = -394$ kJ/mol
$CO_2(g) \rightarrow C_{diamond}(s) + O_2(g)$	$\Delta H_c^0 = 396$ kJ/mol
$C_{graphite}(s) \rightarrow C_{diamond}(s)$	$\Delta H_e^0 = 2$ kJ/mol

ATE, Additional Sample Problems 17–2, p. 522

1. Given: $2N_2(g) + 5O_2(g) \rightarrow 2N_2O_5(g)$
Unknown: ΔH

$2H_2O(l) \rightarrow 2H_2(g) + 2(\frac{1}{2})O_2(g)$	$\Delta H_f^0 = 2(285.8) = 571.6$ kJ/mol
$(2)2HNO_3(l) \rightarrow 2N_2O_5(g) + 2H_2O(l)$	$\Delta H^0 = 2(76.6) = 153.2$ kJ/mol
$(4)\frac{1}{2}N_2(g) + (4)\frac{3}{2}O_2(g) + (4)\frac{1}{2}H_2(g) \rightarrow 4HNO_3(l)$	$\Delta H_f^0 = 4(-174.1)$
	$= -696.4$ kJ/mol
$2N_2 + 5O_2 \rightarrow 2N_2O_5$	$\Delta H = 28.4$ kJ/mol

Practice, p. 524

1. Unknown: ΔH_f of butane (C_4H_{10})

Balanced equation:

$4C(s) + 5H_2(g) \rightarrow C_4H_{10}(g)$

$4C(s) + 4O_2(g) \rightarrow 4CO_2(g)$	$\Delta H_f^0 = 4(-393.5)$
	$= -1574$ kJ/mol
$5H_2(g) + \frac{5}{2}O_2(g) \rightarrow 5H_2O(l)$	$\Delta H_f^0 = 5(-285.8)$
	$= -1429$ kJ/mol
$4CO_2(g) + 5H_2O(l) \rightarrow C_4H_{10}(g) + \frac{13}{2}O_2(g)$	$\Delta H_c^0 = +2877.6$ kJ/mol
$4C(s) + 5H_2(g) \rightarrow C_4H_{10}(g)$	$\Delta H_f^0 = -125.4$ kJ/mol

2. Unknown: ΔH_c of 1 mol of N_2 to form NO_2

Balanced equation:

$N_2(g) + 2O_2(g) \rightarrow 2NO_2(g)$

$\frac{1}{2}N_2(g) + O_2(g) \rightarrow NO_2(g)$	$\Delta H_f^0 = 33.2$ kJ/mol
$(2)\frac{1}{2}N_2(g) + (2)O_2(g) \rightarrow (2)NO_2(g)$	$\Delta H_f^0 = 2(33.2$ kJ/mol$)$
	$= 66.4$ kJ/mol

3. Unknown: ΔH_f of SO_2 from S and O

Balanced equation:

$S(s) + O_2(g) \rightarrow SO_2(g)$

$S(s) + \frac{3}{2}O_2(g) \rightarrow SO_3(g)$	$\Delta H_c^0 = -395.2$ kJ/mol
$(\frac{1}{2})2SO_3(g) \rightarrow (\frac{1}{2})2SO_2(g) + (\frac{1}{2})O_2(g)$	$\Delta H^0 = (\frac{1}{2})(+198.2)$ kJ/mol
	$= 99.1$ kJ/mol
$S(s) + O_2(g) \rightarrow SO_2(g)$	$\Delta H_f = -296.1$ kJ/mol

Section Review, p. 524

5. Given: $m = 75$ g
$\Delta T = 301$ K
$- 295$ K $= 6$ K

Unknown: q

c_p of iron $= 0.449$ J/(g·K)

$$c_p = \frac{q}{m \times \Delta T}$$

$$q = c_p \times m \times \Delta T$$

$$= (0.449 \text{ J/(g·K)})(75 \text{ g})(6 \text{ K})$$

$$= 2.0 \times 10^2 \text{ J}$$

6. Given: CH_4 $\Delta H_c = -890$ kJ/mol
$m = 3.2$ g

Unknown: ΔH

$$(3.2 \text{ g } CH_4)\left(\frac{\text{mol } CH_4}{16 \text{ g } CH_4}\right) = 0.20 \text{ mol } CH_4$$

$$(0.20 \text{ mol})(-890 \text{ kJ/mol}) = -180 \text{ kJ}$$

ATE, Additional Example Problems, p. 529

Unknown: whether ΔS_0 will be > 0, < 0, or $= 0$

1. $3H_2(g) + N_2(g) \rightarrow 2NH_3(g)$

4 mol(g) → 2 mol(g)

S^0 decreases; $\Delta S^0 < 0$

2. $2Mg(s) + O_2(g) \rightarrow 2MgO(s)$

3 mol → 2 mol

S^0 decreases; $\Delta S^0 < 0$

3. $C_6H_{12}O_6(s) + 6O_2(g) \rightarrow 6CO_2(g) + 6H_2O(g)$

7 mol → 12 mol

S^0 increases; $\Delta S^0 > 0$

4. $KNO_3(s) \rightarrow K^+(aq) + NO_3^-(aq)$

1 mol → 2 mol

S^0 increases; $\Delta S^0 > 0$

Practice, p. 530

1. Given: $\Delta H^0 = 31.0$ kJ/mol
$\Delta S^0 = 0.093$ kJ/mol·K

Unknown: temperature at which reaction is spontaneous

$\Delta G^0 = \Delta H^0 - T\Delta S$

Assume $\Delta G^0 < 0$ (ΔG^0 must be negative for a spontaneous reaction)

$0 > \Delta H^0 - T\Delta S$

$$T > \frac{\Delta H^0}{\Delta S^0}$$

$$> \frac{31.0 \text{ kJ/mol}}{0.093 \text{ kJ/mol·K}} = 333 \text{ K}$$

Temperature would need to be > 333K

ATE, Additional Sample Problem 17–4, p. 530

1. Given: $Cu_2S(s) + S(s) \rightarrow 2CuS(s)$
$\Delta H^0 = -26.7$ kJ/mol
$\Delta S^0 = 0.0197$ kJ/(mol·K)
T = 298 K

Unknown: whether reaction will be spontaneous

$\Delta G^0 = \Delta H^0 - T\Delta S^0$
$= -26.7$ kJ/mol $- (298$ K$)(-0.0197$ kJ/(mol·K)$)$
$= -26.7$ kJ/mol $- (-5.87$ kJ/mol$)$
$= -20.8$ kJ/mol

yes; reaction will be spontaneous (ΔG^0 is negative).

Section Review, p. 530

8. Unknown: sign of ΔS^0

a. $CaCO_3(s) \rightarrow CaO(s) + CO_2(g)$

1 mol $\rightarrow$ 2 mol

ΔS^0 increases; $\Delta S^0 > 0 (+\Delta S^0)$

b. $2SO_2(g) + O_2(g) \rightarrow 2SO_3(g)$

3 mol $\rightarrow$ 2 mol

ΔS^0 decreases; $\Delta S^0 < 0 (-\Delta S^0)$

Practice, p. 537

1. Unknown: $\Delta E_{forward}$
$\Delta E_{reverse}$
E_a
E_a'

$\Delta E_{forward}$ = energy of products – energy of reactants
$= -150$ kJ/mol $- 0 = -150$ kJ/mol

$\Delta E_{reverse}$ = energy of reactants – energy of products
$= 0 - (-150$ kJ/mol$) = +150$ kJ/mol

E_a = energy of activated complex – energy of reactants
$= 100$ kJ/mol $- 0 = 100$ kJ/mol

E_a' = energy of activated complex – energy of products
$= 100$ kJ/mol $- (-150$ kJ/mol$) = +250$ kJ/mol

ATE, Additional Sample Problem 17–5, p. 536

1. Unknown: $\Delta E_{forward}$
$\Delta E_{reverse}$
E_a'

$\Delta E_{forward}$ = energy of products – energy of reactants
$= 30$ kJ/mol $- 0$ kJ/mol $= +30$ kJ/mol

$\Delta E_{reverse}$ = energy of reactants – energy of products
$= 0 - 30$ kJ/mol $= -30$ kJ/mol

E_a' = energy of activated complex – energy of products
$= 40$ kJ/mol $- 30$ kJ/mol
$= 10$ kJ/mol

Review Problems

16. Given: $m = 55$ g
$\Delta T = 94.6°C - 22.4°C = 72.2°C = 72.2$ K
c_p of aluminum = 0.897 J/(g·K)
Unknown: q

$c_p = \dfrac{q}{m \times \Delta T}$

$q = c_p \times m \times \Delta T$
$= (0.897 \text{ J/(g·K)})(55 \text{ g})(72.2 \text{ K})$
$= 3.5 \times 10^3$ J

---

17. Given: $q = 3500$ J
$m = 28.2$ g
$T_i = 20°C = 293$ K
Unknown: T_f

c_p of iron = 0.449 J/(g·K)

$c_p = \dfrac{q}{m \times \Delta T}$

$\Delta T = \dfrac{q}{m \times c_p} = \dfrac{3500 \text{ J}}{(28.2 \text{ g})(0.449 \text{ J/(g·K)})}$

$= 280$ K

$\Delta T = T_f - T_i$

$T_f = \Delta T + T_i$
$= 280$ K $+ 293$ K $= 573$ K

---

21. Unknown: ΔH

a. Given: $CaCO_3(s) \rightarrow CaO(s) + CO_2(g)$

$CaCO_3(s) \rightarrow Ca(s) + C(s) + \frac{3}{2}O_2(g)$	$\Delta H = 1206.9$ kJ/mol
$Ca(s) + \frac{1}{2}O_2(g) \rightarrow CaO(s)$	$\Delta H = -634.9$ kJ/mol
$C(s) + O_2(g) \rightarrow CO_2(g)$	$\Delta H = -393.51$ kJ/mol
$CaCO_3(s) \rightarrow CaO(s) + CO_2(g)$	$\Delta H = 178.5$ kJ/mol

$\Delta H = $ [sum of ΔH_f of products] $-$ [sum of ΔH_f of reactants]

$\Delta H = [\Delta H_f^0 \text{ CaO} + \Delta H_f \text{ CO}_2] - [\Delta H_f^0 \text{ CaCO}_3]$

$= [(-634.9 \text{ kJ/mol}) + (-393.51 \text{ kJ/mol})] - (-1206.9 \text{ kJ/mol})$

$= 178.5$ kJ/mol

b. Given: $Ca(OH)_2(s) \rightarrow CaO(s) + H_2O(g)$

$Ca(OH)_2(s) \rightarrow Ca(s) + O_2(g) + H_2(g)$	$\Delta H = 983.2$ kJ/mol
$Ca(s) + \frac{1}{2}O_2(g) \rightarrow CaO(s)$	$\Delta H = -634.9$ kJ/mol
$H_2(g) + \frac{1}{2}O_2(g) \rightarrow H_2O(g)$	$\Delta H = -241.8$ kJ/mol
$Ca(OH)_2(s) \rightarrow CaO(s) + H_2O(g)$	$\Delta H = 106.5$ kJ/mol

$\Delta H = [\Delta H_f^0 \text{ CaO} + \Delta H_f^0 \text{ H}_2\text{O}] - [\Delta H_f^0 \text{ Ca(OH)}_2]$

$= [(-634.9 \text{ kJ/mol}) + (-241.8 \text{ kJ/mol})] - (-983.2 \text{ kJ/mol})$

$= 106.5$ kJ/mol

c. Given: $Fe_2O_3(s) + 3CO(g) \rightarrow 2Fe(s) + 3CO_2(g)$

$Fe_2O_3(s) \rightarrow 2Fe(s) + \frac{3}{2}O_2(g)$	$\Delta H = 825.5$ kJ/mol
$3CO(g) \rightarrow 3C(s) + \frac{3}{2}O_2(g)$	$\Delta H = 3(110.5) = 331.5$ kJ/mol
$3C(s) + 3O_2(g) \rightarrow 3CO_2(g)$	$\Delta H = 3(-393.5) = -1180.5$ kJ/mol

$Fe_2O_3(s) + 3CO(g) \rightarrow 2Fe(s) + 3CO_2(g) \quad \Delta H = -23.5$ kJ/mol

$\Delta H = [2\Delta H_f^0 \text{ Fe} + 3\Delta H_f^0 \text{ CO}_2] - [\Delta H_f^0 \text{ Fe}_2O_3 + 3\Delta H_f^0 \text{ CO}]$

$= [0 + 3(-393.5 \text{ kJ/mol})] - [(-825.5 \text{ kJ/mol}) + 3(110.5 \text{ kJ/mol})]$

$= -23.5$ kJ/mol

---

22. Unknown: ΔH

a. Reaction: $C_2H_6(g) + \frac{7}{2}O_2(g) \rightarrow 2CO_2(g) + 3H_2O(l)$

$C_2H_6(g) \rightarrow 2C(s) + 3H_2(g)$	$\Delta H = 83.8$ kJ/mol
$2C(s) + 2O_2(g) \rightarrow 2CO_2(g)$	$\Delta H = 2(-393.5 \text{ kJ/mol})$
	$= -787$ kJ/mol
$3H_2(g) + \frac{3}{2}O_2(g) \rightarrow 3H_2O(l)$	$\Delta H = 3(-285.8 \text{ kJ/mol})$
	$= -857.4$ kJ/mol

$C_2H_6(g) + \frac{7}{2}O_2(g) \rightarrow 2CO_2(g) + 3H_2O(l) \quad \Delta H = -1560.6$ kJ/mol

$\Delta H = [2\Delta H_f^0 \text{ CO}_2 + 3\Delta H_f^0 \text{ H}_2\text{O}] - [\Delta H_f^0 \text{ C}_2\text{H}_6 + \frac{7}{2}\Delta H_f^0 \text{ O}_2]$

$= [2(-393.5 \text{ kJ/mol}) + 3(-285.8 \text{ kJ/mol})] - [(-83.8 \text{ kJ/mol}) + 0]$

$= -1560.6$ kJ/mol

b. Reaction: $C_6H_6(l) + \frac{15}{2}O_2(g) \rightarrow 6CO_2(g) + 3H_2O(l)$

$C_6H_6(l) \rightarrow 6C(s) + 3H_2(g)$	$\Delta H = -49.080$ kJ/mol
$6C(s) + 6O_2(g) \rightarrow 6CO_2(g)$	$\Delta H = 6(-393.5 \text{ kJ/mol})$
	$= -2361$ kJ/mol
$3H_2(g) + \frac{3}{2}O_2(g) \rightarrow 3H_2O(l)$	$\Delta H = 3(-285.8 \text{ kJ/mol})$
	$= -857.4$ kJ/mol

$C_6H_6(l) + \frac{15}{2}O_2(g) \rightarrow 6CO_2(g) + 3H_2O(l) \quad \Delta H = -3267.5$ kJ/mol

$\Delta H = [6\Delta H_f^0 \text{ CO}_2 + 3\Delta H_f^0 \text{ H}_2\text{O}] - [\Delta H_f^0 \text{ C}_6\text{H}_6 + \frac{15}{2}\Delta H_f^0 \text{ O}_2]$

$= [6(-393.5 \text{ kJ/mol}) + 3(-285.8 \text{ kJ/mol})] - [(+49.08 \text{ kJ/mol} - 0)]$

$= -3267.5$ kJ/mol

---

23. Given: ΔH_f^0 of $C_2H_5OH = -277.0$ kJ/mol

Unknown: ΔH_c^0 of C_2H_5OH

Balanced equation:

$C_2H_5OH(l) + 3O_2(g) \rightarrow 2CO_2(g) + 3H_2O(l)$

$C_2H_5OH \rightarrow 2C + 3H_2 + \frac{1}{2}O_2$	$\Delta H = 277.0$ kJ/mol
$2C + 2O_2 \rightarrow 2CO_2$	$\Delta H = 2(-393.5 \text{ kJ/mol}) = -787.0$ kJ/mol
$3H_2 + \frac{3}{2}O_2 \rightarrow 3H_2O$	$\Delta H = 3(-285.8 \text{ kJ/mol}) = -857.4$ kJ/mol

$C_2H_5OH + 3O_2 \rightarrow 2CO_2 + 3H_2O \quad \Delta H = -1367.4$ kJ/mol

24. Unknown: ΔG; whether reaction will occur spontaneously

    a. Given: $\Delta H = +125$ kJ/mol

$T = 293$ K

$\Delta S = 0.0350$ kJ/(mol·K)

$\Delta G = \Delta H - T\Delta S$

        $= (125 \text{ kJ/mol}) - (293 \text{ K})(0.0350 \text{ kJ/(mol·K)})$

        $= +115$ kJ/mol; not spontaneous

    b. Given: $\Delta H = -85.2$ kJ/mol

$T = 127°C = 400.$ K

$\Delta S = 0.125$ kJ/(mol·K)

$\Delta G = \Delta H - T\Delta S$

        $= (-85.2 \text{ kJ/mol}) - (400. \text{ K})(0.125 \text{ kJ/(mol·K)})$

        $= -135$ kJ/mol; spontaneous

    c. Given: $\Delta H = -275$ kJ/mol

$T = 773$ K

$\Delta S = 0.450$ kJ/(mol·K)

$\Delta G = \Delta H - T\Delta S$

        $= (-275 \text{ kJ/mol}) - (773 \text{ K})(0.450 \text{ kJ/(mol·K)})$

        $= -623$ kJ/mol; spontaneous

---

25. Given: $C(s) + O_2(g) \rightarrow CO_2(g) + 393.51$ kJ
$\Delta S^0 = 0.003\ 00$ kJ/(mol·K)
$T = 298.15$ K

Unknown: ΔG^0; whether reaction will occur spontaneously

$\Delta G = \Delta H - T\Delta S$

$\Delta H = -393.51$ kJ/mol

$\Delta G = (-393.51 \text{ kJ/mol}) - (298.15 \text{ K})(0.003\ 00 \text{ kJ/(mol·K)})$

    $= -394$ kJ/mol; spontaneous

---

26. Unknown: $\Delta E_{forward}$, $\Delta E_{reverse}$, E_a, E_a'

$\Delta E_{forward}$ = energy of products − energy of reactants

$\Delta E_{reverse}$ = energy of reactants − energy of products

E_a = energy of activated complex − energy of reactants

E_a' = energy of activated complex − energy of products

    a. $\Delta E_{forward} = +60 \text{ kJ/mol} - (-20 \text{ kJ/mol}) = +80$ kJ/mol

$\Delta E_{reverse} = -20 \text{ kJ/mol} - 60 \text{ kJ/mol} = -80$ kJ/mol

$E_a = 80 \text{ kJ/mol} - (-20 \text{ kJ/mol}) = 100$ kJ/mol

$E_a' = 80 \text{ kJ/mol} - 60 \text{ kJ/mol} = 20$ kJ/mol

b. $\Delta E_{forward} = -40$ kJ/mol $- 0$ kJ/mol $= -40$ kJ/mol

$\Delta E_{reverse} = 0$ kJ/mol $- (-40$ kJ/mol$) = +40$ kJ/mol

$E_a = 20$ kJ/mol $- 0$ kJ/mol $= 20$ kJ/mol

$E_a' = 20$ kJ/mol $- (-40$ kJ/mol$) = 60$ kJ/mol

c. $\Delta E_{forward} = 10$ kJ/mol $- 0$ kJ/mol $= +10$ kJ/mol

$\Delta E_{reverse} = 0$ kJ/mol $- 10$ kJ/mol $= -10$ kJ/mol

$E_a = 70$ kJ/mol $- 0$ kJ/mol $= 70$ kJ/mol

$E_a' = 70$ kJ/mol $- 10$ kJ/mol $= 60$ kJ/mol

---

27. a. Given: $\Delta E_{forward} = -10$ kJ/mol
$E_a' = 40$ kJ/mol

Unknown: $\Delta E_{reverse}, E_a$

$\Delta E_{reverse}$ = energy of reactants − energy of products

$= 0$ kJ/mol $- (-10$ kJ/mol$) = +10$ kJ/mol

E_a = energy of activated complex − energy of reactants

$= 30$ kJ/mol $- 0$ kJ/mol $= 130$ kJ/mol

b. Given: $\Delta E_{forward} = -95$ kJ/mol
$E_a = 20$ kJ/mol

Unknown: $\Delta E_{reverse}, E_a'$

$\Delta E_{reverse}$ = energy of reactants − energy of products

$= 0 - (-95$ kJ/mol$) = +95$ kJ/mol

E_a' = energy of activated complex − energy of products

$= 20$ kJ/mol $- (-95$ kJ/mol$) = +115$ kJ/mol

c. Given: $\Delta E_{reverse} = -40$ kJ/mol
$E_a' = 30$ kJ/mol

Unknown: $\Delta E_{forward}, E_a$

$\Delta E_{forward}$ = energy of products − energy of reactants

$= 40$ kJ/mol $- 0 = +40$ kJ/mol

E_a = energy of activated complex − energy of reactants

$= 70$ kJ/mol $- 0 = +70$ kJ/mol

---

28. Given: Step 1: $B_2 + B_2 \rightarrow E_3 + D$, slow
Step 2: $E_3 + A \rightarrow B_2 + C_2$, fast

Unknown: balanced equation, rate law

$A + B_2 \rightarrow C_2 + D$

$R = k[B_2]^2$ (from the 2 molecules of B_2 in the rate-determining step)

---

29. Given: $2A + B \rightarrow A_2B$

Unknown: (a) rate law, (b) effect of doubling the concentration of either reactant on reaction rate

a. $R = k[A]^2[B]$

b. If [A] is doubled; the rate will increase fourfold: $R = k[2A]^2[B]$

If [B] is doubled, the rate will double: $R = k[A]^2[2B]$

30. Given: A + 2B → C
(See data table, p. 549 for experimental data.)

a. Unknown: rate law

$R = k[A][B]^2$

b. Unknown: value of k (specific rate constant)

$R = k[A][B]^2$

$k = \dfrac{R}{[A][B]^2}$

$= \dfrac{2.0 \times 10^{-4} \text{ M/min}}{(0.20 \text{ M})(0.20 \text{ M})^2} = 2.5 \times 10^{-2} \text{ min}^{-1} \text{ M}^{-2}$

c. Unknown: initial rate that C is formed at, if initial concentrations of A and B = 0.30 M

$R = k[A][B]^2$

$= (2.5 \times 10^{-2} \text{ min}^{-1} \text{ M}^{-2})(0.30 \text{ M})(0.30 \text{ M})^2$

$= 6.8 \times 10^{-4} \text{ M/min}$

Mixed Review

31. Given: Temperature = 300 K
$\Delta H = -74.8$ kJ/mol
$\Delta S = -0.809$ kJ/(mol·K)

Unknown: if reaction will occur spontaneously

$\Delta G = \Delta H - T\Delta S$

$= (-74.8 \text{ kJ/mol}) - (300 \text{ K})(-0.0809 \text{ kJ/(mol·K)})$

$= -50.5$ kJ/mol

yes; it will occur spontaneously

33. Given: ΔH_f^0 for SO_2 = -0.2968 kJ/(mol·K)

Unknown: ΔH_f^0 for 30.0 g SO_2

$(30.0 \text{ g } SO_2)\left(\dfrac{\text{mol } SO_2}{64.1 \text{ g } SO_2}\right) = 0.468$ mol SO_2

$(0.468 \text{ mol } SO_2)(0.2968 \text{ kJ/mol}) = 0.139$ kJ

34. Given: $T = 298.15$ K
$Fe_2O_3 + 2Al(s) \rightarrow 2Fe(s) + Al_2O_3(s)$
$\Delta H^0 = -851.5$ kJ/mol
$\Delta S^0 = -0.0385$ kJ/(mol·K)

Unknown: ΔG at 448 K

$\Delta G = \Delta H - T\Delta S$

$= (-851.5 \text{ kJ/mol}) - (448 \text{ K})(-0.0385 \text{ kJ/(mol·K)})$

$= -834$ kJ/mol

35a. Given: $\Delta E = +30$ kJ/mol
$E_a' = 20$ kJ/mol
energy of reactants = 0

Unknown: E_a

E_a = energy of activated complex − energy of reactants

= 50 kJ/mol − 0 = +50 kJ/mol

b. Given: $\Delta E = -30$ kJ/mol
$E_a = 20$ kJ/mol

Unknown: E_a'

E_a' = energy of activated complex − energy of products

= 20 kJ/mol − (−30 kJ/mol) = +50 kJ/mol

---

36. Given: $R = k[A][B]^2$

a. Unknown: effect on R if [A] is cut in half

$R = k\frac{1}{2}[A][B]^2$

R is reduced by $\frac{1}{2}$.

b. Unknown: effect on R if [B] is tripled

$R = k[A][3B]^2$

R is increased by a factor of 9. (3^2)

c. Unknown: effect on R if [A] is doubled, but [B] is cut in half

$R = k[2A]\frac{1}{2}[B]^2$

R is reduced by $\frac{1}{2}$. $\left((2)\left(\frac{1}{2}\right)^2 = (2)\left(\frac{1}{4}\right) = \frac{1}{2}\right)$

---

39. Given: $4FeO(s) + O_2(g) \rightarrow 2Fe_2O_3(s)$

(4)FeO → (4)Fe + (4)O

$\Delta H = 4(272.0$ kJ/mol$)$
= 1088.0 kJ/mol

$(2)2Fe + (2)\frac{3}{2}O_2 \rightarrow (2)Fe_2O_3$

$\Delta H = 2(-824.2$ kJ/mol$)$
= −1648.4 kJ/mol

$4FeO + O_2 \rightarrow 2Fe_2O_3$

$\Delta H = -560.4$ kJ/mol

CHAPTER 18
Chemical Equilibrium

Practice, p. 559

1. Given: $[N_2] = 0.602$ mol/L
$[H_2] = 0.420$ mol/L
$[NH_3] = 0.113$ mol/L
$N_2(g) + 3H_2(g) \rightleftharpoons 2NH_3(g)$

Unknown: K

$K = \dfrac{[NH_3]^2}{[N_2][H_2]^3}$

$= \dfrac{(0.113 \text{ mol/L})^2}{(0.602 \text{ mol/L})(0.420 \text{ mol/L})^3}$

$= 0.28$

---

2. Given: $AB_2C = 0.084$ mol
$B_2 = 0.035$ mol
$AC = 0.059$ mol
volume = 5.00 L
$AB_2C(g) \rightleftharpoons B_2(g) + AC(g)$

Unknown: K

$K = \dfrac{[B_2][AC]}{[AB_2C]}$

$= \dfrac{(0.035 \text{ mol/5 L})(0.059 \text{ mol/5 L})}{(0.084 \text{ mol/5 L})}$

$= 4.9 \times 10^{-3}$

---

3. Given: volume = 1.0 L
$[H_2] = 20.0$ mol/L
$[CO_2] = 18.0$ mol/L
$[H_2O] = 12.0$ mol/L
$[CO] = 5.9$ mol/L
$CO_2(g) + H_2(g) \rightleftharpoons CO(g) + H_2O(g)$

Unknown: K

$K = \dfrac{[CO][H_2O]}{[CO_2][H_2]}$

$= \dfrac{(5.9 \text{ mol/L})(12.0 \text{ mol/L})}{(18.0 \text{ mol/L})(20.0 \text{ mol/L})}$

$= 0.20$

---

4. Given: $[SO_2] = 1.50$ mol/L
$[O_2] = 1.25$ mol/L
$[SO_3] = 3.50$ mol/L

Unknown: K

$2SO_2(g) + O_2(g) \rightleftharpoons 2SO_3(g)$

$K = \dfrac{[SO_3]^2}{[SO_2]^2[O_2]}$

$= \dfrac{(3.50 \text{ mol/L})^2}{(1.50 \text{ mol/L})^2(1.25 \text{ mol/L})}$

$= 4.38$

ATE, Additional Sample Problems 18-1, p. 558

1. Given: $[PCl_3] = 6.4 \times 10^{-3}$ mol/L

$[Cl_2] = 2.5 \times 10^{-2}$ mol/L

$[PCl_5] = 4.0 \times 10^{-3}$ mol/L

$PCl_5(g) \rightleftharpoons PCl_3(g) + C_2(g)$

Unknown: K

$K = \dfrac{[PCl_3][Cl_2]}{[PCl_5]}$

$= \dfrac{(6.4 \times 10^{-3} \text{ mol/L})(2.5 \times 10^{-2} \text{ mol/L})}{(4.0 \times 10^{-3} \text{ mol/L})}$

$= 4.0 \times 10^{-2}$

2. Given: volume = 2.0 L

$[H_2] = 0.36$ mol

$[Br_2] = 0.11$ mol

$[HBr] = 37$ mol

$H_2(g) + Br_2(g) \rightleftharpoons 2HBr(g)$

Unknown: K

$K = \dfrac{[HBr]^2}{[H_2][Br_2]}$

$= \dfrac{(37 \text{ mol/2 L})^2}{(0.36 \text{ mol/2 L})(0.11 \text{ mol/2 L})}$

$= 3.5 \times 10^4$

Section Review, p. 559

7. Given: $[HCl] = 0.0625$ mol/L

$[H_2] = 0.0045$ mol/L

$[Cl_2] = 0.0045$ mol/L

$H_2(g) + Cl_2(g) \rightleftharpoons 2HCl(g)$

Unknown: K

$K = \dfrac{[HCl]^2}{[H_2][Cl_2]}$

$= \dfrac{(0.0625 \text{ mol/L})^2}{(0.0045 \text{ mol/L})(0.0045 \text{ mol/L})}$

$= 190$

8. Given: $[H_2] = 1.83 \times 10^{-3}$ mol/L

$[I_2] = 3.13 \times 10^{-3}$ mol/L

$[HI] = 1.77 \times 10^{-2}$ mol/L

$H_2(g) + I_2(g) \rightleftharpoons 2HI(g)$

Unknown: K

$K = \dfrac{[HI]^2}{[H_2][I_2]}$

$= \dfrac{(1.77 \times 10^{-2} \text{ mol/L})^2}{(1.83 \times 10^{-3} \text{ mol/L})(3.13 \times 10^{-3} \text{ mol/L})}$

$= 54.6$

9. Given: $[H_2] = 4.79 \times 10^{-4}$ mol/L

$[I_2] = 4.79 \times 10^{-4}$ mol/L

$K = 54.3$

$H_2(g) + I_2(g) \rightleftharpoons 2HI(g)$

Unknown: $[HI]$

$K = \dfrac{[HI]^2}{[H_2][I_2]}$

$[HI] = \sqrt{[H_2][I_2]K}$

$= \sqrt{(4.79 \times 10^{-4}\ \text{mol/L})(4.79 \times 10^{-4}\ \text{mol/L})(54.3)}$

$= 3.54 \times 10^{-3}$

Practice p. 580

1. Given: solubility of $PbCl_2 = 1.0$ g/100. g H_2O

Unknown: K_{sp}

solubility $= \left(\dfrac{1.0\ \text{g } PbCl_2}{100\ \text{g } H_2O}\right)\left(\dfrac{1\ \text{g } H_2O}{1\ \text{mL}}\right)\left(\dfrac{1000\ \text{mL}}{L}\right)\left(\dfrac{1\ \text{mol } PbCl_2}{278\ \text{g } PbCl_2}\right)$

$= 3.6 \times 10^{-2}$ mol/L $PbCl_2$

$PbCl_2(s) \rightleftharpoons Pb^{2+}(aq) + 2Cl^-(aq)$

$[Pb^{2+}] = 3.6 \times 10^{-2}$

$[Cl^-] = 2(3.6 \times 10^{-2}) = 7.2 \times 10^{-2}$

$K_{sp} = [Pb^{2+}][Cl^-]^2$

$K_{sp} = (3.6 \times 10^{-2})(7.2 \times 10^{-2})^2$

$= 1.9 \times 10^{-4}$

2. Given: solubility of $Ag_2SO_4 = 5$ g/1 L H_2O

Unknown: K_{sp}

solubility $= \left(\dfrac{5\ \text{g } Ag_2SO_4}{L}\right)\left(\dfrac{\text{mol } Ag_2SO_4}{311.77\ \text{g } Ag_2SO_4}\right)$

$= 0.016$ mol/L Ag_2SO_4

$Ag_2SO_4(s) \rightleftharpoons 2Ag^+(aq) + SO_4^{2-}(aq)$

$[Ag^+] = 2(0.016) = 0.032$

$[SO_4^-] = 0.016$

$K_{sp} = [Ag^+]^2[SO_4^-]$

$= 2 \times 10^{-5}$

ATE, Additional Sample Problems 18-2, p. 580

1. Given: solubility of $SnS = 5.2 \times 10^{-12}$ g/100 g H_2O

Unknown: K_{sp}

solubility $= \left(\dfrac{5.2 \times 10^{-12}\ \text{g SnS}}{100\ \text{g } H_2O}\right)\left(\dfrac{1\ \text{g}H_2O}{1\ \text{mL}}\right)\left(\dfrac{1000\ \text{mL}}{L}\right)\left(\dfrac{\text{mol SnS}}{150.78\ \text{g SnS}}\right)$

$= 3.4 \times 10^{-13}$ mol/L Sns

$SnS(s) \rightleftharpoons Sn^2(aq) + S^{2-}(aq)$

$[Sn^{2+}] = 3.4 \times 10^{-13}$

$[S^{2-}] = 3.4 \times 10^{-13}$

$K_{sp} = [Sn^{2+}][S^{2-}]$

$= (3.4 \times 10^{-13})(3.4 \times 10^{-13})$

$= 1.2 \times 10^{-25}$

2. Given: solubility of
$CaCO_3 = 5.3 \times 10^{-5}$ g/L H$_2$O

Unknown: K_{sp}

$$\text{solubility} = \left(\frac{5.3 \times 10^{-5} \text{ g CaCO}_3}{L}\right)\left(\frac{\text{mol CaCO}_3}{100.06 \text{ g CaCO}_3}\right)$$

$= 5.3 \times 10^{-7}$ mol/L CaCO$_3$

$CaCO_3(s) \rightleftharpoons Ca^{2+}(aq) + CO_3^{2-}(aq)$

$[Ca^{2+}] = [CO_3^{2-}] = 5.3 \times 10^{-7}$

$K_{sp} = [Ca^{2+}][CO_3^{2-}]$

$= (5.3 \times 10^{-7})(5.3 \times 10^{-7})$

$= 2.8 \times 10^{-13}$

Practice, p. 582

1. Given: K_{sp} of CdS = 8.0×10^{-27}

Unknown: solubility of CdS in mol/L

$CdS(s) \rightleftharpoons Cd^{2+}(aq) + S^{2-}(aq)$

$K_{sp} = [Cd^{2+}][S^{2-}]$

$[Cd^{2+}] = [S^{2-}] = x$

$K_{sp} = x^2$

$x^2 = 8.0 \times 10^{-27}$

$x = 8.9 \times 10^{-14}$ mol/L = solubility of CdS

2. Given: K_{sp} of SrSO$_4$ = 3.2×10^{-7}

Unknown: $[Sr^{2+}]$

$SrSO_4(s) \rightleftharpoons Sr^{2+}(aq) + SO_4^{2-}(aq)$

$K_{sp} = [Sr^{2+}][SO_4^{2-}]$

$[Sr^{2+}] = [SO_4^{2-}]$

$K_{sp} = x^2 = 3.2 \times 10^{-7}$

$x = \sqrt{3.2 \times 10^{-7}}$

$= 5.7 \times 10^{-4}$ mol/L = $[Sr^{2+}]$

ATE, Additional Sample Problems 18-3, p. 582

1. Given: K_{sp} of MnS = 2.5×10^{-13}

Unknown: solubility of MnS in mol/L

$MnS(s) \rightleftharpoons Mn^{2+}(aq) + S^{2-}(aq)$

$K_{sp} = [Mn^{2+}][S^{2-}]$

$[Mn^{2+}] = [S^{2-}]$

$K_{sp} = x^2$

$x^2 = 2.5 \times 10^{-13}$

$x = 5.0 \times 10^{-7}$ mol/L = solubility of MnS

2. Given: K_{sp} of ZnS = 1.6×10^{-24}

Unknown: $[Zn^2]$

$ZnS(s) \rightleftharpoons Zn^{2+}(aq) + S^{2-}(aq)$

$K_{sp} = [Zn^{2+}][S^{2-}]$

$[Zn^{2+}] = [S^{2-}]$

$K_{sp} = x^2 = 1.6 \times 10^{-24}$

$x = 1.3 \times 10^{-12}$ mol/L = $[Zn^{2+}]$

Practice, p. 584

1. Given: $V\,AgNO_3 = 0.1\text{ L}$

$[AgNO_3] = 0.0025\text{ M}$

$V\,NaBr = 0.150\text{ L}$

$[NaBr] = 0.0020\text{ M}$

Unknown: whether a precipitate forms

$AgNO_3 + NaBr \rightarrow AgBr + NaNO_3$

$AgBr(s) \rightleftharpoons Ag^+(aq) + Br^-(aq)$

$K_{sp} = [Ag^+][Br^-] = 5.0 \times 10^{-13}$

$(0.150\text{ L})\left(\dfrac{0.0020\text{ mol Ag}^+}{\text{L}}\right) = 3.0 \times 10^{-4}\text{ mol Ag}^+$

$(0.100\text{ L})\left(\dfrac{0.0025\text{ mol Br}^-}{\text{L}}\right) = 2.5 \times 10^{-4}\text{ mol Br}^-$

Total volume = $0.150\text{ L} + 0.100\text{ L} = 0.250\text{ L}$

$\dfrac{3.0 \times 10^{-4}\text{ mol Ag}^+}{0.250\text{ L}} = 1.2 \times 10^{-3}\text{ mol/L Ag}^+$

$\dfrac{2.5 \times 10^{-4}\text{ mol Br}^-}{0.250\text{ L}} = 1.0 \times 10^{-3}\text{ mol/L Br}^-$

$[Ag^+][Br^-] = (1.2 \times 10^{-3})(1.0 \times 10^{-3}) = 1.2 \times 10^{-6}$

$1.2 \times 10^{-6} > K_{sp}$; AgBr precipitates

2. Given: $V\,Pb(NO_3)_2 = 0.020\text{ L}$

$[Pb(NO_3)_2] = 0.038\text{ M}$

$V\,KCl = 0.030\text{ L}$

$[KCl] = 0.018\text{ M}$

Unknown: whether a precipitate forms

$Pb(NO_3)_2 + 2KCl \rightarrow PbCl_2 + 2KNO_3$

$PbCl_2(s) \rightleftharpoons Pb^{2+}(aq) + 2Cl^-(aq)$

$K_{sp} = [Pb^{2+}][Cl^-]^2 = 1.6 \times 10^{-5}$

$(0.020\text{ L})\left(\dfrac{0.038\text{ mol Pb}^{2+}}{\text{L}}\right) = 7.6 \times 10^{-4}\text{ mol Pb}^{2+}$

$(0.030\text{ L})\left(\dfrac{0.018\text{ mol Cl}^-}{\text{L}}\right) = 5.4 \times 10^{-4}\text{ mol Cl}^-$

Total volume = $0.020\text{ L} + 0.030\text{ L} = 0.050\text{ L}$

$\dfrac{7.6 \times 10^{-4}\text{ mol Pb}^{2+}}{0.050\text{ L}} = 1.5 \times 10^{-2}\text{ mol/L Pb}^{2+}$

$\dfrac{5.4 \times 10^{-4}\text{ mol Cl}^-}{0.050\text{ L}} = 1.1 \times 10^{-2}\text{ mol/L Cl}^-$

$[Pb^{2+}][Cl^-]^2 = (1.5 \times 10^{-2})(1.1 \times 10^{-2})^2$

$\qquad\qquad\qquad = 1.8 \times 10^{-6}$

$1.8 \times 10^{-6} < K_{sp}$; $PbCl_2$ does not precipitate.

ATE, Additional Sample Problems 18-4, p. 584

1. Given: V NaCl = 0.02 L
[NaCl] = 0.034 M
V CuNO$_3$ = 0.15 L
[CuNO$_3$] = 0.083 M

Unknown: whether a precipitate forms

NaCl + CuNO$_3$ → NaNO$_3$ + CuCl
CuCl(s) ⇌ Cu$^+$(aq) + Cl$^-$(aq)
K_{sp} = [Cu$^+$][Cl$^-$] = 1.2 × 10^{-6}

$(0.15 \text{ L})\left(\dfrac{0.083 \text{ mol Cu}^+}{\text{L}}\right) = 1.2 \times 10^{-2}$ mol Cu$^+$

$(0.02 \text{ L})\left(\dfrac{0.034 \text{ mol Cl}^-}{\text{L}}\right) = 6.8 \times 10^{-4}$ mol Cl$^-$

total volume = 0.02 L + 0.15 L = 0.17 L

$\dfrac{1.2 \times 10^{-2} \text{ mol Cu}^+}{0.17 \text{ L}} = 7.0 \times 10^{-2}$ mol/L Cu$^+$

$\dfrac{6.8 \times 10^{-4} \text{ mol Cl}^-}{0.17 \text{ L}} = 4.0 \times 10^{-3}$ mol/L Cl$^-$

[Cu$^+$][Cl$^-$] = (7.0 × 10^{-2})(4.0 × 10^{-3}) = 2.8 × 10^{-4}

2.8 × 10^{-4} > K_{sp}; CuCl precipitates

2. Given: V Ca(NO$_3$)$_2$ = 0.1 L
[Ca(NO$_3$)$_2$] = 0.0014 M
V Na$_2$SO$_4$ = 0.2 L
[Na$_2$SO$_4$] = 0.000 20 M

Unknown: whether a precipitate forms

Ca(NO$_3$)$_2$ + Na$_2$SO$_4$ → CaSO$_4$ + NaNO$_3$
CaSO$_4$(s) ⇌ Ca^{2+}(aq) + SO$_4^{2-}$(aq)
K_{sp} = [Ca^{2+}][SO$_4^{2-}$] = 9.1 × 10^{-6}

$(0.1 \text{ L})\left(\dfrac{0.0014 \text{ mol Ca}^+}{\text{L}}\right) = 1.4 \times 10^{-4}$ mol Ca^{2+}

$(0.2 \text{ L})\left(\dfrac{0.0020 \text{ mol SO}_4^{2-}}{\text{L}}\right) = 4.0 \times 10^{-5}$ mol SO$_4^{2-}$

Total volume = 0.1 L + 0.2 L = 0.3 L

$\dfrac{1.4 \times 10^{-4} \text{ mol Ca}^{2+}}{0.3 \text{ L}} = 4.7 \times 10^{-4}$ mol/L Ca^{2+}

$\dfrac{4.0 \times 10^{-5} \text{ mol SO}_4^{2-}}{0.3 \text{ L}} = 1.3 \times 10^{-4}$ mol/L SO$_4^{2-}$

[Ca^{2+}][SO$_4^{2-}$] = (4.7 × 10^{-4})(1.3 × 10^{-4})
$= 6.1 \times 10^{-8}$

6.1 × 10^{-8} < 9.1 × 10^{-6}; CaSO$_4$ does not precipitate.

Section Review, p. 584

5. Given: solubility of Ag$_2$SO$_4$ = 5.40 g/1.00 L H$_2$O

Unknown: K_{sp}

solubility = $\left(\dfrac{5.4 \text{ g Ag}_2\text{SO}_4}{\text{L}}\right)\left(\dfrac{\text{mol Ag}_2\text{SO}_4}{311.75 \text{ g Ag}_2\text{SO}_4}\right)$
= 0.0173 mol/L Ag$_2$SO$_4$

Ag$_2$SO$_4$(s) ⇌ 2 Ag$^+$(aq) + SO$_4^{2-}$(aq)

[Ag$^+$] = 2(0.0173) = 0.0346 mol/L

[SO$_4^{2-}$] = 0.0173 mol/L

K_{sp} = [Ag$^+$]^2[SO$_4^{2-}$]
= (0.0346)^2(0.0173)
= 2.07 × 10^{-5}

6. Given: $V\,AgNO_3 = 0.02$ L

[AgNO$_3$] = 1.00×10^{-7} M

V NaCl = 0.02 L

[NaCl] = 2.00×10^{-9} M

Unknown: whether a precipitate forms

$AgNO_3 + NaCl \rightarrow AgCl + NaNO_3$

$AgCl(s) \rightleftharpoons Ag^+(aq) + Cl^-(aq)$

$K_{sp} = [Na^+][Cl^-] = 1.8 \times 10^{-10}$

$(0.02\,L)\left(\dfrac{1.00 \times 10^{-7}\,mol\,Ag^+}{L}\right) = 2.0 \times 10^{-9}\,mol\,Ag^+$

$(0.02\,L)\left(\dfrac{2.00 \times 10^{-9}\,mol\,Cl^-}{L}\right) = 4.0 \times 10^{-11}\,mol\,Cl^-$

Total volume = 0.02 L + 0.02 L = 0.04 L

$\dfrac{2.0 \times 10^{-9}\,mol\,Ag^+}{0.04\,L} = 5.0 \times 10^{-8}\,mol/L\,Ag^+$

$\dfrac{4.0 \times 10^{-11}\,mol\,Cl^-}{0.04\,L} = 1.0 \times 10^{-9}\,mol/L\,Cl^-$

$[Ag^+][Cl^-] = (5.0 \times 10^{-8})(1.0 \times 10^{-9})$

$= 5.0 \times 10^{-17}$

$5.0 \times 10^{-17} < K_{sp}$; AgCl does not precipitate.

Review Problems

24. Unknown: K

$K = \dfrac{[C]}{[A][B]}$

a. Given: [A] = 2.0
[B] = 3.0
[C] = 4.0

A + B $\rightleftharpoons$ C

$= \dfrac{4.0}{(2.0)(3.0)} = 0.67$

b. Given: [D] = 1.5
[E] = 2.0
[F] = 1.8
[G] = 1.2

D + 2E $\rightleftharpoons$ F + 3G

$K = \dfrac{[F][G]^3}{[D][E]^2} = \dfrac{(1.8)(1.2)^3}{(1.5)(2.0)^2} = 0.52$

c. Given: [N$_2$] = 0.45
[H$_2$] = 0.14
[NH$_3$] = 0.62

N$_2$(g) + 3H$_2$(g) $\rightleftharpoons$ 2NH$_3$(g)

$K = \dfrac{[NH_3]^2}{[N_2][H_2]^3} = \dfrac{(0.62)^2}{(0.45)(0.14)^3} = 310$

25. Given: [HCl] = 1.2×10^{-3} mol/L

[O$_2$] = 3.8×10^{-4} mol/L

[H$_2$O] = 5.8×10^{-2} mol/L

[Cl$_2$] = 5.8×10^{-2} mol/L

4HCl(g) + O$_2$(g) $\rightleftharpoons$ 2H$_2$O(g) + 2Cl$_2$(g)

Unknown: K

$K = \dfrac{[H_2O]^2[Cl_2]^2}{[HCl]^4[O_2]}$

$= \dfrac{(5.8 \times 10^{-2})^2(5.8 \times 10^{-2})^2}{(1.2 \times 10^{-3})^4(3.8 \times 10^{-4})}$

$= 1.5 \times 10^{10}$

26. Given: $K = 6.59 \times 10^{-3}$

$[NH_3] = 1.23 \times 10^{-4}$ M

$[H_2] = 2.75 \times 10^{-3}$ M

$N_2(g) + 3H_2(g) \rightleftharpoons 2NH_3(g)$

Unknown: $[N_2]$

$K = \dfrac{[NH_3]^2}{[N_2][H_2]^3}$

$[N_2] = \dfrac{[NH_3]^2}{K[H_2]^3}$

$= \dfrac{(1.23 \times 10^{-4}\ M)^2}{(6.59 \times 10^{-3}\ M)(2.75 \times 10^{-3}\ M)^3}$

$= 110$

---

27. Given: $K = 40.0$

$H_2(g) + I_2(g) \rightleftharpoons 2HI(g)$

Unknown: K for reverse reaction

$2HI(g) \rightleftharpoons H_2(g) + I_2(g)$

K for forward reaction $= 40.0 = \dfrac{[HI]^2}{[H_2][I_2]}$

K for reverse reaction $= \dfrac{[H_2][I_2]}{[HI]^2} = \dfrac{1}{40.0}$

$= 0.0250$

---

28. Given: solubility of EJ = 8.45×10^{-6} mol/L

Unknown: K_{sp}

$EJ \rightleftharpoons E^{2+} + J^{2-}$

$[E^{2+}] = 8.45 \times 10^{-6} = [J^{2-}]$

$K_{sp} = [E^{2+}][J^{2-}]$

$= (8.45 \times 10^{-6})(8.45 \times 10^{-6})$

$= 7.14 \times 10^{-11}$

---

29. Unknown: K_{sp}

a. Given: solubility of $BaSO_4 = 2.4 \times 10^{-4}$ g/100. g H_2O

solubility $= \left(\dfrac{2.4 \times 10^{-4}\ g\ BaSO_4}{100\ g\ H_2O}\right)\left(\dfrac{1\ g\ H_2O}{1\ mL}\right)\left(\dfrac{1000\ mL}{L}\right)\left(\dfrac{1\ mol\ BaSO_4}{233.36\ g\ BaSO_4}\right)$

$= 1.0 \times 10^{-5}$ mol/L $BaSO_4$

$BaSO_4(s) \rightleftharpoons Ba^{2+}(aq) + SO_4^{2-}(aq)$

$[Ba^{2+}] = 1.0 \times 10^{-5} = [SO_4^{2-}]$

$K_{sp} = [Ba^{2+}][SO_4^{2-}]$

$= (1.0 \times 10^{-5})(1.0 \times 10^{-5})$

$= 1.0 \times 10^{-10}$

b. Given: solubility of $Ca(OH)_2 = 0.173$ g/100. g H_2O

solubility $= \left(\dfrac{0.173\ g\ Ca(OH)_2}{100\ g\ H_2O}\right)\left(\dfrac{1\ g\ H_2O}{1\ mL}\right)\left(\dfrac{1000\ mL}{L}\right)\left(\dfrac{1\ mol\ Ca(OH)_2}{74.1\ g\ Ca(OH)_2}\right)$

$= 0.0233$ mol/L $Ca(OH)_2$

$Ca(OH)_2(s) \rightleftharpoons Ca^{2+}(aq) + 2OH^-(aq)$

$[Ca^{2+}] = 0.0233$

$[OH^-] = 2(0.0233) = 0.0466$

$K_{sp} = [Ca^{2+}][OH^-]^2$

$= (0.0233)(0.0466)^2$

$= 5.06 \times 10^{-5}$

---

30. Given: K_{sp} of $MN = 8.1 \times 10^{-6}$

Unknown: solubility of MN

$MN \rightleftharpoons M^{2+} + N^{2-}$

$K_{sp} = [M^{2+}][N^{2-}]$

$[M^{2+}] = [N^{2-}] = x$

$K_{sp} = x^2 = 8.1 \times 10^{-6}$

$x = 2.8 \times 10^{-3}$ mol/L = solubility of MN

31. a. Given: K_{sp} of AgBr = 5.0×10^{-13}

Unknown: solubility of AgBr

$AgBr(s) \rightarrow Ag^+(aq) + Br^-(aq)$

$K_{sp} = [Ag^+][Br^-]$

$[Ag^+] = [Br^-] = x$

$K_{sp} = x^2 = 5.0 \times 10^{-13}$

$x = 7.1 \times 10^{-7}$ mol/L = solubility of AgBr

b. Given: K_{sp} of CoS = 4.0×10^{-21}

Unknown: solubility of CoS

$CoS(s) \rightleftharpoons Co^{2+}(aq) + S^{2-}(aq)$

$K_{sp} = [Co^{2+}][S^{2-}]$

$[Co^{2+}] = [S^{2-}] = x$

$K_{sp} = x^2 = 4.0 \times 10^{-21}$

$x = 6.3 \times 10^{-11}$ mol/L = solubility of CoS

33. Given: $T_3U_2 \rightarrow T^{2+} + U^{3-}$

solubility of T_3U_2 = 3.77×10^{-20} mol/L

Unknown: K_{sp}

$T_3U_2(s) \rightleftharpoons 3T^{2+}(aq) + 2U^{3-}(aq)$

$K_{sp} = [T^{2+}]^3[U^{3-}]^2$

$[T^{2+}] = 3(3.77 \times 10^{-20}) = 1.13 \times 10^{-19}$

$[U^{3-}] = 2(3.77 \times 10^{-20}) = 7.54 \times 10^{-20}$

$K_{sp} = (1.13 \times 10^{-19})^3 (7.54 \times 10^{-20})^2$

$= (1.44 \times 10^{-57})(5.69 \times 10^{-39})$

$= 8.19 \times 10^{-96}$

34. Given: $[Ag^+] = 2.7 \times 10^{-10}$ mol/L

Unknown: $[I^-]$

$AgI(s) \rightleftharpoons Ag^+(aq) + I^-(aq)$

K_{sp} of AgI = 8.3×10^{-17}

$K_{sp} = [Ag^+][I^-]$

$[I^-] = K_{sp}/[Ag^+]$

$= 8.3 \times 10^{-17}/2.7 \times 10^{-10}$ mol/L

$= 3.1 \times 10^{-7}$ mol/L

35. Given: $V\ Ca(NO_3)_2$ = 0.35 L

$[Ca(NO_3)_2]$ = 0.0044 M

V NaOH = 0.17 L

[NaOH] = 0.000 39 M

Unknown: whether a precipitate forms

$Ca(NO_3)_2 + NaOH \rightarrow Ca(OH)_2 + NaNO_3$

$Ca(OH)_2(s) \rightleftharpoons Ca^{2+}(aq) + 2OH^-(aq)$

$K_{sp} = [Ca^{2+}][OH^-]^2 = 5.5 \times 10^{-6}$

$(0.35\ L)\left(\dfrac{0.0044\ mol\ Ca^{2+}}{L}\right) = 1.5 \times 10^{-3}$ mol Ca^{2+}

$(0.17\ L)\left(\dfrac{0.000\ 39\ mol\ OH^-}{L}\right) = 6.6 \times 10^{-5}$ mol OH^-

Total volume = 0.35 L + 0.17 L = 0.52 L

$\dfrac{1.5 \times 10^{-3}\ mol\ Ca^{2+}}{0.52\ L} = 2.9 \times 10^{-3}$ mol/L Ca^{2+}

$\dfrac{6.6 \times 10^{-5}\ mol\ OH^-}{0.52\ L} = 1.3 \times 10^{-4}$ mol/L OH^-

$[Ca^{2+}][OH^-]^2 = (2.9 \times 10^{-3})(1.3 \times 10^{-4})^2$

$= 4.9 \times 10^{-11}$

$4.9 \times 10^{-11} < K_{sp}$; $Ca(OH)_2$ does not precipitate.

36. Given: solubility of
$AgNO_3 = 1.70$ g/200. mL H_2O

solubility of
$NaCl = 14.5$ g/200 mL H_2O

Unknown: whether a precipitate will form

$AgNO_3 + NaCl \rightarrow AgCl + NaNO_3$

$AgCl(s) \rightleftharpoons Ag^+(aq) + Cl^-(aq)$

$K_{sp} = [Ag^+][Cl^-] = 1.8 \times 10^{-10}$

$(1.70 \text{ g AgNO}_3)\left(\dfrac{1 \text{ mol AgNO}_3}{169.85 \text{ g AgNO}_3}\right) = 0.0100 \text{ mol AgNO}_3 = 0.0100 \text{ mol Ag}^+$

$\left(\dfrac{0.0100 \text{ mol Ag}^+}{200. \text{ mL}}\right)\left(\dfrac{1000 \text{ mL}}{\text{L}}\right) = 0.0500 \text{ mol/L Ag}^+$

$(14.5 \text{ g NaCl})\left(\dfrac{1 \text{ mol NaCl}}{58.44 \text{ g NaCl}}\right) = 0.248 \text{ mol NaCl} = 0.248 \text{ mol Cl}^-$

$\left(\dfrac{0.248 \text{ mol Cl}^-}{200. \text{ mL}}\right)\left(\dfrac{1000 \text{ mL}}{\text{L}}\right) = 1.24 \text{ mol/L Cl}^-$

$[Ag^+][Cl^-] = (0.0500)(1.24) = 6.20 \times 10^{-2}$

$6.20 \times 10^{-2} > K_{sp}$; AgCl precipitates.

37. Given: solubility of
$Fe(NO_3)_3 = 2.5$ g/100. mL NaOH

$[NaOH] = 1.0 \times 10^{-20}$ M

Unknown: whether a precipitate will form

$Fe(NO_3)_3 + NaOH \rightarrow Fe(OH)_3 + NaNO_3$

$Fe(OH)_3(s) \rightleftharpoons Fe^{3+}(aq) + 3OH^-(aq)$

$K_{sp} = [Fe^{3+}][OH^-]^3 = 4 \times 10^{-38}$

$(2.5 \text{ g Fe(NO}_3)_3)\left(\dfrac{1 \text{ mol Fe(NO}_3)_3}{241.79 \text{ g Fe(NO}_3)_3}\right) = 1.03 \times 10^{-2} \text{ mol Fe(NO}_3)_3$

$= 1.03 \times 10^{-2} \text{ mol Fe}^{3+}$

$\left(\dfrac{1.03 \times 10^{-2} \text{ mol Fe}^{3+}}{100 \text{ mL}}\right)\left(\dfrac{1000 \text{ mL}}{\text{L}}\right) = 1.03 \times 10^{-1} \text{ mol/L Fe}^{3+}$

$[NaOH] = 1.0 \times 10^{-20}$ M = $[OH^-]$

$[Fe^{3+}][OH^-]^3 = (1.03 \times 10^{-1})(1.0 \times 10^{-20})^3$

$= 1.03 \times 10^{-61}$

$1.03 \times 10^{-61} < K_{sp}$; $Fe(OH)_3$ does not precipitate.

Mixed Review

39. Given: HgS(s) forms a saturated solution

Unknown: **a.** $[Hg^{2+}]$
b. volume of solution that contains one Hg^{2+} ion

a. $HgS(s) \rightleftharpoons Hg^{2+}(aq) + S^{2-}(aq)$

$K_{sp} = [Hg^{2+}][S^{2-}]$

$[Hg^{2+}] = [S^{2-}] = x$

$K_{sp} = x^2 = 1.6 \times 10^{-52}$

$x = 1.3 \times 10^{-26}$ mol/L = $[Hg^{2+}]$

b. $\left(\dfrac{1 \text{ L}}{1.3 \times 10^{-26} \text{ mol}}\right)\left(\dfrac{1 \text{ mol}}{6.022 \times 10^{23} \text{ ions}}\right) = 1.3 \times 10^2$ L/ion

40. Given: $H_2(g) + CO_2(g) \rightleftharpoons H_2O(g) + CO(g)$

$[H_2] = 0.61$ mol/L
$[CO_2] = 1.6$ mol/L
$[H_2O] = 1.1$ mol/L
$[CO] = 1.4$ mol/L

Unknown: K

$K = \dfrac{[H_2O][CO]}{[H_2][CO_2]}$

$= \dfrac{(1.1 \text{ mol/L})(1.4 \text{ mol/L})}{(0.61 \text{ mol/L})(1.6 \text{ mol/L})}$

$= 1.5$

41. Given: $[Ba^{2+}] = 5 \times 10^{-4}$ M

$K_{sp} = 3.4 \times 10^{-23}$

solute = $Ba_3(PO_4)_2$

Unknown: $[PO_4^{3-}]$

$Ba_3(PO_4)_2(s) \rightleftharpoons 3Ba^{2+}(aq) + 2\,PO_4^{3-}(aq)$

$K_{sp} = [Ba^{2+}]^3[PO_4^{3-}]^2$

$[PO_4^{3-}]^2 = K_{sp}/[Ba^{2+}]^3$

$[PO_4^{3-}] = \sqrt{K_{sp}/[Ba^{2+}]^3}$

$= \sqrt{3.4 \times 10^{-23}/(5 \times 10^{-4})^3}$

$= 5 \times 10^{-7}$ mol/L PO_4^{3-}

---

42. Given: $K = 1.7 \times 10^{-13}$

$N_2O(g) + \tfrac{1}{2}O_2(g) \rightleftharpoons 2NO(g)$

$[N_2O] = 0.0035$ mol/L

$[O_2] = 0.0027$ mol/L

Unknown: [NO]

$K = \dfrac{[NO]^2}{[N_2O][O_2]^{1/2}}$

$[NO]^2 = K\,[N_2O][O_2]^{1/2}$

$[NO] = \sqrt{K\,[N_2O][O_2]^{1/2}}$

$= \sqrt{(1.7 \times 10^{-13})(0.0035)(0.0027)^{1/2}}$

$= \sqrt{3.1 \times 10^{-17}}$

$= 5.6 \times 10^{-9}$

---

43. Given: K_{sp} of fluorapatite $(Ca_5(PO_4)_3F) = 1 \times 10^{-60}$

Unknown: solubility of fluorapatite in H_2O

$Ca_5(PO_4)_3F \rightleftharpoons 5Ca^{2+} + 3PO_4^{3-} + F^-(aq)$

$K_{sp} = [Ca^{2+}]^5[PO_4^{3-}]^3[F^-]$

Let $[F^-] = x$

Therefore $[Ca^{2+}] = 5x$ and $[PO_4^{3-}] = 3x$

$K_{sp} = [5x]^5[3x]^3[x]$

$K_{sp} = [3125x^5][27x^3][x] = 84375x^9$

$1 \times 10^{-60} = 84375x^9$

$x^9 = 1.2 \times 10^{-65}$

$x = 6.1 \times 10^{-8}$ mol/L

---

44. Given: amount of $Na_2CO_3 = 0.96$ g/10 L

amount of $BaBr_2 = 0.20$ g/10 L

$K_{sp} = 2.8 \times 10^{-9}$

Unknown: whether a precipitate will form

$Na_2CO_3 + BaBr_2 \rightarrow NaBr + BaCO_3$

$NaBr(s) \rightleftharpoons Na^+(aq) + Br^-(aq)$

$K_{sp} = [Na^+][Br^-]$

$(0.96\text{ g }Na_2CO_3)\left(\dfrac{2\text{ mol }Na^+}{105.96\text{ g }Na_2NO_3}\right) = 1.8 \times 10^{-2}$ mol Na^+

$\dfrac{1.8 \times 10^{-2}\text{ mol }Na^+}{10\text{ L}} = 1.8 \times 10^{-3}$ mol/L Na^+

$(0.20\text{ g }BaBr_2)\left(\dfrac{2\text{ mol }Br^-}{297.13\text{ g }BaBr_2}\right) = 1.3 \times 10^{-3}$ mol Br^-

$\dfrac{1.3 \times 10^{-3}\text{ mol }Br^-}{10\text{ L}} = 1.3 \times 10^{-4}$ mol/L Br^-

$[Na^+][Br^-] = (1.8 \times 10^{-2})(1.3 \times 10^{-4}) = 2.3 \times 10^{-6}$

$2.3 \times 10^{-6} > K_{sp}$; NaBr precipitates.

45. Given: $K = 5.2 \times 10^{-5}$ for NH_3

$[N_2] = 2.00$ M

$[H_2] = 0.80$ M

$N_2(g) + 3H_2(g) \rightleftharpoons 2NH_3(g)$

Unknown: number of grams of NH_3 in 10 L at equilibrium

$K = \dfrac{[NH_3]^2}{[N_2][H_2]^3}$

$[NH_3]^2 = K[N_2][H_2]^3$

$[NH_3] = \sqrt{K[N_2][H_2]^3}$

$= \sqrt{(5.2 \times 10^{-5})(2.00 \text{ M})(0.80 \text{ M})^3}$

$= \sqrt{5.32 \times 10^{-5}} = 7.3 \times 10^{-3}$ M

$\left(\dfrac{7.3 \times 10^{-3} \text{ mol NH}_3}{\text{L}}\right)\left(\dfrac{17.04 \text{ g NH}_3}{\text{mol}}\right)$

$= 1.2 \times 10^{-1}$ g NH_3/L $= 1.2$ g/10 L

CHAPTER 19
Oxidation-Reduction Reactions

Section Review, p. 595

3. Unknown: if reaction is a redox reaction

a. $2KNO_3(s) \rightarrow 2KNO_2(s) + O_2(g)$

$2\overset{+1}{K}\overset{+5}{N}\overset{-2}{O_3} \rightarrow 2\overset{+1}{K}\overset{+3}{N}\overset{-2}{O_2} + \overset{0}{O_2}$

$2\overset{-2}{O} \rightarrow \overset{0}{O_2} + 4e^-$

$2\overset{+5}{N} + 4e^- \rightarrow 2\overset{+3}{N}$

redox

b. $H_2(g) + CuO(s) \rightarrow Cu(s) + H_2O(\ell)$

$\overset{0}{H_2} + \overset{+2}{Cu}\overset{-2}{O} \rightarrow \overset{0}{Cu} + \overset{+1}{H_2}\overset{-2}{O}$

$\overset{0}{H_2} \rightarrow 2\overset{+1}{H} + 2e^-$

$\overset{+2}{Cu} + 2e^- \rightarrow \overset{0}{Cu}$

redox

c. $NaOH(aq) + HCl(aq) \rightarrow NaCl(aq) + H_2O(\ell)$

$\overset{+1}{Na}\overset{-2}{O}\overset{+1}{H} + \overset{+1}{H}\overset{-1}{Cl} \rightarrow \overset{+1}{Na}\overset{-1}{Cl} + \overset{+1}{H_2}\overset{-2}{O}$

not redox

d. $H_2(g) + Cl_2(g) \rightarrow 2HCl(g)$

$\overset{0}{H_2} + \overset{0}{Cl_2} \rightarrow 2\overset{+1}{H}\overset{-1}{Cl}$

$\overset{0}{H_2} \rightarrow 2\overset{+1}{H} + 2e^-$

$\overset{0}{Cl_2} + 2e^- \rightarrow 2\overset{-1}{Cl}$

redox

e. $SO_3(g) + H_2O(\ell) \rightarrow H_2SO_4(aq)$

$\overset{+6}{S}\overset{-2}{O_3} + \overset{+1}{H_2}\overset{-2}{O} \rightarrow \overset{+1}{H_2}\overset{+6}{S}\overset{-2}{O_4}$

not redox

Practice, p. 601

1. Given: $Cu + H_2SO_4 \rightarrow CuSO_4 + SO_2 + H_2O$

Unknown: balanced equation

$Cu \rightarrow Cu^{2+} + 2e^-$

$SO_4^{2-} + 4H^+ + 2e^- \rightarrow SO_2 + 2H_2O$

$\overline{Cu + SO_4^{2-} + 4H^+ \rightarrow Cu^{2+} + SO_2 + 2H_2O}$

$Cu + 2H_2SO_4 \rightarrow CuSO_4 + SO_2 + 2H_2O$

2. Given: $HNO_3 + KI \rightarrow$
$KNO_3 + I_2 +$
$NO + H_2O$

Unknown: balanced equation

$3[2I^- \rightarrow I_2^0 + 2e^-]$
$2[\overset{+5}{N}O_3^- + 4H^+ + 3e^- \rightarrow \overset{+2}{N}O + 2H_2O]$
$\overline{2NO_3^- + 8H^+ + 6I^- \rightarrow 3I_2 + 2NO + 4H_2O}$

$8HNO_3 + 6KI \rightarrow 6KNO_3 + 3I_2 + 2NO + 4H_2O$

3. Given: $Fe + O_2 + H_2O \rightarrow$
$Fe(OH)_3$

Unknown: balanced equation

$4[\overset{0}{Fe} \rightarrow \overset{+3}{Fe}{}^{3+} + 3e^-]$
$3\overset{0}{O_2} + 12e^- + 6H_2O \rightarrow 12\overset{-2}{O}H^-$
$\overline{4Fe + 3O_2 + 6H_2O \rightarrow 4Fe^{3+} + 12OH^-}$

$4Fe + 3O_2 + 6H_2O \rightarrow 4Fe(OH)_3$

ATE, Additional Sample Problems 19-1, p. 601

1. Given: $K_2Cr_2O_7 +$
$C_2H_5OH + HCl \rightarrow$
$CrCl_3 + H_2O +$
$CO_2 + KCl$

Unknown: balanced equation

$\overset{-2}{C_2}H_5OH + 3H_2O \rightarrow 2\overset{+4}{C}O_2 + 12H^+ + 12e^-$
$2[\overset{+6}{Cr_2}O_7^{2-} + 14\overset{+1}{H}{}^+ + 6e^- \rightarrow 2\overset{+3}{Cr}{}^{3+} + 7H_2O]$
$\overline{2Cr_2O_7^{2-} + C_2H_5OH + 3H_2O + 28H^+ \rightarrow 4Cr^{3+} + 14H_2O + 2CO_2 + 12H^+}$

$2K_2Cr_2O_7 + C_2H_5OH + 16\ HCl \rightarrow 4CrCl_3 + 11H_2O + 2CO_2 + 4\ KCl$

2. Given: $XeO_3 + NaI +$
$HNO_3 \rightarrow NaI_3 +$
$Xe + H_2O + NaNO_3$

Unknown: balanced equation

$3[3\overset{+1}{Na}I \rightarrow \overset{+3}{Na}I_3 + 2Na^+ + 2e^-]$
$\overset{+6}{Xe}O_3 + 6H^+ + 6e^- \rightarrow \overset{0}{Xe} + 3H_2O$
$\overline{XeO_3 + 9NaI + 6HNO_3 \rightarrow 3NaI_3 + Xe + 3H_2O + 6NaNO_3}$

Section Review, p. 601

3. Given: $Na_2SnO_2 +$
$Bi(OH)_3 \rightarrow Bi +$
$Na_2SnO_3 + H_2O$

Unknown: balanced equation

$3[Na_2\overset{+2}{Sn}O_2 + H_2O \rightarrow Na_2\overset{+4}{Sn}O_3 + 2H^+ + 2e^-]$
$2[\overset{+3}{Bi}(OH)_3 + 3H^+ + 3e^- \rightarrow \overset{0}{Bi} + 3H_2O]$
$\overline{3Na_2SnO_2 + 3H_2O + 2Bi(OH)_3 + 6H^+ \rightarrow 3Na_2SnO_3 + 6H^+ + 2Bi + 6H_2O}$

$3Na_2SnO_2 + 2Bi(OH)_3 \rightarrow 2Bi + 3Na_2SnO_3 + 3H_2O$

Section Review, p. 616

6. Given: Na^+/Na and K^+/K half-cells

Unknown: a. overall electrochemical reaction
b. E^0 value

Half-reactions:
$K^+ + e^- \rightarrow K \quad E^0 = -2.93$ V
$Na^+ + e^- \rightarrow Na \quad E^0 = -2.71$ V

Anode = K (oxidation)
Cathode = Na (reduction)

Overall reaction:
$K + Na^+ \rightarrow K^+ + Na$

$E^0{}_{anode} = -2.93$ V
$E^0{}_{cathode} = -2.71$ V
$E^0{}_{cell} = E^0{}_{cathode} - E^0{}_{anode}$
$= -2.71$ V $- (-2.93$ V$) = +0.22$ V

Chapter 19 Review Problems, p. 618
Redox Equations

13. Unknown: if reaction is a redox reaction

a. $2NH_4Cl(aq) + Ca(OH)_2(aq) \rightarrow 2NH_3(aq) + 2H_2O(\ell) + CaCl_2(aq)$

Ionic equation:

$2\overset{-3+1}{NH_4^+} + \overset{-1}{Cl^-} + \overset{+2}{Ca^{2+}} + 2\overset{-2+1}{OH^-} \rightarrow 2\overset{-3+1}{NH_3} + 2\overset{+1-2}{H_2O} + \overset{+2}{Ca^{2+}} + 2\overset{-1}{Cl^-}$

nonredox

b. $2HNO_3(aq) + 3H_2S(g) \rightarrow 2NO(g) + 4H_2O(\ell) + 3S(s)$

Ionic equation:

$2\overset{+1}{H^+} + \overset{+5-2}{NO_3^-} + 3\overset{+1-2}{H_2S} \rightarrow 2\overset{+2-2}{NO} + 4\overset{+1-2}{H_2O} + 3\overset{0}{S}$

$2\overset{+5}{NO_3^-} + 6e^- \rightarrow 2\overset{+2}{NO}$

$3\overset{-2}{H_2S} \rightarrow 3\overset{0}{S} + 6e^-$

redox

c. $[Be(H_2O)_4]^{2+}(aq) + H_2O(\ell) \rightarrow H_3O^+(aq) + [Be(H_2O)_3OH]^+(aq)$

Ionic equation:

$\overset{+2\ +1\ -2}{[Be(H_2O)_4]^{2+}} + \overset{+1-2}{H_2O} \rightarrow \overset{+1-2}{H_3O^+} + \overset{+2\ +1\ -2\ -2+1}{[Be(H_2O)_3OH]^+}$

nonredox

14.

Oxidation numbers:

$\overset{0}{Xe}, \overset{+1-1}{XeF}, \overset{+2-1}{XeF_2}, \overset{+4-2-1}{XeOF_2},$

$\overset{+6-2}{XeO_3}, \overset{+1\ +7-1}{CsXeF_8}$

Voltaic and Electrolytic Cells

19. Given: pairs of half-cells

Unknown: electrochemical reaction that proceeds spontaneously

a. $Cu^{2+}/Cu, Ag^+/Ag$

Half-reactions:
$Cu^{2+} + 2e^- \rightarrow Cu \quad E^0 = +0.34$ V
$2(Ag^+ + e^- \rightarrow Ag) \quad E^0 = +0.80$ V

Anode = Cu
Cathode = Ag

Overall reaction:
$Cu + 2Ag^+ \rightarrow Cu^{2+} + 2Ag$

b. $Cd^{2+}/Cd, Co^{2+}/Co$

Half-reactions:
$Cd^{2+} + 2e^- \rightarrow Cd \quad E^0 = -0.40$ V
$Co^{2+} + 2e^- \rightarrow Co \quad E^0 = -0.28$ V

Anode = Cd
Cathode = Co

Overall reaction:
$Co^{2+} + Cd \rightarrow Co + Cd^{2+}$

c. Na^+/Na, Ni^{2+}/Ni

Half-reactions:
$2(Na^+ + e^- \rightarrow Na)$ $E^0 = -2.71$ V
$Ni^{2+} + 2e^- \rightarrow Ni$ $E^0 = -0.26$ V

Anode = Na
Cathode = Ni

Overall reaction:
$2Na + Ni^{2+} \rightarrow 2Na^+ + Ni$

d. I_2/I^-, Br_2/Br^-

Half-reactions:
$I_2 + 2e^- \rightarrow 2I^-$ $E^0 = +0.54$ V
$Br_2 + 2e^- \rightarrow 2Br^-$ $E^0 = +1.07$ V

Anode = I_2
Cathode = Br_2

Overall reaction:
$Br_2 + 2I^- \rightarrow 2Br^- + I_2$

---

20. Unknown: E^0 values for the cells in #19

a. $E^0_{cell} = +0.80$ V $- +0.34$ V $= +0.46$ V

b. $E^0_{cell} = -0.28$ V $- (-0.40$ V$) = +0.12$ V

c. $E^0_{cell} = -0.26$ V $- (-2.71$ V$) = +2.45$ V

d. $E^0_{cell} = +1.07$ V $- +0.54$ V $= +0.53$ V

---

21. Given: $I_2 + 2e^- \rightarrow 2I^-$
new $E^0 = +0.54$ V $-$ 0.54 V $= 0$ V

Unknown: E^0_{cell}

a. $Br_2 + 2e^- \rightarrow 2Br^-$ $E^0 = +1.07$ V

$E^0_{cell} = +1.07$ V $- +0.54$ V $= +0.53$ V

b. $Al^{3+} + 3e^- \rightarrow Al$ $E^0 = -1.66$ V

Anode = Al
Cathode = I_2

$E^0_{cell} = E^0_{cathode} - E^0_{anode}$
$= +0.54$ V $-(-1.66$ V$) = +2.20$ V

c. original: $+1.07$ V $- +0.54$ V $= +0.53$ V

new: $+0.53$ V $- 0 = +0.53$ V

No change would be observed.

Mixed Review

29. Unknown: if reaction is a redox reaction

a. $\overset{0}{Mg}(s) + \overset{+2\ -1}{ZnCl_2}(aq) \rightarrow$
$\overset{0}{Zn}(s) + \overset{+2\ -1}{MgCl_2}(aq)$

$\overset{0}{Mg} \rightarrow Mg^{2+} + 2e^-$

$Zn^{2+} + 2e^- \rightarrow \overset{0}{Zn}$

redox

b. $\overset{0}{H_2}(g) + \overset{+2-1}{OF_2}(g) \rightarrow$
$\overset{+1\ -2}{H_2O}(g) + \overset{+1-1}{HF}(g)$

$\overset{0}{H_2} \rightarrow 2\overset{+1}{H} + 2e^-$
$\overset{+2}{O} + 4e^- \rightarrow \overset{-2}{O}$

redox

c. $2\overset{+1-1}{KI}(aq) + \overset{+2\ +5-2}{Pb(NO_3)_2}(aq) \rightarrow$
$\overset{+2-1}{PbI_2}(s) + 2\overset{+1+5-2}{KNO_3}(aq)$

nonredox

d. $\overset{+2-2}{CaO}(s) + \overset{+1\ -2}{H_2O}(\ell) \rightarrow$
$\overset{+2\ -2+1}{Ca(OH)_2}(aq)$

nonredox

e. $3\overset{+2\ -1}{CuCl_2}(aq) +$
$2\overset{-3+1\ +5-2}{(NH_4)_3PO_4}(aq) \rightarrow$
$6\overset{-3+1\ -1}{NH_4Cl}(aq) +$
$\overset{+2\ +5-2}{Cu_3(PO_4)_2}(s)$

nonredox

f. $\overset{-4+1}{CH_4}(g) + 2\overset{0}{O_2}(g) \rightarrow$
$\overset{+4-2}{CO_2}(g) + 2\overset{+1\ -2}{H_2O}(g)$

$\overset{-4}{C} \rightarrow \overset{+4}{C} + 8e^-$
$\overset{0}{O_2} + 4e^- \rightarrow 2\overset{-2}{O}$

redox

33. $\overset{+5-2}{NO_3},\ \overset{+4\ -2}{N_2O_4},\ \overset{+1\ -2}{N_2O},\ \overset{0}{N_2},$
$\overset{-2\ +1}{N_2H_4},\ \overset{-3+1}{NH_3}$

34. a. $SbCl_5 + KI \rightarrow KCl +$
$I_2 + SbCl_3$

$2I^- \rightarrow I_2 + 2e^-$
$\underline{SbCl_5 + 2e^- \rightarrow SbCl_3 + 2Cl^-}$
$SbCl_5 + 2I^- \rightarrow SbCl_3 + 2Cl^- + I_2$

$SbCl_5 + 2KI \rightarrow 2KCl + I_2 + SbCl_3$

b. $Ca(OH)_2 + NaOH +$
$ClO_2 + C \rightarrow NaClO_2 +$
$CaCO_3 + H_2O$

$\overset{0}{C} + 6OH^- \rightarrow \overset{+4}{CO_3^{2-}} + 3H_2O + 4e^-$
$4[\overset{+4}{ClO_2} + 1e^- \rightarrow \overset{+3}{ClO_2^-}]$
$C + 6OH^- + 4ClO_2 \rightarrow CO_3^{2-} + 3H_2O + 4ClO_2^-$

$Ca(OH)_2 + 4NaOH + 4ClO_2 + C \rightarrow 4NaClO_2 + CaCO_3 + 3H_2O$

CHAPTER 20

Carbon and Hydrocarbons

Chapter 20 Review Problems, Calculations with Carbon Compounds, p. 660

45. Given: 1 carat = 0.200 g
density of diamond = 3.51 g/cm^3

Unknown: volume of a 1.00 carat diamond

$$V = \frac{m}{D} = \frac{0.200 \text{ g}}{3.51 \text{ g/cm}^3} = 0.0570 \text{ cm}^3$$

---

46. Given: mass of butadiene (C_4H_6) = 100.0 g

a. Unknown: number of moles

$$(100.0 \text{ g } C_4H_6)\left(\frac{\text{mol } C_4H_6}{54.10 \text{ g } C_4H_6}\right) = 1.85 \text{ mol } C_4H_6$$

b. Unknown: number of molecules

$$(1.85 \text{ mol } C_4H_6)\left(\frac{6.02 \times 10^{23} \text{ molecules } C_4H_6}{\text{mol } C_4H_6}\right) = 1.11 \times 10^{24} \text{ molecules } C_4H_6$$

---

47. Given: molecular formula of alkene = $C_{12}H_{24}$

Unknown: percent composition

molar mass of $C_{12}H_{24}$ = 168.36

$$\% \text{ C} = \frac{144.12 \text{ g C}}{168.36 \text{ g } C_{12}H_{24}} \times 100 = 85.6\% \text{ C}$$

$$\% \text{ H} = \frac{24.24 \text{ g H}}{168.36 \text{ g } C_{12}H_{24}} \times 100 = 14.4\% \text{ H}$$

---

48. Given: volume of propane (C_3H_8) = 15.0 L

Unknown: volume of CO_2 produced by combustion of propane

$C_3H_8 + 5O_2 \rightarrow 3CO_2 + 4H_2O$
1 mol   3 mol

$$(15.0 \text{ L } C_3H_8)\left(\frac{3 \text{ mol } CO_2}{1 \text{ mol } C_3H_8}\right) = 45.0 \text{ L } CO_2$$

---

49. Given: density of isooctane = 0.692 g/mL

Unknown: mass of 12.0 gal of isooctane in kg

1 gal = (3.78 L)(1000 mL/L) = 3780 mL

$$(12.0 \text{ gal})\left(\frac{3780 \text{ mL}}{1 \text{ gal}}\right)\left(\frac{0.692 \text{ g}}{1 \text{ mL}}\right)\left(\frac{1 \text{ kg}}{1000 \text{ g}}\right) = 31.4 \text{ kg}$$

CHAPTER 21

Other Organic Compounds

Calculations with Organic Compounds, p. 697

57. Unknown: molecular mass of trichlorofluoromethane (CCl_3F)

Carbon: $12.01 \times 1 = 12.01$
Chlorine: $35.45 \times 3 = 106.35$
Fluorine: $19.00 \times 1 = \underline{19.00}$
137.36 amu

58. Given: Percent composition is 54.5% C, 9.1% H, 36.4% O

a. Unknown: simplest formula

$(54.5 \text{ g C})\left(\dfrac{1 \text{ mol C}}{12.01 \text{ g C}}\right) = 4.54 \text{ mol C}$

$(9.1 \text{ g H})\left(\dfrac{1 \text{ mol H}}{1.01 \text{ g H}}\right) = 9.01 \text{ mol H}$

$(36.4 \text{ g O})\left(\dfrac{1 \text{ mol O}}{16.00 \text{ g O}}\right) = 2.28 \text{ mol O}$

Ratio = 4.54 mol C : 9.01 mol H : 2.28 mol O

Smallest whole-number mole ratio = $\dfrac{4.54 \text{ mol C}}{2.28} : \dfrac{9.01 \text{ mol H}}{2.28} : \dfrac{2.28 \text{ mol O}}{2.28}$

$= 1.99 \text{ mol C} : 3.95 \text{ mol H} : 1 \text{ mol O}$

$= 2 \text{ mol C} : 4 \text{ mol H} : 1 \text{ mol O}$

Simplest formula = C_2H_4O

b. Given: molecular mass = 88.1 g

Unknown: molecular formula

molecular mass of $C_2H_4O = 44.05$

molecular mass of compound $(C_2H_4O)_x = 88.1$ g

$x = \dfrac{88.1 \text{ g}}{44.05 \text{ g}} = 2$

molecular formula = $C_4H_8O_2$

59. Given: [H_3O^+] in 0.05 M acetic acid = 9.4×10^{-4} mol/L

Unknown: pH

pH = $-\log (9.4 \times 10^{-4}) = 3.03$

60. Given: [C_2H_5OH] = 0.750 M
volume of solution = 500. mL
density of ethanol = 0.789 g/mL

Unknown: volume pure C_2H_5OH to prepare 500. mL solution

$\left(\dfrac{0.750 \text{ mol } C_2H_5OH}{L}\right)(500 \text{ mL})\left(\dfrac{1 \text{ L}}{1000 \text{ mL}}\right)\left(\dfrac{46.07 \text{ g } C_2H_5OH}{1 \text{ mol } C_2H_5OH}\right) = 17.3 \text{ g } C_2H_5OH$

$V = m/D = \dfrac{17.3 \text{ g}}{0.789 \text{ g/mL}} = 21.9 \text{ mL } C_2H_5OH$

61. Given: density of ethylene glycol [$C_2H_4(OH)_2$] = 1.432 g/mL

a. Unknown: theoretical freezing point of H_2O in a 50% solution of ethylene glycol

$m = DV = (1.432 \text{ g/mL})(1000 \text{ mL}) = 1432 \text{ g}$

number of moles = $(1432 \text{ g } C_2H_4(OH)_2)\left(\dfrac{1 \text{ mol } C_2H_4(OH)_2}{62.06 \text{ g } C_2H_4(OH)_2}\right)$

$= 23.07 \text{ mol } C_2H_4(OH)_2$

molality $(m) = \dfrac{23.1 \text{ mol } C_2H_4(OH)_2}{1 \text{ kg } H_2O} = 23.1 \, m$

$\Delta t_f = (-1.86°\text{C}/m)(23.07 \, m) = -42.9°\text{C}$

freezing point of solution = $0°\text{C} + (-42.9°\text{C}) = -42.9°\text{C}$

CHAPTER 22

Nuclear Chemistry

ATE, Additional Example Problems, p. 702

1. Given: measured atomic mass of $^{32}_{16}\text{S}$ = 31.972 070 amu

Unknown: binding energy of $^{32}_{16}\text{S}$

16 protons: $(16 \times 1.007\,276 \text{ amu}) = 16.116\,416$ amu
16 neutrons: $(16 \times 1.008\,665 \text{ amu}) = 16.138\,64$ amu
16 electrons: $(16 \times 0.000\,5486 \text{ amu}) = \underline{0.008\,7776 \text{ amu}}$
total combined mass: $32.263\,834$ amu

mass defect = 32.263 834 amu − 31.972 070 amu
 = 0.291 764 amu

$$= (0.291\,764 \text{ amu})\left(\frac{1.6605 \times 10^{-27} \text{ kg}}{1 \text{ amu}}\right)$$

$$= 4.8447 \times 10^{-28} \text{ kg}$$

$$E = mc^2$$

$$E = (4.8447 \times 10^{-28} \text{ kg})(3.00 \times 10^8 \text{ m/s})^2$$

$$= 4.36 \times 10^{-11} \text{ kg} \cdot \text{m}^2/\text{s}^2 = 4.36 \times 10^{-11} \text{ J}$$

2. Given: measured atomic mass of $^{16}_{8}\text{O}$ = 15.994 915 amu

Unknown: binding energy for 1 mole of $^{16}_{8}\text{O}$

8 protons: $(8 \times 1.007\,276 \text{ amu}) = 8.058\,8208$ amu
8 neutrons: $(8 \times 1.008\,665 \text{ amu}) = 8.069\,32$ amu
8 electrons: $(8 \times 0.000\,5486 \text{ amu}) = \underline{0.004\,3888 \text{ amu}}$
total combined mass: $16.131\,917$ amu

mass defect = 16.131 917 amu − 15.994 915 amu = 0.137 0018 amu

$$= (0.137\,0018 \text{ amu})\left(\frac{1.6605 \times 10^{-27} \text{ kg}}{1 \text{ amu}}\right) = 2.2749 \times 10^{-28} \text{ kg}$$

$$E = mc^2$$

$$E = (2.2749 \times 10^{-28} \text{ kg})(3.00 \times 10^8 \text{ m/s})^2$$

$$= 2.047 \times 10^{-11} \text{ kg} \cdot \text{m}^2/\text{s}^2 = 2.047 \times 10^{-11} \text{ J}$$

binding energy/mol = $(2.047 \times 10^{-11} \text{ J/atom})(6.022 \times 10^{-23} \text{ atoms/mol})$
 = 1.23×10^{13} J/mol

3. Given: measured atomic mass of $^{55}_{25}$Mn = 54.938 047 amu

Unknown: binding energy per nucleon of a $^{55}_{25}$Mn atom

25 protons: (25 × 1.007 276 amu) = 25.1819 amu
30 neutrons: (30 × 1.008 665 amu) = 30.259 95 amu
25 electrons: (25 × 0.000 5486 amu) = $\underline{0.0137\ 15\ \text{amu}}$
total combined mass: 55.455 115 amu

mass defect = 55.455 115 amu − 54.938 047 amu = 0.517 068 amu

$$= (0.517\ 068\ \text{amu})\left(\frac{1.6605 \times 10^{-27}\ \text{kg}}{1\ \text{amu}}\right) = 8.5859 \times 10^{-28}\ \text{kg}$$

$$E = mc^2$$

$$E = (88.5859 \times 10^{-28}\ \text{kg})(3.00 \times 10^8\ \text{m/s})^2$$

$$= 7.727 \times 10^{-11}\ \text{kg} \cdot \text{m}^2/\text{s}^2 = 7.727 \times 10^{-11}\ \text{J}$$

$$\text{binding energy/nucleon} = \frac{7.727 \times 10^{-11}\ \text{J}}{55\ \text{nucleons}} = 1.41 \times 10^{-12}\ \text{J/nucleon}$$

Practice, p. 704

1. Given: $^{218}_{84}\text{Po} \rightarrow\ ^4_2\text{He} + ?$

mass number: 218 − 4 = 214

atomic number: 84 − 2 = 82

$? =\ ^{214}_{82}\text{Pb}$

2. Given: $^{253}_{99}\text{Es} +\ ^4_2\text{He} \rightarrow\ ^1_0n + ?$

mass number: 253 + 4 − 1 = 256

atomic number: 99 + 2 − 0 = 101

$? =\ ^{256}_{101}\text{Md}$

3. Given: $^{142}_{61}\text{Pm} + ? \rightarrow\ ^{142}_{60}\text{Nd}$

mass number: 142 − 142 = 0

atomic number: 60 − 61 = −1

$? =\ ^{\ \ 0}_{-1}e$

ATE, Additional Sample Problems 22–1, p. 704

1. Given: $^{238}_{92}\text{U} \rightarrow ? +\ ^{234}_{90}\text{Th}$

mass number: 238 − 234 = 4

atomic number: 92 − 90 = 2

$? =\ ^4_2\text{He}$

2. Given: $^{37}_{18}\text{Ar} + ? \rightarrow\ ^{37}_{17}\text{Cl}$

mass number: 37 − 37 = 0

atomic number: 17 − 18 = −1

$? =\ ^{\ \ 0}_{-1}e$

Section Review, p. 704

4. a. Given: $^{187}_{75}\text{Re} + ? \rightarrow$
$^{188}_{75}\text{Re} + ^{1}_{1}\text{H}$

mass number: $188 + 1 - 187 = 2$

atomic number: $75 + 1 - 75 = 1$

$? = ^{2}_{1}\text{H}$

b. $^{9}_{4}\text{Be} + ^{4}_{2}\text{He} \rightarrow ? + ^{1}_{0}n$

mass number: $9 + 4 - 1 = 12$

atomic number: $4 + 2 - 0 = 6$

$? = ^{12}_{6}\text{C}$

c. $^{22}_{11}\text{Na} + ? \rightarrow ^{22}_{10}\text{Ne}$

mass number: $22 - 22 = 0$

atomic number: $10 - 11 = -1$

$? = ^{0}_{-1}e$

Practice, p. 709

1. Given: orig. mass of ^{210}Po = 2.0 mg
half-life of ^{210}Po = 1388.4 days
time elapsed = 415.2 days

Unknown: mass of ^{210}Po remaining after 415.2 days

number of half-lives = $(415.2 \text{ days})\left(\dfrac{1 \text{ half-life}}{138.4 \text{ days}}\right) = 3$ half-lives

mass of ^{210}Po remaining = $2.0 \text{ mg} \times \left(\dfrac{1}{2}\right)^3 = 0.25$ mg

2. Given: half-life of radium-226 = 1599 years

Unknown: number of years needed for decay of $\dfrac{15}{16}$ of a given amount of radium-226

amount remaining = $\dfrac{1}{16} = 0.0625 = \left(\dfrac{1}{2}\right)^4$; 4 half-lives

years needed for decay of $\dfrac{15}{16} = (1599 \text{ years})(4) = 6396$ years

3. Given: half-life of radon-222 = 3.824 days

Unknown: time needed for $\dfrac{1}{4}$ of a given amount of radon-222 to remain

amount remaining = $\dfrac{1}{4} = 0.25 = \left(\dfrac{1}{2}\right)^2$; 2 half-lives

days needed for $\dfrac{3}{4}$ to decay = $(3.824 \text{ days})(2) = 7.648$ days

4. Given: starting mass of ^{60}Co = 10.0 mg
half-life of ^{60}Co = 10.47 min
time elapsed = 104.7 min

Unknown: mass of ^{60}Co remaining after 104.7 min

number of half-lives = $(104.7 \text{ min})\left(\dfrac{1 \text{ half-life}}{10.47 \text{ min}}\right)$ = 10 half-lives

mass remaining after 10 half-lives = $\left(\dfrac{1}{2}\right)^{10}(10.0 \text{ mg})$ = 0.009 77 mg

---

5. Given: mass of ^{238}U 4.46×10^9 years ago = 4.0 mg
half-life of ^{238}U = 4.46×10^9 years

Unknown: mass of uranium-238 remaining today

number of half-lives = $(4.46 \times 10^9 \text{ years})\left(\dfrac{1 \text{ half-life}}{4.46 \times 10^9 \text{ years}}\right)$ = 1 half-life

mass of uranium-238 remaining = $4.0 \text{ mg} \times \dfrac{1}{2}$ = 2.0 mg

---

6. Given: original mass of ^{218}Po = 16 mg
half-life of ^{218}Po = 3.0 min

Unknown: length of time before 1.0 mg of ^{218}Po remains

amount remaining = $\dfrac{1}{16}$ = 0.0625 = $\left(\dfrac{1}{2}\right)^4$; 4 half-lives

time until 1.0 mg remains = $(3.0 \text{ min})(4)$ = 12 min

ATE, Additional Sample Problems 22-2, p. 709

1. Given: original mass of radon-222 = 4.38 μg
half-life of radon-222 = 3.8 days

Unknown: mass of radon-222 remaining after 1.52 days

number of half-lives = $(15.2 \text{ days})\left(\dfrac{1 \text{ half-life}}{3.8 \text{ days}}\right)$ = 4 half-lives

mass of radon-222 remaining = $4.38 \text{ μg} \times \left(\dfrac{1}{2}\right)^4$ = 0.274 μg

---

2. Given: half-life of uranium-238 = 4.46×10^9 years

Unknown: time needed for $\dfrac{7}{8}$ of sample to decay

amount remaining = $\dfrac{1}{8}$ = 0.125 = $\left(\dfrac{1}{2}\right)^3$; 3 half-lives

time needed for decay = $(4.46 \times 10^9 \text{ years})(3)$ = 1.34×10^{10} years

3. Given: half-life of carbon-14 = 5715 years

Unknown: time needed for $\frac{1}{2}$ of sample to remain

time needed for $\frac{1}{2}$ of sample to decay = 1 half-life = 5715 years

4. Given: half-life of iodine-131 = 8.040 days

Unknown: percentage of a sample remaining after 40.2 days

number of half-lives = $(40.2 \text{ days})\left(\dfrac{1 \text{ half-life}}{8.040 \text{ days}}\right)$ = 5 half-lives

percent of sample remaining after 5 half-lives = $\left(\dfrac{1}{2}\right)^5 \times 100 = 3.13\%$

Chapter Review Problems, p. 723
Mass Defect

25. Given: measured atomic mass of $^{20}_{10}\text{Ne}$ = 19.992 44 amu

Unknown: mass defect

10 protons: (10 × 1.007 276 amu) = 10.072 76 amu
10 neutrons: (10 × 1.008 665 amu) = 10.086 65 amu
10 electrons: (10 × 0.000 5486 amu) = $\underline{0.005\ 486\ \text{amu}}$
total combined mass: 20.164 896 amu

mass defect = 20.164 896 amu − 19.992 44 amu = 0.172 46 amu

26. Given: measured atomic mass of $^{7}_{3}\text{Li}$ = 7.016 00 amu

Unknown: mass defect

3 protons: (3 × 1.007 276 amu) = 3.021 828 amu
4 neutrons: (4 × 1.008 665 amu) = 4.034 66 amu
3 electrons: (3 × 0.000 5486 amu) = $\underline{0.001\ 646\ \text{amu}}$
total combined mass: 7.058 1338 amu

mass defect = 7.058 133 8 amu − 7.016 00 amu = 0.042 13 amu

Nuclear Binding Energy

27. Given: measured atomic mass of $^{6}_{3}\text{Li}$ = 6.015 amu

Unknown: nuclear binding energy

3 protons: (3 × 1.007 276 amu) = 3.021 828 amu
3 neutrons: (3 × 1.008 665 amu) = 3.025 995 amu
3 electrons: (3 × 0.000 5486 amu) = $\underline{0.001\ 646\ \text{amu}}$
total combined mass: 6.049 4688 amu

mass defect = 6.049 469 amu − 6.015 amu = 0.034 469 amu

$= (0.034\ 469\ \text{amu})\left(\dfrac{1.6605 \times 10^{-27} \text{ kg}}{1 \text{ amu}}\right) = 5.7235 \times 10^{-29}$ kg

$E = mc^2$

$E = (5.7235 \times 10^{-29} \text{ kg})(3.00 \times 10^8 \text{ m/s})^2$

$= 5.15 \times 10^{-12}$ kg·m²/s² $= 5.2 \times 10^{-12}$ J

28. Given: measured atomic mass of $^{35}_{19}K$ = 34.988 011 amu
measured atomic mass of $^{23}_{11}Na$ = 22.989 757 amu

Unknown: nuclear binding energy of both nuclei; which nucleus releases more energy when formed

a. $^{35}_{19}K$

b. $^{23}_{11}Na$

$^{35}_{19}K$: 19 protons: $(19 \times 1.007\ 276\ \text{amu}) = 19.138\ 244$ amu

16 neutrons: $(16 \times 1.008\ 665\ \text{amu}) = 16.138\ 64$ amu

19 electrons: $(19 \times 1.000\ 5486\ \text{amu}) = \underline{0.010\ 4234\ \text{amu}}$

total combined mass: 35.287 307 amu

mass defect = 35.287 307 amu − 34.988 011 amu = 0.299 296 amu

$$= (0.299\ 296\ \text{amu})\left(\frac{1.6605 \times 10^{-27}\ \text{kg}}{1\ \text{amu}}\right) = 4.9698 \times 10^{-28}\ \text{kg}$$

$E = mc^2$

$E = (4.9698 \times 10^{-28}\ \text{kg})(3.00 \times 10^8\ \text{m/s})^2$

$= 4.47 \times 10^{-11}\ \text{kg} \cdot \text{m}^2/\text{s}^2 = 4.47 \times 10^{-11}$ J

$^{23}_{11}Na$: 11 protons: $(11 \times 1.007\ 276\ \text{amu}) = 11.080\ 036$ amu

12 neutrons: $(12 \times 1.008\ 665\ \text{amu}) = 12.103\ 98$ amu

11 electrons: $(11 \times 0.000\ 5486\ \text{amu}) = \underline{0.060\ 346\ \text{amu}}$

total combined mass: 23.190 051 amu

mass defect = 23.190 151 amu − 22.989 767 amu = 0.200 2836 amu

$$= (0.200\ 2836\ \text{amu})\left(\frac{1.6605 \times 10^{-27}\ \text{kg}}{1\ \text{amu}}\right) = 3.3257 \times 10^{-28}\ \text{kg}$$

$E = mc^2$

$E = (3.3257 \times 10^{-28}\ \text{kg})(3.00 \times 10^8\ \text{m/s})^2$

$= 2.99 \times 10^{-11}\ \text{kg} \cdot \text{m}^2/\text{s}^2 = 2.99 \times 10^{-11}$ J

The nucleus in part **a** ($^{35}_{19}K$) releases more energy.

---

29. a. Unknown: binding energy per nucleon for each nucleus in problem #28

b. Unknown: which nucleus is more stable

$^{35}_{19}K$: $\dfrac{4.5 \times 10^{-11}\ \text{J}}{35\ \text{nucleons}} = 1.28 \times 10^{-12}$ J/nucleon

$^{23}_{11}Na$: $\dfrac{2.99 \times 10^{-11}\ \text{J}}{23\ \text{nucleons}} = 1.30 \times 10^{-12}$ J/nucleon

The greater the nuclear binding energy, the greater the stability. Therefore, $^{23}_{11}Na$ is more stable.

30. Given: measured atomic mass of $^{7}_{3}$Li = 7.016 00 amu

Unknown: binding energy per nucleon

3 protons: (3 × 1.007 276 amu) = 3.021 828 amu
4 neutrons: (4 × 1.008 665 amu) = 4.034 66 amu
3 electrons: (3 × 0.000 5486 amu) = 0.016 46 amu
total combined mass: 7.058 1338 amu

mass defect = 7.058 1338 amu − 7.016 00 amu = 0.042 13 amu

$$= (0.042\ 13\ \text{amu})\left(\frac{1.6605 \times 10^{-27}\ \text{kg}}{\text{amu}}\right) = 6.9956 \times 10^{-29}\ \text{kg}$$

$E = mc^2$

$E = (6.9956 \times 10^{-29}\ \text{kg})(3.00 \times 10^8\ \text{m/s})^2$

$= 6.30 \times 10^{-12}\ \text{kg} \cdot \text{m}^2/\text{s}^2 = 6.30 \times 10^{-29}\ \text{J}$

$$\frac{6.30 \times 10^{-29}\ \text{J}}{7\ \text{nucleons}} = 9.00 \times 10^{-13}\ \text{J/nucleon}$$

Neutron-Proton Ratio

31. Unknown: neutron-proton ratios

a. $^{12}_{6}$C = 6 protons, 6 neutrons

Ratio = $\frac{6}{6}$ = 1 : 1

b. $^{3}_{1}$H = 1 proton, 2 neutrons

Ratio = 2 : 1

c. $^{206}_{82}$Pb = 82 protons, 124 neutrons

Ratio = $\frac{124}{82}$ = 1.51 : 1

d. $^{134}_{50}$Sn = 50 protons, 84 neutrons

Ratio = $\frac{84}{50}$ = 1.68 : 1

Nuclear Equations

33. a. $^{43}_{19}$K → $^{43}_{20}$Ca + ?

mass number: 43 − 43 = 0

atomic number: 19 − 20 = −1

? = $^{0}_{-1}\beta$

b. $^{233}_{92}$U → $^{229}_{90}$Th + ?

mass number: 233 − 229 = 4

atomic number: 92 − 90 = 2

? = $^{4}_{2}$He

c. $^{11}_{6}$C + ? → $^{11}_{5}$B

mass number: 11 − 11 = 0

atomic number: 5 − 6 = 1

? = $^{0}_{-1}e$

d. $^{13}_{7}$N → $^{0}_{+1}\beta$ + ?

mass number: 13 − 0 = 13

atomic number: 7 − 1 = 6

? = $^{13}_{6}$C

Half-Life

36. Given: original mass of ^{239}Pu = 100 g

Unknown: mass of ^{239}Pu remaining after 96 440 years

number of half-lives = (96 440 years)$\left(\dfrac{1 \text{ half-life}}{24\ 110 \text{ years}}\right)$ = 4 half-lives

mass of plutonium-239 remaining after 4 half-lives = (100 g)$\left(\dfrac{1}{2}\right)^4$ = 6.25 g

---

37. Given: half-life of ^{227}Th = 18.72 days

Unknown: length of time for $\frac{3}{4}$ of a given amount to decay

amount remaining = $\dfrac{1}{4} = \left(\dfrac{1}{2}\right)^2$; 2 half-lives

length of time for decay = (18.72 days)(2) = 37.44 days

---

38. Given: half-life of ^{234}Pa = 6.69 hours

Unknown: fraction of a given amount remaining after 26.76 hours

number of half-lives = (26.76 hours)$\left(\dfrac{1 \text{ half-life}}{6.69 \text{ hours}}\right)$ = 4 half-lives

amount remaining after 4 half-lives = $\left(\dfrac{1}{2}\right)^4 = \dfrac{1}{16}$

---

39. Given: original mass of ^{226}Ra = 15.0 mg
half-life of ^{226}Ra = 1599 years

Unknown: mass remaining after 6396 years

number of half-lives = (6396 years)$\left(\dfrac{1 \text{ half-life}}{1\ 599 \text{ years}}\right)$ = 4 half-lives

mass remaining = 15.0 mg × $\left(\dfrac{1}{2}\right)^4$ = 0.938 mg

Mixed Review

40. a. $^{239}_{93}\text{Np} \rightarrow\ ^{0}_{-1}\beta + ?$ mass number: 239 − 0 = 239

atomic number: 93 − (−1) = 94

$? =\ ^{239}_{94}\text{Pu}$

b. $^{9}_{4}\text{Be} +\ ^{4}_{2}\text{He} \rightarrow ?$ mass number: 9 + 4 = 13

atomic number: 4 + 2 = 6

$? =\ ^{13}_{6}\text{C}$

c. $^{32}_{15}\text{P} + ? \rightarrow\ ^{33}_{15}\text{P}$ mass number: 33 − 32 = 1

atomic number: 15 − 15 = 0

$? =\ ^{1}_{0}n$

d. $^{236}_{92}\text{U} \rightarrow\ ^{94}_{36}\text{Kr} + ? + 3\ ^{1}_{0}n$ mass number: 236 − 94 = 139

atomic number: 92 − 36 − 0 = 56

$? =\ ^{139}_{56}\text{Ba}$

---

41. Given: half-life of ^{226}Ra = 1599 years
original mass of ^{226}Ra = 0.250 g

Unknown: mass remaining after 4797 years

number of half-lives = $(4797 \text{ years})\left(\dfrac{1 \text{ half-life}}{1599}\right)$ = 3 half-lives

mass remaining = $0.250 \text{ g} \times \left(\dfrac{1}{2}\right)^3 = 0.0313 \text{ g}$

---

43. Given: half-life of ^{225}Ra = 3.66 days
mass remaining after 7.32 days = 0.0500 g

Unknown: original mass of radium-224

number of half-lives = $(7.32 \text{ days})\left(\dfrac{1 \text{ half-life}}{3.66 \text{ days}}\right)$ = 2 half-lives

x = original mass

$(x)\left(\dfrac{1}{2}\right)^2 = 0.0500 \text{ g}$

$x = (0.0500 \text{ g})\left(\dfrac{2}{1}\right)^2 = 0.200 \text{ g}$

---

44. Unknown: neutron-proton ratios; location in relation to band of stability

a. $^{235}_{92}\text{U}$: 92 protons, 143 neutrons   Ratio = $\dfrac{143}{92}$ = 1.55 : 1 (outside)

b. $^{16}_{8}\text{O}$: 8 protons, 8 neutrons   Ratio = $\dfrac{8}{8}$ = 1 : 1 (within)

c. $^{56}_{26}\text{Fe}$: 26 protons, 30 neutrons   Ratio = $\dfrac{30}{26}$ = 1.15 : 1 (within)

d. $^{156}_{60}\text{Nd}$: 60 protons, 96 neutrons   Ratio = $\dfrac{96}{60}$ = 1.6 : 1 (outside)

45. Given: measured atomic mass of $^{238}_{92}U$ = 238.050 784 amu

Unknown: binding energy per nucleon

92 protons: (92 × 1.007 276 amu) = 92.669 392 amu
146 neutrons: (146 × 1.008 665 amu) = 147.265 09 amu
92 electrons: (92 × 0.000 5486 amu) = $\underline{0.050\ 4712\ \text{amu}}$
total combined mass: 239.984 95 amu

mass defect = 239.984 95 amu − 238.050 784 amu = 1.933 72 amu

$$= (1.933\ 72\ \text{amu})\left(\frac{1.6605 \times 10^{-27}\ \text{kg}}{1\ \text{amu}}\right) = 3.210 \times 10^{-27}\ \text{kg}$$

$E = mc^2$

$E = (3.2109 \times 10^{-27}\ \text{kg})(3.00 \times 10^8\ \text{m/s})^2$

$= 2.889 \times 10^{-10}\ \text{kg} \cdot \text{m}^2/\text{s}^2 = 2.889 \times 10^{-10}\ \text{J}$

energy per nucleon $= \dfrac{2.889 \times 10^{-10}\ \text{J}}{238\ \text{nucleons}} = 1.21 \times 10^{-12}$ J/nucleon

---

46. Given: nuclear binding energy of $^{56}_{26}\text{Fe}$ = 7.89 × 10^{-11} J

Unknown: mass lost in kg

$E = mc^2$

$m = E/C^2$

$= \dfrac{7.89 \times 10^{-11}\ \text{J}}{(3.00 \times 10^8\ \text{m/s})^2} = 8.77 \times 10^{-28}\ \text{kg}$

---

47. Given: measured atomic mass of deuterium = 2.0140 amu

Unknown: binding energy for 1 mole of deuterium atoms

deuterium = ^2_1H

1 proton: (1 × 1.007 276 amu) = 1.007 276 amu
1 neutron: (1 × 1.008 665 amu) = 1.008 665 amu
1 electron: (1 × 0.000 5486 amu) = $\underline{0.000\ 5486\ \text{amu}}$
total combined mass: 2.016 4896 amu

mass defect = 2.016 4896 amu − 2.0140 amu = 0.002 4896 amu

$$= (0.002\ 489\ \text{amu})\left(\frac{1.6605 \times 10^{-27}\ \text{kg}}{1\ \text{amu}}\right) = 4.1339 \times 10^{-30}\ \text{kg}$$

$E = mc^2$

$E = (4.1339 \times 10^{-30}\ \text{kg})(3.00 \times 10^8\ \text{m/s})^2$

$= 3.72 \times 10^{-13}\ \text{kg} \cdot \text{m}^2/\text{s}^2 = 3.72 \times 10^{-13}\ \text{J}$

energy per mole $= (3.72 \times 10^{-13}\ \text{J/atom})(6.022 \times 10^{23}\ \text{atoms/mol})$
$= 2.24 \times 10^{11}$ J/mol

APPENDIX D

Problem Bank

1. a. Given: 5.2 cm
Unknown: length in millimeters

$$5.2 \text{ cm} \times \frac{10 \text{ mm}}{\text{cm}} = 52 \text{ mm}$$

b. Given: 0.049 kg
Unknown: mass in grams

$$0.049 \text{ kg} \times \frac{1000 \text{ g}}{\text{kg}} = 49 \text{ g}$$

c. Given: 1.60 mL
Unknown: volume in microliters

$$1.60 \text{ mL} \times \frac{1000 \text{ μL}}{\text{mL}} = 1600 \text{ μL}$$

d. Given: 0.0025 g
Unknown: mass in micrograms

$$0.0025 \text{ g} \times \frac{1000\ 000 \text{ μg}}{\text{g}} = 2500 \text{ μg}$$

e. Given: 0.020 kg
Unknown: mass in milligrams

$$0.020 \text{ kg} \times \frac{1000\ 000 \text{ mg}}{\text{kg}} = 20\ 000 \text{ mg}$$

f. Given: 3 kL
Unknown: volume in liters

$$3 \text{ kL} \times \frac{1000 \text{ L}}{\text{kL}} = 3000 \text{ L}$$

2. a. Given: 150 mg
Unknown: mass in grams

$$150 \text{ mg} \times \frac{1 \text{ g}}{1000 \text{ mg}} = 0.15 \text{ g}$$

b. Given: 2500 mL
Unknown: volume in liters

$$2500 \text{ mL} \times \frac{1 \text{ L}}{1000 \text{ mL}} = 2.5 \text{ L}$$

c. Given: 0.5 g
Unknown: mass in kilograms

$$0.5 \text{ g} \times \frac{1 \text{ kg}}{1000 \text{ g}} = 0.0005 \text{ kg}$$

d. Given: 55 L
Unknown: volume in kiloliters

$$55 \text{ L} \times \frac{1 \text{ kL}}{1000 \text{ L}} = 0.055 \text{ kL}$$

e. Given: 35 mm
Unknown: length in cm

$$35 \text{ mm} \times \frac{1 \text{ cm}}{10 \text{ mm}} = 3.5 \text{ cm}$$

f. Given: 8740 m
Unknown: length in kilometers

$$8740 \text{ m} \times \frac{1 \text{ km}}{1000 \text{ m}} = 8.74 \text{ km}$$

g. Given: 209 nm
Unknown: length in millimeters

$$209 \text{ nm} \times \frac{1 \text{ mm}}{1\,000\,000 \text{ nm}} = 0.000\,209 \text{ mm}$$

h. Given: 500 000 µg
Unknown: mass in kilograms

$$500\,000 \text{ µg} \times \frac{1 \text{ kg}}{1\,000\,000\,000 \text{ µg}} = 0.0005 \text{ kg}$$

3. Given: 152 million km
Unknown: length in megameters

$$152\,000\,000 \text{ km} \times \frac{1 \text{ Mm}}{1000 \text{ km}} = 152\,000 \text{ Mm}$$

4. Given: 1.87 L in a 2.00 L bottle
Unknown: volume in milliliters needed to fill the bottle

$$2.00 \text{ L} - 1.87 \text{ L} = 0.13 \text{ L}$$

$$0.13 \text{ L} \times \frac{1000 \text{ mL}}{\text{L}} = 130 \text{ mL}$$

5. Given: a wire 150 cm long
Unknown: length in millimeters; number of 50 mm segments in the wire

$$150 \text{ cm} \times \frac{10 \text{ mm}}{\text{cm}} = 1500 \text{ mm}$$

$$1500 \text{ mm} \times \frac{1 \text{ piece}}{50 \text{ mm}} = 30 \text{ pieces}$$

6. Given: 8500 kg to fill a ladle
646 metric tons to make rails
1 metric ton = 1000 kg
Unknown: number of ladlefuls to make rails

$$646 \text{ metric tons} \times \frac{1000 \text{ kg}}{\text{metric ton}} = 646\,000 \text{ kg}$$

$$646\,000 \text{ kg} \times \frac{1 \text{ ladleful}}{8500 \text{ kg}} = 76 \text{ ladlefuls}$$

7. a. Given: 310 000 cm^3
Unknown: volume in cubic meters

$$310\,000 \text{ cm}^3 \times \frac{1 \text{ m}^3}{1\,000\,000 \text{ cm}^3} = 0.31 \text{ m}^3$$

b. Given: 6.5 m^2
Unknown: area in square centimeters

$$6.5 \text{ m}^2 \times \frac{10\,000 \text{ cm}^2}{\text{m}^2} = 65\,000 \text{ cm}^2$$

c. Given: 0.035 m^3
Unknown: volume in cubic centimeters

$$0.035 \text{ m}^3 \times \frac{1\,000\,000 \text{ cm}^3}{\text{m}^3} = 35\,000 \text{ cm}^3$$

d. Given: 0.49 cm^2
Unknown: area in square millimeters

$$0.49 \text{ cm}^2 \times \frac{100 \text{ mm}^2}{\text{cm}^2} = 49 \text{ mm}^2$$

e. Given 1200 dm^3
Unknown: volume in cubic meters

$$1200 \text{ dm}^3 \times \frac{1 \text{ m}^3}{1000 \text{ dm}^3} = 1.2 \text{ m}^3$$

f. Given: 87.5 mm^3
Unknown: volume in cubic centimeters

$$87.5 \text{ mm}^3 \times \frac{1 \text{ cm}^3}{1000 \text{ mm}^3} = 0.0875 \text{ cm}^3$$

g. Given: 250 000 cm^2
Unknown: area in square meters

$$250\,000 \text{ cm}^2 \times \frac{1 \text{ m}^2}{10\,000 \text{ cm}^2} = 25 \text{ m}^2$$

8. Given: volume of a cell = 0.0147 mm^3
Unknown: number of cells that fit into a volume of 1.0 cm^3

$$1.0 \text{ cm}^3 \times \frac{1000 \text{ mm}^3}{\text{cm}^3} = 1000.0 \text{ mm}^3$$

$$1000.00 \text{ mm}^3 \times \frac{1 \text{ cell}}{0.0147 \text{ mm}^3} = 68\,027 \text{ cells}$$

9. a. Given: 12.75 Mm
Unknown: length in kilometers

$$12.75 \text{ Mm} \times \frac{1000 \text{ km}}{\text{Mm}} = 12\,750 \text{ km}$$

b. Given: 277 cm
Unknown: length in meters

$$277 \text{ cm} \times \frac{1 \text{ m}}{100 \text{ cm}} = 2.77 \text{ m}$$

c. Given: 30 560 m^2
1 ha = 10 000 m^2
Unknown: area in hectares

$$30\,560 \text{ m}^2 \times \frac{1 \text{ ha}}{10\,000 \text{ m}^2} = 3.056 \text{ ha}$$

d. Given: 81.9 cm²
Unknown: area in square meters

$$81.9 \text{ cm}^2 \times \frac{1 \text{ m}^2}{10\,000 \text{ cm}^2} = 0.00819 \text{ m}^2$$

e. Given: 300 000 km
Unknown: length in megameters

$$300\,000 \text{ km} \times \frac{1 \text{ Mm}}{1000 \text{ km}} = 300 \text{ Mm}$$

10. a. Given: 0.62 km
Unknown: length in meters

$$0.62 \text{ km} \times \frac{1000 \text{ m}}{\text{km}} = 620 \text{ m}$$

b. Given: 3857 g
Unknown: mass in milligrams

$$3857 \text{ g} \times \frac{1000 \text{ mg}}{\text{g}} = 3857\,000 \text{ mg}$$

c. Given: 0.0036 mL
Unknown: volume in microliters

$$0.0036 \text{ mL} \times \frac{1000 \text{ μL}}{\text{mL}} = 3.6 \text{ μL}$$

d. Given: 0.342 metric tons
1 metric ton = 1000 kg
Unknown: mass in kilograms

$$0.342 \text{ metric tons} \times \frac{1000 \text{ kg}}{\text{metric ton}} = 342 \text{ kg}$$

e. Given: 68.71 kL
Unknown: volume in liters

$$68.71 \text{ kL} \times \frac{1000 \text{ L}}{\text{kL}} = 68\,710 \text{ L}$$

11. a. Given: 856 mg
Unknown: mass in kilograms

$$856 \text{ mg} \times \frac{1 \text{ kg}}{1\,000\,000 \text{ mg}} = 0.000\,856 \text{ kg}$$

b. Given: 1 210 000 μg
Unknown: mass in kilograms

$$1\,210\,000 \text{ μg} \times \frac{1 \text{ kg}}{1\,000\,000\,000 \text{ μg}} = 0.00121 \text{ kg}$$

c. Given: 6598 μL
1 mL = 1 cm³
Unknown: volume in cm³

$$6598 \text{ μL} \times \frac{1 \text{ mL}}{1000 \text{ μL}} \times \frac{1 \text{ cm}^3}{\text{mL}} = 6.598 \text{ cm}^3$$

d. Given: 80 600 nm
Unknown: length in millimeters

$$80\,600 \text{ nm} \times \frac{1 \text{ mm}}{1\,000\,000 \text{ nm}} = 0.0806 \text{ mm}$$

e. Given: 10.74 cm³
Unknown: volume in liters

$$10.74 \text{ cm}^3 \times \frac{1 \text{ mL}}{\text{cm}^3} \times \frac{1 \text{ L}}{1000 \text{ mL}} = 0.010\ 74 \text{ L}$$

12. a. Given: 7.93 L
Unknown: volume in cubic centimeters

$$7.93 \text{ L} \times \frac{1000 \text{ mL}}{\text{L}} \times \frac{1 \text{ cm}^3}{\text{mL}} = 7930 \text{ cm}^3$$

b. Given: 0.0059 km
Unknown: length in centimeters

$$0.0059 \text{ km} \times \frac{100\ 000 \text{ cm}}{\text{km}} = 590 \text{ cm}$$

c. Given: 4.19 L
Unknown: volume in cubic decimeters

$$4.19 \text{ L} \times \frac{1 \text{ dm}^3}{\text{L}} = 4.19 \text{ dm}^3$$

d. Given: 7.48 m²
Unknown: area in square centimeters

$$7.48 \text{ m}^2 \times \frac{10\ 000 \text{ cm}^2}{\text{m}^2} = 74\ 800 \text{ cm}^2$$

e. Given: 0.197 m³
Unknown: volume in liters

$$0.197 \text{ m}^3 \times \frac{1000 \text{ dm}^3}{\text{m}^3} \times \frac{1 \text{ L}}{\text{dm}^3} = 197 \text{ L}$$

13. Given: 0.05 mL oil used per kilometer
Unknown: volume in liters of oil used for 20 000 km

$$20\ 000 \text{ km} \times \frac{0.05 \text{ mL}}{\text{km}} \times \frac{1 \text{ L}}{1000 \text{ mL}} = 1 \text{ L}$$

14. Given: 370 mm³
Unknown: volume in microliters

$$370 \text{ mm}^3 \times \frac{1 \text{ cm}^3}{1000 \text{ mm}^3} \times \frac{1 \text{ mL}}{\text{cm}^3} \times \frac{1000 \text{ µL}}{\text{mL}} = 370 \text{ µL}$$

15. Given: 1.5 tsp vanilla per cake
1 tsp = 5 mL
Unknown: volume of vanilla in liters for 800 cakes

$$800 \text{ cakes} \times \frac{1.5 \text{ tsp}}{\text{cake}} \times \frac{5 \text{ mL}}{\text{tsp}} \times \frac{1 \text{ L}}{1000 \text{ mL}} = 6 \text{ L}$$

16. Given: eight 300 mL glasses of water/day
1 yr = 365 days
D_{water} = 1.00 kg/L

Unknown: volume in liters of water consumed in a year; mass in kilograms of this volume

$$\frac{8 \text{ glasses}}{\text{day}} \times \frac{365 \text{ days}}{\text{yr}} \times \frac{300 \text{ mL}}{\text{glass}} \times \frac{1 \text{ L}}{1000 \text{ mL}} = 876 \text{ L per year}$$

$$876 \text{ L} \times \frac{1 \text{ kg}}{\text{L}} = 876 \text{ kg}$$

---

17. a. Given: 465 m/s
Unknown: velocity in kilometers per hour

$$\frac{465 \text{ m}}{\text{s}} \times \frac{1 \text{ km}}{1000 \text{ m}} \times \frac{3600 \text{ s}}{\text{h}} = 1674 \text{ km/h}$$

b. Given: 465 m/s
Unknown: velocity in kilometers per day

$$\frac{1674 \text{ km}}{\text{h}} \times \frac{24 \text{ h}}{\text{day}} = 40\ 176 \text{ km/day}$$

---

18. Given: 130 g/student; 60 students
Unknown: total mass in kilograms

$$\frac{130 \text{ g}}{\text{student}} \times 60 \text{ students} \times \frac{1 \text{ kg}}{1000 \text{ g}} = 7.8 \text{ kg}$$

---

19. Given: 750 mm/student; 60 students
Unknown: total length in meters

$$\frac{750 \text{ mm}}{\text{student}} \times 60 \text{ students} \times \frac{1 \text{ m}}{1000 \text{ mm}} = 45 \text{ m}$$

---

20. a. Given: 550 μL/h
Unknown: rate in milliliters per day

$$\frac{550 \text{ μL}}{\text{h}} \times \frac{1 \text{ mL}}{1000 \text{ μL}} \times \frac{24 \text{ h}}{\text{day}} = 13.2 \text{ mL/day}$$

b. Given: 9.00 metric tons/h
Unknown: rate in kilograms per minute

$$\frac{9.00 \text{ metric tons}}{\text{h}} \times \frac{1000 \text{ kg}}{\text{metric ton}} \times \frac{1 \text{ h}}{60 \text{ min}} = 150 \text{ kg/min}$$

c. Given: 3.72 L/h
Unknown: rate in cubic centimeters per minute

$$\frac{3.72 \text{ L}}{\text{h}} \times \frac{1000 \text{ mL}}{\text{L}} \times \frac{1 \text{ cm}^3}{\text{mL}} \times \frac{1 \text{ h}}{60 \text{ min}} = 62 \text{ cm}^3/\text{min}$$

d. Given: 6.12 km/h
Unknown: rate in meters per second

$$\frac{6.12 \text{ km}}{\text{h}} \times \frac{1000 \text{ m}}{\text{km}} \times \frac{1 \text{ h}}{3600 \text{ s}} = 1.7 \text{ m/s}$$

21. a. Given: 2.97 kg/L
Unknown: rate in g/cm^3

$$\frac{2.97 \text{ kg}}{\text{L}} \times \frac{1000 \text{ g}}{\text{kg}} \times \frac{1 \text{ L}}{1000 \text{ mL}} \times \frac{1 \text{ mL}}{\text{cm}^3} = 2.97 \text{ g/cm}^3$$

b. Given: 4128 g/dm^2
Unknown: mass per area in kilograms per square centimeters

$$\frac{4128 \text{ g}}{\text{dm}^2} \times \frac{1 \text{ kg}}{1000 \text{ g}} \times \frac{1 \text{ dm}^2}{100 \text{ cm}^2} = 0.04128 \text{ kg/cm}^2$$

c. Given: 5.27 g/cm^3
Unknown: density as kilograms per cubic decimeter

$$\frac{5.27 \text{ g}}{\text{cm}^3} \times \frac{1 \text{ kg}}{1000 \text{ g}} \times \frac{1000 \text{ cm}^3}{\text{dm}^3} = 5.27 \text{ kg/dm}^3$$

d. Given: 6.91 kg/m^3
Unknown: density as milligrams per cubic millimeter

$$\frac{6.91 \text{ kg}}{\text{m}^3} \times \frac{1\,000\,000 \text{ mg}}{\text{kg}} \times \frac{1 \text{ m}^3}{1\,000\,000\,000 \text{ mm}^3} = 0.00691 \text{ mg/mm}^3$$

22. a. Given: density of 5.56 g/L
Unknown: volume in milliliters occupied by 4.17 g

$$4.17 \text{ g} \times \frac{1 \text{ L}}{5.56 \text{ g}} \times \frac{1000 \text{ mL}}{\text{L}} = 750 \text{ mL}$$

b. Given: density of 5.56 g/L
Unknown: mass in kilograms of 1 m^3

$$1 \text{ m}^3 \times \frac{1000 \text{ dm}^3}{\text{m}^3} \times \frac{1 \text{ L}}{\text{dm}^3} \times \frac{5.56 \text{ g}}{\text{L}} \times \frac{1 \text{ kg}}{1000 \text{ g}} = 5.56 \text{ kg}$$

23. Given: mass per area of 0.10 g/cm^2
Unknown: mass in kilometers per 0.125 ha

$$0.125 \text{ ha} \times \frac{10\,000 \text{ m}^2}{\text{ha}} \times \frac{10\,000 \text{ cm}^2}{\text{m}^2} \times \frac{0.10 \text{ g}}{\text{cm}^2} \times \frac{1 \text{ kg}}{1000 \text{ g}} = 1250 \text{ kg}$$

24. a. Given: length of book
= 250. mm
width of book
= 224 mm
thickness of book =
50.0 mm
Unknown: volume in cubic meters

$$250.\text{ mm} \times 224\text{ mm} \times 50.0\text{ mm} = 2\,800\,000\text{ mm}^3 \times \frac{1\text{ m}^3}{1\,000\,000\,000\text{ mm}^3}$$
$$= 0.0028\text{ m}^3$$

b. Given: length of book
= 250. mm
width of book
= 224 mm
thickness of book =
50.0 mm
mass of book
= 2.94 kg
Unknown: density in grams per cubic centimeter

$$\frac{2.94\text{ kg}}{0.0028\text{ m}^3} \times \frac{1000\text{ g}}{\text{kg}} \times \frac{1\text{ m}^3}{1\,000\,000\text{ cm}^3} = 1.05\text{ g/cm}^3$$

c. Given: length of book
= 250. mm
width of book
= 224 mm
thickness of book =
50.0 mm
Unknown: area of front cover in square meters

$$250\text{ mm} \times 224\text{ mm} \times \frac{1\text{ m}^2}{1\,000\,000\text{ mm}^2} = 0.056\text{ m}^2$$

---

25. a. Given: 25 drops = 1.00 mL
Unknown: volume of one drop in milliliters

$$\frac{1.00\text{ mL}}{25\text{ drops}} = 0.04\text{ mL}$$

b. Given: 25 drops = 1.00 mL
Unknown: volume in milliliters of 37 drops

$$37\text{ drops} \times \frac{0.04\text{ mL}}{\text{drop}} = 1.48\text{ mL}$$

c. Given: 25 drops = 1.00 mL
Unknown: number of drops in 0.68 L

$$0.68\text{ L} \times \frac{1000\text{ mL}}{\text{L}} \times \frac{1\text{ drop}}{0.04\text{ mL}} = 17\,000\text{ drops}$$

26. a. Given: 504 700 mg
Unknown: mass in kilograms and grams

$$504\ 700\ \text{mg} \times \frac{1\ \text{kg}}{1000\ 000\ \text{mg}} = 0.5047\ \text{kg}$$

$$504\ 700\ \text{mg} \times \frac{1\ \text{g}}{1000\ \text{mg}} = 504.7\ \text{g}$$

b. Given: 9 200 000 μg
Unknown: mass in kilograms and grams

$$9\ 200\ 000\ \mu\text{g} \times \frac{1\ \text{kg}}{1000\ 000\ 000\ \mu\text{g}} = 0.0092\ \text{kg}$$

$$9\ 200\ 000\ \mu\text{g} \times \frac{1\ \text{g}}{1000\ 000\ \mu\text{g}} = 9.2\ \text{kg}$$

c. Given: 122 mg
Unknown: mass in kilograms and grams

$$122\ \text{mg} \times \frac{1\ \text{kg}}{1000\ 000\ \text{mg}} = 0.000\ 122\ \text{kg}$$

$$122\ \text{mg} \times \frac{1\ \text{g}}{1000\ \text{mg}} = 0.122\ \text{g}$$

d. Given: 7195 cg
Unknown: mass in kilograms and grams

$$7195\ \text{cg} \times \frac{1\ \text{kg}}{100\ 000\ \text{cg}} = 0.07195\ \text{kg}$$

$$7195\ \text{cg} \times \frac{1\ \text{g}}{100\ \text{cg}} = 71.95\ \text{g}$$

27. a. Given: 582 cm³
Unknown: volume in liters and milliliters

$$582\ \text{cm}^3 \times \frac{1\ \text{mL}}{\text{cm}^3} \times \frac{1\ \text{L}}{1000\ \text{mL}} = 0.582\ \text{L}$$

$$582\ \text{cm}^3 \times \frac{1\ \text{mL}}{\text{cm}^3} = 582\ \text{mL}$$

b. Given: 0.0025 m³
Unknown: volume in liters and milliliters

$$0.0025\ \text{m}^3 \times \frac{1000\ \text{dm}^3}{\text{m}^3} \times \frac{1\ \text{L}}{\text{dm}^3} = 2.5\ \text{L}$$

$$0.0025\ \text{m}^3 \times \frac{1000\ 000\ \text{cm}^3}{\text{m}^3} \times \frac{1\ \text{mL}}{\text{cm}^3} = 2500\ \text{mL}$$

c. Given: 1.18 dm³
Unknown: volume in liters and milliliters

$$1.18\ \text{dm}^3 \times \frac{1\ \text{L}}{\text{dm}^3} = 1.18\ \text{L}$$

$$1.18\ \text{dm}^3 \times \frac{1\ \text{L}}{\text{dm}^3} \times \frac{1000\ \text{mL}}{\text{L}} = 1180\ \text{mL}$$

d. Given: 32 900 μL
Unknown: volume in liters and milliliters

$$32\ 900\ \mu\text{L} \times \frac{1\ \text{L}}{1000\ 000\ \mu\text{L}} = 0.0329\ \text{L}$$

$$32\ 900\ \mu\text{L} \times \frac{1\ \text{mL}}{1000\ \mu\text{L}} = 32.9\ \text{mL}$$

28. a. Given: 1.37 g/cm³
Unknown: density in grams per liter and kilograms per cubic meter

$$\frac{1.37\ \text{g}}{\text{cm}^3} \times \frac{1\ \text{cm}^3}{\text{mL}} \times \frac{1000\ \text{mL}}{\text{L}} = 1370\ \text{g/L}$$

$$\frac{1.37\ \text{g}}{\text{cm}^3} \times \frac{1\ \text{kg}}{1000\ \text{g}} \times \frac{1000\ 000\ \text{cm}^3}{\text{m}^3} = 1370\ \text{kg/m}^3$$

b. Given: 0.692 kg/dm^3
Unknown: density in grams per liter and kilograms per cubic meter

$$\frac{0.692 \text{ kg}}{\text{dm}^3} \times \frac{1000 \text{ g}}{\text{kg}} \times \frac{1 \text{ dm}^3}{\text{L}} = 692 \text{ g/L}$$

$$\frac{0.692 \text{ kg}}{\text{dm}^3} \times \frac{1000 \text{ dm}^3}{\text{m}^3} = 692 \text{ kg/m}^3$$

c. Given: 5.2 kg/L
Unknown: density in gams per liter and kilograms per cubic meter

$$\frac{5.2 \text{ kg}}{\text{L}} \times \frac{1000 \text{ g}}{\text{kg}} = 5200 \text{ g/L}$$

$$\frac{5.2 \text{ kg}}{\text{L}} \times \frac{1 \text{ L}}{\text{dm}^3} \times \frac{1000 \text{ dm}^3}{\text{m}^3} = 5200 \text{ kg/m}^3$$

d. Given: $38\,000 \text{ g/m}^3$
Unknown: density in grams per liter and kilograms per cubic meter

$$\frac{38\,000 \text{ g}}{\text{m}^3} \times \frac{1 \text{ m}^3}{1000 \text{ dm}^3} \times \frac{1 \text{ dm}^3}{\text{L}} = 38 \text{ g/L}$$

$$\frac{38\,000 \text{ g}}{\text{m}^3} \times \frac{1 \text{ kg}}{1000 \text{ g}} = 38 \text{ kg/m}^3$$

e. Given: 5.79 mg/mm^3
Unknown: density in grams per liter and kilograms per cubic meter

$$\frac{5.79 \text{ mg}}{\text{mm}^3} \times \frac{1 \text{ g}}{1000 \text{ mg}} \times \frac{1000\,000 \text{ mm}^3}{\text{dm}^3} \times \frac{1 \text{ dm}^3}{\text{L}} = 5790 \text{ g/L}$$

$$\frac{5.79 \text{ mg}}{\text{mm}^3} \times \frac{1 \text{ kg}}{1000\,000 \text{ mg}} \times \frac{1000\,000\,000 \text{ mm}^3}{\text{m}^3} = 5790 \text{ kg/m}^3$$

f. Given: 1.1 μg/ml
Unknown: density in grams per liter and kilograms per cubic meter

$$\frac{1.1 \text{ μg}}{\text{mL}} \times \frac{1 \text{ g}}{1000\,000 \text{ μg}} \times \frac{1000 \text{ mL}}{\text{L}} = 0.0011 \text{ g/L}$$

$$\frac{1.1 \text{ μg}}{\text{mL}} \times \frac{1 \text{ kg}}{1000\,000\,000 \text{ μg}} \times \frac{1 \text{ mL}}{\text{cm}^3} \times \frac{1000\,000 \text{ cm}^3}{\text{m}^3} = 0.0011 \text{ kg/m}^3$$

---

29. a. Given: $648 \text{ kg}/30.0 \text{ h}$
Unknown: rate in grams per minute

$$\frac{648 \text{ kg}}{30.0 \text{ h}} \times \frac{1000 \text{ g}}{\text{kg}} \times \frac{1 \text{ h}}{60 \text{ min}} = 360 \text{ g/min}$$

b. Given: $648 \text{ kg}/30.0 \text{ h}$
Unknown: rate in kilograms per day

$$\frac{648.0 \text{ kg}}{30.0 \text{ h}} \times \frac{24 \text{ h}}{\text{day}} = 518.4 \text{ kg/day}$$

c. Given: $648 \text{ kg}/30.0 \text{ h}$
Unknown: rate in milligrams per millisecond

$$\frac{648.0 \text{ kg}}{30.0 \text{ h}} \times \frac{1000\,000 \text{ mg}}{\text{kg}} \times \frac{1 \text{ h}}{3600 \text{ s}} \times \frac{1 \text{ s}}{1000 \text{ ms}} = 6 \text{ mg/ms}$$

30. Given: 100 km/h
Unknown: rate in meters per second

$$\frac{100.\text{ km}}{\text{h}} \times \frac{1000\text{ m}}{\text{km}} \times \frac{1\text{ h}}{3600\text{ s}} = 27.8 \text{ m/s}$$

31. Given: 330 kJ/min
1 cal = 4.184 J
Unknown: rate in kilocalories per hour

$$\frac{330\text{ kJ}}{\text{min}} \times \frac{60\text{ min}}{\text{h}} \times \frac{1\text{ kcal}}{4.184\text{ kJ}} = 4732 \text{ kcal/h}$$

32. Given: 62 g/m²
1 ha = 10 000 m²
Unknown: mass in kilograms required for 1.0 ha

$$\frac{62\text{ g}}{\text{m}^2} \times \frac{10\,000\text{ m}^2}{\text{ha}} \times \frac{1\text{ kg}}{1000\text{ g}} \times \frac{1.0\text{ ha}}{1000\text{ g}} = 620 \text{ kg}$$

33. Given: 3.9 mL/h
1 year = 365 days
Unknown: volume in liters per year

$$\frac{3.9\text{ mL}}{\text{h}} \times \frac{1\text{ L}}{1000\text{ mL}} \times \frac{24\text{ h}}{\text{day}} \times \frac{365\text{ days}}{\text{yr}} = 34 \text{ L/yr}$$

34. Given: 50 μL/dose
2.0 mL/bottle
Unknown: number of doses in a bottle

$$2.0\text{ mL} \times \frac{1\text{ dose}}{50\text{ μL}} \times \frac{1000\text{ μL}}{\text{mL}} = 40 \text{ doses}$$

35. a. Given: 640 cm³
Unknown: number of signicant figures

2; the zero is not significant

b. Given: 200.0 mL
Unknown: number of significant figures

4; all digits are significant

c. Given: 0.5200 g
Unknown: number of significant figures

4; all digits to the right of the decimal point are significant

d. Given: 1.005 kg
Unknown: number of significant figures

4; all digits are significant

e. Given: 10 000 L
Unknown: number of significant figures

1; the zeros are placeholders

f. Given: 20.900 cm
Unknown: number of significant figures

5; all digits are significant

g. Given: 0.000 000 56 g/L
Unknown: number of significant figures

2; all the zeros are placeholders

h. Given: 0.040 02 kg/m^3
Unknown: number of significant figures

4; the two initial zeros are placeholders

i. Given: 790 001 cm^2
Unknown: number of significant figures

6; all digits are significant

j. Given: 665.000 kg·m/s^2
Unknown: number of significant figures

6; all digits are significant

38. a. Given: 0.0120 m
Unknown: number of significant figures

3; the two initial zeros are placeholders

b. Given: 100.5 mL
Unknown: number of significant figures

4; all digits are significant

c. Given: 101 g
Unknown: number of significant figures

3; all digits are significant

d. Given: 350 cm^2
Unknown: number of significant figures

2; zero is a placeholder

e. Given: 0.97 km
Unknown: number of significant figures

2; zero is a placeholder

f. Given: 1000 kg
Unknown: number of significant figures

1; zeros are placeholders

g. Given: 180. mm
Unknown: number of significant figures

3; all digits are significant

h. Given: 0.4936 L
Unknown: number of significant figures

4; zero is a placeholder

i. Given: 0.020 700 s
Unknown: number of significant figures

5; initial zeros are placeholders

39. a. Given: 5 487 129 m
Unknown: value expressed to 3 significant figures

5 490 000 m; the digit following the last digit to be retained is greater than 5

b. Given: 0.013 479 265 mL
Unknown: value expressed to 6 significant figures

0.013 479 3; initial zeros are placeholders; the digit following the last digit to be retained is greater than 5

c. Given: 31 947.972 cm^2

Unknown: value expressed to 4 significant figures

31 950 cm^2; the digit following the last digit to be retained is greater than 5

d. Given: 192.6739 m^2

Unknown: value expressed to 5 significant figures

192.67 m^2; the digit following the last digit to be retained is less than 5

e. Given: 786.9164 cm

Unknown: value expressed to 2 significant figures

790 cm; the digit following the last digit to be retained is greater than 5

f. Given: 389 277 600 J

Unknown: value expressed to 6 significant figures

389 278 000 J; the digit following the last digit to be retained is greater than 5; the zeros are placeholders

g. Given: 225 834.762 cm^3

Unknown: value expressed to 7 significant figures

225 834.8 cm^3; the digit following the last to be retained is greater than 5

---

42. a. Given: dimensions of 87.59 cm × 35.1 mm

Unknown: area in cm^2

$$87.59 \text{ cm} \times 35.1 \text{ mm} \times \frac{1 \text{ cm}}{10 \text{ mm}} = 307 \text{ cm}^2$$

b. Given: dimensions of 87.59 cm × 35.1 mm

Unknown: area in mm^2

$$87.59 \text{ cm} \times \frac{10 \text{ mm}}{1 \text{ cm}} \times 35.1 \text{ mm} = 30\ 700. \text{ mm}^2$$

c. Given: dimensions of 87.59 cm × 35.1 mm

Unknown: area in m^2

$$87.59 \text{ cm} \times \frac{1 \text{ m}}{100 \text{ cm}} \times 35.1 \text{ mm} \times \frac{1 \text{ m}}{1000 \text{ mm}} = 0.0307 \text{ m}^2$$

43. a. Given: dimensions of 900. mm × 31.5 mm × 6.3 cm

Unknown: volume in cm^3

$$900.\ \text{mm} \times \frac{10\ \text{cm}}{10\ \text{mm}} \times 31.5\ \text{mm} \times \frac{1\ \text{cm}}{10\ \text{mm}} \times 6.3\ \text{cm} = 1800\ \text{cm}^3$$

b. Given: dimensions of 900. mm × 31.5 mm × 6.3 cm

Unknown: volume in m^3

$$900.\ \text{mm} \times \frac{1\ \text{m}}{1000\ \text{mm}} \times 31.5\ \text{mm} \times \frac{1\ \text{m}}{1000\ \text{mm}} \times 6.3\ \text{cm} \times \frac{1\ \text{m}}{100\ \text{cm}}$$
$$= 0.0018\ \text{m}^3$$

c. Given: dimensions of 900. mm × 31.5 mm × 6.3 cm

Unknown: volume in mm^3

$$900.\ \text{mm} \times 31.5\ \text{mm} \times 6.3\ \text{cm} \times \frac{10\ \text{mm}}{1\ \text{cm}} = 1\ 800\ 000\ \text{mm}^3$$

---

44. a. Given: 0.16 kg/125 mL

Unknown: density in kg/m^3

$$\frac{0.16\ \text{kg}}{125\ \text{mL}} \times \frac{1\ \text{mL}}{1\ \text{cm}^3} \times \frac{1\ 000\ 000\ \text{cm}^3}{1\ \text{m}^3} = 1300\ \text{kg/m}^3$$

b. Given: 0.16 kg/125 mL

Unknown: density in g/mL

$$\frac{0.16\ \text{kg}}{125\ \text{mL}} \times \frac{1000\ \text{g}}{1\ \text{kg}} = 1.3\ \text{g/mL}$$

c. Given: 0.16 kg/125 mL

Unknown: density in kg/dm^3

$$\frac{0.16\ \text{kg}}{125\ \text{mL}} \times \frac{1000\ \text{mL}}{1\ \text{L}} \times \frac{1\ \text{L}}{1\ \text{dm}^3} = 1.3\ \text{kg/dm}^3$$

---

45. a. Given: numbers with 4, 3, and 2 significant figures, respectively

Unknown: product to 2 significant figures

$$13.75\ \text{mm} \times 10.1\ \text{mm} \times 0.91\ \text{mm} = 130\ \text{mm}^3$$

b. Given: numbers with 3 and 2 significant figures, respectively

Unknown: product to 2 significant figures

$$89.4\ \text{cm}^2 \times 4.8\ \text{cm} = 430\ \text{cm}^3$$

c. Given: numbers with 3 and 2 significant figures, respectively

Unknown: quotient to 2 significant figures

$14.9 \text{ m}^3 \div 3.0 \text{ m}^2 = 5.0 \text{ m}$

d. Given: numbers with 4, 1, and 3 significant figures, respectively

Unknown: product to 1 significant figure

$6.975 \text{ m} \times 30 \text{ m} \times 21.5 \text{ m} = 4000 \text{ m}^3$

48. Given: dimensions of 30.5 mm × 202 mm × 153 mm; mass empty = 0.30 kg; mass full = 1.33 kg

Unknown: density of the liquid in kg/L

$V = 30.5 \text{ mm} \times 202 \text{ mm} \times 153 \text{ mm} = 943\,000 \text{ mm}^3$

$m = 1.33 \text{ kg} - 0.30 \text{ kg} = 1.03 \text{ kg}$

$D = \dfrac{1.03 \text{ kg}}{943\,000 \text{ mm}^3} \times \dfrac{1.000\,000 \text{ mm}^3}{1 \text{ dm}^3} \times \dfrac{1 \text{ dm}^3}{1 \text{ L}} = 1.09 \text{ kg/L}$

49. Given: 3.3 kg/7.76 km

Unknown: mass in g/m; length with a mass of 1.0 g

$\dfrac{3.3 \text{ kg}}{7.76 \text{ km}} \times \dfrac{1000 \text{ g}}{1 \text{ kg}} \times \dfrac{1 \text{ km}}{1000 \text{ m}} = 0.43 \text{ g/m}$

$1.0 \text{ g} \times \dfrac{1 \text{ m}}{0.43 \text{ g}} = 2.3 \text{ m}$

50. Given: rate of 52 kg/ha; container holds 10 kg; 1 ha = 10 000 m²

Unknown: area in m² covered by full container

$\dfrac{10 \text{ kg}}{1 \text{ container}} \times \dfrac{1 \text{ ha}}{52 \text{ kg}} \times \dfrac{10\,000 \text{ m}^2}{1 \text{ ha}} = 2000 \text{ m}^2$

51. Given: 974 550 kJ/37.0 min

Unknown: rate in kJ/min and kJ/s

$\dfrac{974\,550 \text{ kJ}}{37.0 \text{ min}} = 26\,300 \text{ kJ/min}$

$\dfrac{974\,550 \text{ kJ}}{37.0 \text{ min}} \times \dfrac{1 \text{ min}}{60 \text{ s}} = 439 \text{ kJ/s}$

52. a. Given: dimensions of 189 cm × 307 cm × 272 cm

Unknown: volume in cubic meters

$189 \text{ cm} \times 307 \text{ cm} \times 272 \text{ cm} \times \dfrac{1 \text{ m}^3}{1\,000\,000 \text{ cm}^3} = 15.8 \text{ m}^3$

b. Given: dimensions of 189 cm × 307 cm × 272 cm; fill time of 97 s

Unknown: rate in liters per minute

$$\frac{189 \text{ cm} \times 307 \text{ cm} \times 272 \text{ cm}}{97 \text{ s}} \times \frac{1 \text{ dm}^3}{1000 \text{ cm}^3} \times \frac{60 \text{ s}}{1 \text{ min}} \times \frac{1 \text{ L}}{1 \text{ dm}^3} = 9800 \text{ L/min}$$

c. Given: dimensions of 189 cm × 307 cm × 272 cm; fill time of 97 s

Unknown: rate in cubic meters per hour

$$\frac{189 \text{ cm} \times 307 \text{ cm} \times 272 \text{ cm}}{97 \text{ s}} \times \frac{1 \text{ m}^3}{1\,000\,000 \text{ cm}^3} \times \frac{3600 \text{ s}}{1 \text{ h}} = 590 \text{ m}^3/\text{h}$$

55. a. Given: lengths expressed in scientific notation

Unknown: sum expressed in scientific notation to hundredths place

4.74×10^4 km + 7.71×10^3 km + 1.05×10^3 km
= 4.74×10^4 km + 0.771×10^4 km + 0.105×10^4 km
= 5.62×10^4 km

b. Given: lengths expressed in scientific notation

Unknown: sum expressed in scientific notation to thousandths place

2.75×10^{-4} m + 8.03×10^{-5} m + 2.122×10^{-3} m
= 0.275×10^{-3} m + 0.0803×10^{-3} m + 2.122×10^{-3} m
= 2.477×10^{-3} m

c. Given: volume expressed in scientific notation

Unknown: answer expressed in scientific notation to tenths place

4.0×10^{-5} m^3 + 6.85×10^{-6} m^3 − 1.05×10^{-5} m^3
= 4.0×10^{-5} m^3 + 0.685×10^{-5} m^3 − 1.05×10^{-5} m^3
= 3.6×10^{-5} m^3

d. Given: masses expressed in scientific notation

Unknown: sum expressed in scientific notation to hundredths place

3.15×10^2 mg $+ 3.15 \times 10^3$ mg $+ 3.15 \times 10^4$ mg
$= 0.0315 \times 10^4$ mg $+ 0.315 \times 10^4$ mg $+ 3.15 \times 10^4$ mg
$= 3.50 \times 10^4$ mg

e. Given: number of atoms expressed in scientific notation

Unknown: sum expressed in scientific notation to hundredths place

3.01×10^{22} atoms $+ 1.19 \times 10^{23}$ atoms $+ 9.80 \times 10^{21}$ atoms
$= 0.301 \times 10^{23}$ atoms $+ 1.19 \times 10^{23}$ atoms $+ 0.0980 \times 10^{23}$ atoms
$= 1.59 \times 10^{23}$ atoms

f. Given: lengths expressed in scientific notation

Unknown: answer expressed in scientific notation to thousandths place

6.85×10^7 nm $+ 4.0229 \times 10^8$ nm $- 8.38 \times 10^6$ nm
$= 0.685 \times 10^8$ nm $+ 4.0229 \times 10^8$ nm $- 0.0838 \times 10^8$ nm
$= 4.624 \times 10^8$ nm

65. a. Given: 7.11×10^{24} molecules per 100.0 cm^3

Unknown: number of molecules per 1.09 cm^3

$$\frac{7.11 \times 10^{24} \text{ molecules}}{100.0 \text{ cm}^3} \times 1.09 \text{ cm}^3 = 7.75 \times 10^{22} \text{ molecules}$$

b. Given: 7.11×10^{24} molecules per 100.0 cm^3

Unknown: number of molecules in 2.24×10^4 cm^3

$$\frac{7.11 \times 10^{24} \text{ molecules}}{100.0 \text{ cm}^3} \times 2.24 \times 10^4 \text{ cm}^3 = 0.159 \times 10^{28} \text{ molecules}$$

$= 1.59 \times 10^{27}$ molecules

c. Given: 7.11×10^{24} molecules per 100.0 cm^3

Unknown: number of molecules in $9.01 \times 10^{-6} \text{ cm}^3$

$$\frac{7.11 \times 10^{24} \text{ molecules}}{100.0 \text{ cm}^3} \times 9.01 \times 10^{-6} \text{ cm}^3$$

$= 0.641 \times 10^{18}$ molecules $= 6.41 \times 10^{17}$ molecules

66. a. Given: 3 518 000 transistors per $9.5 \text{ mm} \times 8.2 \text{ mm}$

Unknown: area per transistor

$$\frac{9.5 \text{ mm} \times 8.2 \text{ mm}}{3\,578\,000 \text{ transistors}} = 0.000\,022 \text{ mm}^2/\text{transistor}$$

$= 2.2 \times 10^{-5} \text{ mm}^2/\text{transistor}$

b. Given: $2.2 \times 10^{-5} \text{ mm}^2/$transistor

Unknown: number of transistors on $353 \text{ mm} \times 265 \text{ mm}$

$$353 \text{ mm} \times 265 \text{ mm} \times \frac{1 \text{ transistor}}{2.2 \times 10^{-5} \text{ mm}^2} = 43\,000 \times 10^5 \text{ transistors}$$

$= 4.3 \times 10^9$ transistors

67. Given: 0.0501 g per 1.00 L

Unknown: concentration in grams per microliter

$$\frac{0.0501 \text{ g}}{1.00 \text{ L}} \times \frac{1 \text{ L}}{1 \times 10^6 \text{ μL}} = 0.0501 \times 10^{-6} \text{ g/μL} = 5.01 \times 10^{-8} \text{ g/μL}$$

68. Given: 5.30×10^{-10} m/Cs atom

Unknown: number of Cs atoms in 2.54 cm

$$\frac{1 \text{ Cs atom}}{5.30 \times 10^{-10} \text{ m}} \times \frac{1 \text{ m}}{10^2 \text{ cm}} \times 2.54 \text{ cm} = 4.79 \times 10^7 \text{ Cs atoms}$$

69. Given: $V_{\text{neutron}} = 1.4 \times 10^{-44} \text{ m}^3$
$M_{\text{neutron}} = 1.675 \times 10^{-24}$ g

Unknown: D_{neutron} in g/m^3 mass of 1.0 cm^3 of neutrons in kg

$$D_{\text{neutron}} = \frac{1.675 \times 10^{-24} \text{ g}}{1.4 \times 10^{-44} \text{ m}^3} = 1.2 \times 10^{20} \text{ g/m}^3$$

$$\frac{1.675 \times 10^{-24} \text{ g}}{1.4 \times 10^{-44} \text{ m}^3} \times \frac{1 \text{ kg}}{10^3 \text{ g}} \times \frac{1 \text{ m}^3}{10^6 \text{ cm}^3} \times 1.0 \text{ m}^3 = 1.2 \times 10^{11} \text{ kg}$$

70. Given: 1.6×10^{-8} m per pit

Unknown: number of pits in 0.305 m

$$0.305 \text{ m} \times \frac{1 \text{ pit}}{1.6 \times 10^{-8} \text{ m}} = 1.9 \times 10^7 \text{ pits}$$

71. a. Given: 6.022×10^{23} O_2 molecules per 22 400 mL at 0°C and standard atmospheric pressure

Unknown: number of O_2 molecules in 0.100 mL

$$\frac{6.022 \times 10^{23} \text{ } O_2 \text{ molecules}}{22\ 400 \text{ mL}} \times 0.100 \text{ mL} = 0.000\ 026\ 9 \times 10^{23} \text{ } O_2 \text{ molecules}$$

$$= 2.69 \times 10^{18} \text{ } O_2 \text{ molecules}$$

b. Given: 6.022×10^{23} O_2 molecules per 22 400 mL at 0°C and standard atmospheric pressure

Unknown: number of O_2 molecules in 1.00 L

$$\frac{6.022 \times 10^{23} \text{ } O_2 \text{ molecules}}{22\ 400 \text{ mL}} \times \frac{10^3 \text{ mL}}{1 \text{ L}} \times 1.00 \text{ L}$$

$$= 0.000\ 269 \times 10^{26} \text{ } O_2 \text{ molecules} = 2.69 \times 10^{22} \text{ } O_2 \text{ molecules}$$

c. Given: 6.022×10^{23} O_2 molecules per 22 400 mL at 0°C and standard atmospheric pressure

Unknown: average space in milliliters occupied by one oxygen molecule

$$\frac{22\ 400 \text{ mL}}{6.022 \times 10^{23} \text{ } O_2 \text{ molecules}} = 3720 \times 10^{-23} \text{ mL/}O_2 \text{ molecule}$$

$$= 3.72 \times 10^{-20} \text{ mL/}O_2 \text{ molecule}$$

72. a. Given: $m = 5.136 \times 10^{18}$ kg; 6 500 000 000 people

Unknown: mass in kg per person

$$\frac{5.136 \times 10^{18} \text{ kg}}{6.5 \times 10^9 \text{ people}} = 7.9 \times 10^8 \text{ kg/person}$$

b. Given: $m = 5.136 \times 10^{18}$ kg; 6 500 000 000 people

Unknown: mass in metric tons per person

$$\frac{5.136 \times 10^{18} \text{ kg}}{6.5 \times 10^9 \text{ people}} \times \frac{1 \text{ metric ton}}{10^3 \text{ kg}} = 0.79 \times 10^6 \text{ metric ton/person}$$

$$= 7.9 \times 10^5 \text{ metric ton/person}$$

c. Given: $m = 5.136 \times 10^{18}$ kg; 9 500 000 000 people

Unknown: mass in kg per person

$$\frac{5.136 \times 10^{18} \text{ kg}}{9.5 \times 10^9 \text{ people}} = 0.54 \times 10^9 \text{ kg/person} = 5.4 \times 10^8 \text{ kg/person}$$

73. Given: $m_{sun} = 1.989 \times 10^{30}$ kg
$m_{earth} = 5.974 \times 10^{24}$ kg

Unknown: number of Earths to equal mass of sun

$$\frac{1.989 \times 10^{30} \text{ kg}}{5.974 \times 10^{24} \text{ kg}} = 0.3329 \times 10^6 = 3.329 \times 10^5 \text{ Earths}$$

74. c. Given: landfill dimensions of 2.3 km × 1.4 km × 0.15 km; 250 000 000 objects, each 0.060 m³, per year

Unknown: how many years to fill landfill

$$\frac{2.3 \text{ km} \times 1.4 \text{ km} \times 0.15 \text{ km}}{2.5 \times 10^8 \text{ objects/yr} \times 6.0 \times 10^{-2} \text{ m}^3/\text{objects}} \times \frac{10^9 \text{ m}^3}{1 \text{ km}^3}$$
$$= 0.032 \times 10^3 \text{ yr} = 32 \text{ yr}$$

75. Given: 1 C = 1000 cal
intake of 2400 C per day
1 cal = 4.184 J

Unknown: intake in joules per day

$$\frac{2400 \text{ C}}{1 \text{ day}} \times \frac{1000 \text{ cal}}{1 \text{ C}} \times \frac{4.184 \text{ J}}{1 \text{ cal}} = 10\,000\,000 \text{ J/day} = 1.0 \times 10^7 \text{ J/day}$$

76. Given: $D = 0.73$ g/cm³
$m_{automobile} = 1271$ kg

Unknown: volume in L of gasoline to raise mass of car to 1305 kg

$$\frac{1305 \text{ kg} - 1271 \text{ kg}}{0.73 \text{ g/cm}^3} \times \frac{1 \text{ L}}{10^3 \text{ mL}} \times \frac{10^3 \text{ g}}{1 \text{ kg}} \times \frac{1 \text{ mL}}{1 \text{ cm}^3} = 47 \text{ L}$$

77. Given: pool dimensions of 9.0 m × 3.5 m × 1.75 m
D_{water} = 0.997 g/cm³
1 metric ton = 1000 kg
Unknown: mass of water in pool in metric tons

$$9.0 \text{ m} \times 3.5 \text{ m} \times 1.75 \text{ m} \times \frac{0.997 \text{ g}}{1 \text{ cm}^3} \times \frac{10^6 \text{ cm}^3}{1 \text{ m}^3} \times \frac{1 \text{ kg}}{10^3 \text{ g}} \times \frac{1 \text{ metric ton}}{10^3 \text{ kg}}$$
= 55 metric tons

78. Given: m = 250 g; dimensions of 7.0 cm × 17.0 cm × 19.0 cm
Unknown: density in kilograms per liter

$$\frac{250 \text{ g}}{7.0 \text{ cm} \times 17.0 \text{ cm} \times 19.0 \text{ cm}} \times \frac{1 \text{ kg}}{10^3 \text{ g}} \times \frac{10 \text{ cm}^3}{1 \text{ mL}} \times \frac{10^3 \text{ mL}}{1 \text{ L}} = 0.11 \text{ kg/L}$$

79. Given: area of 18.5 m²; mass of 1275 g; density of 2.7 g/cm³
Unknown: thickness in millimeters

$$\frac{1275 \text{ g}}{18.5 \text{ m}^2} \times \frac{1 \text{ cm}^3}{2.7 \text{ g}} \times \frac{1 \text{ m}^3}{10^6 \text{ cm}^3} \times \frac{10^3 \text{ mm}}{1 \text{ m}} = 2.6 \times 10^{-2} \text{ mm}$$

80. Given: density of 1.17 g/cm³; mass of 3.75 kg
Unknown: volume in liters

$$3.75 \text{ kg} \times \frac{1 \text{ cm}^3}{1.17 \text{ g}} \times \frac{10^3 \text{ g}}{1 \text{ kg}} \times \frac{1 \text{ mL}}{1 \text{ cm}^3} \times \frac{1 \text{ L}}{10^3 \text{ mL}} = 3.21 \text{ L}$$

81. Given: dimensions of 28 cm × 21 cm × 44.5 mm; mass of 2090 g
Unknown: density in g/cm³

$$\frac{2090 \text{ g}}{28 \text{ cm} \times 21 \text{ cm} \times 44.5 \text{ mm}} \times \frac{10 \text{ mm}}{1 \text{ cm}} = 0.80 \text{ g/cm}^3$$

82. Given: mass of 6.58 g; triangle with base of 36.4 mm, height of 30.1 mm, and thickness of 0.560 mm
Unknown: density in g/cm³

$$\frac{6.58 \text{ g}}{0.5(36.4 \text{ mm} \times 30.1 \text{ mm}) \times 0.560 \text{ mm}} \times \frac{10^3 \text{ mm}^3}{1 \text{ cm}^3} = 21.4 \text{ g/cm}^3$$

83. Given: crate dimensions of 0.40 m × 0.40 m × 0.25 m; box dimensions of 22.0 cm × 12.0 cm × 5.0 cm
Unknown: number of boxes to fill the crate

$$\frac{0.40 \text{ m} \times 0.40 \text{ m} \times 0.25 \text{ m}}{1 \text{ crate}} \times \frac{1 \text{ box}}{22.0 \text{ cm} \times 12.0 \text{ cm} \times 5.0 \text{ cm}} \times \frac{10^6 \text{ cm}^3}{1 \text{ m}^3}$$
= 30 boxes/crate

84. a. Given: $V_{cube} = l \times l \times l$
$D = 2.27$ g/cm^3
$m = 3.93$ kg
Unknown: volume in liters; dimensions of the cube

$3.93 \text{ kg} \times \dfrac{1 \text{ cm}^3}{2.27 \text{ g}} \times \dfrac{10^3 \text{ g}}{1 \text{ kg}} \times \dfrac{1 \text{ mL}}{1 \text{ cm}^3} \times \dfrac{1 \text{ L}}{10^3 \text{ mL}} = 1.73 \text{ L}$

$1.73 \text{ L} \times \dfrac{1 \text{ dm}^3}{1 \text{ L}} \times \dfrac{1 \text{ m}^3}{10^3 \text{ dm}^3} = 1.73 \times 10^{-3} \text{ m}^3 = 0.00173 \text{ m}^3$

$\sqrt[3]{0.00173 \text{ m}^3} = 0.120$ m; dimensions $= 0.120$ m $\times$ 0.120 m $\times$ 0.120 m

b. Given: $V_{rectangle} = l \times w \times h$
$D = 1.85$ g/cm^3
dimensions of 33 mm $\times$ 21 mm $\times$ 7.2 mm
Unknown: mass in grams volume in cm^3

$V = 33 \text{ mm} \times 21 \text{ mm} \times 7.2 \text{ mm} \times \dfrac{1 \text{ cm}^3}{10^3 \text{ mm}^3} = 5.0 \text{ cm}^3$

$m = DV = \dfrac{1.85 \text{ g}}{1 \text{ cm}^3} \times 5.0 \text{ cm}^3 = 9.2 \text{ g}$

c. Given: $V_{sphere} = \tfrac{4}{3}\pi r^3$;
$D = 3.21$ g/L;
diameter $= 3.30$ m
Unknown: mass in kilograms, volume in dm^3

$V = \tfrac{4}{3}\pi r^3 = \tfrac{4}{3} \times 3.14 \times \left(\dfrac{3.30 \text{ m}}{2}\right)^3 \times \dfrac{10^2 \text{ dm}^3}{1 \text{ m}^3} = 18.8 \times 10^3 \text{ dm}^3$
$= 1.88 \times 10^4 \text{ dm}^3$

$m = DV = \dfrac{3.21 \text{ g}}{1 \text{ L}} \times 1.88 \times 10^4 \text{ dm}^3 \times \dfrac{1 \text{ L}}{1 \text{ dm}^3} \times \dfrac{1 \text{ kg}}{10^3 \text{ g}} = 60.3 \text{ kg}$

d. Given: $V_{cylinder} = \pi r^2 \times h$;
mass = 497 g;
dimensions of cylinder: 7.5 cm diameter $\times$ 12 cm
Unknown: density in g/cm^3, volume in m^3

$V = \pi r^2 \times h = 3.14 \left(\dfrac{7.5 \text{ cm}}{2}\right)^2 \times 12 \text{ cm} \times \dfrac{1 \text{ m}^3}{10^6 \text{ cm}^3} = 5.3 \times 10^{-4} \text{ m}^3$

$D = \dfrac{m}{V} = \dfrac{497 \text{ g}}{5.3 \times 10^{-4} \text{ m}^3} \times \dfrac{1 \text{ m}^3}{10^6 \text{ cm}^3} = 0.94 \text{ g/cm}^3$

e. Given: $V_{rectangle} = l \times w \times h$;
$D = 0.92$ g/cm^3;
dimensions of 3.5 m $\times$ 1.2 m $\times$ 0.65 m
Unknown: mass in kilograms, volume in cm^3

$V = l \times w \times h = 3.5 \text{ m} \times 1.2 \text{ m} \times 0.65 \text{ m} \times \dfrac{10^6 \text{ cm}^3}{1 \text{ m}^3} = 2.7 \times 10^6 \text{ cm}^3$

$m = DV = \dfrac{0.92 \text{ g}}{1 \text{ cm}^3} \times (2.7 \times 10^6 \text{ cm}^3) \times \dfrac{1 \text{ kg}}{10^3 \text{ g}} = 2.5 \times 10^3 \text{ kg}$

85. Given: mass of 9.65 g; initial V_{water} = 16.0 mL; final V_{water} = 19.5 mL
Unknown: Density in g/cm^3

$V = 19.5 \text{ mL} - 16.0 \text{ mL} = 3.5 \text{ mL}$

$D = \dfrac{m}{V} = \dfrac{9.65 \text{ g}}{3.5 \text{ mL}} \times \dfrac{1 \text{ mL}}{1 \text{ cm}^3} = 2.8 \text{ g/cm}^3$

---

86. a. Given: $m = 50.$ kg; area = 3620 m^2; $D = 19.3$ g/cm^3
Unknown: thickness of the gold in micrometers

$D = \dfrac{m}{\text{area} \times \text{thickness}}$; thickness $= \dfrac{m}{\text{area} \times D}$

$= \dfrac{50. \text{ kg}}{3620 \text{ m}^2} \times \dfrac{1 \text{ cm}^3}{19.3 \text{ g}} \times \dfrac{1 \text{ m}^2}{10^4 \text{ cm}^2} \times \dfrac{10^4 \text{ μm}}{\text{cm}} \times \dfrac{10^3 \text{ g}}{1 \text{ kg}} = 0.000\,72 \times 10^3$ μm

$= 0.72$ μm

b. Given: thickness = 0.72 μm; atom radius = 1.44 × 10^{-10} m
Unknown: number of atoms

diameter = 2 × radius = 2(1.44 × 10^{-10} m) = 2.88 × 10^{-10} m

$0.72 \text{ μm} \times \dfrac{1 \text{ atom}}{2.88 \times 10^{-10} \text{ m}} \times \dfrac{1 \text{ m}}{10^6 \text{ μm}} = 2.5 \times 10^3$ atoms

---

87. Given: fill time = 238 s; cylinder diameter = 1.2 m; cylinder height = 4.6 m
Unknown: flow rate in L/min

$V_{\text{cylinder}} = \pi r^2 \times h = 3.14 \left(\dfrac{1.2 \text{ m}}{2}\right)^2 \times 4.6 \text{ m} = 5.2 \text{ m}^3$

$\dfrac{5.2 \text{ m}^3}{238 \text{ s}} \times \dfrac{10^3 \text{ dm}^3}{1 \text{ m}^3} \times \dfrac{1 \text{ L}}{1 \text{ dm}^3} \times \dfrac{60 \text{ s}}{1 \text{ min}} = 1300$ L/min

---

88. Given: 2.8 g produces 1.0 J/s; 1 cal = 4.184 J; dimensions = 4.5 cm × 3.05 cm × 15 cm; $D = 19.86$ g/cm^3
Unknown: calories generated per hour

$V = 4.5 \text{ cm} \times 3.05 \text{ cm} \times 15 \text{ cm} = 210 \text{ cm}^3$

$m = DV = \dfrac{19.86 \text{ g}}{1 \text{ cm}^3} \times 210 \text{ cm}^3 = 4200$ g

$4200 \text{ g} \times \dfrac{1.0 \text{ J}}{2.8 \text{ g} \cdot \text{s}} \times \dfrac{1 \text{ cal}}{4.184 \text{ J}} \times \dfrac{3600 \text{ s}}{1 \text{ h}} = 1.3 \times 10^6$ cal/h

---

89. Given: $m = 5.974 \times 10^{24}$ kg; sphere diameter $= 1.28 \times 10^4$ km
Unknown: density in g/cm^3

$V = \dfrac{4}{3}\pi r^3 = \dfrac{4}{3} \times 3.14 \left(\dfrac{1.28 \times 10^4 \text{ km}}{2}\right)^3 \times \dfrac{10^{15} \text{ cm}^3}{1 \text{ km}^3} = 1.10 \times 10^{27}$ cm^3

$D = \dfrac{m}{V} = \dfrac{5.974 \times 10^{24} \text{ kg}}{1.10 \times 10^{27} \text{ cm}^3} \times \dfrac{10^3 \text{ g}}{1 \text{ kg}} = 5.43$ g/cm^3

90. Given: D_{Mg} = 1.74 g/cm³
 D_{Pt} = 21.45 g/cm³

Unknown: volume of Mg in cm³ with the same mass as 1.82 dm³ of Pt

$m_{Pt} = 1.82 \text{ dm}^3 \times \dfrac{21.45 \text{ g}}{1 \text{ cm}^3} \times \dfrac{10^3 \text{ cm}^3}{1 \text{ dm}^3} = 3.90 \times 10^4 \text{ g} = m_{Mg}$

$V_{Mg} = \dfrac{m_{Mg}}{D_{Mg}} = 3.90 \times 10^4 \text{ g} \times \dfrac{1 \text{ cm}^3}{1.74 \text{ g}} = 2.24 \times 10^4 \text{ cm}^3$

---

91. Given: 66 m/roll; 5.0 cm/use

Unknown: number of uses in 24 rolls

$\dfrac{66 \text{ m}}{1 \text{ roll}} \times 24 \text{ rolls} \times \dfrac{1 \text{ use}}{5.0 \text{ cm}} \times \dfrac{100 \text{ cm}}{1 \text{ m}} = 32\,000 \text{ uses}$

---

92. Given: 38 km/4.0 L gasoline; driven 75% of year; 86 km/day; 1 yr = 365 days

Unknown: volume of gasoline in liters per year

365 days × 0.75 = 274 days driven

274 days driven × $\dfrac{86 \text{ km}}{1 \text{ day}}$ = 24 000 km driven

24 000 km × $\dfrac{4.0 \text{ L}}{38 \text{ km}}$ = 2500 L

---

93. Given: fill time of 97 h; pool dimensions = 9.0 m × 3.5 m × 1.75 m

Unknown: rate of fill in L/min

$\dfrac{9.0 \text{ m} \times 3.5 \text{ m} \times 1.75 \text{ m}}{97 \text{ h}} \times \dfrac{10^3 \text{ dm}^3}{1 \text{ m}^3} \times \dfrac{1 \text{ L}}{1 \text{ dm}^3} \times \dfrac{1 \text{ h}}{60 \text{ min}} = 9.5 \text{ L/min}$

---

94. Given: $D_{H_2SO_4}$ = 1.285 g/cm³; 38% sulfuric acid in battery

Unknown: mass of sulfuric acid in 500. mL battery acid

$500. \text{ mL} \times 0.38 \times \dfrac{1.285 \text{ g}}{1 \text{ cm}^3} \times \dfrac{1 \text{ cm}^3}{1 \text{ mL}} = 244 \text{ g H}_2\text{SO}_4$

---

95. a. Given: 64.1 g of Al

Unknown: number of moles Al

$64.1 \text{ g Al} \times \dfrac{1 \text{ mol Al}}{26.98 \text{ g Al}} = 2.38 \text{ mol Al}$

b. Given: 28.1 g of Si

Unknown: number of moles Si

$28.1 \text{ g Si} \times \dfrac{1 \text{ mol Si}}{28.09 \text{ g Si}} = 1.00 \text{ mol Si}$

c. Given: 0.255 g of S

Unknown: number of moles S

$0.255 \text{ g S} \times \dfrac{1 \text{ mol S}}{32.07 \text{ g S}} = 7.95 \times 10^{-3} \text{ mol S}$

d. Given: 850.5 g of Zn
Unknown: number of moles Zn

$$850.5 \text{ g Zn} \times \frac{1 \text{ mol Zn}}{65.39 \text{ g Zn}} = 13.01 \text{ mol Zn}$$

96. a. Given: 1.22 mol Na
Unknown: mass Na

$$1.22 \text{ mol Na} \times \frac{22.99 \text{ g N}}{1 \text{ mol Na}} = 28.0 \text{ g Na}$$

b. Given: 14.5 mol Cu
Unknown: mass Cu

$$14.5 \text{ mol Cu} \times \frac{63.55 \text{ g Cu}}{1 \text{ mol Cu}} = 921 \text{ g Cu}$$

c. Given: 0.275 mol Hg
Unknown: mass Hg

$$0.275 \text{ mol Hg} \times \frac{200.59 \text{ g Hg}}{1 \text{ mol Hg}} = 55.2 \text{ g Hg}$$

d. Given: 9.37×10^{-3} mol Mg
Unknown: mass Mg

$$9.37 \times 10^{-3} \text{ mol Mg} \times \frac{24.31 \text{ g Mg}}{1 \text{ mol Mg}} = 0.228 \text{ g Mg}$$

97. a. Given: 3.01×10^{23} atoms Rb
Unknown: amount in moles

$$3.01 \times 10^{23} \text{ atoms Rb} \times \frac{1 \text{ mol}}{6.022 \times 10^{23} \text{ atoms}} = 0.500 \text{ mol Rb}$$

b. Given: 8.08×10^{22} atoms Kr
Unknown: amount in moles

$$8.08 \times 10^{22} \text{ atoms Kr} \times \frac{1 \text{ mol}}{6.022 \times 10^{23} \text{ atoms}} = 0.134 \text{ mol Kr}$$

c. Given: 5 700 000 000 atoms of Pb
Unknown: amount in moles

$$5.7 \times 10^{9} \text{ atoms Pb} \times \frac{1 \text{ mol}}{6.022 \times 10^{23} \text{ atoms}} = 9.5 \times 10^{-15} \text{ mol Pb}$$

d. Given: 2.997×10^{25} atoms of V
Unknown: amount in moles

$$2.997 \times 10^{25} \text{ atoms V} \times \frac{1 \text{ mol}}{6.022 \times 10^{23} \text{ atoms}} = 49.77 \text{ mol V}$$

98. a. Given: 1.004 mol Bi
Unknown: number of atoms

$$1.004 \text{ mol Bi} \times \frac{6.022 \times 10^{23} \text{ atoms}}{1 \text{ mol}} = 6.046 \times 10^{23} \text{ atoms Bi}$$

b. Given: 2.5 mol Mn
Unknown: number of atoms

$$2.5 \text{ mol Mn} \times \frac{6.022 \times 10^{23} \text{ atoms}}{1 \text{ mol}} = 1.5 \times 10^{24} \text{ atoms Mn}$$

c. Given: 0.000 000 2 mol He
Unknown: number of atoms

$$2 \times 10^{-7} \text{ mol He} \times \frac{6.022 \times 10^{23} \text{ atoms}}{1 \text{ mol}} = 1 \times 10^{17} \text{ atoms He}$$

d. Given: 32.6 mol Sr
Unknown: number of atoms

$$32.6 \text{ mol Sr} \times \frac{6.022 \times 10^{23} \text{ atoms}}{1 \text{ mol}} = 1.96 \times 10^{25} \text{ atoms Sr}$$

99. a. Given: 54.0 g Al
Unknown: number of of atoms Al

$$54.0 \text{ g Al} \times \frac{1 \text{ mol Al}}{26.98 \text{ g Al}} \times \frac{6.022 \times 10^{23} \text{ atoms}}{1 \text{ mol}} = 1.21 \times 10^{24} \text{ atoms Al}$$

b. Given: 69.45 g La
Unknown: number of atoms La

$$69.45 \text{ g La} \times \frac{1 \text{ mol La}}{138.91 \text{ g La}} \times \frac{6.022 \times 10^{23} \text{ atoms}}{1 \text{ mol}} = 3.011 \times 10^{23} \text{ atoms La}$$

c. Given: 0.697 g Ga
Unknown: number of atoms Ga

$$0.697 \text{ g Ga} \times \frac{1 \text{ mol Ga}}{69.72 \text{ g Ga}} \times \frac{6.022 \times 10^{23} \text{ atoms}}{1 \text{ mol}} = 6.02 \times 10^{21} \text{ atoms Ga}$$

d. Given: 0.000 000 020 g Be
Unknown: number of atoms Be

$$2.0 \times 10^{-8} \text{ g Be} \times \frac{1 \text{ mol Be}}{9.01 \text{ g Be}} \times \frac{6.022 \times 10^{23} \text{ atoms}}{1 \text{ mol}} = 1.3 \times 10^{15} \text{ atoms Be}$$

100. a. Given: 6.022×10^{24} atoms Ta
Unknown: mass Ta

$$6.022 \times 10^{24} \text{ atoms Ta} \times \frac{1 \text{ mol}}{6.022 \times 10^{23} \text{ atoms}} \times \frac{180.95 \text{ g Ta}}{1 \text{ mol Ta}} = 1810 \text{ g Ta}$$

b. Given: 3.01×10^{21} atoms Co
Unknown: mass Co

$$3.01 \times 10^{21} \text{ atoms Co} \times \frac{1 \text{ mol}}{6.022 \times 10^{23} \text{ atoms}} \times \frac{58.93 \text{ g Co}}{1 \text{ mol Co}} = 0.295 \text{ g Co}$$

c. Given: 1.506×10^{24} atoms Ar
Unknown: mass Ar

$$1.506 \times 10^{24} \text{ atoms Ar} \times \frac{1 \text{ mol}}{6.022 \times 10^{23} \text{ atoms}} \times \frac{39.95 \text{ g Ar}}{1 \text{ mol Ar}} = 99.91 \text{ g Ar}$$

d. Given: 1.20×10^{25} atoms He
Unknown: mass He

$$1.20 \times 10^{25} \text{ atoms He} \times \frac{1 \text{ mol}}{6.022 \times 10^{23} \text{ atoms}} \times \frac{4.00 \text{ g He}}{1 \text{ mol He}} = 79.7 \text{ g He}$$

101. a. Given: 3.00 g BBr$_3$
Unknown: number of moles BBr$_3$

$$\text{formula mass} = 1 \text{ atom B} \times \frac{10.81 \text{ amu}}{1 \text{ atom B}} + 3 \text{ atoms Br} \times \frac{79.90 \text{ amu}}{1 \text{ atom Br}}$$

$$= 10.81 \text{ amu} + 239.70 \text{ amu} = 250.51 \text{ amu}$$

$$3.00 \text{ g BBr}_3 \times \frac{1 \text{ mol BBr}_3}{250.51 \text{ g BBr}_3} = 0.0120 \text{ mol BBr}_3$$

b. Given: 0.472 g NaF
Unknown: number of moles NaF

$$\text{formula mass NaF} = 1 \text{ atom Na} \times \frac{22.99 \text{ amu}}{1 \text{ atom Na}} + 1 \text{ atom F} \times \frac{19.00 \text{ amu}}{1 \text{ atom F}}$$

$$= 41.99 \text{ amu}$$

$$0.472 \text{ g NaF} \times \frac{1 \text{ mol NaF}}{41.99 \text{ g NaF}} = 0.0112 \text{ mol NaF}$$

c. Given: 7.50×10^2 g CH_3OH
Unknown: number of moles CH_3OH

formula mass CH_3OH = 1 atom C $\times \dfrac{12.01 \text{ amu}}{1 \text{ atom C}}$ + 4 atoms H $\times \dfrac{1.01 \text{ amu}}{1 \text{ atom H}}$

+ 1 atom O $\times \dfrac{16.00 \text{ amu}}{1 \text{ atom O}}$ = 32.05 amu

7.50×10^2 g $CH_3OH \times \dfrac{1 \text{ mol } CH_3OH}{32.05 \text{ g } CH_3OH}$ = 23.4 mol CH_3OH

d. Given: 50.0 g $Ca(ClO_3)_2$
Unknown: number of moles $Ca(ClO_3)_2$

formula mass $Ca(ClO_3)_2$ = 1 atom Ca $\times \dfrac{40.08 \text{ amu}}{1 \text{ atom Ca}}$ + 2 atoms Cl

$\times \dfrac{35.45 \text{ amu}}{1 \text{ atom Cl}}$ + 6 atoms O $\times \dfrac{16.00 \text{ amu}}{1 \text{ atom O}}$

= 206.98 amu

50.0 g $Ca(ClO_3)_2 \times \dfrac{1 \text{ mol } Ca(ClO_3)_2}{206.98 \text{ g } Ca(ClO_3)_2}$ = 0.242 mol $Ca(ClO_3)_2$

102. a. Given: 1.366 mol NH_3
Unknown: mass NH_3

formula mass NH_3 = 1 atom N $\times \dfrac{14.01 \text{ amu}}{1 \text{ atom N}}$ + 3 atoms H $\times \dfrac{1.01 \text{ amu}}{1 \text{ atom H}}$

= 17.04 amu

1.366 mol $NH_3 \times \dfrac{17.04 \text{ g } NH_3}{1 \text{ mol } NH_3}$ = 23.28 g NH_3

b. Given: 0.120 mol $C_6H_{12}O_6$
Unknown: mass $C_6H_{12}O_6$

formula mass $C_6H_{12}O_6$ = 6 atoms C $\times \dfrac{12.01 \text{ amu}}{1 \text{ atom C}}$ + 12 atoms H

$\times \dfrac{1.01 \text{ amu}}{1 \text{ atom H}}$ + 6 atoms O $\times \dfrac{16.00 \text{ amu}}{1 \text{ atom O}}$

= 180.18 amu

0.120 mol $C_6H_{12}O_6 \times \dfrac{180.18 \text{ g } C_6H_{12}O_6}{1 \text{ mol } C_6H_{12}O_6}$ = 21.6 g $C_6H_{12}O_6$

c. Given: 6.94 mol $BaCl_2$
Unknown: mass $BaCl_2$

formula mass $BaCl_2$ = 1 atom Ba $\times \dfrac{137.33 \text{ amu}}{1 \text{ atom Ba}}$ + 2 atoms Cl $\times \dfrac{35.45 \text{ amu}}{1 \text{ atom Cl}}$

= 208.23 amu

6.94 mol $BaCl_2 \times \dfrac{208.23 \text{ g } BaCl_2}{1 \text{ mol } BaCl_2}$ = 1450 g $BaCl_2$

d. Given: 0.005 mol C_3H_8
Unknown: mass C_3H_8

formula mass C_3H_8 = 3 atoms C $\times \dfrac{12.01 \text{ amu}}{1 \text{ atom C}}$ + 8 atoms H $\times \dfrac{1.01 \text{ amu}}{1 \text{ atom H}}$

= 44.11 amu

0.005 mol $C_3H_8 \times \dfrac{44.11 \text{ g } C_3H_8}{1 \text{ mol } C_3H_8}$ = 0.2 g C_3H_8

103. a. Given: 4.99 mol CH_4
Unknown: number of molecules

4.99 mol $CH_4 \times \dfrac{6.022 \times 10^{23} \text{ molecules}}{1 \text{ mol}}$ = 3.00×10^{24} molecules CH_4

b. Given: 0.005 20 mol N_2
Unknown: number of molecules

5.20×10^{-3} mol $N_2 \times \dfrac{6.022 \times 10^{23} \text{ molecules}}{1 \text{ mol}} = 3.13 \times 10^{21}$ molecules N_2

c. Given: 1.05 mol PCl_3
Unknown: number of molecules

1.05 mol $PCl_3 \times \dfrac{6.022 \times 10^{23} \text{ molecules}}{1 \text{ mol}} = 6.32 \times 10^{23}$ molecules PCl_3

d. Given: 3.5×10^{-5} mol $C_6H_8O_6$
Unknown: number of molecules

3.5×10^{-5} mol $C_6H_8O_6 \times \dfrac{6.022 \times 10^{23} \text{ molecules}}{1 \text{ mol}}$

$= 2.1 \times 10^{19}$ molecules $C_6H_8O_6$

104. a. Given: 1.25 mol KBr
Unknown: number of formula units

1.25 mol KBr $\times \dfrac{6.022 \times 10^{23} \text{ formula units}}{1 \text{ mol}}$

$= 7.53 \times 10^{23}$ formula units KBr

b. Given: 5.00 mol $MgCl_2$
Unknown: number of formula units

5.00 mol $MgCl_2 \times \dfrac{6.022 \times 10^{23} \text{ formula units}}{1 \text{ mol}}$

$= 3.01 \times 10^{24}$ formula units $MgCl_2$

c. Given: 0.025 mol Na_2CO_3
Unknown: number of formula units

0.025 mol $Na_2CO_3 \times \dfrac{6.022 \times 10^{23} \text{ formula units}}{1 \text{ mol}}$

$= 1.5 \times 10^{22}$ formula units Na_2CO_3

d. Given: 6.82×10^{-6} mol $Pb(NO_3)_2$
Unknown: number of formula units

6.82×10^{-6} mol $Pb(NO_3)_2 \times \dfrac{6.022 \times 10^{23} \text{ formula units}}{1 \text{ mol}}$

$= 4.11 \times 10^{18}$ formula units $Pb(NO_3)_2$

105. a. Given: 3.34×10^{34} formula units $Cu(OH)_2$
Unknown: amount in moles

3.34×10^{34} formula units $Cu(OH)_2 \times \dfrac{1 \text{ mol}}{6.022 \times 10^{23} \text{ formula units}}$

$= 5.55 \times 10^{10}$ mol $Cu(OH)_2$

b. Given: 1.17×10^{16} molecules of H_2S
Unknown: amount in moles

1.17×10^{16} molecules $H_2S \times \dfrac{1 \text{ mol}}{6.022 \times 10^{23} \text{ molecules}} = 1.94 \times 10^{-8}$ mol H_2S

c. Given: 5.47×10^{21} formula units of $NiSO_4$
Unknown: amount in moles

5.47×10^{21} formula units $NiSO_4 \times \dfrac{1 \text{ mol}}{6.022 \times 10^{23} \text{ formula units}}$

$= 9.08 \times 10^{-3}$ mol $NiSO_4$

d. Given: 7.66×10^{19} molecules of H_2O_2
Unknown: amount in moles

7.66×10^{19} molecules $H_2O_2 \times \dfrac{1 \text{ mol}}{6.022 \times 10^{23} \text{ molecules}}$

$= 1.27 \times 10^{-4}$ mol H_2O_2

106. a. Given: 2.41×10^{24} molecules H_2
Unknown: mass H_2

formula mass $H_2 = 2$ atoms $H \times \dfrac{1.01 \text{ amu}}{1 \text{ atom H}} = 2.02$ amu

2.41×10^{24} molecules $H_2 \times \dfrac{1 \text{ mol}}{6.022 \times 10^{23} \text{ molecules}} \times \dfrac{2.02 \text{ g } H_2}{1 \text{ mol } H_2}$

$= 8.08$ g H_2

b. Given: 5.00×10^{21} formula units $Al(OH)_3$
Unknown: mass $Al(OH)_3$

formula mass $Al(OH)_3 = 1$ atom $Al \times \dfrac{26.98 \text{ amu}}{1 \text{ atom Al}} + 3$ atoms O

$\times \dfrac{16.00 \text{ amu}}{1 \text{ atom O}} + 3$ atoms $H \times \dfrac{1.01 \text{ amu}}{1 \text{ atom H}} = 78.01$ amu

5.00×10^{21} formula units $Al(OH)_3 \times \dfrac{1 \text{ mol}}{6.022 \times 10^{23} \text{ formula units}}$

$\times \dfrac{78.01 \text{ g } Al(OH)_3}{1 \text{ mol } Al(OH)_3} = 0.648$ g $Al(OH)_3$

c. Given: 8.25×10^{22} molecules BrF_5
Unknown: mass BrF_5

formula mass $BrF_5 = 1$ atom $Br \times \dfrac{79.90 \text{ amu}}{1 \text{ atom Br}} + 5$ atoms $F \times \dfrac{19.00 \text{ amu}}{1 \text{ atom F}}$

$= 174.90$ amu

8.25×10^{22} molecules $BrF_5 \times \dfrac{1 \text{ mol}}{6.022 \times 10^{23} \text{ molecules}} \times \dfrac{174.90 \text{ g } BrF_5}{1 \text{ mol } BrF_5}$

$= 24.0$ g BrF_5

d. Given: 1.20×10^{23} formula units $Na_2C_2O_4$
Unknown: mass $Na_2C_2O_4$

formula mass $Na_2C_2O_4 = 2$ atoms $Na \times \dfrac{22.99 \text{ amu}}{1 \text{ atom Na}} + 2$ atoms C

$\times \dfrac{12.01 \text{ amu}}{1 \text{ atom C}} + 4$ atoms $O \times \dfrac{16.00 \text{ amu}}{1 \text{ atom O}} = 134$ amu

1.20×10^{23} formula units $Na_2C_2O_4 \times \dfrac{1 \text{ mol}}{6.022 \times 10^{23} \text{ formula units}}$

$\times \dfrac{134 \text{ g } Na_2C_2O_4}{1 \text{ mol } Na_2C_2O_4} = 26.7$ g $Na_2C_2O_4$

107. a. Given: 22.9 g Na_2S
Unknown: number of formula units Na_2S

formula mass $Na_2S = 2$ atoms $Na \times \dfrac{22.99 \text{ amu}}{1 \text{ atom Na}} + 1$ atom $S \times \dfrac{32.01 \text{ amu}}{1 \text{ atom S}}$

$= 77.99$ amu

22.9 g $Na_2S \times \dfrac{1 \text{ mol } Na_2S}{77.99 \text{ g } Na_2S} \times \dfrac{6.022 \times 10^{23} \text{ formula units}}{1 \text{ mol}}$

$= 1.77 \times 10^{23}$ formula units Na_2S

b. Given: 0.272 g Ni(NO$_3$)$_2$
Unknown: number of formula units Ni(NO$_3$)$_2$

formula mass Ni(NO$_3$)$_2$ = 1 atom Ni × $\frac{58.69 \text{ amu}}{1 \text{ atom Ni}}$ + 2 atoms N

× $\frac{14.01 \text{ amu}}{1 \text{ atom N}}$ + 6 atoms O × $\frac{16.00 \text{ amu}}{1 \text{ atom O}}$ = 182.71 amu

0.272 g Ni(NO$_3$)$_2$ × $\frac{1 \text{ mol Ni(NO}_3)_2}{182.71 \text{ g Ni(NO}_3)_2}$ × $\frac{6.022 \times 10^{23} \text{ formula units}}{1 \text{ mol}}$

= 8.96 × 10^{20} formula units Ni(NO$_3$)$_2$

c. Given: 260 mg CH$_2$CHCN
Unknown: number of molecules CH$_2$CHCN

formula mass CH$_2$CHCN = 3 atoms C × $\frac{12.01 \text{ amu}}{1 \text{ atom C}}$ + 3 atoms H

× $\frac{1.01 \text{ amu}}{1 \text{ atom H}}$ + 1 atom N × $\frac{14.01 \text{ amu}}{1 \text{ atom N}}$ = 53.07 amu

260 mg CH$_2$CHCN × $\frac{1 \text{ mol CH}_2\text{CHCN}}{53.07 \text{ g CH}_2\text{CHCN}}$ × $\frac{1 \text{ g}}{1000 \text{ mg}}$

× $\frac{6.022 \times 10^{23} \text{ molecules}}{1 \text{ mol}}$ = 3.0 × 10^{21} molecules CH$_2$CHCN

108. a. Given: 0.039 g Pd
Unknown: number of moles Pd

0.039 g Pd × $\frac{1 \text{ mol Pd}}{106.42 \text{ g Pd}}$ = 3.7 × 10^{-4} mol Pd

b. Given: 8200 g Fe
Unknown: number of moles Fe

8200 g Fe × $\frac{1 \text{ mol Fe}}{55.85 \text{ g Fe}}$ = 150 mol Fe

c. Given: 0.0073 kg Ta
Unknown: number of moles Ta

0.0073 kg Ta × $\frac{1000 \text{ g}}{1 \text{ kg}}$ × $\frac{1 \text{ mol Ta}}{180.95 \text{ g Ta}}$ = 0.040 mol Ta

d. Given: 0.006 55 g Sb
Unknown: number of moles Sb

0.006 55 g Sb × $\frac{1 \text{ mol Sb}}{121.76 \text{ g Sb}}$ = 5.38 × 10^{-5} mol Sb

e. Given: 5.64 kg Ba
Unknown: number of moles Ba

5.64 kg Ba × $\frac{1 \text{ mol Ba}}{137.33 \text{ g Ba}}$ × $\frac{1000 \text{ g}}{1 \text{ kg}}$ = 41.1 mol Ba

f. Given: 3.37 × 10^{-6} g Mo
Unknown: number of moles Mo

3.37 × 10^{-6} g Mo × $\frac{1 \text{ mol Mo}}{95.94 \text{ g Mo}}$ = 3.51 × 10^{-8} mol Mo

109. a. Given: 1.002 mol Cr
Unknown: mass Cr

1.002 mol Cr × $\frac{52.00 \text{ g Cr}}{1 \text{ mol Cr}}$ = 52.10 g Cr

b. Given: 550 mol Al
Unknown: mass Al

550 mol Al × $\frac{26.98 \text{ g Al}}{1 \text{ mol Al}}$ = 1.5 × 10^4 g Al

c. Given: 4.08×10^{-8} mol Ne
Unknown: mass Ne

4.08×10^{-8} mol Ne $\times \dfrac{20.18 \text{ g Ne}}{1 \text{ mol Ne}} = 8.23 \times 10^{-7}$ g Ne

d. Given: 7 mol Ti
Unknown: mass Ti

7 mol Ti $\times \dfrac{47.88 \text{ g Ti}}{1 \text{ mol Ti}} = 3 \times 10^2$ g Ti

e. Given: 0.0086 mol Xe
Unknown: mass Xe

0.0086 mol Xe $\times \dfrac{131.29 \text{ g Xe}}{1 \text{ mol Xe}} = 1.1$ g Xe

f. Given: 3.29×10^4 mol Li
Unknown: mass Li

3.29×10^4 mol Li $\times \dfrac{6.94 \text{ g Li}}{1 \text{ mol Li}} = 2.28 \times 10^5$ g Li

110. a. Given: 17.0 mol Ge
Unknown: number of atoms

17.0 mol Ge $\times \dfrac{6.022 \times 10^{23} \text{ atoms}}{1 \text{ mol}} = 1.02 \times 10^{25}$ atoms Ge

b. Given: 0.6144 mol Cu
Unknown: number of atoms

0.6144 mol Cu $\times \dfrac{6.022 \times 10^{23} \text{ atoms}}{1 \text{ mol}} = 3.700 \times 10^{23}$ atoms Cu

c. Given: 3.02 mol Sn
Unknown: number of atoms

3.02 mol Sn $\times \dfrac{6.022 \times 10^{23} \text{ atoms}}{1 \text{ mol}} = 1.82 \times 10^{24}$ atoms Sn

d. Given: 2.0×10^6 mol C
Unknown: number of atoms

2.0×10^6 mol C $\times \dfrac{6.022 \times 10^{23} \text{ atoms}}{1 \text{ mol}} = 1.2 \times 10^{30}$ atoms C

e. Given: 0.0019 mol Zr
Unknown: number of atoms

0.0019 mol Zr $\times \dfrac{6.022 \times 10^{23} \text{ atoms}}{1 \text{ mol}} = 1.1 \times 10^{21}$ atoms Zr

f. Given: 3.227×10^{-10} mol K
Unknown: number of atoms

3.227×10^{-10} mol K $\times \dfrac{6.022 \times 10^{23} \text{ atoms}}{1 \text{ mol}} = 1.943 \times 10^{14}$ atoms K

111. a. Given: 6.022×10^{24} atoms Co
Unknown: number of moles Co

6.022×10^{24} atoms Co $\times \dfrac{1 \text{ mol}}{6.022 \times 10^{23} \text{ atoms}} = 10.00$ mol Co

b. Given: 1.06×10^{23} atoms W
Unknown: number of moles W

1.06×10^{23} atoms W $\times \dfrac{1 \text{ mol}}{6.022 \times 10^{23} \text{ atoms}} = 0.176$ mol W

c. Given: 3.008×10^{19} atoms Ag
Unknown: number of moles Ag

$$3.008 \times 10^{19} \text{ atoms Ag} \times \frac{1 \text{ mol}}{6.022 \times 10^{23} \text{ atoms}} = 4.995 \times 10^{-5} \text{ mol Ag}$$

d. Given: 950 000 000 atoms Pu
Unknown: number of moles Pu

$$9.5 \times 10^{8} \text{ atoms Pu} \times \frac{1 \text{ mol}}{6.022 \times 10^{23} \text{ atoms}} = 1.6 \times 10^{-15} \text{ mol Pu}$$

e. Given 4.61×10^{17} atoms Rn
Unknown: number of moles Rn

$$4.61 \times 10^{17} \text{ atoms Rn} \times \frac{1 \text{ mol}}{6.022 \times 10^{23} \text{ atoms}} = 7.66 \times 10^{-7} \text{ mol Rn}$$

f. Given: 8 trillion atoms Ce
Unknown: number of moles Ce

$$8 \times 10^{12} \text{ atoms Ce} \times \frac{1 \text{ mol}}{6.022 \times 10^{23} \text{ atoms}} = 1 \times 10^{-11} \text{ mol Ce}$$

112. a. Given: 0.0082 g Au
Unknown: number of atoms Au

$$0.0082 \text{ g Au} \times \frac{1 \text{ mol Au}}{196.97 \text{ g Au}} \times \frac{6.022 \times 10^{23} \text{ atoms}}{1 \text{ mol}} = 2.5 \times 10^{19} \text{ atoms Au}$$

b. Given: 812 g Mo
Unknown: number of atoms Mo

$$812 \text{ g Mo} \times \frac{1 \text{ mol Mo}}{95.94 \text{ g Mo}} \times \frac{6.022 \times 10^{23} \text{ atoms}}{1 \text{ mol}} = 5.10 \times 10^{24} \text{ atoms Mo}$$

c. Given: 2.00×10^{2} mg Am
Unknown: number of atoms Am

$$2.00 \times 10^{2} \text{ mg Am} \times \frac{1 \text{ g}}{1000 \text{ mg}} \times \frac{1 \text{ mol Am}}{243.06 \text{ g Am}} \times \frac{6.022 \times 10^{23} \text{ atoms}}{1 \text{ mol}}$$
$$= 4.96 \times 10^{20} \text{ atoms Am}$$

d. Given: 10.09 kg Ne
Unknown: number of atoms Ne

$$10.09 \text{ kg Ne} \times \frac{1000 \text{ g}}{1 \text{ kg}} \times \frac{1 \text{ mol Ne}}{20.18 \text{ g Ne}} \times \frac{6.022 \times 10^{23} \text{ atoms}}{1 \text{ mol}}$$
$$= 3.011 \times 10^{26} \text{ atoms Ne}$$

e. Given: 0.705 mg Bi
Unknown: number of atoms Bi

$$0.705 \text{ mg Bi} \times \frac{1 \text{ g}}{1000 \text{ mg}} \times \frac{1 \text{ mol Bi}}{208.98 \text{ g Bi}} \times \frac{6.022 \times 10^{23} \text{ atoms}}{1 \text{ mol}}$$
$$= 2.03 \times 10^{18} \text{ atoms Bi}$$

f. Given: 37 μg U
Unknown: number of atoms U

$$37 \text{ μg U} \times \frac{1 \text{ g}}{10^{6} \text{ μg}} \times \frac{1 \text{ mol U}}{238.03 \text{ g U}} \times \frac{6.022 \times 10^{23} \text{ atoms}}{1 \text{ mol}} = 9.4 \times 10^{16} \text{ atoms U}$$

113. a. Given: 8.22×10^{23} atoms Rb
Unknown: mass Rb

$$8.22 \times 10^{23} \text{ atoms Rb} \times \frac{1 \text{ mol}}{6.022 \times 10^{23} \text{ atoms}} \times \frac{85.47 \text{ g Rb}}{1 \text{ mol Rb}} = 117 \text{ g Rb}$$

b. Given: 4.05 Avogadro's constants of Mn atoms
Unknown: mass Mn

$$4.05 \times 6.022 \times 10^{23} \text{ atoms Mn} \times \frac{1 \text{ mol}}{6.022 \times 10^{23} \text{ atoms}} \times \frac{54.94 \text{ g Mn}}{1 \text{ mol Mn}}$$
$$= 223 \text{ g Mn}$$

c. Given: 9.96×10^{26} atoms Te
Unknown: mass Te

$$9.96 \times 10^{26} \text{ atoms Te} \times \frac{1 \text{ mol}}{6.022 \times 10^{23} \text{ atoms}} \times \frac{127.60 \text{ g Te}}{1 \text{ mol Te}} = 2.11 \times 10^5 \text{ g Te}$$

d. Given: 0.000 025 Avogadro's constants of Rh atoms
Unknown: mass Rh

$$2.5 \times 10^{-5} \times 6.022 \times 10^{23} \text{ atoms Rh} \times \frac{1 \text{ mol}}{6.022 \times 10^{23} \text{ atoms}} \times \frac{102.91 \text{ g Rh}}{1 \text{ mol Rh}}$$
$$= 2.6 \times 10^{-3} \text{ g Rh}$$

e. Given: 88 300 000 000 000 atoms Ra
Unknown: mass Ra

$$8.83 \times 10^{13} \text{ atoms Ra} \times \frac{1 \text{ mol}}{6.022 \times 10^{23} \text{ atoms}} \times \frac{226.03 \text{ g Ra}}{1 \text{ mol Ra}}$$
$$= 3.31 \times 10^{-8} \text{ g Ra}$$

f. Given: 2.94×10^{17} atoms Hf
Unknown: mass Hf

$$2.94 \times 10^{17} \text{ atoms Hf} \times \frac{1 \text{ mol}}{6.022 \times 10^{23} \text{ atoms}} \times \frac{178.49 \text{ g Hf}}{1 \text{ mol Hf}} = 8.71 \times 10^{-5} \text{ g Hf}$$

114. a. Given: 45.0 g CH_3COOH
Unknown: moles CH_3COOH

formula mass CH_3COOH = 2 atoms C × $\frac{12.01 \text{ amu}}{1 \text{ atom C}}$

+ 4 atoms H × $\frac{1.01 \text{ amu}}{1 \text{ atom H}}$ + 2 atoms O × $\frac{16.00 \text{ amu}}{1 \text{ atom O}}$ = 60.06 amu

$$45.0 \text{ g } CH_3COOH \times \frac{1 \text{ mol } CH_3COOH}{60.06 \text{ g } CH_3COOH} = 0.749 \text{ mol } CH_3COOH$$

b. Given: 7.04 g $Pb(NO_3)_2$
Unknown: moles $Pb(NO_3)_2$

formula mass $Pb(NO_3)_2$ = 1 atom Pb × $\frac{207.2 \text{ amu}}{1 \text{ atom Pb}}$ + 2 atoms N

× $\frac{14.01 \text{ amu}}{1 \text{ atom N}}$ + 6 atoms O × $\frac{16.00 \text{ amu}}{1 \text{ atom O}}$ = 331.22 amu

$$7.04 \text{ g } Pb(NO_3)_2 \times \frac{1 \text{ mol } Pb(NO_3)_2}{331.22 \text{ g } Pb(NO_3)_2} = 0.0213 \text{ mol } Pb(NO_3)_2$$

c. Given: 5000 kg Fe_2O_3
Unknown: moles Fe_2O_3

formula mass Fe_2O_3 = 2 atoms Fe × $\frac{55.85 \text{ amu}}{1 \text{ atom Fe}}$ + 3 atoms O × $\frac{16.00 \text{ amu}}{1 \text{ atom O}}$

= 159.70 amu

$$5000 \text{ kg } Fe_2O_3 \times \frac{1000 \text{ g}}{1 \text{ kg}} \times \frac{1 \text{ mol } Fe_2O_3}{159.70 \text{ g } Fe_2O_3} = 3 \times 10^4 \text{ mol } Fe_2O_3$$

d. Given: 12.0 mg $C_2H_5NH_2$
Unknown: moles $C_2H_5NH_2$

formula mass $C_2H_5NH_2$ = 2 atoms C × $\dfrac{12.01 \text{ amu}}{1 \text{ atom C}}$ + 7 atoms H × $\dfrac{1.01 \text{ amu}}{1 \text{ atom H}}$

+ 1 atom N × $\dfrac{14.01 \text{ amu}}{1 \text{ atom N}}$ = 45.10 amu

12.0 mg $C_2H_5NH_2$ × $\dfrac{1 \text{ g}}{1000 \text{ mg}}$ × $\dfrac{1 \text{ mol } C_2H_5NH_2}{45.10 \text{ g } C_2H_5NH_2}$

= 2.66×10^{-4} mol $C_2H_5NH_2$

e. Given: 0.003 22 g $C_{17}H_{35}COOH$
Unknown: moles $C_{17}H_{35}COOH$

formula mass $C_{17}H_{35}COOH$ = 18 atoms C × $\dfrac{12.01 \text{ amu}}{1 \text{ atom C}}$

+ 36 atoms H × $\dfrac{1.01 \text{ amu}}{1 \text{ atom H}}$ + 2 atoms O × $\dfrac{16.00 \text{ amu}}{1 \text{ atom O}}$ = 284.54 amu

3.22×10^{-3} g $C_{17}H_{35}COOH$ × $\dfrac{1 \text{ mol } C_{17}H_{35}COOH}{284.54 \text{ g } C_{17}H_{35}COOH}$

= 1.13×10^{-5} mol $C_{17}H_{35}COOH$

f. Given: 50.0 kg $(NH_4)_2SO_4$
Unknown: moles $(NH_4)_2SO_4$

formula mass $(NH_4)_2SO_4$ = 2 atoms N × $\dfrac{14.01 \text{ amu}}{1 \text{ atom N}}$ + 8 atoms H

× $\dfrac{1.01 \text{ amu}}{1 \text{ atom H}}$ + 1 atom S × $\dfrac{32.07 \text{ amu}}{1 \text{ atom S}}$ + 4 atoms O × $\dfrac{16.00 \text{ amu}}{1 \text{ atom O}}$

= 132.17 amu

50.0 kg $(NH_4)_2SO_4$ × $\dfrac{1000 \text{ g}}{1 \text{ kg}}$ × $\dfrac{1 \text{ mol }(NH_4)_2SO_4}{132.17 \text{ g }(NH_4)_2SO_4}$

= 378 mol $(NH_4)_2SO_4$

115. a. Given: 3.00 mol $SeOBr_2$
Unknown: mass $SeOBr_2$

formula mass $SeOBr_2$ = 1 atom Se × $\dfrac{78.96 \text{ amu}}{1 \text{ atom Se}}$

+ 1 atom O × $\dfrac{16.00 \text{ amu}}{1 \text{ atom O}}$ + 2 atoms Br × $\dfrac{79.90 \text{ amu}}{1 \text{ atom Br}}$

= 254.76 amu

3.00 mol $SeOBr_2$ × $\dfrac{254.76 \text{ g } SeOBr_2}{1 \text{ mol } SeOBr_2}$ = 764 g $SeOBr_2$

b. Given: 488 mol $CaCO_3$
Unknown: mass $CaCO_3$

formula mass $CaCO_3$ = 1 atom Ca × $\dfrac{40.08 \text{ amu}}{1 \text{ atom Ca}}$ + 1 atom C × $\dfrac{12.01 \text{ amu}}{1 \text{ atom C}}$

+ 3 atoms O × $\dfrac{16.00 \text{ amu}}{1 \text{ atom O}}$ = 100.09 amu

488 mol $CaCO_3$ × $\dfrac{100.09 \text{ g } CaCO_3}{1 \text{ mol } CaCO_3}$ = 4.88×10^4 g $CaCO_3$

c. Given: 0.0091 mol $C_{20}H_{28}O_2$
Unknown: mass $C_{20}H_{28}O_2$

formula mass $C_{20}H_{28}O_2$ = 20 atoms C $\times \dfrac{12.01 \text{ amu}}{1 \text{ atom C}}$
+ 28 atoms H $\times \dfrac{1.01 \text{ amu}}{1 \text{ atom H}}$ + 2 atoms O $\times \dfrac{16.00 \text{ amu}}{1 \text{ atom O}}$ = 300.48 amu

0.0091 mol $C_{20}H_{28}O_2 \times \dfrac{300.48 \text{ g } C_{20}H_{28}O_2}{1 \text{ mol } C_{20}H_{28}O_2}$ = 2.7 g $C_{20}H_{28}O_2$

d. Given: 6.00×10^{-8} mol $C_{10}H_{14}N_2$
Unknown: mass $C_{10}H_{14}N_2$

formula mass $C_{10}H_{14}N_2$ = 10 atoms C $\times \dfrac{12.01 \text{ amu}}{1 \text{ atom C}}$ + 14 atoms H $\times \dfrac{1.01 \text{ amu}}{1 \text{ atom H}}$ + 2 atoms N $\times \dfrac{14.01 \text{ amu}}{1 \text{ atom N}}$ = 162.26 amu

6.00×10^{-8} mol $C_{10}H_{14}N_2 \times \dfrac{162.26 \text{ g } C_{10}H_{14}N_2}{1 \text{ mol } C_{10}H_{14}N_2}$ = 9.74×10^{-6} g $C_{10}H_{14}N_2$

e. Given: 2.50 mol $Sr(NO_3)_2$
Unknown: mass $Sr(NO_3)_2$

formula mass $Sr(NO_3)_2$ = 1 atom Sr $\times \dfrac{87.62 \text{ amu}}{1 \text{ atom Sr}}$
+ 2 atoms N $\times \dfrac{14.01 \text{ amu}}{1 \text{ atom N}}$ + 6 atoms O $\times \dfrac{16.00 \text{ amu}}{1 \text{ atom O}}$
= 211.64 amu

2.50 mol $Sr(NO_3)_2 \times \dfrac{211.64 \text{ g } Sr(NO_3)_2}{1 \text{ mol } Sr(NO_3)_2}$ = 529 g $Sr(NO_3)_2$

f. Given: 3.50×10^{-6} mol UF_6
Unknown: mass UF_6

formula mass UF_6 = 1 atom U $\times \dfrac{238.03 \text{ amu}}{1 \text{ atom U}}$ + 6 atoms F $\times \dfrac{19.00 \text{ amu}}{1 \text{ atom F}}$
= 352.03 amu

3.50×10^{-6} mol $UF_6 \times \dfrac{352.03 \text{ g } UF_6}{1 \text{ mol } UF_6}$ = 1.23×10^{-3} g UF_6

116. a. Given: 4.72 mol WO_3
Unknown: number of formula units

4.27 mol $WO_3 \times \dfrac{6.022 \times 10^{23} \text{ formula units}}{1 \text{ mol}}$
= 2.57×10^{24} formula units WO_3

b. Given: 0.003 00 mol $Sr(NO_3)_2$
Unknown: number of formula units

3.00×10^{-3} mol $Sr(NO_3)_2 \times \dfrac{6.022 \times 10^{23} \text{ formula units}}{1 \text{ mol}}$
= 1.81×10^{21} formula units $Sr(NO_3)_2$

c. Given 72.5 mol $C_6H_5CH_3$
Unknown: number of molecules

72.5 mol $C_6H_5CH_3 \times \dfrac{6.022 \times 10^{23} \text{ molecules}}{1 \text{ mol}}$
= 4.37×10^{25} molecules $C_6H_5CH_3$

d. Given: 5.11×10^{-7} mol $C_{29}H_{50}O_2$
Unknown: number of molecules

5.11×10^{-7} mol $C_{29}H_{50}O_2 \times \dfrac{6.022 \times 10^{23} \text{ molecules}}{1 \text{ mol}}$

$= 3.08 \times 10^{17}$ molecules $C_{29}H_{50}O_2$

e. Given: 1500 mol N_2H_4
Unknown: number of molecules

1500 mol $N_2H_4 \times \dfrac{6.022 \times 10^{23} \text{ molecules}}{1 \text{ mol}} = 9.0 \times 10^{26}$ molecules N_2H_4

f. Given: 0.989 mol $C_6H_5NO_2$
Unknown: number of molecules

0.989 mol $C_6H_5NO_2 \times \dfrac{6.022 \times 10^{23} \text{ molecules}}{1 \text{ mol}}$

$= 5.96 \times 10^{23}$ molecules $C_6H_5NO_2$

117. a. Given: 285 g $FePO_4$
Unknown: number of formula units $FePO_4$

formula mass $FePO_4$ = 1 atom Fe $\times \dfrac{55.85 \text{ amu}}{1 \text{ atom Fe}}$ + 1 atom P $\times \dfrac{30.97 \text{ amu}}{1 \text{ atom P}}$

+ 4 atoms O $\times \dfrac{16.00 \text{ amu}}{1 \text{ atom O}} = 150.82$ amu

285 g $FePO_4 \times \dfrac{1 \text{ mol } FePO_4}{150.82 \text{ g } FePO_4} \times \dfrac{6.022 \times 10^{23} \text{ formula units}}{1 \text{ mol}}$

$= 1.14 \times 10^{24}$ formula units $FePO_4$

b. Given: 0.0084 g C_5H_5N
Unknown: number of molecules C_5H_5N

formula mass C_5H_5N = 5 atoms C $\times \dfrac{12.01 \text{ amu}}{1 \text{ atom C}}$ + 5 atoms H $\times \dfrac{1.01 \text{ amu}}{1 \text{ atom H}}$

+ 1 atom N $\times \dfrac{14.01 \text{ amu}}{1 \text{ atom N}} = 79.11$ amu

0.0084 g $C_5H_5N \times \dfrac{1 \text{ mol } C_5H_5N}{79.11 \text{ g } C_5H_5N} \times \dfrac{6.022 \times 10^{23} \text{ molecules}}{1 \text{ mol}}$

$= 6.4 \times 10^{19}$ molecules C_5H_5N

c. Given: 85 mg $(CH_3)_2CHCH_2OH$
Unknown: number of molecules $(CH_3)_2CHCH_2OH$

formula mass $(CH_3)_2CHCH_2OH$ = 4 atoms C $\times \dfrac{12.01 \text{ amu}}{1 \text{ atom C}}$

+ 10 atoms H $\times \dfrac{1.01 \text{ amu}}{1 \text{ atom H}}$ + 1 atom O $\times \dfrac{16.00 \text{ amu}}{1 \text{ atom O}} = 74.14$ amu

85 mg $(CH_3)_2CHCH_2OH \times \dfrac{1 \text{ g}}{1000 \text{ mg}} \times \dfrac{1 \text{ mol } (CH_3)_2CHCH_2OH}{74.14 \text{ g } (CH_3)_2CHCH_2OH}$

$\times \dfrac{6.022 \times 10^{23} \text{ molecules}}{1 \text{ mol}} = 6.9 \times 10^{20}$ molecules $(CH_3)_2CHCH_2OH$

d. Given: 4.6×10^{-4} g $Hg(C_2H_3O_2)_2$

Unknown: number of formula units $Hg(C_2H_3O_2)_2$

formula mass $Hg(C_2H_3O_2)_2$ = 1 atom Hg $\times \dfrac{200.59 \text{ amu}}{1 \text{ atom Hg}}$ + 4 atoms C $\times \dfrac{12.01 \text{ amu}}{1 \text{ atom C}}$ + 6 atoms H $\times \dfrac{1.01 \text{ amu}}{1 \text{ atom H}}$ + 4 atoms O $\times \dfrac{16.00 \text{ amu}}{1 \text{ atom O}}$

= 318.69 amu

4.6×10^{-4} g $Hg(C_2H_3O_2)_2 \times \dfrac{1 \text{ mol } Hg(C_2H_3O_2)_2}{318.69 \text{ g } Hg(C_2H_3O_2)_2}$

$\times \dfrac{6.022 \times 10^{23} \text{ formula units}}{1 \text{ mol}} = 8.7 \times 10^{17}$ formula units $Hg(C_2H_3O_2)_2$

e. Given: 0.0067 g Li_2CO_3

Unknown: number of formula units Li_2CO_3

formula mass Li_2CO_3 = 2 atoms Li $\times \dfrac{6.94 \text{ amu}}{1 \text{ atom Li}}$ + 1 atom C $\times \dfrac{12.01 \text{ amu}}{1 \text{ atom C}}$ + 3 atoms O $\times \dfrac{16.00 \text{ amu}}{1 \text{ atom O}}$ = 73.89 amu

6.7×10^{-3} g $Li_2CO_3 \times \dfrac{1 \text{ mol } Li_2CO_3}{73.89 \text{ g } Li_2CO_3} \times \dfrac{6.022 \times 10^{23} \text{ formula units}}{1 \text{ mol}}$

= 5.5×10^{19} formula units Li_2CO_3

118. a. Given: 8.39×10^{23} molecules F_2

Unknown: mass F_2

formula mass F_2 = 2 atoms F $\times \dfrac{19.00 \text{ amu}}{1 \text{ atom F}}$ = 38.00 amu

8.39×10^{23} molecules $F_2 \times \dfrac{1 \text{ mol}}{6.022 \times 10^{23} \text{ molecules}} \times \dfrac{38.00 \text{ g } F_2}{1 \text{ mol } F_2} = 52.9$ g F_2

b. Given: 6.82×10^{24} formula units $BeSO_4$

Unknown: mass $BeSO_4$

formula mass $BeSO_4$ = 1 atom Be $\times \dfrac{9.01 \text{ amu}}{1 \text{ atom Be}}$ + 1 atom S $\times \dfrac{32.07 \text{ amu}}{1 \text{ atom S}}$ + 4 atoms O $\times \dfrac{16.00 \text{ amu}}{1 \text{ atom O}}$ = 105.08 amu

6.82×10^{24} formula units $BeSO_4 \times \dfrac{1 \text{ mol}}{6.022 \times 10^{23} \text{ formula units}}$

$\times \dfrac{105.08 \text{ g } BeSO_4}{1 \text{ mol } BeSO_4} = 1190$ g $BeSO_4$

c. Given: 7.004×10^{26} molecules of $CHCl_3$

Unknown: mass $CHCl_3$

formula mass $CHCl_3$ = 1 atom C $\times \dfrac{12.01 \text{ amu}}{1 \text{ atom C}}$ + 1 atom H $\times \dfrac{1.01 \text{ amu}}{1 \text{ atom H}}$ + 3 atoms Cl $\times \dfrac{35.45 \text{ amu}}{1 \text{ atom Cl}}$ = 119.37 amu

7.004×10^{26} molecules $CHCl_3 \times \dfrac{1 \text{ mol}}{6.022 \times 10^{23} \text{ molecules}}$

$\times \dfrac{119.37 \text{ g } CHCl_3}{1 \text{ mol } CHCl_3} = 1.388 \times 10^5$ g $CHCl_3$

d. Given: 31 billion formula units Cr(CHO$_2$)$_3$
Unknown: mass Cr(CHO$_2$)$_3$

formula mass Cr(CHO$_2$)$_3$ = 1 atom Cr × $\frac{52.00 \text{ amu}}{1 \text{ atom Cr}}$

+ 3 atoms C × $\frac{12.01 \text{ amu}}{1 \text{ atom C}}$ + 3 atoms H × $\frac{1.01 \text{ amu}}{1 \text{ atom H}}$ + 6 atoms O × $\frac{16.00 \text{ amu}}{1 \text{ atom O}}$

= 187.06 amu

3.1 × 10^{10} formula units Cr(CHO$_2$)$_3$ × $\frac{1 \text{ mol}}{6.022 \times 10^{23} \text{ formula units}}$

× $\frac{187.06 \text{ g Cr(CHO}_2)_3}{1 \text{ mol Cr(CHO}_2)_3}$ = 9.6 × 10^{-12} g Cr(CHO$_2$)$_3$

e. Given: 6.3 × 10^{18} molecules HNO$_3$
Unknown: mass HNO$_3$

formula mass HNO = 1 atom H × $\frac{1.01 \text{ amu}}{1 \text{ atom H}}$ + 1 atom N × $\frac{14.01 \text{ amu}}{1 \text{ atom N}}$

+ 3 atoms O × $\frac{16.00 \text{ amu}}{1 \text{ atom O}}$ = 63.02 amu

6.3 × 10^{18} molecules HNO$_3$ × $\frac{1 \text{ mol}}{6.022 \times 10^{23} \text{ molecules}}$ × $\frac{63.02 \text{ g HNO}_3}{1 \text{ mol HNO}_3}$

= 6.6 × 10^{-4} g HNO$_3$

f. Given: 8.37 × 10^{25} molecules C$_2$Cl$_2$F$_4$
Unknown: mass C$_2$Cl$_2$F$_4$

formula mass C$_2$Cl$_2$F$_4$ = 2 atoms C × $\frac{12.01 \text{ amu}}{1 \text{ atom C}}$

+ 2 atoms Cl × $\frac{35.45 \text{ amu}}{1 \text{ atom Cl}}$ + 4 atoms F × $\frac{19.00 \text{ amu}}{1 \text{ atom F}}$ = 170.92 amu

8.37 × 10^{25} molecules C$_2$Cl$_2$F$_4$ × $\frac{1 \text{ mol}}{6.022 \times 10^{23} \text{ molecules}}$

× $\frac{170.92 \text{ g C}_2\text{Cl}_2\text{F}_4}{1 \text{ mol C}_2\text{Cl}_2\text{F}_4}$ = 2.38 × 10^4 g C$_2$Cl$_2$F$_4$

119. Given: 1 troy ounce = 31.1 g
Unknown: moles in a troy ounce of Au, Pt, Ag

31.1 g Au × $\frac{1 \text{ mol Au}}{196.97 \text{ g Au}}$ = 0.158 mol Au

31.1 g Pt × $\frac{1 \text{ mol Pt}}{195.08 \text{ g Pt}}$ = 0.159 mol Pt

31.1 g Ag × $\frac{1 \text{ mol Ag}}{107.87 \text{ g Ag}}$ = 0.288 mol Ag

120. Given: 22.0 g C$_6$H$_5$OH
Unknown: moles C$_6$H$_5$OH

formula mass C$_6$H$_5$OH = 6 atoms C × $\frac{12.01 \text{ amu}}{1 \text{ atom C}}$

+ 6 atoms H × $\frac{1.01 \text{ amu}}{1 \text{ atom H}}$ + 1 atom O × $\frac{16.00 \text{ amu}}{1 \text{ atom O}}$ = 94.12 amu

22.0 g C$_6$H$_5$OH × $\frac{1 \text{ mol C}_6\text{H}_5\text{OH}}{94.12 \text{ g C}_6\text{H}_5\text{OH}}$ = 0.234 mol C$_6$H$_5$OH

121. Given: 0.015 mol I_2
Unknown: mass I_2

formula mass I_2 = 2 atoms I $\times \dfrac{126.90 \text{ amu}}{1 \text{ atom I}}$ = 253.80 amu

0.015 mol $I_2 \times \dfrac{253.80 \text{ g } I_2}{1 \text{ mol } I_2}$ = 3.8 g I_2

---

122. Given: 1 carat = 200 mg
Unknown: number of C atoms in 1.00 carat

1 carat $\times \dfrac{200. \text{ mg}}{1 \text{ carat}} \times \dfrac{1 \text{ g}}{1000 \text{ mg}} \times \dfrac{1 \text{ mol C}}{12.01 \text{ g C}} \times \dfrac{6.022 \times 10^{23} \text{ atoms}}{1 \text{ mol}}$

= 1.00×10^{22} atoms C

---

123. a. Given: 8.00 g $CaCl_2$; 1.000 kg water
Unknown: moles of $CaCl_2$ and water

formula mass $CaCl_2$ = 1 atom Ca $\times \dfrac{40.08 \text{ amu}}{1 \text{ atom Ca}}$

+ 2 atoms Cl $\times \dfrac{35.45 \text{ amu}}{1 \text{ atom Cl}}$ = 110.98 amu

8.00 g $CaCl_2 \times \dfrac{1 \text{ mol } CaCl_2}{110.98 \text{ g } CaCl_2}$ = 0.0721 mol $CaCl_2$

formula mass H_2O = 2 atoms H $\times \dfrac{1.01 \text{ amu}}{1 \text{ atom H}}$ + 1 atom O $\times \dfrac{16.00 \text{ amu}}{1 \text{ atom O}}$

= 18.02 amu

1.000 kg $H_2O \times \dfrac{1000 \text{ g}}{1 \text{ kg}} \times \dfrac{1 \text{ mol } H_2O}{18.02 \text{ g } H_2O}$ = 55.49 mol H_2O

b. Given: 0.0721 mol $CaCl_2$ from part **a**
Unknown: moles of Ca^{2+} and Cl^- ions

0.0721 mol $CaCl_2 \times \dfrac{1 \text{ mol } Ca^{2+}}{1 \text{ mol } CaCl_2}$ = 0.0721 mol Ca^{2+}

0.0721 mol $CaCl_2 \times \dfrac{2 \text{ mol } Cl^-}{1 \text{ mol } CaCl_2}$ = 0.144 mol Cl^-

---

124. a. Given: 453.6 g $C_{12}H_{22}O_{11}$
Unknown: moles $C_{12}H_{22}O_{11}$

formula mass $C_{12}H_{22}O_{11}$ = 12 atoms C $\times \dfrac{12.01 \text{ amu}}{1 \text{ atom C}}$

+ 22 atoms H $\times \dfrac{1.01 \text{ amu}}{1 \text{ atom H}}$ + 11 atoms O $\times \dfrac{16.00 \text{ amu}}{1 \text{ atom O}}$ = 342.34 amu

453.6 g $C_{12}H_{22}O_{11} \times \dfrac{1 \text{ mol } C_{12}H_{22}O_{11}}{342.34 \text{ g } C_{12}H_{22}O_{11}}$ = 1.325 mol $C_{12}H_{22}O_{11}$

b. Given: 453.6 g NaCl
Unknown: moles NaCl

formula mass NaCl = 1 atom Na $\times \dfrac{22.99 \text{ amu}}{1 \text{ atom Na}}$ + 1 atom Cl $\times \dfrac{35.45 \text{ amu}}{1 \text{ atom Cl}}$

= 58.44 amu

453.6 g NaCl $\times \dfrac{1 \text{ mol NaCl}}{58.44 \text{ g NaCl}}$ = 7.762 mol NaCl

125. Given: 10.7 g NH_4Cl
Unknown: moles of ions

formula mass NH_4Cl = 1 atom N $\times \dfrac{14.01 \text{ amu}}{1 \text{ atom N}}$ + 4 atoms H $\times \dfrac{1.01 \text{ amu}}{1 \text{ atom H}}$

+ 1 atom Cl $\times \dfrac{35.45 \text{ amu}}{1 \text{ atom Cl}}$ = 53.50 amu

10.7 g $NH_4Cl \times \dfrac{1 \text{ mol } NH_4Cl}{53.50 \text{ g } NH_4Cl} \times \dfrac{2 \text{ mol ions}}{1 \text{ mol } NH_4Cl}$ = 0.400 mol ions

126. Given: 2.41×10^{24} atoms Cr; 1.51×10^{23} atoms Ni; 3.01×10^{23} atoms Cr
Unknown: total moles

2.41×10^{24} atoms + 0.15×10^{24} atoms + 0.301×10^{24} atoms
= 2.86×10^{24} atoms

2.86×10^{24} atoms $\times \dfrac{1 \text{ mol}}{6.022 \times 10^{23} \text{ atoms}}$ = 4.75 mol

127. a. Given: 250.0 mL H_2O; D_{H_2O} = 0.997 g/mL
Unknown: mass H_2O

250.0 mL $H_2O \times \dfrac{0.997 \text{ g } H_2O}{1 \text{ mL } H_2O}$ = 249 g H_2O

b. Given: 249 g H_2O (from part **a**)
Unknown: moles H_2O

formula mass H_2O = 2 atoms H $\times \dfrac{1.01 \text{ amu}}{1 \text{ atom H}}$ + 1 atom O $\times \dfrac{16.00 \text{ amu}}{1 \text{ atom O}}$

= 18.02 amu

249 g $H_2O \times \dfrac{1 \text{ mol } H_2O}{18.02 \text{ g } H_2O}$ = 13.8 mol H_2O

c. Given: D_{H_2O} = 0.997 g/mL; 2000 mol H_2O
Unknown: volume

2.000 mol $H_2O \times \dfrac{18.02 \text{ g } H_2O}{1 \text{ mol } H_2O} \times \dfrac{1 \text{ mL } H_2O}{0.997 \text{ g } H_2O}$ = 36.1 mL H_2O

d. Given: 2.000 mol H_2O
Unknown: mass

2.000 mol $H_2O \times \dfrac{18.02 \text{ g } H_2O}{1 \text{ mol } H_2O}$ = 36.04 g H_2O

129. Given: 6.35 g Cd
Unknown: mass of same number of Al atoms

6.35 g Cd $\times \dfrac{1 \text{ mol}}{112.41 \text{ g Cd}} \times \dfrac{26.98 \text{ g Al}}{1 \text{ mol}}$ = 1.52 g Al

130. Given: Oxygen in cylinder: initial = 1027.8 g; final = 1023.2 g
Unknown: moles of O_2 used

formula mass O_2 = 2 atoms O $\times \dfrac{16.00 \text{ amu}}{1 \text{ atom O}}$ = 32.00 amu

mass of O_2 used = 1027.8 g − 1023.2 g = 4.6 g

4.6 g $O_2 \times \dfrac{1 \text{ mol } O_2}{32.00 \text{ g } O_2}$ = 0.14 mol O_2

131. a. Given: 0.250 mol Ag_2S
Unknown: moles of Ag and S

0.250 mol $Ag_2S \times \dfrac{2 \text{ mol Ag}}{1 \text{ mol } Ag_2S}$ = 0.500 mol Ag

0.250 mol $Ag_2S \times \dfrac{1 \text{ mol S}}{1 \text{ mol } Ag_2S}$ = 0.250 mol S

b. Given: 38.8 g Ag_2S
Unknown: moles Ag_2S, Ag, and S

formula mass Ag_2S = 2 atoms Ag × $\dfrac{107.87 \text{ Amu}}{1 \text{ atom Ag}}$ + 1 atom S × $\dfrac{32.07 \text{ amu}}{1 \text{ atom S}}$

= 247.81 amu

38.8 g Ag_2S × $\dfrac{1 \text{ mol } Ag_2S}{247.81 \text{ g } Ag_2S}$ = 0.157 mol Ag_2S

0.157 mol Ag_2S × $\dfrac{2 \text{ mol Ag}}{1 \text{ mol } Ag_2S}$ = 0.314 mol Ag

0.157 mol Ag_2S × $\dfrac{1 \text{ mol S}}{1 \text{ mol } Ag_2S}$ = 0.157 mol S

c. Given: 0.314 mol Ag; 0.157 mol S
Unknown: masses of Ag and S

0.314 mol Ag × $\dfrac{107.87 \text{ g Ag}}{1 \text{ mol Ag}}$ = 133.9 g Ag

0.157 mol S × $\dfrac{32.07 \text{ g S}}{1 \text{ mol S}}$ = 5.03 g S

---

132. a. Given: $Na_2C_2O_4$
Unknown: percentage composition

molar mass $Na_2C_2O_4$ = 2 mol Na × $\dfrac{22.99 \text{ g Na}}{1 \text{ mol Na}}$ + 2 mol C × $\dfrac{12.01 \text{ g C}}{1 \text{ mol C}}$

+ 4 mol O × $\dfrac{16.00 \text{ g O}}{1 \text{ mol O}}$ = 134.00 g

$\dfrac{2 \times 22.99 \text{ g Na}}{134.00 \text{ g } Na_2C_2O_4}$ × 100 = 34.31% Na

$\dfrac{2 \times 12.01 \text{ g C}}{134.00 \text{ g } Na_2C_2O_4}$ × 100 = 17.93% C

$\dfrac{4 \times 16.00 \text{ g O}}{134.00 \text{ g } Na_2C_2O_4}$ × 100 = 47.76% O

b. Given: C_2H_5OH
Unknown: percentage composition

molar mass C_2H_5OH = 2 mol C × $\dfrac{12.01 \text{ g C}}{1 \text{ mol C}}$ + 6 mol H × $\dfrac{1.01 \text{ g H}}{1 \text{ mol H}}$

+ 1 mol O × $\dfrac{16.00 \text{ g O}}{1 \text{ mol O}}$ = 46.08 g

$\dfrac{2 \times 12.01 \text{ g C}}{46.08 \text{ g } C_2H_5OH}$ × 100 = 52.13% C

$\dfrac{6 \times 1.01 \text{ g H}}{46.08 \text{ g } C_2H_5OH}$ × 100 = 13.15% H

$\dfrac{1 \times 16.00 \text{ g O}}{46.08 \text{ g } C_2H_5OH}$ × 100 = 34.72% O

c. Given: Al_2O_3
Unknown: percentage composition

molar mass Al_2O_3 = 2 mol Al × $\dfrac{26.98 \text{ g Al}}{1 \text{ mol Al}}$ + 3 mol O × $\dfrac{16.00 \text{ g O}}{1 \text{ mol O}}$ = 101.96 g

$\dfrac{2 \times 26.98 \text{ Al}}{101.96 \text{ } Al_2O_3}$ × 100 = 52.92% Al

$\dfrac{3 \times 16.00 \text{ g O}}{101.96 \text{ g } Al_2O_3}$ × 100 = 47.08% O

d. Given: K_2SO_4
Unknown: percentage composition

molar mass K_2SO_4 = 2 mol K × $\dfrac{39.10 \text{ g K}}{1 \text{ mol K}}$ + 1 mol S × $\dfrac{32.07 \text{ g S}}{1 \text{ mol S}}$

+ 4 mol O × $\dfrac{16.00 \text{ g O}}{1 \text{ mol O}}$ = 174.27 g

$\dfrac{2 \times 39.10 \text{ g K}}{174.27 \text{ g K}_2\text{SO}_4} \times 100 = 44.87\% \text{ K}$

$\dfrac{1 \times 32.07 \text{ g S}}{174.27 \text{ g K}_2\text{SO}_4} \times 100 = 18.40\% \text{ S}$

$\dfrac{4 \times 16.00 \text{ O}}{174.27 \text{ g K}_2\text{SO}_4} \times 100 = 36.72\% \text{ O}$

133. Given: percentage composition
Unknown: identity of the compound

100.0 g compound contains 42.59 g Na, 12.02 g C, and 44.99 g O.

$\dfrac{42.59 \text{ g Na}}{22.99 \text{ g/mol}} = 1.853 \text{ mol Na} \cong 2 \text{ mol Na}$

$\dfrac{12.02 \text{ g C}}{12.01 \text{ g/mol}} = 1.001 \text{ mol C} \cong 1 \text{ mol C}$

$\dfrac{44.99 \text{ g O}}{16.00 \text{ g/mol}} = 2.812 \text{ mol O} \cong 3 \text{ mol O}$

Na_2CO_3, or sodium carbonate

134. a. Given: 50.0 g KBr
Unknown: mass Br

molar mass KBr = 1 mol K × $\dfrac{39.10 \text{ g K}}{1 \text{ mol K}}$ + 1 mol Br × $\dfrac{79.90 \text{ g Br}}{1 \text{ mol Br}}$ = 119.00 g

50.0 g KBr × $\dfrac{79.90 \text{ g Br}}{119.00 \text{ g KBr}}$ = 33.6 g Br

b. Given: 1.00 kg $Na_2Cr_2O_7$
Unknown: mass Cr

molar mass $Na_2Cr_2O_7$ = 2 mol Na × $\dfrac{22.99 \text{ g Na}}{1 \text{ mol Na}}$ + 2 mol Cr × $\dfrac{52.00 \text{ g Cr}}{1 \text{ mol Cr}}$

+ 7 mol O × $\dfrac{16.00 \text{ g O}}{1 \text{ mol O}}$ = 261.98 g

1.00 kg $Na_2Cr_2O_7$ × $\dfrac{1000 \text{ g}}{1 \text{ kg}}$ × $\dfrac{2 \times 52.00 \text{ g Cr}}{261.98 \text{ g Na}_2\text{Cr}_2\text{O}_7}$ = 397 g Cr

c. Given: 85.0 mg $C_6H_{14}N_2O_2$
Unknown: mass N

molar mass $C_6H_{14}N_2O_2$ = 6 mol C × $\dfrac{12.01 \text{ g C}}{1 \text{ mol C}}$ + 14 mol H × $\dfrac{1.01 \text{ g H}}{1 \text{ mol H}}$

+ 2 mol N × $\dfrac{14.01 \text{ g N}}{1 \text{ mol N}}$ + 2 mol O × $\dfrac{16.00 \text{ g O}}{1 \text{ mol O}}$

= 146.22 g

0.085 g $C_6H_{14}N_2O_2$ × $\dfrac{28.02 \text{ g N}}{146.22 \text{ g C}_6\text{H}_{14}\text{N}_2\text{O}_2}$ = 0.0163 g = 16.3 mg N

d. Given: 2.84 g Co($C_2H_3O_2$)$_2$

Unknown: mass of Co

molar mass Co($C_2H_3O_2$)$_2$ = 1 mol Co × $\dfrac{58.93 \text{ g Co}}{1 \text{ mol Co}}$ + 4 mol C × $\dfrac{12.01 \text{ g C}}{1 \text{ mol C}}$

+ 6 mol H × $\dfrac{1.01 \text{ g H}}{1 \text{ mol H}}$ + 4 mol O × $\dfrac{16.00 \text{ g O}}{1 \text{ mol O}}$ = 177.03 g

2.84 g Co($C_2H_3O_2$)$_2$ × $\dfrac{58.93 \text{ g Co}}{177.03 \text{ g Co}(C_2H_3O_2)_2}$ = 0.945 g Co

---

135. a. Given: $Na_2CO_3 \cdot 10H_2O$

Unknown: percentage of water

molar mass $Na_2CO_3 \cdot 10H_2O$ = 2 mol Na × $\dfrac{22.99 \text{ g Na}}{1 \text{ mol Na}}$

+ 1 mol C × $\dfrac{12.01 \text{ g C}}{1 \text{ mol C}}$ + 13 mol O × $\dfrac{16.00 \text{ g O}}{1 \text{ mol O}}$ + 20 mol H × $\dfrac{1.01 \text{ g H}}{1 \text{ mol H}}$ = 286.19 g

molar mass H_2O = 2 mol H × $\dfrac{1.01 \text{ g H}}{1 \text{ mol H}}$ + 1 mol O × $\dfrac{16.00 \text{ g O}}{1 \text{ mol O}}$ = 18.02 g

$\dfrac{10(18.02) \text{ g } H_2O}{286.19 \text{ g } Na_2CO_3 \cdot 10 \, H_2O}$ × 100 = 62.97% H_2O

b. Given: $NiI_2 \cdot 6H_2O$

Unknown: percentage of water

molar mass $NiI_2 \cdot 6H_2O$ = 1 mol Ni × $\dfrac{58.69 \text{ g Ni}}{1 \text{ mol Ni}}$ + 2 mol I × $\dfrac{126.90 \text{ g I}}{1 \text{ mol I}}$

+ 12 mol H × $\dfrac{1.01 \text{ g H}}{1 \text{ mol H}}$ + 6 mol O × $\dfrac{16.00 \text{ g O}}{1 \text{ mol O}}$ = 420.61 g

$\dfrac{6(18.02) \text{ g } H_2O}{420.61 \text{ g } NiI_2 \cdot 6H_2O}$ × 100 = 25.71% H_2O

c. Given: $(NH_4)_2Fe(CN)_6 \cdot 3H_2O$

Unknown: percentage of water

molar mass $(NH_4)_2Fe(CN)_6 \cdot 3H_2O$ = 8 mol N × $\dfrac{14.01 \text{ g N}}{1 \text{ mol N}}$

+ 14 mol H × $\dfrac{1.01 \text{ H}}{1 \text{ mol H}}$ + 1 mol Fe × $\dfrac{55.85 \text{ g Fe}}{1 \text{ mol Fe}}$ + 6 mol C × $\dfrac{12.01 \text{ g C}}{1 \text{ mol C}}$

+ 3 mol O × $\dfrac{16.00 \text{ g O}}{1 \text{ mol O}}$ = 302.13 g

$\dfrac{3(18.02) \text{ g } H_2O}{302.13 \text{ g } (NH_4)_2Fe(CN)_6 \cdot 3H_2O}$ × 100 = 17.89% H_2O

d. Given: $AlBr_3 \cdot 6H_2O$
Unknown: percentage of water

molar mass $AlBr_3 \cdot 6H_2O = 1 \text{ mol Al} \times \dfrac{26.98 \text{ g Al}}{1 \text{ mol Al}}$

$+ \, 3 \text{ mol Br} \times \dfrac{79.90 \text{ g Br}}{1 \text{ mol Br}} + 12 \text{ mol H} \times \dfrac{1.01 \text{ g H}}{1 \text{ mol H}}$

$+ \, 6 \text{ mol O} \times \dfrac{16.00 \text{ g O}}{1 \text{ mol O}} = 374.80 \text{ g}$

$\dfrac{6(18.02) \text{ g H}_2\text{O}}{374.80 \text{ g AlBr}_3 \cdot 6\text{H}_2\text{O}} \times 100 = 28.85\% \text{ H}_2\text{O}$

---

136. a. Given: nitric acid
Unknown: formula; percentage composition

formula is HNO_3

molar mass $HNO_3 = 1 \text{ mol H} \times \dfrac{1.01 \text{ g H}}{1 \text{ mol H}} + 1 \text{ mol N} \times \dfrac{14.01 \text{ g N}}{1 \text{ mol N}}$

$+ \, 3 \text{ mol O} \times \dfrac{16.00 \text{ g O}}{1 \text{ mol O}} = 63.02 \text{ g}$

$\dfrac{1.01 \text{ g H}}{63.02 \text{ g HNO}_3} \times 100 = 1.60\% \text{ H}$

$\dfrac{14.01 \text{ g N}}{63.02 \text{ g HNO}_3} \times 100 = 22.23\% \text{ N}$

$\dfrac{3 \times 16.00 \text{ g O}}{63.02 \text{ g HNO}_3} \times 100 = 76.17\% \text{ O}$

b. Given: ammonia
Unknown: formula; percentage composition

formula is NH_3

molar mass $NH_3 = 1 \text{ mol N} \times \dfrac{14.01 \text{ g N}}{1 \text{ mol N}} + 3 \text{ mol H} \times \dfrac{1.01 \text{ g H}}{1 \text{ mol H}} = 17.04 \text{ g}$

$\dfrac{14.01 \text{ g N}}{17.04 \text{ g NH}_3} \times 100 = 82.22\% \text{ N}$

$\dfrac{3 \times 1.01 \text{ g H}}{17.04 \text{ g NH}_3} \times 100 = 17.78\% \text{ H}$

c. Given: mercury (II) sulfate
Unknown: formula; percentage composition

formula is $HgSO_4$

molar mass $HgSO_4 = 1 \text{ mol Hg} \times \dfrac{200.59 \text{ g Hg}}{1 \text{ mol Hg}} + 1 \text{ mol S} \times \dfrac{32.07 \text{ g S}}{1 \text{ mol S}}$

$+ \, 4 \text{ mol O} \times \dfrac{16.00 \text{ g O}}{1 \text{ mol O}} = 296.66 \text{ g}$

$\dfrac{200.59 \text{ g Hg}}{296.66 \text{ g HgSO}_4} \times 100 = 67.616\% \text{ Hg}$

$\dfrac{32.07 \text{ g S}}{296.66 \text{ g HgSO}_4} \times 100 = 10.81\% \text{ S}$

$\dfrac{4 \times 16.00 \text{ g O}}{296.66 \text{ g HgSO}_4} \times 100 = 21.57\% \text{ O}$

d. Given: antimony (V) flouride
Unknown: formula; percentage composition

formula is SbF_5

$$\text{molar mass } SbF_5 = 1 \text{ mol Sb} \times \frac{121.76 \text{ g Sb}}{1 \text{ mol Sb}} + 5 \text{ mol F} \times \frac{19.00 \text{ g F}}{1 \text{ mol F}} = 216.76 \text{ g}$$

$$\frac{121.76 \text{ g Sb}}{216.76 \text{ g } SbF_5} \times 100 = 56.173\% \text{ Sb}$$

$$\frac{5 \times 19.00 \text{ g F}}{216.76 \text{ g } SbF_5} \times 100 = 43.83\% \text{ F}$$

---

137. a. Given: LiBr
Unknown: percentage composition

$$\text{molar mass LiBr} = 1 \text{ mol Li} \times \frac{6.94 \text{ g Li}}{1 \text{ mol Li}} + 1 \text{ mol Br} \times \frac{79.90 \text{ g Br}}{1 \text{ mol Br}} = 86.84 \text{ g}$$

$$\frac{6.94 \text{ g Li}}{86.84 \text{ g LiBr}} \times 100 = 7.99\% \text{ Li}$$

$$\frac{79.90 \text{ g Br}}{86.84 \text{ LiBr}} \times 100 = 92.01\% \text{ Br}$$

b. Given: $C_{14}H_{10}$
Unknown: percentage composition

$$\text{molar mass } C_{14}H_{10} = 14 \text{ mol C} \times \frac{12.01 \text{ g C}}{1 \text{ mol C}} + 10 \text{ mol H} \times \frac{1.01 \text{ g H}}{1 \text{ mol H}}$$
$$= 178.24 \text{ g}$$

$$\frac{14 \times 12.01 \text{ g C}}{178.24 \text{ g } C_{14}H_{10}} \times 100 = 94.33\% \text{ C}$$

$$\frac{10 \times 1.01 \text{ g H}}{178.24 \text{ g } C_{14}H_{10}} \times 100 = 5.67\% \text{ H}$$

c. Given: NH_4NO_3
Unknown: percentage composition

$$\text{molar mass } NH_4NO_3 = 2 \text{ mol N} \times \frac{14.01 \text{ g N}}{1 \text{ mol N}} + 4 \text{ mol H} \times \frac{1.01 \text{ g H}}{1 \text{ mol H}}$$
$$+ 3 \text{ mol O} \times \frac{16.00 \text{ g O}}{1 \text{ mol O}} = 80.06 \text{ g}$$

$$\frac{2 \times 14.01 \text{ g N}}{80.06 \text{ g } NH_4NO_3} \times 100 = 35.00\% \text{ N}$$

$$\frac{4 \times 1.01 \text{ g H}}{80.06 \text{ g } NH_4NO_3} \times 100 = 5.05\% \text{ H}$$

$$\frac{3 \times 16.00 \text{ g O}}{80.06 \text{ g } NH_4NO_3} \times 100 = 59.96\% \text{ O}$$

d. Given: HNO_2
Unknown: percentage composition

$$\text{molar mass } HNO_2 = 1 \text{ mol H} \times \frac{1.01 \text{ g H}}{1 \text{ mol H}} + 1 \text{ mol N} \times \frac{14.01 \text{ g N}}{1 \text{ mol N}}$$
$$+ 2 \text{ mol O} \times \frac{16.00 \text{ g O}}{1 \text{ mol O}} = 47.02 \text{ g}$$

$$\frac{1.01 \text{ g H}}{47.02 \text{ g } HNO_2} \times 100 = 2.15\% \text{ H}$$

$$\frac{14.01 \text{ g N}}{47.02 \text{ g } HNO_2} \times 100 = 29.80\% \text{ N}$$

$$\frac{2 \times 16.00 \text{ g O}}{47.02 \text{ g } HNO_2} \times 100 = 68.06\% \text{ O}$$

e. Given: Ag_2S
Unknown: percentage composition

molar mass Ag_2S = 2 mol Ag × $\dfrac{107.87 \text{ g Ag}}{1 \text{ mol Ag}}$ + 1 mol S × $\dfrac{32.07 \text{ g S}}{1 \text{ mol S}}$ = 247.81 g

$\dfrac{2 \times 107.87 \text{ g Ag}}{247.81 \text{ g Ag}_2\text{S}} \times 100 = 87.059\% \text{ Ag}$

$\dfrac{32.07 \text{ g S}}{247.81 \text{ g Ag}_2\text{S}} \times 100 = 12.94\% \text{ S}$

f. Given: $Fe(SCN)_2$
Unknown: percentage composition

molar mass $Fe(SCN)_2$ = 1 mol Fe × $\dfrac{55.85 \text{ g Fe}}{1 \text{ mol Fe}}$ + 2 mol S × $\dfrac{32.07 \text{ g S}}{1 \text{ mol S}}$

+ 2 mol C × $\dfrac{12.01 \text{ g C}}{1 \text{ mol C}}$ + 2 mol N × $\dfrac{14.01 \text{ g N}}{1 \text{ mol N}}$ = 172.03 g

$\dfrac{55.85 \text{ g Fe}}{172.03 \text{ g Fe(SCN)}_2} \times 100 = 32.47\% \text{ Fe}$

$\dfrac{2 \times 32.00 \text{ g S}}{172.03 \text{ g Fe(SCN)}_2} \times 100 = 37.28\% \text{ S}$

$\dfrac{2 \times 12.01 \text{ g C}}{172.03 \text{ g Fe(SCN)}_2} \times 100 = 13.96\% \text{ C}$

$\dfrac{2 \times 14.01 \text{ g N}}{172.03 \text{ g Fe(SCN)}_2} \times 100 = 16.29\% \text{ N}$

g. Given: lithium acetate
Unknown: percentage composition

molar mass $LiC_2H_3O_2$ = 1 mol Li × $\dfrac{6.94 \text{ g Li}}{1 \text{ mol Li}}$ + 2 mol C × $\dfrac{12.01 \text{ g C}}{1 \text{ mol C}}$

+ 3 mol H × $\dfrac{1.01 \text{ g H}}{1 \text{ mol H}}$ + 2 mol O × $\dfrac{16.00 \text{ g O}}{1 \text{ mol O}}$ = 65.99 g

$\dfrac{6.94 \text{ g Li}}{65.99 \text{ g LiC}_2\text{H}_3\text{O}_2} \times 100 = 10.52\% \text{ Li}$

$\dfrac{2 \times 12.01 \text{ g C}}{65.99 \text{ g LiC}_2\text{H}_3\text{O}_2} \times 100 = 36.40\% \text{ C}$

$\dfrac{3 \times 1.01 \text{ g H}}{65.99 \text{ g LiC}_2\text{H}_3\text{O}_2} \times 100 = 4.59\% \text{ H}$

$\dfrac{2 \times 16.00 \text{ g O}}{65.99 \text{ g LiC}_2\text{H}_3\text{O}_2} \times 100 = 48.49\% \text{ O}$

h. Given: nickel (II) formate
Unknown: percentage composition

molar mass $Ni(CHO_2)_2 = 1 \text{ mol Ni} \times \dfrac{58.69 \text{ g Ni}}{1 \text{ mol Ni}} + 2 \text{ mol C} \times \dfrac{12.01 \text{ g C}}{1 \text{ mol C}}$

$+ 2 \text{ mol H} \times \dfrac{1.01 \text{ g H}}{1 \text{ mol H}} + 4 \text{ mol O} \times \dfrac{16.00 \text{ g O}}{1 \text{ mol O}} = 148.73 \text{ g}$

$\dfrac{58.69 \text{ g Ni}}{148.73 \text{ g Ni(CHO}_2)_2} \times 100 = 39.46\% \text{ Ni}$

$\dfrac{2 \times 12.01 \text{ g C}}{148.73 \text{ g Ni(CHO}_2)_2} \times 100 = 16.15\% \text{ C}$

$\dfrac{2 \times 1.01 \text{ g H}}{148.73 \text{ Ni(CHO}_2)_2} \times 100 = 1.36\% \text{ H}$

$\dfrac{3 \times 16.00 \text{ g O}}{148.73 \text{ g Ni(CHO}_2)_2} \times 100 = 43.03\% \text{ O}$

138. a. Given: NH_2CONH_2
Unknown: percentage of nitrogen

molar mass $NH_2CONH_2 = 2 \text{ mol N} \times \dfrac{14.01 \text{ g N}}{1 \text{ mol N}} + 1 \text{ mol C} \times \dfrac{12.01 \text{ g C}}{1 \text{ mol C}}$

$+ 4 \text{ mol H} \times \dfrac{1.01 \text{ g H}}{1 \text{ mol H}} + 1 \text{ mol O} \times \dfrac{16.00 \text{ g O}}{1 \text{ mol O}} = 60.07 \text{ g}$

$\dfrac{2 \times 14.01 \text{ g N}}{60.07 \text{ g NH}_2CONH_2} \times 100 = 46.65\% \text{ N}$

b. Given: SO_2Cl_2
Unknown: percentage of sulfur

molar mass $SO_2Cl_2 = 1 \text{ mol S} \times \dfrac{32.07 \text{ g S}}{1 \text{ mol S}} + 2 \text{ mol O} \times \dfrac{16.00 \text{ g O}}{1 \text{ mol O}}$

$+ 2 \text{ mol Cl} \times \dfrac{35.45 \text{ g Cl}}{1 \text{ mol Cl}} = 134.97 \text{ g}$

$\dfrac{32.07 \text{ g S}}{134.97 \text{ g SO}_2Cl_2} \times 100 = 23.76\% \text{ S}$

c. Given: Tl_2O_3
Unknown: percentage of thallium

molar mass $Tl_2O_3 = 2 \text{ mol Tl} \times \dfrac{204.38 \text{ g Tl}}{1 \text{ mol Tl}} + 3 \text{ mol O} \times \dfrac{16.00 \text{ g O}}{1 \text{ mol O}} = 456.76 \text{ g}$

$\dfrac{2 \times 204.38 \text{ g Tl}}{456.76 \text{ g Tl}_2O_3} \times 100 = 89.491\% \text{ Tl}$

d. Given: $KClO_3$
Unknown: percentage of oxygen

molar mass $KClO_3 = 1 \text{ mol K} \times \dfrac{39.10 \text{ g K}}{1 \text{ mol K}} + 1 \text{ mol Cl} \times \dfrac{35.45 \text{ g Cl}}{1 \text{ mol Cl}}$

$+ 3 \text{ mol O} \times \dfrac{16.00 \text{ g O}}{1 \text{ mol O}} = 122.55 \text{ g}$

$\dfrac{3 \times 16.00 \text{ g O}}{122.55 \text{ g KClO}_3} \times 100 = 39.17\% \text{ O}$

e. Given: $CaBr_2$
Unknown: percentage of bromine

molar mass $CaBr_2 = 1 \text{ mol Ca} \times \dfrac{40.08 \text{ g Ca}}{1 \text{ mol Ca}} + 2 \text{ mol Br} \times \dfrac{79.90 \text{ g Br}}{1 \text{ mol Br}} = 199.88 \text{ g}$

$\dfrac{2 \times 79.9 \text{ g Br}}{199.88 \text{ g CaBr}_2} \times 100 = 79.95\% \text{ Br}$

f. Given: SnO_2
Unknown: percentage of tin

molar mass $SnO_2 = 1 \text{ mol Sn} \times \dfrac{118.71 \text{ g Sn}}{1 \text{ mol Sn}} + 2 \text{ mol O} \times \dfrac{16.00 \text{ g O}}{1 \text{ mol O}}$
$= 150.71 \text{ g}$

$\dfrac{118.71 \text{ g Sn}}{150.71 \text{ g SnO}_2} \times 100 = 78.767\% \text{ Sn}$

139. a. Given: 4.00 g MnO_2
Unknown: mass of oxygen

molar mass $MnO_2 = 1 \text{ mol Mn} \times \dfrac{54.94 \text{ g Mn}}{1 \text{ mol Mn}} + 2 \text{ mol O} \times \dfrac{16.00 \text{ g O}}{1 \text{ mol O}} = 86.94 \text{ g}$

$\dfrac{4.00 \text{ g MnO}_2 \times 2 \times 16.00 \text{ g O}}{86.94 \text{ g MnO}_2} = 1.47 \text{ g O}$

b. Given: 50.0 metric tons Al_2O_3
Unknown: mass of aluminum

molar mass $Al_2O_3 = 2 \text{ mol Al} \times \dfrac{26.98 \text{ g Al}}{1 \text{ mol Al}} + 3 \text{ mol O} \times \dfrac{16.00 \text{ g O}}{1 \text{ mol O}} = 101.96 \text{ g}$

$50.0 \text{ metric tons Al}_2O_3 \times \dfrac{2 \times 26.98 \text{ g Al}}{101.96 \text{ g Al}_2O_3} = 26.5 \text{ metric tons Al}$

c. Given: 325 g AgCN
Unknown: mass of silver

molar mass $AgCN = 1 \text{ mol Ag} \times \dfrac{107.87 \text{ g Ag}}{1 \text{ mol Ag}} + 1 \text{ mol C} \times \dfrac{12.01 \text{ g}}{1 \text{ mol C}}$
$+ 1 \text{ mol N} \times \dfrac{14.01 \text{ g N}}{1 \text{ mol N}} = 133.89 \text{ g}$

$325 \text{ g AgCN} \times \dfrac{107.87 \text{ g Ag}}{133.89 \text{ g AgCN}} = 262 \text{ g Ag}$

d. Given: 0.780 g Au_2Se_3
Unknown: mass of gold

molar mass $Au_2Se_3 = 2 \text{ mol Au} \times \dfrac{196.97 \text{ g Au}}{1 \text{ mol Au}} + 3 \text{ mol Se} \times \dfrac{78.96 \text{ g Se}}{1 \text{ mol Se}}$
$= 630.82 \text{ g}$

$0.780 \text{ g Au}_2Se_3 \times \dfrac{2 \times 196.97 \text{ g Au}}{630.82 \text{ g Au}_2Se_3} = 0.487 \text{ g Au}$

e. Given: 683 g Na_2SeO_3
Unknown: mass of selenium

molar mass $Na_2SeO_3 = 2 \text{ mol Na} \times \dfrac{22.99 \text{ g Na}}{1 \text{ mol Na}} + 1 \text{ mol Se} \times \dfrac{78.96 \text{ g Se}}{1 \text{ mol Se}}$
$+ 3 \text{ mol O} \times \dfrac{16.00 \text{ g O}}{1 \text{ mol O}} = 172.94 \text{ g}$

$683 \text{ g Na}_2SeO_3 \times \dfrac{78.96 \text{ g Se}}{172.94 \text{ g Na}_2SeO_3} = 312 \text{ g Se}$

f. Given: 5.0×10^4 g $CHCl_2CH_2CH_3$
Unknown: mass of chlorine

molar mass $CHCl_2CH_2CH_3$ = 3 mol C × $\dfrac{12.01 \text{ g C}}{1 \text{ mol C}}$ + 6 mol H × $\dfrac{1.01 \text{ g H}}{1 \text{ mol H}}$

+ 2 mol Cl × $\dfrac{35.45 \text{ g Cl}}{1 \text{ mol Cl}}$ = 112.99 g

5.0×10^4 g $CHCl_2CH_2CH_3$ × $\dfrac{2 \times 35.45 \text{ g Cl}}{112.99 \text{ g } CHCl_2CH_2CH_3}$ = 3.1×10^4 g Cl

---

140. a. Given: $SrCl_2 \cdot 6H_2O$
Unknown: percentage of water

molar mass $SrCl_2 \cdot 6H_2O$ = 1 mol Sr × $\dfrac{87.62 \text{ g Sr}}{1 \text{ mol Sr}}$ + 2 mol Cl × $\dfrac{35.45 \text{ g Cl}}{1 \text{ mol Cl}}$

+ 12 mol H × $\dfrac{1.01 \text{ g H}}{1 \text{ mol H}}$ + 6 mol O × $\dfrac{16.00 \text{ g O}}{1 \text{ mol O}}$ = 266.64 g

$\dfrac{6(18.02) \text{ g } H_2O}{266.64 \text{ g } SrCl_2 \cdot 6H_2O}$ × 100 = 40.55% H_2O

b. Given: $ZnSO_4 \cdot 7H_2O$
Unknown: percentage of water

molar mass $ZnSO_4 \cdot 7H_2O$ = 1 mol Zn × $\dfrac{65.39 \text{ g Zn}}{1 \text{ mol Zn}}$ + 1 mol S × $\dfrac{32.07 \text{ g S}}{1 \text{ mol S}}$

+ 11 mol O × $\dfrac{16.00 \text{ g O}}{1 \text{ mol O}}$ + 14 mol H × $\dfrac{1.01 \text{ g H}}{1 \text{ mol H}}$ = 287.60 g

$\dfrac{7(18.02) \text{ g } H_2O}{287.60 \text{ g } ZnSO_4 \cdot 7H_2O}$ × 100 = 43.86% H_2O

c. Given: $CaFPO_3 \cdot 2H_2O$
Unknown: percentage of water

molar mass $CaFPO_3 \cdot 2H_2O$ = 1 mol Ca × $\dfrac{40.08 \text{ g Ca}}{1 \text{ mol Ca}}$

+ 1 mol F × $\dfrac{19.00 \text{ g F}}{1 \text{ mol F}}$ + 1 mol P × $\dfrac{30.97 \text{ g P}}{1 \text{ mol P}}$ + 5 mol O × $\dfrac{16.00 \text{ g O}}{1 \text{ mol O}}$

+ 4 mol H × $\dfrac{1.01 \text{ g H}}{1 \text{ mol H}}$ = 174.09 g

$\dfrac{2(18.02) \text{ g } H_2O}{174.09 \text{ g } CaFPO_3 \cdot 2H_2O}$ × 100 = 20.70% H_2O

d. Given: $Be(NO_3)_2 \cdot 3H_2O$
Unknown: percentage of water

molar mass $Be(NO_3)_2 \cdot 3H_2O$ = 1 mol Be × $\dfrac{9.01 \text{ g Be}}{1 \text{ mol Be}}$

+ 2 mol N × $\dfrac{14.01 \text{ g N}}{1 \text{ mol N}}$ × 9 mol O × $\dfrac{16.00 \text{ g O}}{1 \text{ mol O}}$

+ 6 mol H × $\dfrac{1.01 \text{ g H}}{1 \text{ mol H}}$ = 187.09 g

$\dfrac{3(18.02) \text{ g } H_2O}{187.09 \text{ g } Be(NO_3)_2 \cdot 3H_2O}$ × 100 = 28.90% H_2O

141. a. Given: nickel (II) acetate tetrahydrate
Unknown: formula; percentage of nickel

formula is $Ni(C_2H_3O_2)_2 \cdot 4H_2O$

molar mass $Ni(C_2H_3O_2)_2 \cdot 4H_2O$ = 1 mol Ni × $\dfrac{58.69 \text{ g Ni}}{1 \text{ mol Ni}}$

+ 4 mol C × $\dfrac{12.01 \text{ g C}}{1 \text{ mol C}}$ + 14 mol H × $\dfrac{1.01 \text{ g H}}{1 \text{ mol H}}$ + 8 mol O × $\dfrac{16.00 \text{ g O}}{1 \text{ mol O}}$

= 248.87 g

$\dfrac{58.69 \text{ g Ni}}{248.87 \text{ g Ni}(C_2H_3O_2)_2 \cdot 4H_2O} \times 100 = 23.58\%$ Ni

b. Given: sodium chromate tetrahydrate
Unknown: formula; percentage of chromium

formula is $Na_2CrO_4 \cdot 4H_2O$

molar mass $Na_2CrO_4 \cdot 4H_2O$ = 2 mol Na × $\dfrac{22.99 \text{ g Na}}{1 \text{ mol Na}}$

+ 1 mol Cr × $\dfrac{52.00 \text{ g Cr}}{1 \text{ mol Cr}}$ + 8 mol O × $\dfrac{16.00 \text{ g O}}{1 \text{ mol O}}$ + 8 mol H × $\dfrac{1.01 \text{ g H}}{1 \text{ mol H}}$

= 234.06 g

$\dfrac{52.00 \text{ g Cr}}{234.06 \text{ g Na}_2CrO_4 \cdot 4H_2O} \times 100 = 22.22\%$ Cr

c. Given: cerium (IV) sulfate tetrahydrate
Unknown: percentage of cerium

formula is $Ce(SO_4)_2 \cdot 4H_2O$

molar mass $Ce(SO_4)_2 \cdot 4H_2O$ = 1 mol Ce × $\dfrac{140.12 \text{ g Ce}}{1 \text{ mol Ce}}$

+ 2 mol S × $\dfrac{32.07 \text{ g S}}{1 \text{ mol S}}$ + 12 mol O × $\dfrac{16.00 \text{ g O}}{1 \text{ mol O}}$ + 8 mol H × $\dfrac{1.01 \text{ g H}}{1 \text{ mol H}}$

= 404.34 g

$\dfrac{140.12 \text{ g Ce}}{404.34 \text{ g Ce}(SO_4)_2 \cdot 4H_2O} \times 100 = 34.65\%$ Ce

142. Given: 50.0 kg HgS
Unknown: mass of mercury

molar mass HgS = 1 mol Hg × $\dfrac{200.59 \text{ g Hg}}{1 \text{ mol Hg}}$ + 1 mol S × $\dfrac{32.07 \text{ g S}}{1 \text{ mol S}}$ = 232.68 g

50.0 kg HgS × $\dfrac{200.59 \text{ g Hg}}{232.68 \text{ g HgS}}$ = 43.1 kg Hg

143. Given: 1.00×10^3 kg of each of $Ca_2(OH)_2CO_3$ and $CuFeS_2$

Unknown: mass of copper for each; which has more Cu

molar mass $Cu_2(OH)_2CO_3 = 2 \text{ mol Cu} \times \dfrac{63.55 \text{ g Cu}}{1 \text{ mol Cu}} + 5 \text{ mol O} \times \dfrac{16.00 \text{ g O}}{1 \text{ mol O}}$

$+ 2 \text{ mol H} \times \dfrac{1.01 \text{ g H}}{1 \text{ mol H}} + 1 \text{ mol C} \times \dfrac{12.01 \text{ g C}}{1 \text{ mol C}} = 221.13 \text{ g}$

$1.00 \times 10^3 \text{ kg } Cu_2(OH)_2CO_3 \times \dfrac{2 \times 63.55 \text{ g Cu}}{221.13 \text{ g } Cu_2(OH)_2CO_3} = 575 \text{ kg Cu}$

molar mass $CuFeS_2 = 1 \text{ mol Cu} \times \dfrac{63.55 \text{ g Cu}}{1 \text{ mol Cu}} + 1 \text{ mol Fe} \times \dfrac{55.85 \text{ g Fe}}{1 \text{ mol F}}$

$+ 2 \text{ mol S} \times \dfrac{32.07 \text{ g S}}{1 \text{ mol S}} = 183.54 \text{ g}$

$1.00 \times 10^3 \text{ kg } CuFeS_2 \times \dfrac{63.55 \text{ g Cu}}{183.54 \text{ g } CuFeS_2} = 346 \text{ kg Cu}$

Malachite, $Cu_2(OH)_2CO_3$, has more copper.

---

144. a. Given: $VOSO_4 \cdot 2H_2O$

Unknown: percentage of vanadium

molar mass $VOSO_4 \cdot 2H_2O = 1 \text{ mol V} \times \dfrac{50.94 \text{ g V}}{1 \text{ mol V}} + 7 \text{ mol O} \times \dfrac{16.00 \text{ g O}}{1 \text{ mol O}}$

$+ 1 \text{ mol S} \times \dfrac{32.07 \text{ g S}}{1 \text{ mol S}} + 4 \text{ mol H} \times \dfrac{1.01 \text{ g H}}{1 \text{ mol H}} = 199.05 \text{ g}$

$\dfrac{50.94 \text{ g V}}{199.05 \text{ g } VOSO_4 \cdot 2 H_2O} \times 100 = 25.59\% \text{ V}$

b. Given: $K_2SnO_3 \cdot 3H_2O$

Unknown: percentage of tin

molar mass $K_2SnO_3 \cdot 3H_2O = 2 \text{ mol K} \times \dfrac{39.10 \text{ g K}}{1 \text{ mol K}}$

$+ 1 \text{ mol Sn} \times \dfrac{118.71 \text{ g Sn}}{1 \text{ mol Sn}} + 6 \text{ mol O} \times \dfrac{16.00 \text{ g O}}{1 \text{ mol O}} + 6 \text{ mol H} \times \dfrac{1.01 \text{ g H}}{1 \text{ mol H}} = 298.97 \text{ g}$

$\dfrac{118.71 \text{ g Sn}}{298.97 \text{ g } K_2SnO_3 \cdot 3H_2O} \times 100 = 39.71\% \text{ Sn}$

c. Given: $CaClO_3 \cdot 2H_2O$

Unknown: percentage of chlorine

molar mass $CaClO_3 \cdot 2H_2O = 1 \text{ mol Ca} \times \dfrac{40.08 \text{ g Ca}}{1 \text{ mol Ca}}$

$+ 1 \text{ mol Cl} \times \dfrac{35.45 \text{ g Cl}}{1 \text{ mol Cl}} + 5 \text{ mol O} \times \dfrac{16.00 \text{ g O}}{1 \text{ mol O}} + 4 \text{ mol H} \times \dfrac{1.01 \text{ g H}}{1 \text{ mol H}}$

$= 159.57 \text{ g}$

$\dfrac{35.45 \text{ g Cl}}{159.57 \text{ g } CaClO_3 \cdot 2H_2O} \times 100 = 22.22\% \text{ Cl}$

145. Given: 500.0 g CuSO$_4$ · 5H$_2$O

Unknown: mass of anhydrous CuSO$_4$

molar mass CuSO$_4$ = 1 mol Cu × $\dfrac{63.55 \text{ g Cu}}{1 \text{ mol Cu}}$ + 1 mol S × $\dfrac{32.07 \text{ g}}{1 \text{ mol S}}$

+ 4 mol O × $\dfrac{16.00 \text{ g O}}{1 \text{ mol O}}$ = 159.62 g

molar mass CuSO$_4$ · 5H$_2$O = 159.62 g + 5 mol H$_2$O × $\dfrac{18.02 \text{ g H}_2\text{O}}{1 \text{ mol H}_2\text{O}}$ = 249.72 g

500.0 g CuSO$_4$ · 5H$_2$O × $\dfrac{159.62 \text{ g CuSO}_4}{249.72 \text{ g CuSO}_4 \cdot 5\text{H}_2\text{O}}$ = 319.6 g CuSO$_4$

---

146. Given: 1.00 g Ag

Unknown: mass of AgNO$_3$

molar mass AgNO$_3$ = 1 mol Ag × $\dfrac{107.87 \text{ g Ag}}{1 \text{ mol Ag}}$ + 1 mol N × $\dfrac{14.01 \text{ g N}}{1 \text{ mol N}}$

+ 3 mol O × $\dfrac{16.00 \text{ g O}}{1 \text{ mol O}}$ = 169.88 g

1.00 g Ag × $\dfrac{169.88 \text{ g AgNO}_3}{107.87 \text{ g Ag}}$ = 1.57 g AgNO$_3$

---

147. Given: 62.4 g Ag$_2$S

Unknown: mass of Ag and S

molar mass Ag$_2$S = 2 mol Ag × $\dfrac{107.87 \text{ g Ag}}{1 \text{ mol Ag}}$ + 1 mol S × $\dfrac{32.07 \text{ g S}}{1 \text{ mol S}}$ = 247.81 g

62.4 g Ag$_2$S × $\dfrac{2(107.87) \text{ g Ag}}{247.81 \text{ g Ag}_2\text{S}}$ = 54.3 g Ag

62.4 g Ag$_2$S × $\dfrac{32.07 \text{ g S}}{247.81 \text{ g Ag}_2\text{S}}$ = 8.08 g S

---

148. Given: MgSO$_4$ · 7H$_2$O; 11.8 g H$_2$O

Unknown: mass of MgSO$_4$ · 7H$_2$O

molar mass MgSO$_4$ · 7H$_2$O = 1 mol Mg × $\dfrac{24.31 \text{ g Mg}}{1 \text{ mol Mg}}$

+ 1 mol S × $\dfrac{32.07 \text{ g S}}{1 \text{ mol S}}$ + 11 mol O × $\dfrac{16.00 \text{ g O}}{1 \text{ mol O}}$ + 14 mol H × $\dfrac{1.01 \text{ g H}}{1 \text{ mol H}}$

= 246.52 g

11.8 g H$_2$O × $\dfrac{246.52 \text{ g MgSO}_4 \cdot 7\text{H}_2\text{O}}{7(18.02) \text{ g H}_2\text{O}}$ = 23.1 g MgSO$_4$ · 7H$_2$O

---

149. Given: 1.00 kg H$_2$SO$_4$

Unknown: mass of sulfur

molar mass H$_2$SO$_4$ = 2 mol H × $\dfrac{1.01 \text{ g H}}{1 \text{ mol H}}$ + 1 mol S × $\dfrac{32.07 \text{ S}}{1 \text{ mol S}}$

+ 4 mol O × $\dfrac{16.00 \text{ g O}}{1 \text{ mol O}}$ = 98.09 g

1.00 kg H$_2$SO$_4$ × $\dfrac{32.07 \text{ g S}}{98.09 \text{ g H}_2\text{SO}_4}$ × $\dfrac{1000 \text{ g}}{1 \text{ kg}}$ = 3.27 × 10^2 g S

---

150. a. Given: 28.4% Cu; 71.6% Br

Unknown: empirical formula of compound

100.0 g of compound contains 28.4 g Cu and 71.6 g Br.

28.4 g Cu × $\dfrac{1 \text{ mol Cu}}{63.55 \text{ g Cu}}$ = 0.447 mol Cu

71.6 g Br × $\dfrac{1 \text{ mol Br}}{79.90 \text{ g Br}}$ = 0.896 mol Br

$\dfrac{0.447 \text{ mol Cu}}{0.447}$: $\dfrac{0.896 \text{ mol Br}}{0.447}$ = 1.00 mol Cu : 2.00 mol Br → CuBr$_2$

b. Given: 39.0% K;
12.0% C;
1.01% H;
47.9% O

Unknown: empirical formula of compound

100.0 g of compound contains 39.0 g K, 12.0 g C, 1.01 g H, and 47.9 g O.

$$39.0 \text{ g K} \times \frac{1 \text{ mol K}}{39.10 \text{ g K}} = 0.997 \text{ mol K}$$

$$12.0 \text{ g C} \times \frac{1 \text{ mol C}}{12.01 \text{ g C}} = 0.999 \text{ mol C}$$

$$1.01 \text{ g H} \times \frac{1 \text{ mol H}}{1.01 \text{ g H}} = 1.00 \text{ mol H}$$

$$47.9 \text{ g O} \times \frac{1 \text{ mol O}}{16.00 \text{ g O}} = 2.99 \text{ mol O}$$

$$\frac{0.997 \text{ mol K}}{0.997} : \frac{0.999 \text{ mol C}}{0.997} : \frac{1.00 \text{ mol H}}{0.997} : \frac{2.99 \text{ mol O}}{0.997} = 1.00 \text{ mol K} : 1.00 \text{ mol C} : 1.00 \text{ mol H} : 3.00 \text{ mol O} \rightarrow KHCO_3$$

c. Given: 77.3% Ag;
7.4% P;
15.3% O

Unknown: empirical formula of compound

100.0 g of compound contains 77.3 g Ag, 7.4 g P, and 15.3 g O.

$$77.3 \text{ g Ag} \times \frac{1 \text{ mol Ag}}{107.87 \text{ g Ag}} = 0.717 \text{ mol Ag}$$

$$7.4 \text{ g P} \times \frac{1 \text{ mol P}}{30.97 \text{ g P}} = 0.24 \text{ mol P}$$

$$15.3 \text{ g O} \times \frac{1 \text{ mol O}}{16.00 \text{ g O}} = 0.956 \text{ mol O}$$

$$\frac{0.717 \text{ mol Ag}}{0.24} : \frac{0.24 \text{ mol P}}{0.24} : \frac{0.956 \text{ mol O}}{0.24} = 3.0 \text{ mol Ag} : 1.0 \text{ mol P} : 4.0 \text{ mol O} \rightarrow Ag_3PO_4$$

d. Given: 0.57% H;
72.1% I;
27.3% O

Unknown: empirical formula of compound

100.0 g of compound contains 0.57 g H, 72.1 g I, and 27.3 g O

$$0.57 \text{ g H} \times \frac{1 \text{ mol H}}{1.01 \text{ g H}} = 0.56 \text{ mol H}$$

$$72.1 \text{ g I} \times \frac{1 \text{ mol I}}{126.90 \text{ g I}} = 0.57 \text{ mol I}$$

$$27.3 \text{ g O} \times \frac{1 \text{ mol O}}{16.00 \text{ g O}} = 1.7 \text{ mol O}$$

$$\frac{0.56 \text{ mol H}}{0.56} : \frac{0.57 \text{ mol I}}{0.56} : \frac{1.7 \text{ mol O}}{0.56} = 1.0 \text{ mol H} : 1.0 \text{ mol I} : 3.0 \text{ mol O}$$

$$\rightarrow HIO_3$$

151. a. Given: 36.2% Al;
63.8% S

Unknown: empirical formula of compound

100.0 g of compound contains 36.2 g Al and 63.8 g S.

$$36.2 \text{ g Al} \times \frac{1 \text{ mol Al}}{26.98 \text{ g Al}} = 1.34 \text{ mol Al}$$

$$63.8 \text{ g S} \times \frac{1 \text{ mol S}}{32.07 \text{ g S}} = 1.99 \text{ mol S}$$

$$\frac{1.34 \text{ mol Al}}{1.34} : \frac{1.99 \text{ mol S}}{1.34} = 1.0 \text{ mol Al} : 1.5 \text{ mol S}$$

$$2(1.0 \text{ mol Al} : 1.5 \text{ mol S}) = 2.0 \text{ mol Al} : 3.0 \text{ mol S} \rightarrow Al_2S_3$$

b. Given: 93.5% Nb; 6.50% O
Unknown: empirical formula of compound

100.0 g of compound contains 93.5 g Nb and 6.50 g O.

$$93.5 \text{ g Nb} \times \frac{1 \text{ mol Nb}}{92.91 \text{ g Nb}} = 1.01 \text{ mol Nb}$$

$$6.50 \text{ g O} \times \frac{1 \text{ mol O}}{16.00 \text{ g O}} = 0.406 \text{ mol O}$$

$$\frac{1.01 \text{ mol Nb}}{0.406} : \frac{0.406 \text{ mol O}}{0.406} = 2.49 \text{ mol Nb} : 1.00 \text{ mol O}$$

$$2(2.49 \text{ mol Nb} : 1.00 \text{ mol O}) = 4.98 \text{ mol Nb} : 2 \text{ mol O} \rightarrow Nb_5O_2$$

c. Given: 57.6% Sr; 13.8% P; 28.6% O
Unknown: empirical formula of compound

100.0 g of compound contains 57.6 g Sr, 13.8 g P, and 28.6 g O

$$57.6 \text{ g Sr} \times \frac{1 \text{ mol Sr}}{87.62 \text{ g Sr}} = 0.657 \text{ mol Sr}$$

$$13.8 \text{ g P} \times \frac{1 \text{ mol P}}{30.97 \text{ g P}} = 0.446 \text{ mol P}$$

$$28.6 \text{ g O} \times \frac{1 \text{ mol O}}{16.00 \text{ g O}} = 1.79 \text{ mol O}$$

$$\frac{0.657 \text{ mol Sr}}{0.446} : \frac{0.446 \text{ mol P}}{0.446} : \frac{1.79 \text{ mol O}}{0.446} = 1.47 \text{ mol Sr} : 1.00 \text{ mol P} : 4.01 \text{ mol O}$$

$$2(1.47 \text{ mol Sr} : 1.00 \text{ mol P} : 4.01 \text{ mol O}) = 2.94 \text{ mol Sr} : 2.00 \text{ mol P} : 8.02 \text{ mol O} \rightarrow Sr_3P_2O_8$$

d. Given: 28.5% Fe; 48.6% O; 22.9% S
Unknown: empirical formula of compound

100.0 g of compound contains 28.5 g Fe, 48.6 g O, and 22.9 g S

$$28.5 \text{ g Fe} \times \frac{1 \text{ mol Fe}}{55.85 \text{ g Fe}} = 0.510 \text{ mol Fe}$$

$$48.6 \text{ g O} \times \frac{1 \text{ mol O}}{16.00 \text{ g O}} = 3.04 \text{ mol O}$$

$$22.9 \text{ g S} \times \frac{1 \text{ mol S}}{32.07 \text{ g S}} = 0.714 \text{ mol S}$$

$$\frac{0.510 \text{ mol Fe}}{0.510} : \frac{3.04 \text{ mol O}}{0.510} : \frac{0.714 \text{ mol S}}{0.510} = 1.00 \text{ mol Fe} : 5.96 \text{ mol O} : 1.40 \text{ mol S}$$

$$2(1.00 \text{ mol Fe} : 5.96 \text{ mol O} : 1.40 \text{ mol S}) = 2.00 \text{ mol Fe} : 11.92 \text{ mol O} : 2.80 \text{ mol S} \rightarrow Fe_2S_3O_{12}$$

152. a. Given: empirical formula: CH_2; molar mass = 28 g/mol
Unknown: molecular formula

$$\text{molar mass } CH_2 = 1 \text{ mol C} \times \frac{12.01 \text{ g C}}{1 \text{ mol C}} + 2 \text{ mol H} \times \frac{1.01 \text{ g H}}{1 \text{ mol H}} = 14.03 \text{ g}$$

$$x = \frac{28 \text{ g}}{14.03 \text{ g}} = 2.0 \qquad C_xH_{2x} = C_2H_4$$

b. Given: empirical formula: B_2H_5; molar mass = 54 g/mol
Unknown: molecular formula

$$\text{molar mass } B_2H_5 = 2 \text{ mol B} \times \frac{10.81 \text{ g B}}{1 \text{ mol B}} + 5 \text{ mol H} \times \frac{1.01 \text{ g H}}{1 \text{ mol H}} = 26.67 \text{ g}$$

$$x = \frac{54 \text{ g}}{26.67 \text{ g}} = 2.0 \qquad B_{2x}H_{5x} = B_4H_{10}$$

c. Given: empirical formula: C_2HCl; molar mass = 179 g/mol
Unknown: molecular formula

molar mass $C_2HCl = 2 \text{ mol C} \times \dfrac{12.01 \text{ g C}}{1 \text{ mol C}} + 1 \text{ mol H} \times \dfrac{1.01 \text{ g H}}{1 \text{ mol C}}$

$+ 1 \text{ mol Cl} \times \dfrac{35.45 \text{ g Cl}}{1 \text{ mol Cl}} = 60.48 \text{ g}$

$x = \dfrac{179 \text{ g}}{60.48 \text{ g}} = 2.96 \cong 3 \quad C_{2x}H_xCl_x = C_6H_3Cl_3$

d. Given: empirical formula: C_6H_8O; molar mass = 290 g/mol
Unknown: molecular formula

molar mass $C_6H_8O = 6 \text{ mol C} \times \dfrac{12.01 \text{ g C}}{1 \text{ mol C}} + 8 \text{ mol H} \times \dfrac{1.01 \text{ g H}}{1 \text{ mol H}}$

$+ 1 \text{ mol O} \times \dfrac{16.00 \text{ g O}}{1 \text{ mol O}} = 96.14 \text{ g}$

$x = \dfrac{290 \text{ g}}{96.14 \text{ g}} = 3.0 \quad C_{6x}H_{8x}O_x = C_{18}H_{24}O_3$

e. Given: empirical formula: C_3H_2O; molar mass = 216 g/mol
Unknown: molecular formula

molar mass $C_3H_2O = 3 \text{ mol C} \times \dfrac{12.01 \text{ g C}}{1 \text{ mol C}} + 2 \text{ mol H} \times \dfrac{1.01 \text{ g H}}{1 \text{ mol H}}$

$+ 1 \text{ mol O} \times \dfrac{16.00 \text{ g O}}{1 \text{ mol O}} = 54.05 \text{ g}$

$x = \dfrac{216 \text{ g}}{54.05} = 4.0 \quad C_{3x}H_{2x}O_x = C_{12}H_8O_4$

153. a. Given: 66.0% Ba; 34.0% Cl
Unknown: empirical formula of compound

100.0 g of compound contains 66.0 g Ba and 34.0 g Cl

$66.0 \text{ g Ba} \times \dfrac{1 \text{ mol Ba}}{137.33 \text{ g Ba}} = 0.481 \text{ mol Ba}$

$34.0 \text{ g Cl} \times \dfrac{1 \text{ mol Cl}}{35.45 \text{ g Cl}} = 0.959 \text{ mol Cl}$

$\dfrac{0.481 \text{ mol Ba}}{0.481} : \dfrac{0.959 \text{ mol Cl}}{0.481} = 1.00 \text{ mol Ba} : 1.99 \text{ mol Cl} \rightarrow BaCl_2$

b. Given: 80.38% Bi; 18.46% O; 1.16% H
Unknown: empirical formula of compound

100.00 g of compound contains 80.38 g Bi, 18.46 g O, and 1.16 g H

$80.38 \text{ g Bi} \times \dfrac{1 \text{ mol Bi}}{28.98 \text{ g Bi}} = 0.3846 \text{ mol Bi}$

$18.46 \text{ g O} \times \dfrac{1 \text{ mol O}}{16.00 \text{ g O}} = 1.154 \text{ mol O}$

$1.16 \text{ g H} \times \dfrac{1 \text{ mol H}}{1.01 \text{ g H}} = 1.149 \text{ mol H}$

$\dfrac{0.3846 \text{ mol Bi}}{0.3846} : \dfrac{1.154 \text{ mol O}}{0.3846} : \dfrac{1.149 \text{ mol H}}{0.3846} = 1.000 \text{ mol Bi} : 3.001 \text{ mol O} :$

$2.964 \text{ mol H} \rightarrow BiO_3H_3$

c. Given: 12.67% Al; 19.73% N; 67.60% O

Unknown: empirical formula of compound

100.00 g of compound contains 12.67 g Al, 19.73 g N, and 67.60 g O.

$$12.67 \text{ g Al} \times \frac{1 \text{ mol Al}}{26.98 \text{ g Al}} = 0.4696 \text{ mol Al}$$

$$19.73 \text{ g N} \times \frac{1 \text{ mol N}}{14.01 \text{ g N}} = 1.408 \text{ mol N}$$

$$67.60 \text{ g O} \times \frac{1 \text{ mol O}}{16.00 \text{ g O}} = 4.225 \text{ mol O}$$

$$\frac{0.4696 \text{ mol Al}}{0.4696} : \frac{1.408 \text{ mol N}}{0.4696} : \frac{4.225 \text{ mol O}}{0.4696} = 1.000 \text{ mol Al} : 2.998 \text{ mol N} : 8.997 \text{ mol O} \rightarrow \text{AlN}_3\text{O}_9$$

d. Given: 35.64% Zn; 26.18% C; 34.88% O; 3.30% H

Unknown: empirical formula of compound

100.00 g of compound contains 35.64 g Zn, 26.18 g C, 34.88 g O, and 3.30 g H.

$$35.64 \text{ g Zn} \times \frac{1 \text{ mol Zn}}{65.39 \text{ g Zn}} = 0.5450 \text{ mol Zn}$$

$$26.18 \text{ g C} \times \frac{1 \text{ mol C}}{12.01 \text{ g C}} = 2.180 \text{ mol C}$$

$$34.88 \text{ g O} \times \frac{1 \text{ mol O}}{16.00 \text{ g O}} = 2.180 \text{ mol O}$$

$$3.30 \text{ g H} \times \frac{1 \text{ mol H}}{1.01 \text{ g H}} = 3.267 \text{ mol H}$$

$$\frac{0.5450 \text{ mol Zn}}{0.5450} : \frac{2.180 \text{ mol C}}{0.5450} : \frac{2.180 \text{ mol O}}{0.5450} : \frac{3.267 \text{ mol H}}{0.5450} = 1.000 \text{ mol Zn} : 4.000 \text{ mol C} : 4.000 \text{ mol O} : 5.994 \text{ mol H} \rightarrow \text{ZnC}_4\text{H}_6\text{O}_4$$

e. Given: 2.8% H; 9.8% N; 20.5% Ni; 44.5% O; 22.4% S

Unknown: empirical formula of compound

100.0 g of compound contains 2.8 g H, 9.8 g N, 20.5 g Ni, 44.5 g O, and 22.4 g S.

$$2.8 \text{ g H} \times \frac{1 \text{ mol H}}{1.01 \text{ g H}} = 2.8 \text{ mol H}$$

$$9.8 \text{ g N} \times \frac{1 \text{ mol N}}{14.01 \text{ g N}} = 0.70 \text{ mol N}$$

$$20.5 \text{ g Ni} \times \frac{1 \text{ mol Ni}}{58.69 \text{ g Ni}} = 0.35 \text{ mol Ni}$$

$$44.5 \text{ g O} \times \frac{1 \text{ mol O}}{16.00 \text{ g O}} = 2.8 \text{ mol O}$$

$$22.4 \text{ g S} \times \frac{1 \text{ mol S}}{32.07 \text{ g S}} = 0.70 \text{ mol S}$$

$$\frac{2.8 \text{ mol H}}{0.35} : \frac{0.70 \text{ mol N}}{0.35} : \frac{0.35 \text{ mol Ni}}{0.35} : \frac{2.8 \text{ mol O}}{0.35} : \frac{0.70 \text{ mol S}}{0.35} = 8.0 \text{ mol H} : 2.0 \text{ mol N} : 1.0 \text{ mol Ni} : 8.0 \text{ mol O} : 2.0 \text{ mol S} \rightarrow \text{NiN}_2\text{S}_2\text{H}_8\text{O}_8$$

f. Given: 8.09% C; 0.34% H; 10.78% O; 80.78% Br

Unknown: empirical formula of compound

100.00 g of compound contains 8.09 g C, 0.34 g H, 10.78 g O, and 80.78 g Br.

$$8.09 \text{ g C} \times \frac{1 \text{ mol C}}{12.01 \text{ g C}} = 0.674 \text{ mol C}$$

$$0.34 \text{ g H} \times \frac{1 \text{ mol H}}{1.01 \text{ g H}} = 0.34 \text{ mol H}$$

$$10.78 \text{ g O} \times \frac{1 \text{ mol O}}{16.00 \text{ g O}} = 0.674 \text{ mol O}$$

$$80.78 \text{ g Br} \times \frac{1 \text{ mol Br}}{79.90 \text{ g Br}} = 1.011 \text{ mol Br}$$

$$\frac{0.674 \text{ mol C}}{0.34} : \frac{0.34 \text{ mol H}}{0.34} : \frac{0.674 \text{ mol O}}{0.34} : \frac{1.011 \text{ mol Br}}{0.34} = 2.0 \text{ mol C} :$$

$$1.00 \text{ mol H} : 2.0 \text{ mol O} : 3.0 \text{ mol Br} \rightarrow C_2HBrO_2$$

---

154. a. Given: 0.537 g Cu; 0.321 g F

Unknown: empirical formula of compound

$$0.537 \text{ g Cu} \times \frac{1 \text{ mol Cu}}{63.55 \text{ g Cu}} = 0.00845 \text{ mol Cu}$$

$$0.321 \text{ g F} \times \frac{1 \text{ mol F}}{19.00 \text{ g F}} = 0.0169 \text{ mol F}$$

$$\frac{0.00845 \text{ mol Cu}}{0.00845} : \frac{0.0169 \text{ Mol F}}{0.00845} = 1.00 \text{ mol Cu} : 2.00 \text{ mol F} \rightarrow CuF_2$$

b. Given: 9.48 g Ba; 1.66 g C; 1.93 g N

Unknown: empirical formula of compound

$$9.48 \text{ g Ba} \times \frac{1 \text{ mol Ba}}{137.33 \text{ g Ba}} = 0.0690 \text{ mol Ba}$$

$$1.66 \text{ g C} \times \frac{1 \text{ mol C}}{12.01 \text{ g C}} = 0.138 \text{ mol C}$$

$$1.93 \text{ g N} \times \frac{1 \text{ mol N}}{14.01 \text{ g N}} = 0.138 \text{ mol N}$$

$$\frac{0.0690 \text{ mol Ba}}{0.0690} : \frac{0.138 \text{ mol C}}{0.0690} : \frac{0.138 \text{ mol N}}{0.0690} = 1.00 \text{ mol Ba} : 2.00 \text{ mol C} :$$

$$2.00 \text{ mol N} \rightarrow BaC_2N_2 = Ba(CN)_2$$

c. Given: 0.0091 g Mn; 0.0106 g O; 0.0053 g S

Unknown: empirical formula of compound

$$0.0091 \text{ g Mn} \times \frac{1 \text{ mol Mn}}{54.94 \text{ Mn}} = 1.7 \times 10^{-4} \text{ mol Mn}$$

$$0.0106 \text{ g O} \times \frac{1 \text{ mol O}}{16.00 \text{ g O}} = 6.63 \times 10^{-4} \text{ mol O}$$

$$0.0053 \text{ g S} \times \frac{1 \text{ mol S}}{32.07 \text{ g S}} = 1.7 \times 10^{-4} \text{ mol S}$$

$$\frac{1.7 \times 10^{-4} \text{ mol Mn}}{1.7 \times 10^{-4}} : \frac{6.63 \times 10^{-4} \text{ mol O}}{1.7 \times 10^{-4}} : \frac{1.7 \times 10^{-4} \text{ mol S}}{1.7 \times 10^{-4}}$$

$$= 1.0 \text{ mol Mn} : 3.9 \text{ mol O} : 1.0 \text{ mol S} \rightarrow MnO_4S = MnSO_4$$

155. a. Given: 0.0015 g Ni; 0.0067 g I
Unknown: empirical formula of compound

$$0.0015 \text{ g Ni} \times \frac{1 \text{ mol Ni}}{58.69 \text{ g Ni}} = 2.6 \times 10^{-5} \text{ mol Ni}$$

$$0.0067 \text{ g I} \times \frac{1 \text{ mol I}}{126.90 \text{ g I}} = 5.3 \times 10^{-5} \text{ mol I}$$

$$\frac{2.6 \times 10^{-5} \text{ mol Ni}}{2.6 \times 10^{-5}} : \frac{5.3 \times 10^{-5} \text{ mol I}}{2.6 \times 10^{-5}} = 1.0 \text{ mol Ni} : 2.0 \text{ mol I} \rightarrow \text{NiI}_2$$

b. Given: 0.144 g Mn; 0.074 g N; 0.252 g O
Unknown: empirical formula of compound

$$0.144 \text{ g Mn} \times \frac{1 \text{ mol Mn}}{54.94 \text{ g Mn}} = 2.62 \times 10^{-3} \text{ mol Mn}$$

$$0.074 \text{ g N} \times \frac{1 \text{ mol N}}{14.01 \text{ g N}} = 5.3 \times 10^{-3} \text{ mol N}$$

$$0.252 \text{ g O} \times \frac{1 \text{ mol O}}{16.00 \text{ g O}} = 1.58 \times 10^{-2} \text{ mol O}$$

$$\frac{2.62 \times 10^{-3} \text{ mol Mn}}{2.62 \times 10^{-3}} : \frac{5.3 \times 10^{-3} \text{ mol N}}{2.62 \times 10^{-3}} : \frac{1.58 \times 10^{-2} \text{ mol O}}{2.62 \times 10^{-3}} =$$

$$1.00 \text{ mol Mn} : 2.0 \text{ mol N} : 6.0 \text{ mol O} \rightarrow \text{MnN}_2\text{O}_6 = \text{Mn(NO}_3)_2$$

c. Given: 0.691 g Mg; 1.824 g S; 1.365 g O
Unknown: empirical formula of compound

$$0.691 \text{ g Mg} \times \frac{1 \text{ mol Mg}}{24.31 \text{ g Mg}} = 0.0284 \text{ mol Mg}$$

$$1.824 \text{ g S} \times \frac{1 \text{ mol S}}{32.07 \text{ g S}} = 0.05688 \text{ mol S}$$

$$1.365 \text{ g O} \times \frac{1 \text{ mol O}}{16.00 \text{ g O}} = 0.08531 \text{ mol O}$$

$$\frac{0.0284 \text{ mol Mg}}{0.0284} : \frac{0.05688 \text{ mol S}}{0.0284} : \frac{0.08531 \text{ mol O}}{0.0284} = 1.00 \text{ mol Mg} :$$

$$2.00 \text{ mol S} : 3.00 \text{ mol O} \rightarrow \text{MgS}_2\text{O}_3$$

d. Given: 14.77 g K; 9.06 g O; 2.42 g Sn
Unknown: empirical formula of compound

$$14.77 \text{ g K} \times \frac{1 \text{ mol K}}{39.10 \text{ g K}} = 0.3777 \text{ mol K}$$

$$9.06 \text{ g O} \times \frac{1 \text{ mol O}}{16.00 \text{ g O}} = 0.566 \text{ mol O}$$

$$22.42 \text{ g Sn} \times \frac{1 \text{ mol Sn}}{118.71 \text{ g Sn}} = 0.1889 \text{ mol Sn}$$

$$\frac{0.3777 \text{ mol K}}{0.3777} : \frac{0.566 \text{ mol O}}{0.3777} : \frac{0.1889 \text{ mol Sn}}{0.3777} = 1.000 \text{ mol K} : 1.50 \text{ mol O} :$$

0.5001 mol Sn

2(1.0 mol K : 1.5 mol O : 0.5 mol Sn) = 2 mol K : 3 mol O : 1 mol Sn
$\rightarrow \text{K}_2\text{O}_3\text{Sn} = \text{K}_2\text{SnO}_3$

156. a. Given: 60.9% As; 39.1% S
Unknown: empirical formula of compound

100.0 g of compound contains 60.9 g As and 39.1 g S.

$60.9 \text{ g As} \times \dfrac{1 \text{ mol As}}{74.92 \text{ g As}} = 0.813 \text{ mol As}$

$39.1 \text{ g S} \times \dfrac{1 \text{ mol S}}{32.07 \text{ g S}} = 1.22 \text{ mol S}$

$\dfrac{0.813 \text{ mol As}}{0.813} : \dfrac{1.22 \text{ mol S}}{0.813} = 1.00 \text{ mol As} : 1.50 \text{ mol S}$

$2(1.00 \text{ mol As} : 1.50 \text{ mol S}) = 2 \text{ mol As} : 3 \text{ mol S} \rightarrow As_2S_3$

b. Given: 76.89% Re; 23.12% O
Unknown: empirical formula of compound

100.00 g of compound contains 76.89 g Re and 23.12 g O.

$76.89 \text{ g Re} \times \dfrac{1 \text{ mol Re}}{186.21 \text{ g Re}} = 0.4129 \text{ mol Re}$

$23.12 \text{ g O} \times \dfrac{1 \text{ mol O}}{16.00 \text{ g O}} = 1.445 \text{ mol O}$

$\dfrac{0.4129 \text{ mol Re}}{0.4129} : \dfrac{1.445 \text{ mol O}}{0.4129} = 1.000 \text{ mol Re} : 3.500 \text{ mol O}$

$2(1.0 \text{ mol Re} : 3.5 \text{ mol O}) = 2 \text{ mol Re} : 7 \text{ mol O} \rightarrow Re_2O_7$

c. Given: 5.04% H; 35.00% N; 59.96% O
Unknown: empirical formula of compound

100.00 g of compound contains 5.04 g H, 35.00 g N, and 59.96 g O

$5.04 \text{ g H} \times \dfrac{1 \text{ mol H}}{1.01 \text{ g H}} = 4.99 \text{ mol H}$

$35.00 \text{ g N} \times \dfrac{1 \text{ mol N}}{14.01 \text{ g N}} = 2.498 \text{ mol N}$

$59.96 \text{ g O} \times \dfrac{1 \text{ mol O}}{16.00 \text{ g O}} = 3.748 \text{ mol O}$

$\dfrac{4.99 \text{ mol H}}{2.498} : \dfrac{2.498 \text{ mol N}}{2.498} : \dfrac{3.748 \text{ mol O}}{2.498} = 2.00 \text{ mol H} : 1.000 \text{ mol N} : 1.500 \text{ mol O}$

$2(2.0 \text{ mol H} : 1.0 \text{ mol N} : 1.5 \text{ mol O}) = 4 \text{ mol H} : 2 \text{ mol N} : 3 \text{ mol O} \rightarrow H_4N_2O_3 = NH_4NO_3$

d. Given: 24.3% Fe; 33.9% Cr; 41.8% O
Unknown: empirical formula of compound

100.0 g of compound contains 24.3 g Fe, 33.9 g Cr and 41.8 g O

$24.3 \text{ g Fe} \times \dfrac{1 \text{ mol Fe}}{55.85 \text{ g Fe}} = 0.435 \text{ mol Fe}$

$33.9 \text{ g Cr} \times \dfrac{1 \text{ mol Cr}}{52.00 \text{ g Cr}} = 0.652 \text{ mol Cr}$

$41.8 \text{ g O} \times \dfrac{1 \text{ mol O}}{16.00 \text{ g O}} = 2.61 \text{ mol O}$

$\dfrac{0.435 \text{ mol Fe}}{0.435} : \dfrac{0.652 \text{ mol Cr}}{0.435} : \dfrac{2.61 \text{ mol O}}{0.435} = 1.00 \text{ mol Fe} : 1.50 \text{ mol Cr} : 6.00 \text{ mol O}$

$2(1.0 \text{ mol Fe} : 1.5 \text{ mol Cr} : 6.0 \text{ mol O}) = 2 \text{ mol Fe} : 3 \text{ mol Cr} : 12 \text{ mol O} \rightarrow Fe_2Cr_3O_{12} = Fe_2(CrO_4)_3$

e. Given: 54.03% C; 37.81% N; 8.16% H
Unknown: empirical formula of compound

100.00 g of compound contains 54.03 g C, 37.81 g N, and 8.16 g H.

$$54.03 \text{ g C} \times \frac{1 \text{ mol C}}{12.01 \text{ g C}} = 4.499 \text{ mol C}$$

$$37.81 \text{ g N} \times \frac{1 \text{ mol N}}{14.01 \text{ g N}} = 2.699 \text{ mol N}$$

$$8.16 \text{ g H} \times \frac{1 \text{ mol H}}{1.01 \text{ g H}} = 8.08 \text{ mol H}$$

$$\frac{4.499 \text{ mol C}}{2.699} : \frac{2.699 \text{ mol N}}{2.699} : \frac{8.08 \text{ mol H}}{2.699} = 1.67 \text{ mol C} : 1.00 \text{ mol N} : 2.99 \text{ mol H}$$

$$3(1.67 \text{ mol C} : 1.00 \text{ mol N} : 2.99 \text{ mol H}) = 5 \text{ mol C} : 3 \text{ mol N} : 9 \text{ mol H} \rightarrow C_5N_3H_9 = C_5H_9N_3$$

f. Given: 55.81% C; 3.90% H; 29.43% F; 10.85% N
Unknown: empirical formula of compound

100.00 g of compound contains 55.81 g C, 3.90 g H, 29.43 g F, and 10.85 g N.

$$55.81 \text{ g C} \times \frac{1 \text{ mol C}}{12.01 \text{ g C}} = 4.647 \text{ mol C}$$

$$3.90 \text{ g H} \times \frac{1 \text{ mol H}}{1.01 \text{ g H}} = 3.86 \text{ mol H}$$

$$29.43 \text{ g F} \times \frac{1 \text{ mol F}}{19.00 \text{ g F}} = 1.549 \text{ mol F}$$

$$10.85 \text{ g N} \times \frac{1 \text{ mol N}}{14.01 \text{ g N}} = 0.7744 \text{ mol N}$$

$$\frac{4.647 \text{ mol C}}{0.7744} : \frac{3.86 \text{ mol H}}{0.7744} : \frac{1.549 \text{ mol F}}{0.7744} : \frac{0.7744 \text{ mol N}}{0.7744} = 6.000 \text{ mol C} : 4.98 \text{ mol H} : 2.000 \text{ mol F} : 1.000 \text{ mol N} \rightarrow C_6H_5F_2N = C_6H_3F_2NH_2$$

157. a. Given: empirical formula: C_2H_4S; molar mass = 179
Unknown: molecular formula of compound

$$\text{molar mass } C_2H_4S = 2 \text{ mol C} \times \frac{12.01 \text{ g C}}{1 \text{ mol C}} + 4 \text{ mol H} \times \frac{1.01 \text{ g H}}{1 \text{ mol H}} + 1 \text{ mol S} \times \frac{32.07 \text{ g S}}{1 \text{ mol S}} = 60.13 \text{ g}$$

$$x = \frac{179 \text{ g}}{60.13 \text{ g}} = 2.98 \cong 3; C_{2x}H_{4x}S_x = C_6H_{12}S_3$$

b. Given: empirical formula: C_2H_4O; molar mass = 176
Unknown: molecular formula of compound

$$\text{molar mass } C_2H_4O = 2 \text{ mol C} \times \frac{12.01 \text{ g C}}{1 \text{ mol C}} + 4 \text{ mol H} \times \frac{1.01 \text{ g H}}{1 \text{ mol H}} + 1 \text{ mol O} \times \frac{16.00 \text{ g O}}{1 \text{ mol O}} = 44.06 \text{ g}$$

$$x = \frac{176 \text{ g}}{44.06 \text{ g}} = 3.99 \cong 4; C_{2x}H_{4x}O_x = C_8H_{16}O_4$$

c. Given: empirical formula: $C_2H_3O_2$; molar mass = 119
Unknown: molecular formula of compound

$$\text{molar mass } C_2H_3O_2 = 2 \text{ mol C} \times \frac{12.01 \text{ g C}}{1 \text{ mol C}} + 3 \text{ mol H} \times \frac{1.01 \text{ g H}}{1 \text{ mol H}} + 2 \text{ mol O} \times \frac{16.00 \text{ g O}}{1 \text{ mol O}} = 59.058$$

$$x = \frac{119 \text{ g}}{59.05 \text{ g}} = 2.02 \cong 2; C_{2x}H_{3x}O_{2x} = C_4H_6O_4$$

d. Given: empirical formula: C_2H_2O; molar mass = 254

Unknown: molecular formula of compound

molar mass C_2H_2O = 2 mol C × $\dfrac{12.01 \text{ g C}}{1 \text{ mol C}}$ + 2 mol H × $\dfrac{1.01 \text{ g H}}{1 \text{ mol H}}$

+ 1 mol O × $\dfrac{16.00 \text{ g O}}{1 \text{ mol O}}$ = 42.04 g

$x = \dfrac{254 \text{ g}}{42.04 \text{ g}} = 6.04 \cong 6$; $C_{2x}H_{2x}O_x = C_{12}H_{12}O_6$

---

158. a. Given: percent composition; molar mass = 116.07

Unknown: molecular formula of compound

100.00 g of compound contains 41.39 g C, 3.47 g H, and 55.14 g O.

41.39 g C × $\dfrac{1 \text{ mol C}}{12.01 \text{ g C}}$ = 3.446 mol C

3.47 g H × $\dfrac{1 \text{ mol H}}{1.01 \text{ g H}}$ = 3.44 mol H

55.14 g O × $\dfrac{1 \text{ mol O}}{16.00 \text{ g O}}$ = 3.446 mol O

$\dfrac{3.446 \text{ mol C}}{3.44} : \dfrac{3.44 \text{ mol H}}{3.44} : \dfrac{3.446 \text{ mol O}}{3.44}$ = 1.00 mol C : 1.00 mol H : 1.00 mol

empirical formula = CHO

molar mass CHO = 1 mol C × $\dfrac{12.01 \text{ g C}}{1 \text{ mol C}}$ + 1 mol H × $\dfrac{1.01 \text{ g H}}{1 \text{ mol H}}$

+ 1 mol O × $\dfrac{16.00 \text{ g O}}{1 \text{ mol O}}$ = 29.02 g

$x = \dfrac{116.07 \text{ g}}{29.02 \text{ g}} = 4.000$; $C_xH_xO_x = C_4H_4O_4$

b. Given: percent composition; molar mass = 88

Unknown: molecular formula of compound

100.00 g of compound contains 54.53 g C, 9.15 g H, and 36.32 g O.

54.53 g C × $\dfrac{1 \text{ mol C}}{12.01 \text{ g C}}$ = 4.540 mol C

9.15 g H × $\dfrac{1 \text{ mol H}}{1.01 \text{ g H}}$ = 9.06 mol H

36.32 g O × $\dfrac{1 \text{ mol O}}{16.00 \text{ g O}}$ = 2.270 mol O

$\dfrac{4.540 \text{ mol C}}{2.270} : \dfrac{2.06 \text{ mol H}}{2.270} : \dfrac{2.270 \text{ mol O}}{2.270}$ = 2.000 mol C : 3.99 mol H : 1.000 mol O

empirical formula = C_2H_4O

molar mass C_2H_4O = 2 mol C × $\dfrac{12.01 \text{ g C}}{1 \text{ mol C}}$ + 4 mol H × $\dfrac{1.01 \text{ g H}}{1 \text{ mol H}}$

+ 1 mol O × $\dfrac{16.00 \text{ g O}}{1 \text{ mol O}}$ = 44.06 g

$x = \dfrac{88 \text{ g}}{44.06 \text{ g}} = 2.0$; $C_{2x}H_{4x}O_x = C_4H_8O_2$

c. Given: percent composition; molar mass = 168.19

Unknown: molecular formula of compound

100.00 g of compound contains 64.27 g C; 7.19 g H, and 28.54 g O.

$$64.27 \text{ g C} \times \frac{1 \text{ mol C}}{12.01 \text{ g C}} = 5.351 \text{ mol C}$$

$$7.19 \text{ g H} \times \frac{1 \text{ mol H}}{1.01 \text{ g H}} = 7.12 \text{ mol H}$$

$$28.54 \text{ g O} \times \frac{1 \text{ mol O}}{16.00 \text{ g O}} = 1.784 \text{ mol O}$$

$$\frac{5.351 \text{ mol C}}{1.784} : \frac{7.12 \text{ mol H}}{1.784} : \frac{1.784 \text{ mol O}}{1.784} = 2.999 \text{ mol C} : 3.99 \text{ mol H} : 1.000 \text{ mol O}$$

empirical formula = C_3H_4O

$$\text{molar mass } C_3H_4O = 3 \text{ mol C} \times \frac{12.01 \text{ g C}}{1 \text{ mol C}} + 4 \text{ mol H} \times \frac{1.01 \text{ g H}}{1 \text{ mol H}} + 1 \text{ mol O} \times \frac{16.00 \text{ g O}}{1 \text{ mol O}} = 56.07 \text{ g}$$

$$x = \frac{168.19 \text{ g}}{56.07 \text{ g}} = 3.000; C_{3x}H_{4x}O_x = C_9H_{12}O_3$$

159. Given: 0.141 g K; 0.115 g S; 0.144 g O

Unknown: empirical formula of compound

$$0.141 \text{ g K} \times \frac{1 \text{ mol K}}{39.10 \text{ g K}} = 3.61 \times 10^{-3} \text{ mol K}$$

$$0.115 \text{ g S} \times \frac{1 \text{ mol S}}{32.07 \text{ g S}} = 3.59 \times 10^{-3} \text{ mol S}$$

$$0.144 \text{ g O} \times \frac{1 \text{ mol O}}{16.00 \text{ g O}} = 9.00 \times 10^{-3} \text{ mol O}$$

$$\frac{3.61 \times 10^{-3} \text{ mol K}}{3.59 \times 10^{-3}} : \frac{3.59 \times 10^{-3} \text{ mol S}}{3.59 \times 10^{-3}} : \frac{9.00 \times 10^{-3} \text{ mol O}}{3.59 \times 10^{-3}}$$

= 1.01 mol K : 1.00 mol S : 2.51 mol O

2(1.01 mol K : 1.00 mol S : 2.51 mol O) = 2 mol K : 2 mol S : 5 mol O → $K_2S_2O_5$

160. a. Given: 9.65 g Pb; 0.99 g O

Unknown: empirical formula of compound

$$9.65 \text{ g Pb} \times \frac{1 \text{ mol Pb}}{207.2 \text{ g Pb}} = 0.0466 \text{ mol Pb}$$

$$0.99 \text{ g O} \times \frac{1 \text{ mol O}}{16.00 \text{ g O}} = 0.062 \text{ mol O}$$

$$\frac{0.466 \text{ mol Pb}}{0.0466} : \frac{0.062 \text{ mol O}}{0.0466} = 1.0 \text{ mol Pb} : 1.3 \text{ mol O}$$

3 (1.0 mol Pb : 1.3 mol 0) = 3 mol Pb : 4 mol O → Pb_3O_4

161. Given: 0.70 g Cr; 0.65 g S; 1.30 g O; molar mass = 392.2

Unknown: molecular formula of compound

$0.70 \text{ g Cr} \times \dfrac{1 \text{ mol Cr}}{52.00 \text{ g Cr}} = 0.013 \text{ mol Cr}$

$0.65 \text{ g S} \times \dfrac{1 \text{ mol S}}{32.07 \text{ g S}} = 0.0203 \text{ mol S}$

$1.30 \text{ g O} \times \dfrac{1 \text{ mol O}}{16.00 \text{ g O}} = 0.812 \text{ mol O}$

$\dfrac{0.0135 \text{ mol Cr}}{0.0135} : \dfrac{0.0203 \text{ mol S}}{0.0135} : \dfrac{0.0812 \text{ mol O}}{0.0135} = 1.0 \text{ mol Cr} : 1.5 \text{ mol S} : 6.0 \text{ mol O}$

$2(1.0 \text{ mol Cr} : 1.5 \text{ mol S} : 6.0 \text{ mol O}) = 2 \text{ mol Cr} : 3 \text{ mol S} : 12 \text{ mol O} \rightarrow Cr_2S_3O_{12} = Cr_2(SO_4)_3$

162. Given: 60.68% C; 3.40% H; 35.92% O

Unknown: empirical formula of compound

100.00 compound contain 60.68 g C, 3.40 g H, and 35.92 g O

$60.68 \text{ g C} \times \dfrac{1 \text{ mol C}}{12.01 \text{ g C}} = 5.052 \text{ mol C}$

$3.40 \text{ g H} \times \dfrac{1 \text{ mol H}}{1.01 \text{ g H}} = 3.37 \text{ mol H}$

$35.92 \text{ g O} \times \dfrac{1 \text{ mol O}}{16.00 \text{ g O}} = 2.245 \text{ mol O}$

$\dfrac{5.052 \text{ mol C}}{2.245} : \dfrac{3.37 \text{ mol H}}{2.245} : \dfrac{2.245 \text{ mol O}}{2.245} = 2.25 \text{ mol C} : 1.50 \text{ mol H} : 1.00 \text{ mol O}$

$4(2.25 \text{ mol C} : 1.50 \text{ mol H} : 1.00 \text{ mol O}) = 9 \text{ mol C} : 6 \text{ mol H} : 4 \text{ mol O} \rightarrow C_9H_6O_4$

163. Given: 208 mg C; 31 mg H; 146 mg N; molar mass = 111

Unknown: molecular formula of compound

$208 \text{ mg C} \times \dfrac{1 \text{ mol C}}{12.01 \text{ g C}} \times \dfrac{1 \text{ g}}{1000 \text{ mg}} = 0.0173 \text{ mol C}$

$31 \text{ mg H} \times \dfrac{1 \text{ mol H}}{1.01 \text{ g H}} \times \dfrac{1 \text{ g}}{1000 \text{ mg}} = 0.031 \text{ mol H}$

$146 \text{ mg N} \times \dfrac{1 \text{ mol N}}{14.01 \text{ g N}} \times \dfrac{1 \text{ g}}{1000 \text{ mg}} = 0.0104 \text{ mol N}$

$\dfrac{0.0173 \text{ mol C}}{0.0104} : \dfrac{0.031 \text{ mol H}}{0.0104} : \dfrac{0.0104 \text{ mol N}}{0.0104} = 1.66 \text{ mol C} : 3.00 \text{ mol H} : 1.00 \text{ mol N}$

$3(1.66 \text{ mol C} : 3.00 \text{ mol H} : 1.00 \text{ mol N}) = 5 \text{ mol C} : 9 \text{ mol H} : 3 \text{ mol N}$

empirical formula = $C_5H_9N_3$

molar mass $C_5H_9N_3$ = $5 \text{ mol C} \times \dfrac{12.01 \text{ g C}}{1 \text{ mol C}} + 9 \text{ mol H} \times \dfrac{1.01 \text{ g H}}{1 \text{ mol H}}$

$+ 3 \text{ mol N} \times \dfrac{14.01 \text{ g N}}{1 \text{ mol N}} = 111.17 \text{ g}$

$\dfrac{111 \text{ g}}{111.17 \text{ g}} = 0.998 \cong 1$

molecular formula = $C_5H_9N_3$

165. Given: balanced equation; 4.0 mol H_2
Unknown: moles Na

$$4.0 \text{ mol } H_2 \times \frac{2 \text{ mol Na}}{1 \text{ mol } H_2} = 8.0 \text{ mol Na}$$

166. Given: balanced equation; 0.046 mol LiBr
Unknown: moles LiCl

$$0.046 \text{ mol LiBr} \times \frac{2 \text{ mol LiCl}}{2 \text{ mol LiBr}} = 0.046 \text{ mol LiCl}$$

167. a. Given: balanced equation; 18 mol Al
Unknown: moles H_2SO_4

$$18 \text{ mol Al} \times \frac{3 \text{ mol } H_2SO_4}{2 \text{ mol Al}} = 27 \text{ mol } H_2SO_4$$

b. Given: balanced equation; 18 mol Al
Unknown: moles of each product

$$18 \text{ mol Al} \times \frac{1 \text{ mol } Al_2(SO_4)_3}{2 \text{ mol Al}} = 9 \text{ mol } Al_2(SO_4)_3$$

$$18 \text{ mol Al} \times \frac{3 \text{ mol } H_2}{2 \text{ mol Al}} = 27 \text{ mol } H_2$$

168. a. Given: balanced equation; 3.85 mol C_3H_8
Unknown: moles of CO_2 and H_2O

$$3.85 \text{ mol } C_3H_8 \times \frac{3 \text{ mol } CO_2}{1 \text{ mol } C_3H_8} = 11.6 \text{ mol } CO_2$$

$$3.85 \text{ mol } C_3H_8 \times \frac{4 \text{ mol } H_2O}{1 \text{ mol } C_3H_8} = 15.4 \text{ mol } H_2O$$

b. Given: balanced equation; 0.647 mol O_2
Unknown: moles of CO_2, H_2O, and C_3H_8

$$0.647 \text{ mol } O_2 \times \frac{3 \text{ mol } CO_2}{5 \text{ mol } O_2} = 0.388 \text{ mol } CO_2$$

$$0.647 \text{ mol } O_2 \times \frac{4 \text{ mol } H_2O}{5 \text{ mol } O_2} = 0.518 \text{ mol } H_2O$$

$$0.647 \text{ mol } O_2 \times \frac{1 \text{ mol } C_3H_8}{5 \text{ mol } O_2} = 0.129 \text{ mol } C_3H_8$$

169. a. Given: balanced equation; 3.25 mol P_4O_{10}
Unknown: mass of P

$$3.25 \text{ mol } P_4O_{10} \times \frac{4 \text{ mol P}}{1 \text{ mol } P_4O_{10}} \times \frac{30.97 \text{ g P}}{1 \text{ mol P}} = 403 \text{ g P}$$

b. Given: balanced equation; 0.489 mol P
Unknown: mass of O_2 and P_4O_{10}

$$0.489 \text{ mol P} \times \frac{5 \text{ mol } O_2}{4 \text{ mol P}} \times \frac{32.00 \text{ g } O_2}{1 \text{ mol } O_2} = 19.6 \text{ g } O_2$$

$$0.489 \text{ mol P} \times \frac{1 \text{ mol } P_4O_{10}}{4 \text{ mol P}} \times \frac{283.88 \text{ g } P_4O_{10}}{1 \text{ mol } P_4O_{10}} = 34.7 \text{ g } P_4O_{10}$$

170. a. Given: balanced equation; 1.840 mol H_2O_2

Unknown: mass of O_2

$$1.840 \text{ mol } H_2O_2 \times \frac{1 \text{ mol } O_2}{2 \text{ mol } H_2O_2} \times \frac{32.00 \text{ g } O_2}{1 \text{ mol } O_2} = 29.44 \text{ g } O_2$$

b. Given: balanced equation; 5.0 mol O_2

Unknown: mass of water

$$5.0 \text{ mol } O_2 \times \frac{2 \text{ mol } H_2O}{1 \text{ mol } O_2} \times \frac{18.02 \text{ g } H_2O}{1 \text{ mol } H_2O} = 180 \text{ g } H_2O$$

171. a. Given: balanced equation; 100.0 g $NaNO_3$

Unknown: moles of Na_2CO_3

$$100.0 \text{ g } NaNO_3 \times \frac{1 \text{ mol } NaNO_3}{85.01 \text{ g } NaNO_3} \times \frac{1 \text{ mol } Na_2CO_3}{2 \text{ mol } NaNO_3} = 0.5882 \text{ mol } Na_2CO_3$$

b. Given: balanced equation; 7.50 g Na_2CO_3

Unknown: moles of CO_2

$$7.50 \text{ g } Na_2CO_3 \times \frac{1 \text{ mol } Na_2CO_3}{105.99 \text{ g } Na_2CO_3} \times \frac{1 \text{ mol } CO_2}{1 \text{ mol } Na_2CO_3} = 0.0708 \text{ mol } CO_2$$

172. a. Given: balanced equation; 625 g Fe_3O_4

Unknown: moles of H_2

$$625 \text{ g } Fe_3O_4 \times \frac{1 \text{ mol } Fe_3O_4}{231.55 \text{ g } Fe_3O_4} \times \frac{4 \text{ mol } H_2}{1 \text{ mol } Fe_3O_4} = 10.8 \text{ mol } H_2$$

b. Given: balanced equation; 27 g H_2

Unknown: moles of Fe

$$27 \text{ g } H_2 \times \frac{1 \text{ mol } H_2}{2.02 \text{ g } H_2} \times \frac{3 \text{ mol Fe}}{4 \text{ mol } H_2} = 10. \text{ mol Fe}$$

173. a. Given: balanced equation; 22.5 g $AgNO_3$

Unknown: mass of AgBr

$$22.5 \text{ g } AgNO_3 \times \frac{1 \text{ mol } AgNO_3}{169.88 \text{ g } AgNO_3} \times \frac{2 \text{ mol AgBr}}{2 \text{ mol } AgNO_3} \times \frac{187.77 \text{ g AgBr}}{1 \text{ mol AgBr}}$$
$$= 24.9 \text{ g AgBr}$$

174. a. Given: balanced equation; 90. g CaC_2

Unknown: mass C_2H_2

$$90. \text{ g } CaC_2 \times \frac{1 \text{ mol } CaC_2}{64.10 \text{ g } CaC_2} \times \frac{1 \text{ mol } C_2H_2}{1 \text{ mol } CaC_2} \times \frac{26.04 \text{ g } C_2H_2}{1 \text{ mol } C_2H_2} = 37 \text{ g } C_2H_2$$

175. a. Given: balanced equation; 25.0 g Cl_2

Unknown: mass MnO_2

$$25.0 \text{ g } Cl_2 \times \frac{1 \text{ mol } Cl_2}{70.90 \text{ g } Cl_2} \times \frac{1 \text{ mol } MnO_2}{1 \text{ mol } Cl_2} \times \frac{86.94 \text{ g } MnO_2}{1 \text{ mol } MnO_2} = 30.7 \text{ g } MnO_2$$

b. Given: balanced equation; 0.091 g Cl_2

Unknown: mass $MnCl_2$

$$0.091 \text{ g } Cl_2 \times \frac{1 \text{ mol } Cl_2}{70.90 \text{ g } Cl_2} \times \frac{1 \text{ mol } MnCl_2}{1 \text{ mol } Cl_2} \times \frac{125.84 \text{ g } MnCl_2}{1 \text{ mol } MnCl_2} = 0.16 \text{ g } MnCl_2$$

176. Given: balanced equation; 30.0 mol NH_3

Unknown: moles $(NH_4)_2SO_4$

$$30.0 \text{ mol } NH_3 \times \frac{1 \text{ mol } (NH_4)_2SO_4}{2 \text{ mol } NH_3} = 15.0 \text{ mol } (NH_4)_2SO_4$$

177. a. Given: balanced equation; 150 g Fe_2O_3

Unknown: mass Al

$$150 \text{ g } Fe_2O_3 \times \frac{1 \text{ mol } Fe_2O_3}{159.70 \text{ g } Fe_2O_3} \times \frac{2 \text{ mol Al}}{1 \text{ mol } Fe_2O_3} \times \frac{26.98 \text{ g Al}}{1 \text{ mol Al}} = 51 \text{ g Al}$$

b. Given: balanced equation; 0.905 mol Al_2O_3

Unknown: mass Fe

$$0.905 \text{ mol } Al_2O_3 \times \frac{2 \text{ mol Fe}}{1 \text{ mol } Al_2O_3} \times \frac{55.85 \text{ g Fe}}{1 \text{ mol Fe}} = 101 \text{ g Fe}$$

c. Given: balanced equation; 99.0 g Al

Unknown: moles Fe_2O_3

$$99.0 \text{ g Al} \times \frac{1 \text{ mol Al}}{26.98 \text{ g Al}} \times \frac{1 \text{ mol } Fe_2O_3}{2 \text{ mol Al}} = 1.83 \text{ mol } Fe_2O_3$$

178. Given: 1.40 g N_2; balanced equation

Unknown: mass H_2

$$1.40 \text{ g } N_2 \times \frac{1 \text{ mol } N_2}{28.02 \text{ g } N_2} \times \frac{3 \text{ mol } H_2}{1 \text{ mol } N_2} \times \frac{2.02 \text{ g } H_2}{1 \text{ mol } H_2} = 0.303 \text{ g } H_2$$

179. Given: 1.27 g KOH

Unknown: mass of H_2SO_4

$2KOH + H_2SO_4 \rightarrow K_2SO_4 + 2H_2O$

$$1.27 \text{ g KOH} \times \frac{1 \text{ mol KOH}}{56.11 \text{ g KOH}} \times \frac{1 \text{ mol } H_2SO_4}{2 \text{ mol KOH}} \times \frac{98.09 \text{ g } H_2SO_4}{1 \text{ mol } H_2SO_4}$$

$= 1.11 \text{ g } H_2SO_4$

180. a. Given: reactants and products

Unknown: balanced equation

$H_3PO_4 + 2NH_3 \rightarrow (NH_4)_2HPO_4$

b. Given: 10.00 g NH_3

Unknown: moles $(NH_4)_2HPO_4$

$$10.00 \text{ g } NH_3 \times \frac{1 \text{ mol } NH_3}{17.04 \text{ g } NH_3} \times \frac{1 \text{ mol } (NH_4)_2HPO_4}{2 \text{ mol } NH_3} = 0.293 \text{ mol } (NH_4)_2HPO_4$$

c. Given: 2800 kg H_3PO_4

Unknown: mass NH_3

$$2800 \text{ kg } H_3PO_4 \times \frac{1 \text{ mol } H_3PO_4}{98.00 \text{ g } H_3PO_4} \times \frac{2 \text{ mol } NH_3}{1 \text{ mol } H_3PO_4} \times \frac{17.04 \text{ g } NH_3}{1 \text{ mol } NH_3}$$

$= 970 \text{ kg } NH_3$

181. a. Given: balanced equation; 30.0 mol $Zn_3(C_6H_5O_7)_2$

Unknown: moles $ZnCO_3$ and $C_6H_8O_7$

$$30.0 \text{ mol } Zn_3(C_6H_5O_7)_2 \times \frac{3 \text{ mol } ZnCO_3}{1 \text{ mol } Zn_3(C_6H_5O_7)_2} = 90.0 \text{ mol } ZnCO_3$$

$$30.0 \text{ mol } Zn_3(C_6H_5O_7)_2 \times \frac{2 \text{ mol } C_6H_8O_7}{1 \text{ mol } Zn_3(C_6H_5O_7)_2} = 60.0 \text{ mol } C_6H_8O_7$$

b. Given: balanced equation; 500. mol $C_6H_8O_7$

Unknown: mass in kg of H_2O and CO_2

$$500. \text{ mol } C_6H_8O_7 \times \frac{3 \text{ mol } H_2O}{2 \text{ mol } C_6H_8O_7} \times \frac{18.02 \text{ g } H_2O}{1 \text{ mol } H_2O} \times \frac{1 \text{ kg}}{1000 \text{ g}} = 13.5 \text{ kg } H_2O$$

$$500. \text{ mol } C_6H_8O_7 \times \frac{3 \text{ mol } CO_2}{2 \text{ mol } C_6H_8O_7} \times \frac{44.01 \text{ g } CO_2}{1 \text{ mol } CO_2} \times \frac{1 \text{ kg}}{1000 \text{ g}} = 33.0 \text{ kg } CO_2$$

182. a. Given: balanced equation; 52.5 g C_3H_7COOH

Unknown: mass $C_3H_7COOCH_3$

$$52.5 \text{ g } C_3H_7COOH \times \frac{1 \text{ mol } C_3H_7COOH}{88.12 \text{ g } C_3H_7COOH} \times \frac{1 \text{ mol } C_3H_7COOCH_3}{1 \text{ mol } C_3H_7COOH}$$

$$\times \frac{102.15 \text{ g } C_3H_7COOCH_3}{1 \text{ mol } C_3H_7COOCH_3} = 60.9 \text{ g } C_3H_7COOCH_3$$

b. Given: balanced equation; 5800. g CH_3OH

Unknown: mass H_2O

$$5800. \text{ g } CH_3OH \times \frac{1 \text{ mol } CH_3OH}{32.05 \text{ g } CH_3OH} \times \frac{1 \text{ mol } H_2O}{1 \text{ mol } CH_3OH} \times \frac{18.02 \text{ g } H_2O}{1 \text{ mol } H_2O} = 3261 \text{ g } H_2O$$

183. a. Given: balanced equation; 36.0 g NH_4NO_3

Unknown: moles N_2

$$36.0 \text{ g } NH_4NO_3 \times \frac{1 \text{ mol } NH_4NO_3}{80.06 \text{ g } NH_4NO_3} \times \frac{2 \text{ mol } N_2}{2 \text{ mol } NH_4NO_3} = 0.450 \text{ mol } N_2$$

b. Given: balanced equation; 7.35 mol H_2O

Unknown: mass NH_4NO_3

$$7.35 \text{ mol } H_2O \times \frac{2 \text{ mol } NH_4NO_3}{4 \text{ mol } H_2O} \times \frac{80.06 \text{ g } NH_4NO_3}{1 \text{ mol } NH_4NO_3} = 294 \text{ g } NH_4NO_3$$

184. Given: 1.23 mg $Pb(NO_3)_2$

Unknown: mass KNO_3

$$Pb(NO_3)_2 + 2KI \rightarrow PbI_2 + 2KNO_3$$

$$1.23 \text{ mg } Pb(NO_3)_2 \times \frac{1 \text{ mol } Pb(NO_3)_2}{331.22 \text{ g } Pb(NO_3)_2} \times \frac{2 \text{ mol } KNO_3}{1 \text{ mol } Pb(NO_3)_2}$$

$$\times \frac{101.11 \text{ g } KNO_3}{1 \text{ mol } KNO_3} = 0.751 \text{ mg } KNO_3$$

185. Given: balanced equation; 0.34 kg Pb

Unknown: moles $PbSO_4$

$$0.34 \text{ kg Pb} \times \frac{1000 \text{ g}}{1 \text{ kg}} \times \frac{1 \text{ mol Pb}}{207.2 \text{ g Pb}} \times \frac{2 \text{ mol } PbSO_4}{1 \text{ mol Pb}} = 3.3 \text{ mol } PbSO_4$$

186. Given: 20.0 mol CO_2
Unknown: mass H_2O

$CO_2 + 2LiOH \rightarrow H_2O + Li_2CO_3$

$20.0 \text{ mol } CO_2 \times \dfrac{1 \text{ mol } H_2O}{1 \text{ mol } CO_2} \times \dfrac{18.02 \text{ g } H_2O}{1 \text{ mol } H_2O} = 360. \text{ g } H_2O$

---

187. a. Given: balanced equation; 1.00×10^2 g P_4O_{10}
Unknown: mass H_2O

$1.00 \times 10^2 \text{ g } P_4O_{10} \times \dfrac{1 \text{ mol } P_4O_{10}}{283.88 \text{ g } P_4O_{10}} \times \dfrac{6 \text{ mol } H_2O}{1 \text{ mol } P_4O_{10}} \times \dfrac{18.02 \text{ g } H_2O}{1 \text{ mol } H_2O}$

$= 38.1 \text{ g } H_2O$

b. Given: balanced equation; 0.614 mol H_2O
Unknown: mass H_3PO_4

$0.614 \text{ mol } H_2O \times \dfrac{4 \text{ mol } H_3PO_4}{6 \text{ mol } H_2O} \times \dfrac{98.00 \text{ g } H_3PO_4}{1 \text{ mol } H_3PO_4} = 40.1 \text{ g } H_3PO_4$

c. Given: mass of H_2O
Unknown: moles H_2O

$(63.70 \text{ g} - 56.64 \text{ g}) \; H_2O \times \dfrac{1 \text{ mol } H_2O}{18.02 \text{ g } H_2O} = 0.392 \text{ mol } H_2O$

---

188. Given: 95.0 g H_2O
Unknown: mass C_2H_5OH

$C_2H_5OH + 3O_2 \rightarrow 2CO_2 + 3H_2O$

$95.0 \text{ g } H_2O \times \dfrac{1 \text{ mol } H_2O}{18.02 \text{ g } H_2O} \times \dfrac{1 \text{ mol } C_2H_5OH}{3 \text{ mol } H_2O} \times \dfrac{46.08 \text{ g } C_2H_5OH}{1 \text{ mol } C_2H_5OH}$

$= 81.0 \text{ g } C_2H_5OH$

---

189. Given: balanced equation; 50.0 g SO_2
Unknown: mass H_2SO_4 and O_2

$50.0 \text{ g } SO_2 \times \dfrac{1 \text{ mol } SO_2}{64.07 \text{ g } SO_2} \times \dfrac{2 \text{ mol } H_2SO_4}{2 \text{ mol } SO_2} \times \dfrac{98.09 \text{ g } H_2SO_4}{1 \text{ mol } H_2SO_4} = 76.5 \text{ g } H_2SO_4$

$50.0 \text{ g } SO_2 \times \dfrac{1 \text{ mol } SO_2}{64.07 \text{ g } SO_2} \times \dfrac{1 \text{ mol } O_2}{2 \text{ mol } SO_2} \times \dfrac{32.00 \text{ g } O_2}{1 \text{ mol } O_2} = 12.5 \text{ g } O_2$

---

190. Given: 5.00 g $NaHCO_3$
Unknown: mass CO_2

$2NaHCO_3 \rightarrow Na_2CO_3 + H_2O + CO_2$

$5.00 \text{ g } NaHCO_3 \times \dfrac{1 \text{ mol } NaHCO_3}{84.01 \text{ g } NaHCO_3} \times \dfrac{1 \text{ mol } CO_2}{2 \text{ mol } NaHCO_3} \times \dfrac{44.01 \text{ g } CO_2}{1 \text{ mol } CO_2}$

$= 1.31 \text{ g } CO_2$

---

191. c. Given: 20 000 mol N_2H_4
Unknown: mol N_2

$20\,000 \text{ mol } N_2H_4 \times \dfrac{3 \text{ mol } N_2}{2 \text{ mol } N_2H_4} = 30\,000 \text{ mol } N_2$

d. Given: 450. kg N_2O_4
Unknown: mass H_2O

$450. \text{ kg } N_2O_4 \times \dfrac{1000 \text{ g}}{1 \text{ kg}} \times \dfrac{1 \text{ mol } N_2O_4}{92.02 \text{ g } N_2O_4} \times \dfrac{4 \text{ mol } H_2O}{1 \text{ mol } N_2O_4} \times \dfrac{18.02 \text{ g } H_2O}{1 \text{ mol } H_2O}$

$= 3.52 \times 10^5 \text{ g } H_2O$

192. Given: 517.84 g HgO
Unknown: mol O_2

$2HgO \rightarrow 2Hg + O_2$

$517.84 \text{ g HgO} \times \dfrac{1 \text{ mol HgO}}{216.59 \text{ g HgO}} \times \dfrac{1 \text{ mol } O_2}{2 \text{ mol HgO}} = 1.1954 \text{ mol } O_2$

193. Given: 58.0 g Cl_2
Unknown: mass Fe

$2Fe + 3Cl_2 \rightarrow 2FeCl_3$

$58.0 \text{ g } Cl_2 \times \dfrac{1 \text{ mol } Cl_2}{70.90 \text{ g } Cl_2} \times \dfrac{2 \text{ mol Fe}}{3 \text{ mol } Cl_2} \times \dfrac{55.85 \text{ g Fe}}{1 \text{ mol Fe}} = 30.5 \text{ g Fe}$

194. Given: balanced equation; 5.00 mg Na_2S
Unknown: mass CdS in mg

$5.00 \text{ mg } Na_2S \times \dfrac{1 \text{ mol } Na_2S}{78.05 \text{ g } Na_2S} \times \dfrac{1 \text{ mol CdS}}{1 \text{ mol } Na_2S} \times \dfrac{144.48 \text{ g CdS}}{1 \text{ mol CdS}} = 9.26 \text{ mg CdS}$

195. a. Given: balanced equation; 4.44 mol $KMnO_4$
Unknown: mol CO_2

$4.44 \text{ mol } KMnO_4 \times \dfrac{5 \text{ mol } CO_2}{14 \text{ mol } KMnO_4} = 1.59 \text{ mol } CO_2$

b. Given: 5.21 g H_2O
Unknown: mol $C_3H_5(OH)_3$

$5.21 \text{ g } H_2O \times \dfrac{1 \text{ mol } H_2O}{18.02 \text{ g } H_2O} \times \dfrac{4 \text{ mol } C_3H_5(OH)_3}{16 \text{ mol } H_2O} = 0.0723 \text{ mol } C_3H_5(OH)_3$

c. Given: 3.39 mol K_2CO_3
Unknown: mass Mn_2O_3 in grams

$3.39 \text{ mol } K_2CO_3 \times \dfrac{7 \text{ mol } Mn_2O_3}{7 \text{ mol } K_2CO_3} \times \dfrac{157.88 \text{ g } Mn_2O_3}{1 \text{ mol } Mn_2O_3} = 535 \text{ g } Mn_2O_3$

d. Given: 50.0 g $KMnO_4$
Unknown: mass $C_3H_5(OH)_3$ and CO_2 in grams

$50.0 \text{ g } KMnO_4 \times \dfrac{1 \text{ mol } KMnO_4}{158.04 \text{ g } KMnO_4} \times \dfrac{4 \text{ mol } C_3H_5(OH)_3}{14 \text{ mol } KMnO_4}$
$\times \dfrac{92.11 \text{ g } C_3H_5(OH)_3}{1 \text{ mol } C_3H_5(OH)_3} = 8.33 \text{ g } C_3H_5(OH)_3$

$50.0 \text{ g } KMnO_4 \times \dfrac{1 \text{ mol } KMnO_4}{158.04 \text{ g } KMnO_4} \times \dfrac{5 \text{ mol } CO_2}{14 \text{ mol } KMnO_4} \times \dfrac{44.01 \text{ g } CO_2}{1 \text{ mol } CO_2}$
$= 4.97 \text{ g } CO_2$

196. a. Given: balanced equation; 5.00×10^3 kg $CaCl_2$
Unknown: mass HCl

$5.00 \times 10^3 \text{ kg } CaCl_2 \times \dfrac{1 \text{ mol } CaCl_2}{110.98 \text{ g } CaCl_2} \times \dfrac{2 \text{ mol HCl}}{1 \text{ mol } CaCl_2} \times \dfrac{36.46 \text{ g HCl}}{1 \text{ mol HCl}}$
$= 3.29 \times 10^3 \text{ kg HCl}$

b. Given: balanced equation; 750 g $CaCO_3$
Unknown: mass CO_2

$$750 \text{ g } CaCO_3 \times \frac{1 \text{ mol } CaCO_3}{100.09 \text{ g } CaCO_3} \times \frac{1 \text{ mol } CO_2}{1 \text{ mol } CaCO_3} \times \frac{44.01 \text{ g } CO_2}{1 \text{ mol } CO_2} = 330 \text{ g } CO_2$$

---

197. a. Given: balanced equation; 1.50×10^5 g Al
Unknown: mass NH_4ClO_4

$$1.50 \times 10^5 \text{ g Al} \times \frac{1 \text{ mol Al}}{26.98 \text{ g Al}} \times \frac{3 \text{ mol } NH_4ClO_4}{3 \text{ mol Al}} \times \frac{117.50 \text{ g } NH_4ClO_4}{1 \text{ mol } NH_4ClO_4}$$

$$= 6.53 \times 10^5 \text{ g } NH_4ClO_4$$

b. Given: balanced equation; 620 kg NH_4ClO_4
Unknown: mass NO

$$620 \text{ kg } NH_4ClO_4 \times \frac{1 \text{ mol } NH_4ClO_4}{117.50 \text{ g } NH_4ClO_4} \times \frac{3 \text{ mol NO}}{3 \text{ mol } NH_4ClO_4} \times \frac{30.01 \text{ g NO}}{1 \text{ mol NO}}$$

$$= 160 \text{ kg NO}$$

---

198. a. Given: balanced equation; 2.50×10^5 kg H_2SO_4
Unknown: moles H_3PO_4

$$2.50 \times 10^5 \text{ kg } H_2SO_4 \times \frac{1000 \text{ g}}{1 \text{ kg}} \times \frac{1 \text{ mol } H_2SO_4}{98.09 \text{ g } H_2SO_4} \times \frac{2 \text{ mol } H_3PO_4}{3 \text{ mol } H_2SO_4}$$

$$= 1.70 \times 10^6 \text{ mol } H_3PO_4$$

b. Given: balanced equation; 400. kg $Ca_3(PO_4)_2$
Unknown: mass $CaSO_4 \cdot 2H_2O$

$$400 \text{ kg } Ca(PO_4)_2 \times \frac{1 \text{ mol } Ca_3(PO_4)_2}{310.18 \text{ g } Ca_3(PO_4)_2} \times \frac{3 \text{ mol } CaSO_4 \cdot 2H_2O}{1 \text{ mol } Ca_3(PO_4)_2}$$

$$\times \frac{172.19 \text{ g } CaSO_4 \cdot 2H_2O}{1 \text{ mol } CaSO_4 \cdot 2H_2O} = 666 \text{ kg } CaSO_4 \cdot 2H_2O$$

c. Given: balanced equation; 68 metric tons rock; 78.8% $Ca_3(PO_4)_2$
Unknown: mass H_3PO_4 in metric tons

$$68 \text{ metric tons rock} \times \frac{0.788 \text{ g } Ca_3(PO_4)_2}{1 \text{ g rock}} \times \frac{1 \text{ mol } Ca_3(PO_4)_2}{310.18 \text{ g } Ca_3(PO_4)_2}$$

$$\times \frac{2 \text{ mol } H_3PO_4}{1 \text{ mol } Ca_3(PO_4)_2} \times \frac{98.00 \text{ g } H_3PO_4}{1 \text{ mol } H_3PO_4} = 34 \text{ metric tons } H_3PO_4$$

---

199. Given: balanced equation; mass Fe = 3.19% × 1650 kg
Unknown: mass of steel scrap after 1 yr

$0.0319 \times 1650 \text{ kg Fe} = 52.6 \text{ kg Fe}$

$$52.6 \text{ kg Fe} \times \frac{\text{mol Fe}}{55.85 \text{ g Fe}} \times \frac{2 \text{ mol } Fe_2O_3}{4 \text{ mol Fe}} \times \frac{159.70 \text{ g } Fe_2O_3}{1 \text{ mol } Fe_2O_3} = 75.2 \text{ kg } Fe_2O_3$$

$1650 \text{ kg} - 52.6 \text{ kg} + 75.2 \text{ kg} = 1670 \text{ kg}$

---

200. Given: balanced equation; 0.048 mol Al; 0.030 mol O_2

$$0.048 \text{ mol Al} \times \frac{3 \text{ mol } O_2}{4 \text{ mol Al}} = 0.036 \text{ mol } O_2$$

0.036 mol O_2 needed; 0.030 mol O_2 available

limiting reactant: O_2

201. Given: balanced equation; 862 g $ZrSiO_4$; 950. g Cl_2

Unknown: limiting reactant; mass $ZrCl_4$

$$862 \text{ g } ZrSiO_4 \times \frac{1 \text{ mol } ZrSiO_4}{183.31 \text{ g } ZrSiO_4} \times \frac{2 \text{ mol } Cl_2}{1 \text{ mol } ZrSiO_4} \times \frac{70.90 \text{ g } Cl_2}{1 \text{ mol } Cl_2} = 667 \text{ g } Cl_2$$

667 g Cl_2 needed; 950. g Cl_2 available; limiting reactant is $ZrSiO_4$

$$862 \text{ g } ZrSiO_4 \times \frac{1 \text{ mol } ZrSiO_4}{183.31 \text{ g } ZrSiO_4} \times \frac{1 \text{ mol } ZrCl_4}{1 \text{ mol } ZrSiO_4} \times \frac{233.02 \text{ g } ZrCl_4}{1 \text{ mol } ZrCl_4}$$

$= 1.10 \times 10^3$ g $ZrCl_4$

202. Given: 1.72 mol ZnS; 3.04 mol O_2

Unknown: balanced equation; limiting reactant

$2ZnS + 3O_2 \rightarrow 2ZnO + 2SO_2$

$$1.72 \text{ mol ZnS} \times \frac{3 \text{ mol } O_2}{2 \text{ mol ZnS}} = 2.58 \text{ mol } O_2$$

2.58 mol O_2 needed; 3.04 mol O_2 available; limiting reactant is ZnS.

203. a. Given: balanced equation; 0.32 mol Al; 0.26 mol O_2

Unknown: limiting reactant

$$0.32 \text{ mol Al} \times \frac{3 \text{ mol } O_2}{4 \text{ mol Al}} = 0.24 \text{ mol } O_2$$

0.24 mol O_2 needed; 0.26 mol O_2 available; limiting reactant is Al

b. Given: balanced equation; 6.38×10^{-3} mol O_2; 9.15×10^{-3} mol Al

Unknown: moles Al_2O_3

$$6.38 \times 10^{-3} \text{ mol } O_2 \times \frac{4 \text{ mol Al}}{3 \text{ mol } O_2} = 8.51 \times 10^{-3} \text{ mol Al}$$

8.51×10^{-3} mol Al needed; 9.15×10^{-3} mol Al available; limiting reactant: O_2

$$6.38 \times 10^{-3} \text{ mol } O_2 \times \frac{2 \text{ mol } Al_2O_3}{3 \text{ mol } O_2} = 4.25 \times 10^{-3} \text{ mol } Al_2O_3$$

c. Given: balanced equation; 3.17 g Al; 2.55 g O_2

Unknown: limiting reactant

$$3.17 \text{ g Al} \times \frac{1 \text{ mol Al}}{26.98 \text{ g Al}} \times \frac{3 \text{ mol } O_2}{4 \text{ mol Al}} \times \frac{32.00 \text{ g } O_2}{1 \text{ mol } O_2} = 2.82 \text{ g } O_2$$

2.82 g O_2 needed; 2.55 O_2 available; limiting reactant: O_2

204. a. Given: balanced equation; 100. g CuS; 56 g O_2

Unknown: limiting reactant

$$100. \text{ g CuS} \times \frac{1 \text{ mol CuS}}{95.62 \text{ g CuS}} \times \frac{3 \text{ mol } O_2}{2 \text{ mol CuS}} \times \frac{32.00 \text{ g } O_2}{1 \text{ mol } O_2} = 50.2 \text{ g } O_2$$

50.2 g O_2 needed; 56 g O_2 available; limiting reactant: CuS

b. Given: balanced equation; 18.7 g CuS; 12.0 g O_2

Unknown: mass CuO

$$18.7 \text{ g CuS} \times \frac{1 \text{ mol CuS}}{95.62 \text{ g CuS}} \times \frac{3 \text{ mol } O_2}{2 \text{ mol CuS}} \times \frac{32.00 \text{ g } O_2}{1 \text{ mol } O_2} = 9.39 \text{ g } O_2$$

9.39 g O_2 needed; 12.0 g O_2 available; limiting reactant: CuS

$$18.7 \text{ g CuS} \times \frac{1 \text{ mol CuS}}{95.62 \text{ CuS}} \times \frac{2 \text{ mol CuO}}{2 \text{ mol CuS}} \times \frac{79.55 \text{ g CuO}}{1 \text{ mol CuO}} = 15.6 \text{ g CuO}$$

205. Given: balanced equation; 0.092 mol Fe; 0.158 mol CuSO$_4$

Unknown: limiting reactant; moles Cu

$$0.092 \text{ mol Fe} \times \frac{3 \text{ mol CuSO}_4}{2 \text{ mol Fe}} = 0.14 \text{ mol CuSO}_4$$

0.14 mol CuSo$_4$ needed; 0.158 mol CuSO$_4$ available; limiting reactant: Fe

$$0.092 \text{ mol Fe} \times \frac{3 \text{ mol Cu}}{2 \text{ mol Fe}} = 0.14 \text{ mol Cu}$$

206. Given: balanced equation; 55 g BaCO$_3$; 26 g HNO$_3$

Unknown: mass Ba(NO$_3$)$_2$

$$55 \text{ g BaCO}_3 \times \frac{1 \text{ mol BaCO}_3}{197.34 \text{ g BaCO}_3} \times \frac{2 \text{ mol HNO}_3}{1 \text{ mol BaCO}_3} \times \frac{63.02 \text{ g HNO}_3}{1 \text{ mol HNO}_3}$$
$$= 35 \text{ g HNO}_3$$

35 g HNO$_3$ needed; 26 HNO$_3$ available; limiting reactant: HNO$_3$

$$26 \text{ g HNO}_3 \times \frac{1 \text{ mol HNO}_3}{63.02 \text{ g HNO}_3} \times \frac{1 \text{ mol Ba(NO}_3)_2}{2 \text{ mol HNO}_3} \times \frac{261.35 \text{ g Ba(NO}_3)_2}{1 \text{ mol Ba(NO}_3)_2}$$
$$= 54 \text{ g Ba(NO}_3)_2$$

207. a. Given: balanced equation; 560 g MgI$_2$; 360 g Br$_2$

Unknown: excess reactant; remaining mass

$$560 \text{ g MgI}_2 \times \frac{1 \text{ mol MgI}_2}{278.11 \text{ g MgI}_2} \times \frac{1 \text{ mol Br}_2}{1 \text{ mol MgI}_2} \times \frac{159.80 \text{ g Br}_2}{1 \text{ mol Br}_2} = 322 \text{ g Br}_2$$

Br$_2$ is in excess; 360 g Br$_2$ − 322 g Br$_2$ = 38 g Br$_2$ excess

b. Given: balanced equation; 560 g MgI$_2$

Unknown: mass I$_2$

$$560 \text{ g MgI}_2 \times \frac{1 \text{ mol MgI}_2}{278.11 \text{ g MgI}_2} \times \frac{1 \text{ mol I}_2}{1 \text{ mol MgI}_2} \times \frac{253.80 \text{ g I}_2}{1 \text{ mol I}_2} = 510 \text{ g I}_2$$

208. a. Given: balanced equation; 22.9 g Ni; 112 g AgNO$_3$

Unknown: excess reactant

$$22.9 \text{ g Ni} \times \frac{1 \text{ mol Ni}}{58.69 \text{ g Ni}} \times \frac{2 \text{ mol AgNO}_3}{1 \text{ mol Ni}} \times \frac{169.88 \text{ g AgNO}_3}{1 \text{ mol AgNO}_3} = 133 \text{ g AgNO}_3$$

Ni is in excess

b. Given: balanced equation; 112 g AgNO$_3$

Unknown: mass Ni(NO$_3$)$_2$

$$112 \text{ g AgNO}_3 \times \frac{1 \text{ mol AgNO}_3}{169.88 \text{ g AgNO}_3} \times \frac{1 \text{ mol Ni(NO}_3)_2}{2 \text{ mol AgNO}_3} \times \frac{182.71 \text{ g Ni(NO}_3)_2}{1 \text{ mol Ni(NO}_3)_2}$$
$$= 60.2 \text{ g Ni(NO}_3)_2$$

209. Given: 1.60 mol CS$_2$; 5.60 mol O$_2$

Unknown: balanced equation; moles of excess, reactant

$$CS_2(g) + 3O_2(g) \rightarrow 2SO_2(g) + CO_2(g)$$

$$1.60 \text{ mol CS}_2 \times \frac{3 \text{ mol O}_2}{1 \text{ mol CS}_2} = 4.80 \text{ mol O}_2$$

5.60 mol O$_2$ − 4.80 mol O$_2$ = 0.80 mol O$_2$ excess

210. a. Given: balanced equation; 0.91 g HgCl$_2$
Unknown: mass Hg(NH$_2$)Cl

$$0.91 \text{ g HgCl}_2 \times \frac{1 \text{ mol HgCl}_2}{271.49 \text{ g HgCl}_2} \times \frac{1 \text{ mol Hg(NH}_2)\text{Cl}}{1 \text{ mol HgCl}_2}$$

$$\times \frac{252.07 \text{ g Hg(NH}_2)\text{Cl}}{1 \text{ mol Hg(NH}_2)\text{Cl}} = 0.84 \text{ g Hg(NH}_2)\text{Cl}$$

b. Given: balanced equation; 0.91 g HgCl$_2$; 0.15 g NH$_3$
Unknown: mass Hg(NH$_2$)Cl

$$0.91 \text{ g HgCl}_2 \times \frac{1 \text{ mol HgCl}_2}{271.49 \text{ g HgCl}_2} \times \frac{2 \text{ mol NH}_3}{1 \text{ mol HgCl}_2} \times \frac{17.04 \text{ g NH}_3}{1 \text{ mol NH}_3} = 0.11 \text{ g NH}_3$$

0.11 g NH$_3$ needed; 0.15 g NH$_3$ available; limiting reactant: HgCl$_2$

$$0.91 \text{ g HgCl}_2 \times \frac{1 \text{ mol HgCl}_2}{271.49 \text{ g HgCl}_2} \times \frac{1 \text{ mol Hg(NH}_2)\text{Cl}}{1 \text{ mol HgCl}_2}$$

$$\times \frac{252.07 \text{ g Hg(NH}_2)\text{Cl}}{1 \text{ mol Hg(NH}_2)\text{Cl}} = 0.84 \text{ g Hg(NH}_2)\text{Cl}$$

211. b. Given: balanced equation from **a**; 0.57 mol Al; 0.37 mol NaOH; excess water
Unknown: moles H$_2$

$$0.57 \text{ mol Al} \times \frac{2 \text{ mol NaOH}}{2 \text{ mol Al}} = 0.57 \text{ mol NaOH}$$

0.57 mol NaOH needed; 0.37 mol NaOH available; limiting reactant: NaOH

$$0.37 \text{ mol NaOH} \times \frac{3 \text{ mol H}_2}{2 \text{ mol NaOH}} = 0.56 \text{ mol H}_2$$

212. a. Given: balanced equation; 0.0845 mol NO$_2$
Unknown: moles of HNO$_3$ and Cu

$$0.0845 \text{ mol NO}_2 \times \frac{4 \text{ mol HNO}_3}{2 \text{ mol NO}_2} = 0.169 \text{ mol HNO}_3$$

$$0.0845 \text{ mol NO}_2 \times \frac{1 \text{ mol Cu}}{2 \text{ mol NO}_2} = 0.0422 \text{ mol Cu}$$

b. Given: balanced equation; 5.94 g Cu; 23.23 g HNO$_3$
Unknown: excess reactant

$$5.94 \text{ g Cu} \times \frac{1 \text{ mol Cu}}{63.55 \text{ g Cu}} \times \frac{4 \text{ mol HNO}_3}{1 \text{ mol Cu}} \times \frac{63.02 \text{ g HNO}_3}{1 \text{ mol HNO}_3} = 23.6 \text{ g HNO}_3$$

23.6 g HNO$_3$ needed; 23.23 g HNO$_3$ available: Cu is in excess.

213. a. Given: balanced equation; 2.90 mol NH$_3$; 3.75 mol O$_2$
Unknown: moles of NO and H$_2$O

$$2.90 \text{ mol NH}_3 \times \frac{5 \text{ mol O}_2}{4 \text{ mol NH}_3} = 3.62 \text{ mol O}_2; \text{ NH}_3 \text{ is limiting.}$$

$$2.90 \text{ mol NH}_3 \times \frac{4 \text{ mol NO}}{4 \text{ mol NH}_3} = 2.90 \text{ mol NO}$$

$$2.90 \text{ mol NH}_3 \times \frac{6 \text{ mol H}_2\text{O}}{4 \text{ mol NH}_3} = 4.35 \text{ mol H}_2\text{O}$$

b. Given: balanced equation; 4.20×10^4 g NH_3; 1.31×10^5 g O_2

Unknown: limiting reactant

$$4.20 \times 10^4 \text{ g } NH_3 \times \frac{1 \text{ mol } NH_3}{17.04 \text{ g } NH_3} \times \frac{5 \text{ mol } O_2}{4 \text{ mol } NH_3} \times \frac{32.00 \text{ g } O_2}{1 \text{ mol } O_2}$$

$= 9.86 \times 10^4$ g O_2

9.86×10^4 g O_2 needed; 1.31×10^5 g O_2 available; limiting reactant: NH_3

c. Given: balanced equation; 869 kg NH_3; 2480 kg O_2

Unknown: mass NO

$$869 \text{ kg } NH_3 \times \frac{1 \text{ mol } NH_3}{17.04 \text{ g } NH_3} \times \frac{5 \text{ mol } O_2}{4 \text{ mol } NH_3} \times \frac{32.00 \text{ g } O_2}{1 \text{ mol } O_2} = 2040 \text{ kg } O_2$$

2040 kg O_2 needed; 2480 kg O_2 available; limiting reactant: NH_3

$$869 \text{ kg } NH_3 \times \frac{1 \text{ mol } NH_3}{17.04 \text{ g } NH_3} \times \frac{4 \text{ mol NO}}{4 \text{ mol } NH_3} \times \frac{30.01 \text{ g NO}}{1 \text{ mol NO}}$$

$= 1.53 \times 10^3$ kg NO

214. Given: balanced equation; 620 g C_2H_5OH; 1020 g CuO

Unknown: mass CH_3CHO; mass excess reactant left

$$620 \text{ g } C_2H_5OH \times \frac{1 \text{ mol } C_2H_5OH}{46.08 \text{ g } C_2H_5OH} \times \frac{1 \text{ mol CuO}}{1 \text{ mol } C_2H_5OH} \times \frac{79.55 \text{ g CuO}}{1 \text{ mol CuO}}$$

$= 1070$ g CuO

1070 g CuO needed; 1020 g CuO available; limiting reactant: CuO

$$1020 \text{ g CuO} \times \frac{1 \text{ mol CuO}}{79.55 \text{ g CuO}} \times \frac{1 \text{ mol } CH_3CHO}{1 \text{ mol CuO}} \times \frac{44.06 \text{ g } CH_3CHO}{1 \text{ mol } CH_3CHO}$$

$= 565$ g CH_3CHO

$$1020 \text{ g CuO} \times \frac{1 \text{ mol CuO}}{79.55 \text{ g CuO}} \times \frac{1 \text{ mol } C_2H_5OH}{1 \text{ mol CuO}} \times \frac{46.08 \text{ g } C_2H_5OH}{1 \text{ mol } C_2H_5OH}$$

$= 591$ g C_2H_5OH

620 g C_2H_5OH − 591 g C_2H_5OH = 29 g C_2H_5OH excess

215. Given: balanced equation; 250 g SO_2; 650 g Br_2; excess H_2O

Unknown: mass HBr

$$250 \text{ g } SO_2 \times \frac{1 \text{ mol } SO_2}{64.07 \text{ g } SO_2} \times \frac{1 \text{ mol } Br_2}{1 \text{ mol } SO_2} \times \frac{159.80 \text{ g } Br_2}{1 \text{ mol } Br_2} = 624 \text{ g } Br_2$$

624 g Br_2 needed; 650 Br_2 available; limiting reactant: SO_2

$$250 \text{ g } SO_2 \times \frac{1 \text{ mol } SO_2}{64.07 \text{ g } SO_2} \times \frac{2 \text{ mol HBr}}{1 \text{ mol } SO_2} \times \frac{80.91 \text{ g HBr}}{1 \text{ mol HBr}} = 630 \text{ g HBr}$$

216. Given: balanced equation; 25.0 g Na_2SO_3; 22.0 g HCl

Unknown: mass SO_2

$$25.0 \text{ g } Na_2SO_3 \times \frac{1 \text{ mol } Na_2SO_3}{126.05 \text{ g } Na_2SO_3} \times \frac{2 \text{ mol HCl}}{1 \text{ mol } Na_2SO_3} \times \frac{36.46 \text{ g HCl}}{1 \text{ mol HCl}} = 14.5 \text{ g HCl}$$

14.5 g HCl needed; 22.0 g HCl available; limiting reactant: Na_2SO_3

$$25.0 \text{ g } Na_2SO_3 \times \frac{1 \text{ mol } Na_2SO_3}{126.05 \text{ g } Na_2SO_3} \times \frac{1 \text{ mol } SO_2}{1 \text{ mol } Na_2SO_3} \times \frac{64.07 \text{ g } SO_2}{1 \text{ mol } SO_2}$$

$= 12.7$ g SO_2

217. a. Given: balanced equation; 27.5 g TbF$_3$; 6.96 g Ca

Unknown: mass Tb

$$27.5 \text{ g TbF}_3 \times \frac{1 \text{ mol TbF}_3}{215.93 \text{ g TbF}_3} \times \frac{3 \text{ mol Ca}}{2 \text{ mol TbF}_3} \times \frac{40.08 \text{ g Ca}}{1 \text{ mol Ca}} = 7.66 \text{ g Ca}$$

7.66 g Ca needed; 6.96 g Ca available; limiting reactant: Ca

$$6.96 \text{ g Ca} \times \frac{1 \text{ mol Ca}}{40.08 \text{ g Ca}} \times \frac{2 \text{ mol Tb}}{3 \text{ mol Ca}} \times \frac{158.93 \text{ g Tb}}{1 \text{ mol Tb}} = 18.4 \text{ g Tb}$$

b. Given: balanced equation; 6.96 g Ca; 27.5 g TbF$_3$; Ca is limiting

Unknown: mass TbF$_3$ remaining

$$6.96 \text{ g Ca} \times \frac{1 \text{ mol Ca}}{40.08 \text{ g Ca}} \times \frac{2 \text{ mol TbF}_3}{3 \text{ mol Ca}} \times \frac{215.93 \text{ g TbF}_3}{1 \text{ mol TbF}_3} = 25.0 \text{ g TbF}_3$$

27.5 g TbF$_3$ − 25.0 g TbF$_3$ = 2.5 g TbF$_3$ remaining

218. a. Given: theoretical yield = 50.0 g; actual yield = 41.9 g

Unknown: percent yield

$$\frac{41.9 \text{ g}}{50.0 \text{ g}} \times 100 = 83.8\% \text{ yield}$$

b. Given: theoretical yield = 290 kg; actual yield = 270 kg

Unknown: percent yield

$$\frac{270 \text{ kg}}{290 \text{ kg}} \times 100 = 93\% \text{ yield}$$

c. Given: theoretical yield = 6.05 × 10^4 kg; actual yield = 4.18 × 10^4 kg

Unknown: percent yield

$$\frac{4.18 \times 10^4 \text{ kg}}{6.05 \times 10^4 \text{ kg}} \times 100 = 69.1\% \text{ yield}$$

d. Given: theoretical yield = 0.00192 g; actual yield = 0.00089 g

Unknown: percent yield

$$\frac{0.00089 \text{ g}}{0.00192 \text{ g}} \times 100 = 46\% \text{ yield}$$

219. a. Given: balanced equation; 8.87 g As$_2$O$_3$; actual yield As = 5.33 g

Unknown: percent yield

theoretical yield = $8.87 \text{ g As}_2\text{O}_3 \times \frac{1 \text{ mol As}_2\text{O}_3}{197.84 \text{ g As}_2\text{O}_3} \times \frac{4 \text{ mol As}}{2 \text{ mol As}_2\text{O}_3}$

$\times \frac{74.92 \text{ g As}}{1 \text{ mol As}} = 6.72 \text{ g As}$

$\frac{5.33 \text{ g As}}{6.72 \text{ g As}} \times 100 = 79.3\% \text{ yield}$

b. Given: balanced equation; 67 g C, actual yield As = 425 g

Unknown: percent yield

theoretical yield = $67 \text{ g C} \times \frac{1 \text{ mol C}}{12.01 \text{ g C}} \times \frac{4 \text{ mol As}}{3 \text{ mol C}} \times \frac{74.92 \text{ g As}}{1 \text{ mol As}} = 560 \text{ g As}$

$\frac{425 \text{ g As}}{560 \text{ g As}} \times 100 = 76\% \text{ yield}$

220. a. Given: theoretical yield = 68.3 g; actual yield = 43.9 g

Unknown: percent yield

$$\frac{43.9 \text{ g}}{68.3 \text{ g}} \times 100 = 64.3\% \text{ yield}$$

b. Given: theoretical yield = 0.0722 mol; actual yield = 0.0419 mol

Unknown: percent yield

$$\frac{0.0419 \text{ mol}}{0.0722 \text{ mol}} \times 100 = 58.0\% \text{ yield}$$

c. Given: 4.29 mol C_2H_5OH; actual yield $CH_3COOC_2H_5$ = 2.98 mol; balanced equation

Unknown: percent yield

theoretical yield $CH_3COOC_2H_5$ =

$$4.29 \text{ mol } C_2H_5OH \times \frac{1 \text{ mol } CH_3COOC_2H_5}{1 \text{ mol } C_2H_5OH} = 4.29 \text{ mol } CH_3COOC_2H_5$$

$$\frac{2.98 \text{ mol } CH_3COOC_2H_5}{4.29 \text{ mol } CH_3COOC_2H_5} \times 100 = 69.5\% \text{ yield}$$

d. Given: balanced equation; 0.58 mol C_2H_5OH; 0.82 mol CH_3COOH; actual yield $CH_3COOC_2H_5$ = 0.46 mol

Unknown: percent yield

$$0.58 \text{ mol } C_2H_5OH \times \frac{1 \text{ mol } CH_3COOH}{1 \text{ mol } C_2H_5OH} = 0.58 \text{ mol } CH_3COOH$$

C_2H_5OH is limiting.

$$0.58 \text{ mol } C_2H_5OH \times \frac{1 \text{ mol } CH_3COOC_2H_5}{1 \text{ mol } C_2H_5OH} = 0.58 \text{ mol } CH_3COOC_2H_5$$

$$\frac{0.46 \text{ mol}}{0.58 \text{ mol}} \times 100 = 79\% \text{ yield}$$

221. a. Given: balanced equation; 0.0251 mol A; actual yield C = 0.0349 mol

Unknown: percent yield

theoretical yield C = $0.0251 \text{ mol A} \times \frac{4 \text{ mol C}}{2 \text{ mol A}} = 0.0502 \text{ mol C}$

$$\frac{0.0349 \text{ mol}}{0.0502 \text{ mol}} \times 100 = 69.5\% \text{ yield}$$

b. Given: balanced equation; 1.19 mol A; actual yield D = 1.41 mol

Unknown: percent yield

theoretical yield D = $1.19 \text{ mol A} \times \frac{3 \text{ mol D}}{2 \text{ mol A}} = 1.785 \text{ mol D}$

$$\frac{1.41 \text{ mol}}{1.785 \text{ mol}} \times 100 = 79.0\% \text{ yield}$$

c. Given: balanced equation; 189 mol B; actual yield D = 39 mol

Unknown: percent yield

theoretical yield D = $189 \text{ mol B} \times \frac{3 \text{ mol D}}{7 \text{ mol B}} = 81.0 \text{ mol D}$

$$\frac{39 \text{ mol}}{81.0 \text{ mol}} \times 100 = 48\% \text{ yield}$$

d. Given: balanced equation; 3500 mol B; actual yield C = 1700 mol
Unknown: percent yield

theoretical yield C = 3500 mol B × $\dfrac{4 \text{ mol C}}{7 \text{ mol B}}$ = 2.0×10^3 mol C

$\dfrac{1.7 \times 10^3 \text{ mol}}{2.0 \times 10^3 \text{ mol}} \times 100 = 85\%$ C

222. a. Given: balanced equation; 57 mol $Ca_3(PO_4)_2$; actual yield $CaSiO_3$ = 101 mol
Unknown: percent yield

theoretical yield $CaSiO_3$ = 57 mol $Ca_3(PO_4)_2$ × $\dfrac{3 \text{ mol } CaSiO_3}{1 \text{ mol } Ca_3(PO_4)_2}$
= 170 mol $CaSiO_3$

$\dfrac{101 \text{ mol}}{170 \text{ mol}} \times 100 = 59\%$ yield

b. Given: balanced equation; 1280 mol C; actual yield $CaSiO_3$ = 622 mol
Unknown: percent yield

theoretical yield $CaSiO_3$ = 1280 mol C × $\dfrac{3 \text{ mol } CaSiO_3}{5 \text{ mol C}}$ = 768 mol $CaSiO_3$

$\dfrac{622 \text{ mol}}{768 \text{ mol}} \times 100 = 81.0\%$ yield

c. Given: balanced equation; 81.5% yield; 1.4×10^5 mol $Ca_3(PO_4)_2$
Unknown: actual mol P

1.4×10^5 mol $Ca_3(PO_4)_2$ × $\dfrac{2 \text{ mol P}}{1 \text{ mol } Ca_3(PO_4)_2}$ × $0.815 = 2.3 \times 10^5$ mol

223. a. Given: balanced equation; 56.9 g WO_3; actual yield W = 41.4 g
Unknown: percent yield

56.9 g WO_3 × $\dfrac{1 \text{ mol } WO_3}{231.84 \text{ g } WO_3}$ × $\dfrac{1 \text{ mol W}}{1 \text{ mol } WO_3}$ × $\dfrac{183.84 \text{ g W}}{1 \text{ mol W}}$ = 45.1 g W

$\dfrac{41.4 \text{ g}}{45.1 \text{ g}} \times 100 = 91.8\%$ yield

b. Given: balanced equation; 3.72 g WO_3; 92.0% yield
Unknown: moles W

3.72 g WO_3 × $\dfrac{1 \text{ mol } WO_3}{231.84 \text{ g } WO_3}$ × $\dfrac{1 \text{ mol W}}{1 \text{ mol } WO_3}$ × $\dfrac{92.0\%}{100\%}$ = 0.0148 mol W

c. Given: balanced equation; actual yield W = 11.4 g; 89.4% yield
Unknown: mass WO_3

$\dfrac{11.4 \text{ g W}}{x} = \dfrac{89.4 \text{ g}}{100.0 \text{ g}}$; x = 12.8 g W

12.8 g W × $\dfrac{1 \text{ mol W}}{183.84 \text{ g W}}$ × $\dfrac{1 \text{ mol } WO_3}{1 \text{ mol W}}$ × $\dfrac{231.84 \text{ g } WO_3}{1 \text{ mol } WO_3}$ = 16.1 g WO_3

224. a. Given: balanced equation; 410. kg CS_2; actual yield $CCl_4 = 719$ kg
Unknown: percent yield

theoretical yield = 410. kg $CS_2 \times \dfrac{1 \text{ mol } CS_2}{76.15 \text{ g } CS_2} \times \dfrac{1 \text{ mol } CCl_4}{1 \text{ mol } CS_2}$

$\times \dfrac{153.81 \text{ g } CCl_4}{1 \text{ mol } CCl_4} = 828$ kg CCl_4

$\dfrac{719 \text{ kg}}{828 \text{ kg}} \times 100 = 86.8\%$ yield

b. Given: balanced equation; 67.5 g Cl_2; actual yield $S_2Cl_2 = 39.5$ g
Unknown: percent yield

theoretical yield = 67.5 g $Cl_2 \times \dfrac{1 \text{ mol } Cl_2}{70.90 \text{ g } Cl_2} \times \dfrac{1 \text{ mol } S_2Cl_2}{3 \text{ mol } Cl_2}$

$\times \dfrac{135.04 \text{ g } S_2Cl_2}{1 \text{ mol } S_2Cl_2} = 42.8$ g S_2Cl_2

$\dfrac{39.5 \text{ g}}{42.8 \text{ g}} \times 100 = 92.2\%$ yield

c. Given: balanced equation; 83.3% yield; actual yield $CCl_4 = 5.00 \times 10^4$ kg
Unknown: kg CS_2 and S_2Cl_2

5.00×10^4 kg $CCl_4 \times \dfrac{100\%}{83.3\%} \times \dfrac{1 \text{ mol } CCl_4}{153.81 \text{ g } CCl_4} \times \dfrac{1 \text{ mol } CS_2}{1 \text{ mol } CCl_4} \times \dfrac{76.15 \text{ g } CS_2}{1 \text{ mol } CS_2} =$

2.97×10^4 kg CS_2

5.00×10^4 kg $CCl_4 \times \dfrac{83.3\%}{100\%} \times \dfrac{1 \text{ mol } CCl_4}{153.81 \text{ g } CCl_4} \times \dfrac{1 \text{ mol } S_2Cl_2}{1 \text{ mol } CCl_4}$

$\times \dfrac{135.04 \text{ g } S_2Cl_2}{1 \text{ mol } S_2Cl_2} = 3.66 \times 10^4$ kg S_2Cl_2

225. a. Given: balanced equation; 0.38 g NO_2; actual yield $N_2O_5 = 0.36$ g
Unknown: percent yield

0.38 g $NO_2 \times \dfrac{1 \text{ mol } NO_2}{46.01 \text{ g } NO_2} \times \dfrac{1 \text{ mol } N_2O_5}{2 \text{ mol } NO_2} \times \dfrac{108.02 \text{ g } N_2O_5}{1 \text{ mol } N_2O_5} = 0.45$ g N_2O_5

$\dfrac{0.36 \text{ g}}{0.45 \text{ g}} \times 100 = 80.\%$ g yield

b. Given: balanced equation; 6.0 mol NO_2; 61.1% yield
Unknown: mass N_2O_5

6.0 mol $NO_2 \times \dfrac{1 \text{ mol } N_2O_5}{2 \text{ mol } NO_2} \times \dfrac{108.02 \text{ g } N_2O_5}{1 \text{ mol } N_2O_5} = 320$ g N_2O_5

320 g $N_2O_5 \times \dfrac{61.1\%}{100\%} = 2.0 \times 10^2$ g N_2O_5

226. Given: balanced equation; 30.0 g NaCl; 0.250 mol H_2SO_4; actual yield HCl = 14.6 g
Unknown: percent yield

30.0 g NaCl $\times \dfrac{1 \text{ mol NaCl}}{58.44 \text{ g NaCl}} \times \dfrac{1 \text{ mol } H_2SO_4}{2 \text{ mol NaCl}} = 0.257$ mol H_2SO_4

0.250 mol H_2SO_4 available, 0.257 mol H_2SO_4 needed; H_2SO_4 is limiting.

0.250 mol $H_2SO_4 \times \dfrac{2 \text{ mol HCl}}{1 \text{ mol } H_2SO_4} \times \dfrac{36.46 \text{ g HCl}}{1 \text{ mol HCl}} = 18.2$ g HCl

$\dfrac{14.6 \text{ g}}{18.2 \text{ g}} \times 100 = 80.2\%$ yield

227. a. Given: balanced equation; 410 g Au; actual yield NaAu(CN)$_2$ = 540 g

Unknown: percent yield

theoretical yield = 410 g Au $\times \dfrac{1 \text{ mol Au}}{196.97 \text{ g Au}} \times \dfrac{4 \text{ mol NaAu(CN)}_2}{4 \text{ mol Au}}$

$\times \dfrac{272.00 \text{ g NaAu(CN)}_2}{1 \text{ mol NaAu(CN)}_2}$ = 570 g NaAu(CN)$_2$

$\dfrac{540 \text{ g}}{570 \text{ g}} \times 100 = 95\%$ yield

b. Given: balanced equation; 79.6% yield; 1.00 kg NaAu(CN)$_2$

Unknown: mass Au

1.00 kg NaAu(CN)$_2 \times \dfrac{100\%}{79.6\%} \times \dfrac{1 \text{ mol NaAu(CN)}_2}{272.00 \text{ g NaAu(CN)}_2}$

$\times \dfrac{4 \text{ mol Au}}{4 \text{ mol NaAu(CN)}_2} \times \dfrac{196.97 \text{ g Au}}{1 \text{ mol Au}}$ = 0.910 kg Au = 9.10 $\times 10^2$ g Au

c. Given: 0.910 kg Au; ore is 0.001% Au

Unknown: mass of ore

$\dfrac{x}{0.910 \text{ kg Au}} = \dfrac{100\%}{0.001\%}$; x = 9 $\times 10^4$ kg ore

228. a. Given: balanced equation; 2.00 g CO; actual yield I$_2$ = 3.17 g

Unknown: percent yield

theoretical yield = 2.00 g CO $\times \dfrac{1 \text{ mol CO}}{28.01 \text{ g CO}} \times \dfrac{1 \text{ mol I}_2}{5 \text{ mol CO}} \times \dfrac{253.80 \text{ g I}_2}{1 \text{ mol I}_2}$

= 3.63 g I$_2$

$\dfrac{3.17 \text{ g}}{3.63 \text{ g}} \times 100 = 87.3\%$ yield

b. Given: 87.6% yield; 2.00 g CO

Unknown: mass of unreacted CO

100.0% − 87.6% = 12.4% unreacted

0.124 $\times$ 2.00 g CO = 0.248 g CO

229. a. Given: balanced equation; 1.2 kg Cl$_2$; actual yield NaClO = 0.90 kg

Unknown: percent yield

theoretical yield = 1.2 kg Cl$_2 \times \dfrac{1 \text{ mol Cl}_2}{70.90 \text{ g Cl}_2} \times \dfrac{1 \text{ mol NaClO}}{1 \text{ mol Cl}_2}$

$\times \dfrac{74.44 \text{ g NaClO}}{1 \text{ mol NaClO}}$ = 1.3 kg NaClO

$\dfrac{0.90 \text{ kg}}{1.3 \text{ kg}} \times 100 = 69\%$

b. Given: balanced equation; 91.8% yield; 25 metric tons actual yield NaClO

Unknown: metric tons Cl$_2$

25 metric tons NaClO $\times \dfrac{100\%}{91.8\%} \times \dfrac{1 \text{ mol NaClO}}{65.46 \text{ g NaClO}} \times \dfrac{1 \text{ mol Cl}_2}{1 \text{ mol NaClO}}$

$\times \dfrac{70.90 \text{ g Cl}_2}{1 \text{ mol Cl}_2}$ = 29 metric tons Cl$_2$

c. Given: balanced equation; 81.8% yield; 1 mol Cl$_2$

Unknown: mass NaCl

$$1.00 \text{ mol Cl}_2 \times \frac{1 \text{ mol NaCl}}{1 \text{ mol Cl}_2} \times \frac{58.44 \text{ g NaCl}}{1 \text{ mol NaCl}} \times \frac{81.8\%}{100\%} = 47.8 \text{ g NaCl}$$

d. Given: balanced equation; 79.5% yield; actual rate NaClO 370 kg/h

Unknown: rate NaOH in kg/h

$$\frac{370 \text{ kg NaClO}}{1 \text{ h}} \times \frac{100\%}{79.5\%} \times \frac{1 \text{ mol NaClO}}{65.46 \text{ g NaClO}} \times \frac{2 \text{ mol NaOH}}{1 \text{ mol NaClO}}$$

$$\times \frac{40.00 \text{ g NaOH}}{1 \text{ mol NaOH}} = 568 \text{ kg NaOH/h}$$

230. b. Given: theoretical yield = 2.04 g; actual yield = 1.79 g

Unknown: percent yield

$$\frac{1.79 \text{ g}}{2.04 \text{ g}} \times 100 = 87.7\% \text{ yield}$$

d. Given: balanced equation from **c**; 0.097 mol Mg; 0.027 mol Mg$_3$N$_2$

Unknown: percent yield

$$\text{theoretical yield} = 0.097 \text{ mol Mg} \times \frac{1 \text{ mol Mg}_3\text{N}_2}{3 \text{ mol Mg}} = 0.032 \text{ mol Mg}_3\text{N}_2$$

$$\frac{0.027 \text{ mol}}{0.032 \text{ mol}} \times 100 = 84\% \text{ yield}$$

231. a. Given: balanced equation; 0.89 g C$_3$H$_7$OH; actual yield C$_2$H$_5$COOH = 0.88 g

Unknown: percent yield

$$\text{theoretical yield} = 0.89 \text{ g C}_3\text{H}_7\text{OH} \times \frac{1 \text{ mol C}_3\text{H}_7\text{OH}}{60.11 \text{ g C}_3\text{H}_7\text{OH}}$$

$$\times \frac{3 \text{ mol C}_2\text{H}_5\text{COOH}}{3 \text{ mol C}_3\text{H}_7\text{OH}} \times \frac{74.09 \text{ g C}_2\text{H}_5\text{COOH}}{1 \text{ mol C}_2\text{H}_5\text{COOH}} = 1.1 \text{ g C}_2\text{H}_5\text{COOH}$$

$$\frac{0.88 \text{ g}}{1.1 \text{ g}} \times 100 = 80.\% \text{ yield}$$

b. Given: balanced equation; actual yield C$_2$H$_5$COOH = 1.50 mol; 136 g C$_3$H$_7$OH

Unknown: percent yield

$$\text{theoretical yield} = 136 \text{ g C}_3\text{H}_7\text{OH} \times \frac{1 \text{ mol C}_3\text{H}_7\text{OH}}{60.11 \text{ g C}_3\text{H}_7\text{OH}}$$

$$\times \frac{3 \text{ mol C}_2\text{H}_5\text{COOH}}{3 \text{ mol C}_3\text{H}_7\text{OH}} = 2.26 \text{ mol C}_2\text{H}_5\text{COOH}$$

$$\frac{1.50 \text{ mol}}{2.26 \text{ mol}} \times 100 = 66.4\% \text{ yield}$$

c. Given: balanced equation; 116 g Na$_2$Cr$_2$O$_7$; actual yield C$_2$H$_5$COOH = 28.1 g

Unknown: percent yield

$$\text{theoretical yield} = 116 \text{ g Na}_2\text{Cr}_2\text{O}_7 \times \frac{1 \text{ mol Na}_2\text{Cr}_2\text{O}_7}{261.98 \text{ g Na}_2\text{Cr}_2\text{O}_7}$$

$$\times \frac{3 \text{ mol C}_2\text{H}_5\text{COOH}}{2 \text{ mol Na}_2\text{Cr}_2\text{O}_7} \times \frac{74.09 \text{ g C}_2\text{H}_5\text{COOH}}{1 \text{ mol C}_2\text{H}_5\text{COOH}} = 49.2 \text{ g C}_2\text{H}_5\text{COOH}$$

$$\frac{28.1 \text{ g}}{49.2 \text{ g}} \times 100 = 57.1\% \text{ yield}$$

232. Given: unbalanced equation; 850. g C_3H_6; 300. g NH_3; excess O_2; actual yield C_3H_3N = 850. g

Unknown: balanced equation; limiting reactant; percent yield

$2C_3H_6(g) + 2NH_3(g) + 3O_2(g) \rightarrow 2C_3H_3N(g) + 6H_2O(g)$

$850. \text{ g } C_3H_6 \times \dfrac{1 \text{ mol } C_3H_6}{42.09 \text{ g } C_3H_6} \times \dfrac{2 \text{ mol } NH_3}{2 \text{ mol } C_3H_6} \times \dfrac{17.04 \text{ g } NH_3}{1 \text{ mol } NH_3} = 344 \text{ g } NH_3$

300 g NH_3 available, 344 g NH_3 needed; NH_3 is limiting.

$300. \text{ g } NH_3 \times \dfrac{1 \text{ mol } NH_3}{17.04 \text{ g } NH_3} \times \dfrac{2 \text{ mol } C_3H_3N}{2 \text{ mol } NH_3} \times \dfrac{53.07 \text{ g } C_3H_3N}{1 \text{ mol } C_3H_3N}$

$= 934 \text{ g } C_3H_3N$

$\dfrac{850. \text{ g}}{934 \text{ g}} \times 100 = 91.0\% \text{ yield}$

233. a. Given: 430 kg H_2; reactants and product

Unknown: balanced equation; mass CH_3OH

$CO + 2H_2 \rightarrow CH_3OH$

$430. \text{ kg } H_2 \times \dfrac{1 \text{ mol } H_2}{2.02 \text{ g } H_2} \times \dfrac{1 \text{ mol } CH_3OH}{2 \text{ mol } H_2} \times \dfrac{32.05 \text{ g } CH_3OH}{1 \text{ mol } CH_3OH}$

$= 3.41 \times 10^3 \text{ kg } CH_3OH$

b. Given: balanced equation and theoretical yield from **a**; actual yield CH_3OH = 3.12×10^3 kg

Unknown: percent yield

$\dfrac{3.12 \times 10^3 \text{ kg}}{3.41 \times 10^3 \text{ kg}} \times 100 = 91.5\% \text{ yield}$

234. Given: balanced equation; 750. g $C_6H_{10}O_4$; actual yield $C_6H_{16}N_2$ = 578 g

Unknown: percent yield

$750. \text{ g } C_6H_{10}O_4 \times \dfrac{1 \text{ mol } C_6H_{10}O_4}{146.16 \text{ g } C_6H_{10}O_4} \times \dfrac{1 \text{ mol } C_6H_{16}N_2}{1 \text{ mol } C_6H_{10}O_4}$

$\times \dfrac{116.24 \text{ g } C_6H_{16}N_2}{1 \text{ mol } C_6H_{16}N_2} = 596 \text{ g } C_6H_{16}N_2$

$\dfrac{578 \text{ g}}{596 \text{ g}} \times 100 = 97.0\% \text{ yield}$

235. Given: unbalanced equation; 1.37×10^4 g CO_2; 63.4% yield

Unknown: balanced equation; mass O_2

$6CO_2 + 6H_2O \rightarrow C_6H_{12}O_6 + 6O_2$

$1.37 \times 10^4 \text{ g } CO_2 \times \dfrac{1 \text{ mol } CO_2}{44.01 \text{ g } CO_2} \times \dfrac{6 \text{ mol } O_2}{6 \text{ mol } CO_2} \times \dfrac{32.00 \text{ g } O_2}{1 \text{ mol } O_2}$

$= 9.96 \times 10^3 \text{ g } O_2$

$9.96 \times 10^3 \text{ g } O_2 \times \dfrac{63.4\%}{100\%} = 6.32 \times 10^3 \text{ g } O_2$

236. Given: balanced equation; 2.67×10^2 mol $Ca(OH)_2$; 54.3% yield

Unknown: mass CaO in kg

$2.67 \times 10^2 \text{ mol } Ca(OH)_2 \times \dfrac{100\%}{54.3\%} \times \dfrac{1 \text{ mol } CaO}{1 \text{ mol } Ca(OH)_2} \times \dfrac{56.08 \text{ g } CaO}{1 \text{ mol } CaO}$

$\times \dfrac{1 \text{ kg}}{1000 \text{ g}} = 27.6 \text{ kg}$

237. a. Given: $P_1 = 3.0$ atm; $V_1 = 25$ mL; $P_2 = 6.0$ atom
Unknown: V_2

$$V_2 = \frac{P_1 V_1}{P_2} = 3.0 \text{ atm} \times \frac{25 \text{ mL}}{6.0 \text{ atm}} = 13 \text{ mL}$$

b. Given: $P_1 = 99.97$ kPa; $V_1 = 550.$ mL; $V_2 = 275$ mL
Unknown: P_2

$$P_2 = \frac{P_1 V_1}{V_2} = \frac{99.97 \text{ kPa} \times 550. \text{ mL}}{275 \text{ mL}} = 200. \text{ kPa}$$

c. Given: $P_1 = 0.89$ atm; $P_2 = 3.56$ atm; $V_2 = 20.0$ L
Unknown: V_1

$$V_1 = \frac{P_2 V_2}{P_1} = \frac{3.56 \text{ atm} \times 20.0 \text{ L}}{0.89 \text{ atm}} = 80. \text{ L}$$

d. Given: $V_1 = 800.$ mL; $P_2 = 500.$ kPa; $V_2 = 160.$ mL
Unknown: P_1

$$P_1 = \frac{P_2 V_2}{V_1} = \frac{500. \text{ kPa} \times 160. \text{ mL}}{800. \text{ mL}} = 100. \text{ kPa}$$

e. Given: $P_1 = 0.040$ atm; $P_2 = 250$ atm; $V_2 = 1.0 \times 10^{-2}$ L
Unknown: V_1

$$V_1 = \frac{P_2 V_2}{P_1} = \frac{250 \text{ atm} \times 1.0 \times 10^{-2} \text{ L}}{0.040 \text{ atm}} = 63 \text{ L}$$

238. Given: $P_1 = 1.8$ atm; $V_1 = 2.8$ L; $P_2 = 1.2$ atm
Unknown: V_2

$$V_2 = \frac{P_1 V_1}{P_2} = \frac{1.8 \text{ atm} \times 2.8 \text{ L}}{1.2 \text{ atm}} = 4.2 \text{ L}$$

239. Given: $P_1 = 99.3$ kPa; $V_1 = 48.0$ L; $V_2 = 16.0$ L
Unknown: P_2

$$P_2 = \frac{P_1 V_1}{V_2} = \frac{99.3 \text{ kPa} \times 48.0 \text{ L}}{16.0 \text{ L}} = 298 \text{ kPa}$$

240. Given: $P_1 = 0.989$ atm; $V_1 = 59.0$ mL; $P_2 = 0.967$ atm
Unknown: V_2

$$V_2 = \frac{P_1 V_1}{P_2} = \frac{0.989 \text{ atm} \times 59.0 \text{ mL}}{0.967 \text{ atm}} = 60.3 \text{ mL}$$

241. Given: $P_1 = 6.5$ atm; $V_1 = 2.2$ L; $P_2 = 1.15$ atm
Unknown: V_2

$$V_2 = \frac{P_1 V_1}{P_2} = \frac{6.5 \text{ atm} \times 2.2 \text{ L}}{1.15 \text{ atm}} = 12 \text{ L}$$

242. a. Given: $V_1 = 40.0$ mL; $T_1 = 280.$ K; $T_2 = 350.$ K

Unknown: V_2

$$V_2 = \frac{V_1 T_2}{T_1} = \frac{40.0 \text{ mL} \times 350 \text{ K}}{280. \text{ K}} = 50.0 \text{ mL}$$

b. Given: $V_1 = 0.606$ L; $T_1 = 300.$ K; $V_2 = 0.404$ L

Unknown: T_2

$$T_2 = \frac{V_2 T_1}{V_1} = \frac{0.404 \text{ L} \times 300. \text{ K}}{0.606 \text{ L}} = 200. \text{ K}$$

c. Given: $T_1 = 292$ K; $V_2 = 250.$ mL; $T_2 = 365$ K

Unknown: V_1

$$V_1 = \frac{V_2 T_1}{T_2} = \frac{250. \text{ mL} \times 292 \text{ K}}{365 \text{ K}} = 200. \text{ mL}$$

d. Given: $V_1 = 100.$ mL; $V_2 = 125$ mL; $T_2 = 305$ K

Unknown: T_1

$$T_1 = \frac{T_2 V_1}{V_2} = \frac{305 \text{ K} \times 100. \text{ mL}}{125 \text{ mL}} = 244 \text{ K}$$

e. Given: $V_1 = 0.0024$ L; $T_1 = 22°C$; $T_2 = -14°C$; $K = 273 + °C$

Unknown: V_2

$$V_2 = \frac{V_1 T_2}{T_1} = \frac{0.0024 \text{ L} \times (273 - 14) \text{ K}}{(273 + 22) \text{ K}} = \frac{0.0024 \text{ L} \times 259 \text{ K}}{259 \text{ K}} = 0.0021 \text{ L}$$

243. Given: $V_1 = 2.75$ L; $T_1 = 18°C$; $T_2 = 45°C$; $K = 273 + °C$

Unknown: V_2

$$V_2 = \frac{V_1 T_2}{T_1} = \frac{2.75 \text{ L} \times (273 + 45) \text{ K}}{(273 + 18) \text{ K}} = \frac{2.75 \text{ L} \times 318 \text{ K}}{291 \text{ K}} = 3.01 \text{ L}$$

244. Given: $V_1 = 0.43$ mL; $T_1 = 24°C$; $V_2 = 0.57$ mL; $K = 273 + °C$

Unknown: T_2

$$T_2 = \frac{T_1 V_2}{V_1} = \frac{(273 + 24) \text{ K} \times 0.57 \text{ mL}}{0.43 \text{ mL}} = 394 \text{ K}$$

$394 \text{ K} - 273 = 121°C$

245. a. Given: $P_1 = 1.50$ atm; $T_1 = 273$ K; $T_2 = 410$ K

Unknown: P_2

$$P_2 = \frac{P_1 T_2}{T_1} = \frac{1.50 \text{ atm} \times 410 \text{ K}}{273 \text{ K}} = 2.25 \text{ atm}$$

b. Given: $P_1 = 0.208$ atm; $T_1 = 300.$ K; $P_2 = 0.156$ atm

Unknown: T_2

$$T_2 = \frac{T_1 P_2}{P_1} = \frac{300. \text{ K} \times 0.156 \text{ atm}}{0.208 \text{ atm}} = 225 \text{ K}$$

c. Given: $T_1 = 52°C$;
$P_2 = 99.7$ kPa;
$T_2 = 77°C$;
K = 273 + °C

Unknown: P_1

$$P_1 = \frac{P_2 T_1}{T_2} = \frac{99.7 \text{ kPa} \times (52 + 273) \text{ K}}{(77 + 273) \text{ K}} = \frac{99.7 \text{ kPa} \times 325 \text{ K}}{350. \text{ K}} = 92.6 \text{ kPa}$$

d. Given: $P_1 = 5.20$ atm;
$P_2 = 4.16$ atm;
$T_2 = -13°C$;
K = 273 + °C

Unknown: T_1

$$T_1 = \frac{P_1 T_2}{P_1} = \frac{5.20 \text{ atm} \times (273 - 13) \text{ K}}{4.16 \text{ atm}} = \frac{5.20 \text{ atm} \times 260. \text{ K}}{4.16 \text{ atm}} = 325 \text{ K}$$

325 K − 273 = 52°C

e. Given: $P_1 = 8.33 \times 10^{-4}$ atm;
$T_1 = -84°C$;
$P_2 = 3.92 \times 10^{-3}$ atm;
K = 273 + °C

Unknown: T_2

$$T_2 = \frac{T_1 P_2}{P_1} = \frac{(273 - 84) \text{ K} \times 3.92 \times 10^{-3} \text{ atm}}{8.33 \times 10^{-4} \text{ atm}} = \frac{189 \text{ K} \times 3.92 \times 10^{-3}}{8.33 \times 10^{-4}}$$

= 889 K

889 K − 273 = 616°C

246. Given: $P_1 = 4.882$ atm;
$P_2 = 4.690$ atm;
$T_2 = 8°C$;
K = 273 + °C

Unknown: T_1

$$T_1 = \frac{P_1 T_2}{P_2} = \frac{4.882 \text{ atm} \times (273 + 8) \text{ K}}{4.690 \text{ atm}} = \frac{4.882 \text{ atm} \times 281 \text{ K}}{4.690 \text{ atm}} = 293 \text{ K}$$

293 K − 273 = 20.°C

247. Given: $P_1 = 107$ kPa;
$T_1 = 22°C$;
$T_2 = 45°C$;
K = 273 + °C

Unknown: P_2

$$P_2 = \frac{P_1 T_2}{T_1} = \frac{107 \text{ kPa} \times (273 + 45) \text{ K}}{(273 + 22) \text{ K}} = \frac{107 \text{ kPa} \times 318 \text{ K}}{295 \text{ K}} = 115 \text{ kPa}$$

248. a. Given: $P_1 = 99.3$ kPa;
$V_1 = 225$ mL;
$T_1 = 15°C$;
$P_2 = 102.8$ kPa;
$T_2 = 24°C$

Unknown: V_2

$$V_2 = \frac{P_1 V_1 T_2}{T_1 P_2} = \frac{99.3 \text{ kPa} \times 225 \text{ mL} \times (273 + 24) \text{ K}}{(273 + 15) \text{ K} \times 102.8 \text{ kPa}} = 224 \text{ mL}$$

b. Given: $P_1 = 0.959$ atm;
$V_1 = 3.50$ L;
$T_1 = 45°C$;
$V_2 = 3.70$ L;
$T_2 = 37°C$

Unknown: P_2

$$P_2 = \frac{P_1 V_1 T_2}{T_1 V_2} = \frac{0.959 \text{ atm} \times 3.50 \text{ L} \times (273 + 37) \text{ K}}{(273 + 45) \text{ K} \times 3.70 \text{ L}} = 0.884 \text{ atm}$$

c. Given: $P_1 = 0.0036$ atm;
$V_1 = 62$ mL;
$T_1 = 373$ K;
$P_2 = 0.0029$ atm;
$V_2 = 64$ mL

Unknown: T_2

$$T_2 = \frac{T_1 P_2 V_2}{P_1 V_1} = \frac{373 \text{ K} \times 0.0029 \text{ atm} \times 64 \text{ mL}}{0.0036 \text{ atm} \times 62 \text{ mL}} = 310 \text{ K}$$

d. Given: $P_1 = 100.$ kPa; $V_1 = 43.2$ mL; $T_1 = 19°C$; $P_2 = 101.3$ kPa; $T_2 = 0°C$
Unknown: V_2

$$V_2 = \frac{P_1 V_1 T_2}{T_1 P_2} = \frac{100. \text{ kPa} \times 43.2 \text{ mL} \times (273 + 0) \text{ K}}{(273 + 19) \text{ K} \times 101.3 \text{ kPa}} = 39.9 \text{ mL}$$

249. Given: $V_1 = 450.$ mL; $P_1 = 100.$ kPa; $T_1 = 17°C$; $T_2 = 0°C$; $P_2 = 101.3$ kPa
Unknown: V_2

$$V_2 = \frac{P_1 V_1 T_2}{T_1 P_2} = \frac{100. \text{ kPa} \times 450. \text{ mL} \times (273 + 0) \text{ K}}{(273 + 17) \text{ K} \times 101.3 \text{ kPa}} = 418 \text{ mL}$$

250. Given: $T = 27°C$; H_2S gas: $P_T = 207.33$ kPa; $V = 15$ mL
Unknown: P_{H_2O}; P_{H_2S}

Per Table A-8: $P_{H_2O} = 3.57$ kPa

$P_{H_2S} = P_T - P_{H_2O} = 207.33$ kPa $- 3.57$ kPa $= 203.76$ kPa

251. Given: $T = 10°C$; $P_T = 105.5$ kPa; $V = 1.93$ L; $\frac{P_1 V_1}{T_1} = \frac{P_2 V_2}{T_2}$
Unknown: V_{H_2} at STP

Per Table A-8: $P_{H_2O} = 1.23$ kPa

$P_{H_2} = P_T - P_{H_2O} = 105.5$ kPa $- 1.23$ kPa $= 104.3$ kPa

$$V_2 = \frac{P_1 V_1 T_2}{T_1 P_2} = \frac{104.3 \text{ kPa} \times 1.93 \text{ L} \times (273 + 0) \text{ K}}{(273 + 10) \text{ K} \times 101.3 \text{ kPa}} = 1.92 \text{ L}$$

252. Given: $V_1 = 338$ mL CH_4 at $T_1 = 19°C$, $P_1 = 0.9566$ atm; $T_2 = 26°C$, $P_2 = 0.989$
Unknown: V_2 of CH_4

$P_{1(H_2O)} = 2.19$ kPa; $P_{2(H_2O)} = 3.36$ kPa

$$P_{1(CH_4)} = 0.9566 \text{ atm} - \left(2.19 \text{ kPa} \times \frac{1 \text{ atm}}{101.3 \text{ kPa}}\right) = 0.935 \text{ atm}$$

$$P_{2(CH_4)} = 0.989 \text{ atm} - \left(3.36 \text{ kPa} \times \frac{1 \text{ atm}}{101.3 \text{ kPa}}\right) = 0.956 \text{ atm}$$

$$V_2 = \frac{P_1 V_1 T_2}{T_1 P_2} = \frac{0.935 \text{ atm} \times 338 \text{ mL} \times (273 + 26) \text{ K}}{(273 + 19) \text{ K} \times 0.956 \text{ atm}} = 339 \text{ mL}$$

$V_2 > V_1$; Student 2 collected more.

253. T is constant; $P_1 V_1 = P_2 V_2$

a. Given: $P_1 = 127.3$ kPa; $V_1 = 796$ cm^3; $V_2 = 965$ cm^3
Unknown: P_2

$$P_2 = \frac{P_1 V_1}{V_2} = \frac{127.3 \text{ kPa} \times 796 \text{ cm}^3}{965 \text{ cm}^3} = 105 \text{ kPa}$$

b. Given: $P_1 = 7.1 \times 10^2$ atm; $P_2 = 9.6 \times 10^{-1}$ atm; $V_2 = 3.7 \times 10^3$ mL

Unknown: V_1

$V_1 = \dfrac{P_2 V_2}{P_1} = \dfrac{9.6 \times 10^{-1} \text{ atm} \times 3.7 \times 10^3 \text{ mL}}{7.1 \times 10^2 \text{ atm}} = 5.0$ mL

c. Given: $V_1 = 1.77$ L; $P_2 = 30.79$ kPa; $V_2 = 2.44$ L

Unknown: P_1

$P_1 = \dfrac{P_2 V_2}{V_1} = \dfrac{30.79 \text{ kPa} \times 2.44 \text{ L}}{1.77 \text{ L}} = 42.4$ kPa

d. Given: $P_1 = 114$ kPa; $V_1 = 2.93$ dm^3; $P_2 = 4.93 \times 10^4$ kPa

Unknown: V_2

$V_2 = \dfrac{P_1 V_1}{P_2} = \dfrac{114 \text{ kPa} \times 2.93 \text{ dm}^3}{4.93 \times 10^4 \text{ kPa}} = 6.78 \times 10^{-3}$ dm^3

e. Given: $P_1 = 1.00$ atm; $V_1 = 120.$ mL; $V_2 = 97.0$ mL

Unknown: P_2

$P_2 = \dfrac{P_1 V_1}{V_2} = \dfrac{1.00 \text{ atm} \times 120. \text{ mL}}{97.0 \text{ mL}} = 1.24$ atm

f. Given: $P_1 = 0.77$ atm; $V_2 = 3.6$ m^3; $P_2 = 1.90$ atm

Unknown: V_2

$V_2 = \dfrac{P_1 V_1}{P_2} = \dfrac{0.77 \text{ atm} \times 3.6 \text{ m}^3}{1.90 \text{ atm}} = 1.5$ m^3

254. Given: $V_1 = 0.722$ m^3; $P_1 = 10.6$ atm; $P_2 = 0.96$ atm; $P_1 V_1 = P_2 V_2$

Unknown: V_2

$V_2 = \dfrac{P_1 V_1}{P_2} = \dfrac{10.6 \text{ atm} \times 0.722 \text{ m}^3}{0.96 \text{ atm}} = 8.0$ m^3

255. Given: $V_1 = 7.50 \times 10^3$ L; $V_2 = 195$ L; $P_2 = 0.993$ atm; $P_1 V_1 = P_2 V_2$

Unknown: P_1

$P_1 = \dfrac{P_2 V_2}{V_1} = \dfrac{0.993 \text{ atm} \times 195 \text{ L}}{7.50 \times 10^3 \text{ L}} = 0.0258$ atm

256. Given: $V_1 = 5.70 \times 10^{-1}$ dm^3; $P_1 = 1.05$ atm; $P_2 = 7.47$ atm; $P_1 V_1 = P_2 V_2$

Unknown: V_2

$V_2 = \dfrac{P_1 V_1}{P_2} = \dfrac{1.05 \text{ atm} \times 5.70 \times 10^{-1} \text{ dm}^3}{7.47 \text{ atm}} = 8.01 \times 10^{-2}$ dm^3

257. P is constant; $\dfrac{V_1}{T_1} = \dfrac{V_2}{T_2}$;

K = 273 + °C

a. Given: $V_1 = 26.5$ mL;
$V_2 = 32.9$ mL;
$T_2 = 290.$ K

Unknown: T_1

$T_1 = \dfrac{T_2 V_1}{V_2} = \dfrac{290.\text{ K} \times 26.5 \text{ mL}}{32.9 \text{ mL}} = 234$ K

b. Given: $T_1 = 100°$C;
$V_2 = 0.83$ dm^3;
$T_2 = 29°$C

Unknown: V_1

$V_1 = \dfrac{T_1 V_2}{T_2} = \dfrac{(273 + 100)\text{ K} \times 0.83 \text{ dm}^3}{(273 + 29)\text{ K}} = 1.0$ dm^3

c. Given: $V_1 = 7.44 \times 10^4$ mm^3;
$T_1 = 870°$C;
$V_2 = 2.59 \times 10^2$ mm^3

Unknown: T_2 in °C

$T_2 = \dfrac{T_1 V_2}{V_1} = \dfrac{(870. + 273)\text{ K} \times 2.59 \times 10^2 \text{ mm}^3}{7.44 \times 10^4 \text{ mm}^3} = 3.98$ K

3.98 K − 273.15 = −269.17°C

d. Given: $V_1 = 5.63 \times 10^{-2}$ L;
$T_1 = 132$ K;
$T_2 = 190.$ K

Unknown: V_2

$V_2 = \dfrac{V_1 T_2}{T_1} = \dfrac{5.63 \times 10^{-2} \text{ L} \times 190.\text{ K}}{132 \text{ K}} = 8.10 \times 10^{-2}$ L

e. Given: $T_1 = 243$ K;
$V_2 = 819$ cm^3;
$T_2 = 409$ K

Unknown: V_1

$V_1 = \dfrac{V_2 T_1}{T_2} = \dfrac{819 \text{ cm}^3 \times 243 \text{ K}}{409 \text{ K}} = 487$ cm^3

f. Given: $V_1 = 679$ m^3;
$T_1 = -3°$C;
$T_2 = -246°$C

Unknown: V_2

$V_2 = \dfrac{V_1 T_2}{T_1} = \dfrac{679 \text{ m}^3 \times (273 - 246)\text{ K}}{(273 - 3)\text{ K}} = 67.9$ m^3

---

258. Given: $V_1 = 1.15$ cm^3;
$T_1 = 22°$C;
$T_2 = 99°$C;
$\dfrac{V_1}{T_1} = \dfrac{V_2}{T_2}$;
K = 273 + °C

Unknown: V_2

$V_2 = \dfrac{V_1 T_2}{T_1} = \dfrac{1.15 \text{ cm}^3 \times (273 + 99)\text{ K}}{(273 + 22)\text{ K}} = 1.45$ cm^3

---

259. Given: $V_1 = 6.75$ dm^3;
$T_1 = 40.°$C;
$V_2 = 5.03$ dm^3;
$\dfrac{V_1}{T_1} = \dfrac{V_2}{T_2}$;
K = 273 + °C

Unknown: T_2

$T_2 = \dfrac{V_2 T_1}{V_1} = \dfrac{5.03 \text{ dm}^3 \times (273 + 40.)\text{ K}}{6.75 \text{ dm}^3} = 233$ K

233 K − 273 = −40.°C

260. V is constant;
$\dfrac{P_1}{T_1} = \dfrac{P_2}{T_2}$;
K = 273 + °C

   a. Given: $P_1 = 0.777$ atm;
       $P_2 = 5.6$ atm;
       $T_2 = 192°C$
   Unknown: T_1 in °C

$T_1 = \dfrac{T_2 P_1}{P_2} = \dfrac{(273 + 192)\text{ K} \times 0.777 \text{ atm}}{5.6 \text{ atm}} = 64.5 \text{ K}$

64.5 K − 273 = −208°C

   b. Given: $P_1 = 152$ kPa;
       $T_1 = 302$ K;
       $T_2 = 11$ K
   Unknown: P_2

$P_2 = \dfrac{P_1 T_2}{T_1} = \dfrac{152 \text{ kPa} \times 11 \text{ K}}{302 \text{ K}} = 5.5 \text{ kPa}$

   c. Given: $T_1 = -76°C$;
       $P_2 = 3.97$ atm;
       $T_2 = 27°C$
   Unknown: P_1

$P_1 = \dfrac{P_2 T_1}{T_2} = \dfrac{3.97 \text{ atm} \times (273 - 76) \text{ K}}{(273 + 27) \text{ K}} = 2.61 \text{ atm}$

   d. Given: $P_1 = 395$ atm;
       $T_1 = 46°C$;
       $P_2 = 706$ atm
   Unknown: T_2 in °C

$T_2 = \dfrac{T_1 P_2}{P_1} = \dfrac{(273 + 46) \text{ K} \times 706 \text{ atm}}{395 \text{ atm}} = 570. \text{ K}$

570. K − 273 = 297°C

   e. Given: $T_1 = -37°C$;
       $P_2 = 350.$ atm;
       $T_2 = 2050°C$
   Unknown: P_1

$P_1 = \dfrac{P_2 T_1}{T_2} = \dfrac{350. \text{ atm} \times (273 - 37) \text{ K}}{(273 + 2050) \text{ K}} = 35.6 \text{ atm}$

   f. Given: $P_1 = 0.39$ atm;
       $T_1 = 263$ K;
       $P_2 = 0.058$ atm
   Unknown: T_2

$T_2 = \dfrac{T_1 P_2}{P_1} = \dfrac{263 \text{ K} \times 0.058 \text{ atm}}{0.39 \text{ atm}} = 39 \text{ K}$

261. Given: $T_1 = 22°C$;
      $P_1 = 0.982$ atm;
      $T_2 = -3°C$;
      $\dfrac{P_1}{T_1} = \dfrac{P_2}{T_2}$;
      K = 273 + °C
Unknown: P_2

$P_2 = \dfrac{T_2 P_1}{T_1} = \dfrac{(273 - 3) \text{ K} \times 0.982 \text{ atm}}{(273 + 22) \text{ K}} = 0.899 \text{ atm}$

262. Given: $P_1 = 2.50$ atm;
      $T_1 = 33°C$;
      $T_2 = 0°C$;
      $\dfrac{P_1}{T_1} = \dfrac{P_2}{T_2}$;
      K = 273 + °C
Unknown: P_2

$P_2 = \dfrac{P_1 T_2}{T_1} = \dfrac{2.50 \text{ atm} \times (273 + 0) \text{ K}}{(273 + 33) \text{ K}} = 2.23 \text{ atm}$

263. Given: $P_1 = 127.5$ kPa;
      $T_1 = 290.$ K;
      $P_2 = 3.51$ kPa;
      $\dfrac{P_1}{T_1} = \dfrac{P_2}{T_2}$;
Unknown: T_2

$T_2 = \dfrac{T_1 P_2}{P_1} = \dfrac{290. \text{ K} \times 3.51 \text{ kPa}}{127.5 \text{ kPa}} = 7.98 \text{ K}$

264. $\dfrac{V_1 P_1}{T_1} = \dfrac{V_2 P_2}{T_2}$; $V_2 = \dfrac{V_1 P_1 T_2}{T_1 P_2} = \dfrac{1.65 \text{ L} \times 1.03 \text{ atm} \times (273+46) \text{ K}}{(273+19) \text{ K} \times 0.920 \text{ atm}} = 2.02 \text{ L}$

K = 273 + °C

a. Given: $P_1 = 1.03$ atm;
$V_1 = 1.65$ L;
$T_1 = 19°C$;
$P_2 = 0.920$ atm;
$T_2 = 46°C$

Unknown: V_2

b. Given: $P_1 = 107.0$ kPa;
$V_1 = 3.79$ dm^3;
$T_1 = 73°C$;
$V_2 = 7.58$ dm^3;
$T_2 = 217°C$

$P_2 = \dfrac{V_1 P_1 T_2}{T_1 V_2} = \dfrac{3.79 \text{ dm}^3 \times 107.0 \text{ kPa} \times (273+217) \text{ K}}{(273+73) \text{ K} \times 7.58 \text{ dm}^3} = 75.8 \text{ kPa}$

Unknown: P_2

c. Given: $P_1 = 0.029$ atm;
$V_1 = 249$ mL;
$P_2 = 0.098$ atm;
$V_2 = 197$ mL;
$T_2 = 293$ K

$T_1 = \dfrac{V_1 P_1 T_2}{V_2 P_2} = \dfrac{249 \text{ mL} \times 0.029 \text{ atm} \times 293 \text{ K}}{197 \text{ mL} \times 0.098 \text{ atm}} = 110 \text{ K}$

Unknown: T_1

d. Given: $P_1 = 113$ kPa;
$T_1 = 12°C$;
$P_2 = 149$ kPa;
$V_2 = 3.18 \times 10^3$ mm^3;
$T_2 = -18°C$

$V_1 = \dfrac{V_2 P_2 T_1}{T_2 P_1} = \dfrac{3.18 \times 10^3 \text{ mm}^3 \times 149 \text{ kPa} \times (273+12) \text{ K}}{(273-18) \text{ K} \times 113 \text{ kPa}}$

$= 4.69 \times 10^3 \text{ mm}^3$

Unknown: V_1

e. Given: $P_1 = 1.15$ atm;
$V_1 = 0.93$ m^3;
$T_1 = -22°C$;
$P_2 = 1.01$ atm;
$V_2 = 0.85$ m^3

$T_2 = \dfrac{V_2 P_2 T_1}{V_1 P_1} = \dfrac{0.85 \text{ m}^3 \times 1.01 \text{ atm} \times (273-22) \text{ K}}{0.93 \text{ m}^3 \times 1.15 \text{ atm}} = 210 \text{ K}$

201 K − 273 = −72°C

Unknown: T_2

f. Given: $V_1 = 156$ cm^3;
$T_1 = 195$ K;
$P_2 = 2.25$ atm;
$V_2 = 468$ cm^3;
$T_2 = 585$ K

$P_1 = \dfrac{V_2 P_2 T_1}{T_2 V_1} = \dfrac{468 \text{ cm}^3 \times 2.25 \text{ atm} \times 195 \text{ K}}{585 \text{ K} \times 156 \text{ cm}^3} = 2.25 \text{ atm}$

Unknown: P_1

---

265. Given: $\dfrac{P_1 V_1}{T_1} = \dfrac{P_2 V_2}{T_2}$;
K = 273 + °C;
$V_1 = 392$ cm^3;
$P_1 = 0.987$ atm;
$T_1 = 21°C$;
$T_2 = 13°C$;
$P_2 = 0.992$ atm

$V_2 = \dfrac{V_1 P_1 T_2}{T_1 P_2} = \dfrac{392 \text{ cm}^3 \times 0.987 \text{ atm} \times (273+13) \text{ K}}{(273+21) \text{ K} \times 0.992 \text{ atm}} = 379 \text{ cm}^3$

Unknown: V_2

266. Given: P_T = 0.989 atm; T = 17°C
Unknown: P_{H_2}

from Table A-8: P_{H_2O} = 1.94 kPa × $\dfrac{1 \text{ atm}}{101.3 \text{ kPa}}$ = 0.0192 atm

$P_{H_2} = P_T - P_{H_2O}$ = 0.989 atm − 0.0192 atm = 0.970 atm or 98.3 kPa

---

267. Given: P_1 = 1.77 atm;
V_1 = 1.00 L;
V_2 = 1.50 L;
P_2 = 0.487 atm;
$P_1V_1 = P_2V_2$
Unknown: equalized P in V_T

$V_T = V_1 + V_2$ = 1.00 L + 1.50 L = 2.50 L

$\dfrac{P_1V_1}{V_T} = \dfrac{1.77 \text{ atm} \times 1.00 \text{ L}}{2.50 \text{ L}}$ = 0.708 atm

$\dfrac{P_2V_2}{V_T} = \dfrac{0.487 \text{ atm} \times 1.50 \text{ L}}{2.50 \text{ L}}$ = 0.292 atm

P_T = 0.708 atm + 0.292 atm = 1.00 atm

---

268. Given: T_1 = 10.°C;
P_T = 1.02 atm;
V_1 = 293 mL;
$\dfrac{P_1V_1}{T_1} = \dfrac{P_2V_2}{T_2}$;
P_2 = 1.00 atm;
T_2 = 0°C;
K = 273
Unknown: V_{O_2} at STP (V_2)

From Table A-8: P_{H_2O} = 1.23 kPa × $\dfrac{1 \text{ atm}}{101.3 \text{ kPa}}$ = 0.0121 atm

$P_{O_2} = P_T - P_{H_2O}$ = 1.02 atm − 0.0121 atm = 1.01 atm = P_1

$V_2 = \dfrac{P_1V_1T_2}{T_1P_2} = \dfrac{1.01 \text{ atm} \times 293 \text{ mL} \times (273 + 0) \text{ K}}{(273 + 10) \text{ K} \times 1.00 \text{ atm}}$ = 285 mL

---

269. Given: P_1 = 101.3 kPa;
T_1 = 20°C;
V_1 = 325 cm³;
P_2 = 76.24 kPa;
T_2 = 10°C;
gases are air over water;
K = 273 + °C;
$P_T = P_{air} + P_{H_2O}$;
$P_{H_2O(20°C)}$ = 2.34 kPa; $P_{H_2O(10°C)}$ = 1.23 kPa
Unknown: V_2; V of water lost

$P_{A_1} = P_1 - P_{H_2O(20°C)}$ = 101.3 kPa − 2.34 kPa = 99.0 kPa

In sealed bottle on mountain: $P_{A_2} = P_{A_1} + P_{H_2O(10°C)}$

= 99.0 kPa + 1.23 kPa = 100.2 kPa

$V_2 = \dfrac{P_{A_2}V_1T_2}{T_1P_2} = \dfrac{100.2 \text{ kPa} \times 325 \text{ cm}^3 \times (273 + 10) \text{ K}}{(273 + 20) \text{ K} \times 76.24 \text{ kPa}}$ = 413 cm³

V_{H_2O} = 413 cm³ − 325 cm³ = 88 cm³ H_2O

---

270. Given: 1°C change in T produces a change of 0.20 cm³ in V;
$\dfrac{V_1}{T_1} = \dfrac{V_2}{T_2}$;
K = 273 + °C
Unknown: V at 20.°C

$\dfrac{V}{(273 + 20) \text{ K}} = \dfrac{V + 0.20 \text{ cm}^3}{(273 + 21) \text{ K}}$

294 V = 293 V + 293 × 0.20 cm³

V = 59 cm³

---

271. Given: V_1 = 62.25 mL;
T = 22°C;
P_T = 97.7 kPa;
V_2 = 50.00 mL;
P_{H_2O} = 2.64 kPa
Unknown: P_2

$P_{N_2} = P_T - P_{H_2O}$ = 97.7 kPa − 2.64 kPa = 95.1 kPa

$P_2 = \dfrac{P_{N_2}V_1}{V_2} = \dfrac{95.1 \text{ kPa} \times 62.25 \text{ mL}}{50.00 \text{ mL}}$ = 118 kPa

272. Given: $V_1 = 844$ mL;
$T_1 = 0.00°C$;
$P_1 = 1.000$ atm;
$P_T = 1.017$ atm;
$P_{H_2O} = 3.17$ kPa;
$T_2 = 25°C$;
$P_T = 1.017$ atm;
$\dfrac{P_1V_1}{T_1} = \dfrac{P_2V_2}{T_2}$;
K = 273 + °C

Unknown: V_2; V_T at 25°C and 1.017 atm

$P_2 = P_{NF_3} = P_T - P_{H_2O} = 1.017 \text{ atm} - 3.17 \text{ kPa}\left(\dfrac{1 \text{ atm}}{101.3 \text{ kPa}}\right) = 0.9857$ atm

$V_2 = \dfrac{P_1V_1T_2}{T_1P_2} = \dfrac{1.000 \text{ atm} \times 844 \text{ mL} \times (273+25) \text{ K}}{(273+0) \text{ K} \times 0.9857 \text{ atm}} = 935$ mL

273. Given: $V_1 = 2.94$ kL;
$P_1 = 1.06$ atm;
$T_1 = 32°C$;
$P_2 = 0.092$ atm;
$T_2 = -35°C$;
$\dfrac{P_1V_1}{T_1} = \dfrac{P_2V_2}{T_2}$;
K = 273 + °C

Unknown: V_2

$V_2 = \dfrac{P_1V_1T_2}{T_1P_2} = \dfrac{1.06 \text{ atm} \times 2.94 \text{ kL} \times (273-35) \text{ K}}{(273+32) \text{ K} \times 0.092 \text{ atm}} = 26.4$ kL

274. Given: $P_1 = 2.96$ atm;
$T_1 = 17°C$;
$T_2 = 95°C$;
$\dfrac{P_1}{T_1} = \dfrac{P_2}{T_2}$;
K = 273 + °C

Unknown: P_2

$P_2 = \dfrac{P_1T_2}{T_1} = \dfrac{2.96 \text{ atm} \times (273+95) \text{ K}}{(273+17) \text{ K}} = 3.76$ atm

275. Given: $T_1 = 39°C$;
$V_2 = 108$ mL;
$T_2 = 21°C$;
$\dfrac{V_1}{T_1} = \dfrac{V_2}{T_2}$;
K = 273 + °C

Unknown: V_1

$V_1 = \dfrac{V_2T_1}{T_2} = \dfrac{108 \text{ mL} \times (273+39) \text{ K}}{(273+21) \text{ K}} = 115$ mL

276. Given: $V_1 = 624$ L;
$P_1 = 1.40$ atm;
$V_2 = 80.0$ L;
$P_1V_1 = P_2V_2$

Unknown: P_2

$P_2 = \dfrac{P_1V_1}{V_2} = \dfrac{1.40 \text{ atm} \times 624 \text{ L}}{80.0 \text{ L}} = 10.9$ atm

277. a. Given: $P = 1.09$ atm;
$n = 0.0881$ mol;
$T = 302$ K

Unknown: V in L

$V = \dfrac{nRT}{P} = \dfrac{0.0881 \text{ mol} \cdot 0.0821 \dfrac{\text{L} \cdot \text{atm}}{\text{mol} \cdot \text{K}} \times 302 \text{ K}}{1.09 \text{ atm}} = 2.00$ L

b. Given: $P = 94.9$ kPa;
$V = 0.0350$ L;
$T = 55°C$

Unknown: n

$n = \dfrac{PV}{RT} = \dfrac{94.9 \text{ kPa} \times 0.0350 \text{ L}}{8.314 \dfrac{\text{L} \cdot \text{kPa}}{\text{mol} \cdot \text{K}} \times (273+55) \text{ K}} = 1.22 \times 10^{-3}$ mol

c. Given: $V = 15.7$ L;
$n = 0.815$ mol;
$T = -20.°C$
Unknown: P in kPa

$$P = \frac{nRT}{V} = \frac{0.815 \text{ mol} \times 8.314 \frac{\text{L} \cdot \text{kPa}}{\text{mol} \cdot \text{K}} \times (273 - 20) \text{ K}}{15.7 \text{ L}} = 109 \text{ kPa}$$

d. Given: $P = 0.500$ atm;
$V = 629$ mL;
$n = 0.0337$ mol
Unknown: T in K

$$T = \frac{PV}{nR} = \frac{0.500 \text{ atm} \times 629 \text{ mL} \times 1 \text{ L}}{0.0337 \text{ mol} \times 0.0821 \frac{\text{L} \cdot \text{atm}}{\text{mol} \cdot \text{K}} \times 1000 \text{ mL}} = 114 \text{ K}$$

e. Given: $P = 0.950$ atm;
$n = 0.0818$ mol;
$T = 19°C$
Unknown: V in L

$$V = \frac{nRT}{P} = \frac{0.0818 \text{ mol} \times 0.0821 \frac{\text{L} \cdot \text{atm}}{\text{mol} \cdot \text{K}} \times (273 + 19) \text{ K}}{0.950 \text{ atm}} = 2.06 \text{ L}$$

f. Given: $P = 107$ kPa;
$V = 39.0$ mL;
$T = 27°C$
Unknown: n

$$n = \frac{PV}{RT} = \frac{107 \text{ kPa} \times 39.0 \text{ mL} \times 1 \text{ L}}{8.314 \frac{\text{L} \cdot \text{kPa}}{\text{mol} \cdot \text{K}} \times (273 + 27) \text{ K} \times 1000 \text{ mL}} = 1.67 \times 10^{-3} \text{ mol}$$

278. Given: $V = 425$ mL;
$T = 24°C$,
$P = 0.899$ atm
Unknown: n

$$n = \frac{PV}{RT} = \frac{0.899 \text{ atm} \times 425 \text{ mL} \times 1 \text{ L}}{0.0821 \frac{\text{L} \cdot \text{atm}}{\text{mol} \cdot \text{K}} \times (273 + 24) \text{ K} \times 1000 \text{ mL}} = 1.57 \times 10^{-2} \text{ mol}$$

279. Given: $m = 0.116$ g;
$V = 25.0$ mL;
$T = 127°C$;
$P = 155.3$ kPa;
$PV = \frac{mRT}{M}$
Unknown: M

$$M = \frac{mRT}{PV} = \frac{0.116 \text{ g} \times 8.314 \frac{\text{L} \cdot \text{kPa}}{\text{mol} \cdot \text{K}} \times (273 + 127) \text{ K} \times 1000 \text{ mL}}{155.3 \text{ kPa} \times 25.0 \text{ mL} \times 1 \text{ L}} = 99.4 \text{ g/mol}$$

280. Given: CO_2 gas;
$V = 7.10$ L;
$P = 1.11$ atm;
$T = 31°C$;
$PV = \frac{mRT}{M}$
Unknown: m

$M = 1 \text{ atom C} \times 12.01 \text{ amu/atom} + 2 \text{ atoms O} \times 16.00 \text{ amu/atom} = 44.01$ amu; $M = 44.01$ g/mol

$$m = \frac{MPV}{RT} = \frac{44.01 \text{ g/mol} \times 1.11 \text{ atm} \times 7.10 \text{ L}}{0.0821 \frac{\text{L} \cdot \text{atm}}{\text{mol} \cdot \text{K}} \times (273 + 31) \text{ K}} = 13.9 \text{ g}$$

281. Given: $T = 72°C$;
$P = 144.5$ kPa;
SiF_4 gas;
$D = \frac{MP}{RT}$
Unknown: D

$$D = \frac{MP}{RT} = \frac{104.09 \text{ g/mol} \times 144.5 \text{ kPa}}{8.314 \frac{\text{L} \cdot \text{kPa}}{\text{mol} \cdot \text{K}} \times (273 + 72) \text{ K}} = 5.24 \text{ g/L}$$

282. Given: $D = 1.13$ g/L;
$P = 1.09$ atm;
N_2 gas;
$D = \frac{MP}{RT}$
Unknown: T

$$T = \frac{MP}{DR} = \frac{28.02 \text{ g/mol} \times 1.09 \text{ atm}}{1.13 \text{ g/L} \times 0.0821 \frac{\text{L} \cdot \text{atm}}{\text{mol} \cdot \text{K}}} = 329 \text{ K}$$

283. a. Given: $P = 0.0477$ atm;
$V = 15\,200$ L;
$T = -15°C$
Unknown: n

$$n = \frac{PV}{RT} = \frac{0.0477 \text{ atm} \times 15\,200 \text{ L}}{0.0821 \frac{\text{L} \cdot \text{atm}}{\text{mol} \cdot \text{K}} \times (273 - 15) \text{ K}} = 34.2 \text{ mol}$$

b. Given: $V = 0.119$ mL;
$n = 0.000\,350$ mol;
$T = 0°C$
Unknown: P in kPa

$$P = \frac{nRT}{V} = \frac{0.000\,350 \text{ mol} \times 8.314 \frac{\text{L} \cdot \text{kPa}}{\text{mol} \cdot \text{K}} \times (273 + 0) \text{ K} \times 1000 \text{ mL}}{0.119 \text{ mL} \times 1 \text{ L}}$$
$= 6.68 \times 10^3$ kPa

c. Given: $P = 500.0$ kPa;
$V = 250.$ mL;
$n = 0.120$ mol
Unknown: T in °C

$$T = \frac{PV}{nR} = \frac{500.0 \text{ kPa} \times 250. \text{ mL} \times 1 \text{ L}}{0.120 \text{ mol} \times 8.314 \frac{\text{L} \cdot \text{kPa}}{\text{mol} \cdot \text{K}} \times 1000 \text{ mL}} = 125 \text{ K}$$

$125 \text{ K} - 273 = -148°C$

d. Given: $P = 19.5$ atm;
$n = 4.7 \times 10^4$ mol;
$T = 300.$ °C
Unknown: V

$$V = \frac{nRT}{P} = \frac{4.7 \times 10^4 \text{ mol} \times 0.0821 \frac{\text{L} \cdot \text{atm}}{\text{mol} \cdot \text{K}} \times (273 + 300) \text{ K}}{19.5 \text{ atm}} = 1.1 \times 10^5 \text{ L}$$

---

284. Given: $PV = \frac{mRT}{M}$

a. Given: $P = 0.955$ atm;
$V = 3.77$ L;
$m = 8.23$ g;
$T = 25°C$
Unknown: M

$$M = \frac{mRT}{PV} = \frac{8.23 \text{ g} \times 0.0821 \frac{\text{L} \cdot \text{atm}}{\text{mol} \cdot \text{K}} \times (273 + 25) \text{ K}}{0.955 \text{ atm} \times 3.77 \text{ L}} = 55.9 \text{ g/mol}$$

b. Given: $P = 105.0$ kPa;
$V = 50.0$ mL;
$M = 48.02$ g/mol;
$T = 0°C$
Unknown: m

$$m = \frac{PVM}{RT} = \frac{105.0 \text{ kPa} \times 50.0 \text{ mL} \times 48.02 \text{ g/mol} \times 1 \text{ L}}{8.314 \frac{\text{L} \cdot \text{kPa}}{\text{mol} \cdot \text{K}} \times (273 + 0) \text{ K} \times 1000 \text{ mL}} = 0.111 \text{ g}$$

c. Given: $P = 0.782$ atm;
$m = 3.20 \times 10^{-3}$ g;
$M = 2.02$ g/mol;
$T = -5°C$
Unknown: V in L

$$V = \frac{mRT}{PM} = \frac{3.20 \times 10^{-3} \text{ g} \times 0.0821 \frac{\text{L} \cdot \text{atm}}{\text{mol} \cdot \text{K}} \times (273 - 5) \text{ K}}{0.782 \text{ atm} \times 2.02 \text{ g/mol}} = 4.46 \times 10^{-2} \text{ L}$$

d. Given: $V = 2.00$ L;
$m = 7.19$ g;
$M = 159.8$ g/mol;
$T = 185°C$
Unknown: P in atm

$$P = \frac{mRT}{MV} = \frac{7.19 \text{ g} \times 0.0821 \frac{\text{L} \cdot \text{atm}}{\text{mol} \cdot \text{K}} \times (273 + 185) \text{ K}}{159.8 \text{ g/mol} \times 2.00 \text{ L}} = 0.846 \text{ atm}$$

e. Given: $P = 107.2$ kPa;
$V = 26.1$ mL;
$m = 0.414$ g;
$T = 45°C$
Unknown: M

$$M = \frac{mRT}{PV} = \frac{0.414 \text{ g} \times 8.314 \frac{\text{L} \cdot \text{kPa}}{\text{mol} \cdot \text{K}} \times (273 + 45) \text{ K} \times 1000 \text{ mL}}{107.2 \text{ kPa} \times 26.1 \text{ mL} \times 1 \text{ L}}$$

$= 391$ g/mol

285. Given: $n = 1.00$ mol;
$T = 25°C$;
$P = 0.915$ kPa
Unknown: V

$$V = \frac{nRT}{P} = \frac{1.00 \text{ mol} \times 8.314 \frac{\text{L} \cdot \text{kPa}}{\text{mol} \cdot \text{K}} \times (273 + 25) \text{ K}}{0.915 \text{ kPa}} = 2.71 \times 10^3 \text{ L}$$

286. $D = \frac{MP}{RT}$

a. Given: $P = 1.12$ atm;
$D = 2.40$ g/L;
$T = 2°C$
Unknown: M

$$M = \frac{DRT}{P} = \frac{2.40 \text{ g/L} \times 0.0821 \frac{\text{L} \cdot \text{atm}}{\text{mol} \cdot \text{K}} \times (273 + 2) \text{ K}}{1.12 \text{ atm}} = 48.4 \text{ g/mol}$$

b. Given: $P = 7.50$ atm;
$M = 30.07$ g/mol;
$T = 20.°C$
Unknown: D in g/L

$$D = \frac{30.07 \text{ g/mol} \times 7.50 \text{ atm}}{0.0821 \frac{\text{L} \cdot \text{atm}}{\text{mol} \cdot \text{K}} \times (273 + 20.) \text{ K}} = 9.38 \text{ g/L}$$

c. Given: $P = 97.4$ kPa;
$M = 104.09$ g/mol;
$D = 4.37$ g/L
Unknown: T in °C

$$T = \frac{MP}{DR} = \frac{104.09 \text{ g/mol} \times 97.4 \text{ kPa}}{4.37 \text{ g/L} \times 8.314 \frac{\text{L} \cdot \text{kPa}}{\text{mol} \cdot \text{K}}} = 279 \text{ K}$$

$279 \text{ K} - 273 = 6°C$

d. Given: $M = 77.95$ g/mol;
$D = 6.27$ g/L;
$T = 66°C$
Unknown: P in atm

$$P = \frac{DRT}{M} = \frac{6.27 \text{ g/L} \times 0.0821 \frac{\text{L} \cdot \text{atm}}{\text{mol} \cdot \text{K}} \times (273 + 66) \text{ K}}{77.95 \text{ g/mol}} = 2.24 \text{ atm}$$

287. Given: $m = 1.36$ kg;
N_2O gas;
$V = 25.0$ L;
$T = 59°C$
Unknown: P in atm

$$P = \frac{mRT}{MV} = \frac{1.36 \text{ kg} \times 0.0821 \frac{\text{L} \cdot \text{atm}}{\text{mol} \cdot \text{K}} \times (273 + 59) \text{ K} \times 1000 \text{ g}}{44.02 \text{ g/mol} \times 25.0 \text{ L} \times 1 \text{ kg}} = 33.7 \text{ atm}$$

288. Given: $AlCl_3$ vapor;
$T = 225°C$;
$P = 0.939$ atm
Unknown: D

$$D = \frac{MP}{RT} = \frac{133.33 \text{ g/mol} \times 0.939 \text{ atm}}{0.0821 \frac{\text{L} \cdot \text{atm}}{\text{mol} \cdot \text{K}} \times (273 + 225) \text{ K}} = 3.06 \text{ g/L}$$

289. Given: $D = 0.0262$ g/mL;
$P = 0.918$ atm;
$T = 10.°C$
Unknown: M

$$M = \frac{DRT}{P} = \frac{0.0262 \text{ g/mL} \times 0.0821 \frac{\text{L} \cdot \text{atm}}{\text{mol} \cdot \text{K}} \times (273 + 10) \text{ K} \times 1000 \text{ mL}}{0.918 \text{ atm} \times 1 \text{ L}}$$

$= 663$ g/mol

290. Given: $m = 11.7$ g;
He gas;
$P = 0.262$ atm;
$T = -50.°C$
Unknown: V

$$V = \frac{mRT}{MP} = \frac{11.9 \text{ g} \times 0.0821 \frac{\text{L} \cdot \text{atm}}{\text{mol} \cdot \text{K}} \times (273 - 50) \text{ K}}{4.00 \text{ g/mol} \times 0.262 \text{ atm}} = 208 \text{ L}$$

291. Given: $T = 15°C$; $P_{H_2O} = 1.5988$ kPa; $P_T = 100.0$ kPa; C_2H_6 gas; $V = 245$ mL
Unknown: n

$P_{C_2H_6} = P_T - P_{H_2O} = 100.0 \text{ kPa} - 1.5988 \text{ kPa} = 98.4 \text{ kPa}$

$$n = \frac{PV}{RT} = \frac{98.4 \text{ kPa} \times 245 \text{ mL} \times 1 \text{ L}}{8.314 \frac{\text{L} \cdot \text{kPa}}{\text{mol} \cdot \text{K}} \times (273 + 15) \text{ K} \times 1000 \text{ mL}} = 0.0101 \text{ mol}$$

292. Given: $V = 3.75$ L; NO gas; $T = 19°C$; $P = 1.10$ atm
Unknown: m

$$m = \frac{MPV}{RT} = \frac{30.01 \text{ g/mol} \times 1.10 \text{ atm} \times 3.75 \text{ L}}{0.0821 \frac{\text{L} \cdot \text{atm}}{\text{mol} \cdot \text{K}} \times (273 + 19) \text{ K}} = 5.16 \text{ g}$$

293. Given: theoretical yield $NH_3 = 8.83$ g;
actual yield NH_3 10.24 L;
$T_1 = 52°C$;
$P_1 = 105.3$ kPa;
1 mole gas at STP = 22.4 L;
STP = 0°C, 101.3 kPa
Unknown: percent yield

$$V_{STP} = \frac{P_1 V_1 T_{STP}}{T_1 P_{STP}} = \frac{105.3 \text{ kPa} \times 10.24 \text{ L} \times (273 + 0) \text{ K}}{(273 + 52) \text{ K} \times 101.3 \text{ kPa}} = 8.94 \text{ L}$$

$8.94 \text{ L} \times \frac{1 \text{ mol}}{22.4 \text{ L}} \times \frac{17.04 \text{ g}}{1 \text{ mol}} = 6.80 \text{ g actual yield}$

$\frac{6.80 \text{ g}}{8.83 \text{ g}} \times 100 = 77.0\%$ yield

294. Given: $D = 0.405$ g/L;
$P = 0.889$ atm;
$T = 7°C$
Unknown: molar mass

$$M = \frac{DRT}{P} = \frac{0.405 \text{ g/L} \times 0.0821 \frac{\text{L} \cdot \text{atm}}{\text{mol} \cdot \text{K}} \times (273 + 7) \text{ K}}{0.889 \text{ atm}} = 10.5 \text{ g/mol}$$

295. Given: $V = 90.0$ L;
$P = 1780$ kPa;
$T = 18°C$;
mass empty tank = 39.2 kg;
mass tank + gas = 50.5 kg
Unknown: molar mass (M) of gas

$m = 50.5 \text{ kg} - 39.2 \text{ kg} = 11.3 \text{ kg}$

$$M = \frac{mRT}{PV} = \frac{11.3 \text{ kg} \times 8.314 \frac{\text{L} \cdot \text{kPa}}{\text{mol} \cdot \text{K}} \times (273 + 18) \text{ K} \times 1000 \text{ g}}{1780 \text{ kPa} \times 90.0 \text{ L} \times 1 \text{ kg}} = 171 \text{ g/mol}$$

296. Given: $V = 1.20 \times 10^3$ L;
$m = 12.0$ kg; HCl gas; $T = 18°C$
Unknown: P

$$P = \frac{mRT}{MV} = \frac{12.0 \text{ kg} \times 0.0821 \frac{\text{L} \cdot \text{atm}}{\text{mol} \cdot \text{K}} \times (273 + 18) \text{ K} \times 1000 \text{ g}}{36.46 \text{ g/mol} \times 1.20 \times 10^3 \text{ L} \times 1 \text{ kg}} = 6.55 \text{ atm}$$

297. Given: $T = 20.°C$; Ne gas; $D = 2.70$ g/L
Unknown: P in kPa

$$P = \frac{DRT}{M} = \frac{2.70 \text{ g/L} \times 8.314 \frac{\text{L} \cdot \text{kPa}}{\text{mol} \cdot \text{K}} \times (273 + 20) \text{ K}}{20.18 \text{ g/mol}} = 326 \text{ kPa}$$

298. Given: $V = 658$ mL; $m = 1.50$ g; Ne gas; $P = 4.50 \times 10^2$ kPa

Unknown: T

$$T = \frac{MPV}{mR} = \frac{20.18 \text{ g/mol} \times 4.50 \times 10^2 \text{ kPa} \times 658 \text{ mL} \times 1 \text{ L}}{1.50 \text{ g} \times 8.314 \frac{\text{L} \cdot \text{kPa}}{\text{mol} \cdot \text{K}} \times 1000 \text{ mL}} = 479 \text{ K}$$

299. Given: $m = 1.00$ g; H_2 gas; $P = 6.75$ millibars; 1 bar = 100 kPa = 0.9869 atm; $T_1 = -75°C$; $T_2 = -8°C$

Unknown: V at T_1 and T_2

$$V = \frac{TmR}{MP} = \frac{(273-75) \text{ K} \times 1.00 \text{ g} \times 8.314 \frac{\text{L} \cdot \text{kPa}}{\text{mol} \cdot \text{K}}}{2.02 \text{ g/mol} \times 0.675 \text{ kPa}} = 1210 \text{ L at } -75°C$$

$$V = \frac{TmR}{MP} = \frac{(273-8) \text{ K} \times 1.00 \text{ g} \times 8.314 \frac{\text{L} \cdot \text{kPa}}{\text{mol} \cdot \text{K}}}{2.02 \text{ g/mol} \times 0.675 \text{ kPa}} = 1620 \text{ L at } -8°C$$

300. Given: $n = 3.95$ mol; $V = 850.$ mL; $T = 15°C$

Unknown: P in kPa

$$P = \frac{nRT}{V} = \frac{3.95 \text{ mol} \times 8.314 \frac{\text{L} \cdot \text{kPa}}{\text{mol} \cdot \text{K}} \times (273+15) \text{ K} \times 1000 \text{ mL}}{850. \text{ mL} \times 1 \text{ L}} = 1.11 \times 10^4 \text{ kPa}$$

301. Given: $n = 0.00660$ mol; $P = 0.907$ atm; $T = 9°C$

Unknown: V in mL

$$V = \frac{nRT}{P} = \frac{0.00660 \text{ mol} \times 0.0821 \frac{\text{L} \cdot \text{atm}}{\text{mol} \cdot \text{K}} \times (273+9) \text{ K} \times 1000 \text{ mL}}{0.907 \text{ atm} \times 1 \text{ L}} = 168 \text{ mL}$$

302. Given: $m = 8.47$ kg; SO_2 gas; $P = 89.4$ kPa; $T = 40.°C$

Unknown: V

$$V = \frac{mRT}{MP} = \frac{8.47 \text{ kg} \times 8.314 \frac{\text{L} \cdot \text{kPa}}{\text{mol} \cdot \text{K}} \times (273+40) \text{ K} \times 1000 \text{ g}}{64.07 \text{ g/mol} \times 89.4 \text{ kPa} \times 1 \text{ kg}} = 3.85 \times 10^3 \text{ L}$$

303. Given: $m = 908$ g; He gas; $P = 128.3$ kPa; $T = 2°C$

Unknown: V

$$V = \frac{mRT}{MP} = \frac{908 \text{ g} \times 8.314 \frac{\text{L} \cdot \text{kPa}}{\text{mol} \cdot \text{K}} \times (273+2) \text{ K}}{4.00 \text{ g/mol} \times 128.3 \text{ kPa}} = 4.05 \times 10^3 \text{ L}$$

304. Given: $D = 1.162$ g/L; $T = 27°C$; $P = 100.0$ kPa

Unknown: M

$$M = \frac{DRT}{P} = \frac{1.162 \text{ g/L} \times 8.314 \frac{\text{L} \cdot \text{kPa}}{\text{mol} \cdot \text{K}} \times (273+27) \text{ K}}{100.0 \text{ kPa}} = 29.0 \text{ g/mol}$$

305. Given: balanced equation; 2800 L NH_3

Unknown: V of NO; V of O_2

$$2800 \text{ L NH}_3 \times \frac{5 \text{ L O}_2}{4 \text{ L NH}_3} = 3500 \text{ L O}_2$$

$$2800 \text{ L NH}_3 \times \frac{4 \text{ L NO}}{4 \text{ L NH}_3} = 2800 \text{ L NO}$$

306. Given: balanced equation; 3.60×10^4 mL F_2

Unknown: V of O_3; V of HF

$$3.60 \times 10^4 \text{ mL F}_2 \times \frac{1 \text{ mL O}_3}{3 \text{ mL F}_2} = 1.20 \times 10^4 \text{ mL O}_3$$

$$3.60 \times 10^4 \text{ mL F}_2 \times \frac{6 \text{ mL HF}}{3 \text{ mL F}_2} = 7.20 \times 10^4 \text{ mL HF}$$

307. Given: balanced equation; $P_1 = 2.26$ atm; $T_1 = 40°C$; $V_1 = 55.8$ mL
Unknown: V_{CO_2} produced at STP

$$V_2 \text{ (at STP)} = \frac{P_1V_1T_2}{T_1P_2} = \frac{2.26 \text{ atm} \times 55.8 \text{ mL} \times (273+0) \text{ K}}{(273+40) \text{ K} \times 1.00 \text{ atm}} = 110. \text{ mL}$$

$$110. \text{ mL O}_2 \times \frac{2 \text{ mol CO}_2}{3 \text{ mol O}_2} = 73.3 \text{ mL CO}_2$$

308. Given: reactants and products: V_1 (of N_2O_5 at STP) = 5.00 L; $T_2 = 64.5°C$; $P_2 = 1.76$ atm
Unknown: balanced equation; V_2 for N_2O_5; V of NO_2

$2N_2O_5 \rightarrow 4NO_2 + O_2$

$$V_2 = \frac{V_1P_1T_2}{T_1P_2} = \frac{5.00 \text{ L} \times 1.00 \text{ atm} \times (273+64.5) \text{ K}}{(273+0) \text{ K} \times 1.76 \text{ atm}} = 3.51 \text{ L N}_2O_5$$

$$3.51 \text{ L N}_2O_5 \times \frac{4 \text{ L NO}_2}{2 \text{ L N}_2O_5} = 7.02 \text{ NO}_2$$

309. Given: balanced equation

a. Given: excess Al; STP; mass $AlCl_3$ = 7.15 g
Unknown: V of Cl_2

$$7.15 \text{ g AlCl}_3 \times \frac{1 \text{ mol AlCl}_3}{133.33 \text{ g AlCl}_3} \times \frac{3 \text{ mol Cl}_2}{2 \text{ mol AlCl}_3} \times \frac{22.4 \text{ L Cl}_2}{1 \text{ mol Cl}_2} = 1.80 \text{ L Cl}_2$$

b. Given: 19.4 g Al, STP
Unknown: V of Cl_2

$$19.4 \text{ g Al} \times \frac{1 \text{ mol Al}}{26.98 \text{ g Al}} \times \frac{3 \text{ mol Cl}_2}{2 \text{ mol Al}} \times \frac{22.4 \text{ L Cl}_2}{1 \text{ mol Cl}_2} = 24.2 \text{ L Cl}_2$$

c. Given: 1.559 kg Al; $T = 20.°C$; $P = 0.945$ atm
Unknown: V of Cl_2

$$1.559 \text{ kg Al} \times \frac{1 \text{ mol Al}}{26.98 \text{ g Al}} \times \frac{1000 \text{ g}}{1 \text{ kg}} \times \frac{3 \text{ mol Cl}_2}{2 \text{ mol Al}} \times \frac{22.4 \text{ L Cl}_2}{1 \text{ mol Cl}_2} = 1940 \text{ L Cl}_2$$

$$V_2 = \frac{P_1V_1T_2}{T_1P_2} = \frac{1.00 \text{ atm} \times 1940 \text{ L} \times (273+20) \text{ K}}{(273+0) \text{ K} \times 0.945 \text{ atm}} = 2.21 \times 10^3 \text{ L Cl}_2$$

d. Given: excess Al; 920. L Cl_2; STP
Unknown: mass $AlCl_3$ in g

$$920 \text{ L Cl}_2 \times \frac{1 \text{ mol}}{22.4 \text{ L}} \times \frac{2 \text{ mol AlCl}_3}{3 \text{ mol Cl}_2} \times \frac{133.33 \text{ g AlCl}_3}{1 \text{ mol AlCl}_3} = 3.65 \times 10^3 \text{ g AlCl}_3$$

e. Given: $V_1 = 1.049$ mL Cl_2; $T_1 = 37°C$; $P_1 = 5.00$ atm
Unknown: mass Al in g

Find V at STP:

$$V_2 = \frac{P_1V_1T_2}{T_1P_2} = \frac{5.00 \text{ atm} \times 1.049 \text{ mL} \times (273+0) \text{ K}}{(273+37) \text{ K} \times 1.00 \text{ atm}} = 4.62 \text{ mL}$$

$$4.62 \text{ mL Cl}_2 \times \frac{1 \text{ mol Cl}_2}{22\,400 \text{ mL Cl}_2} \times \frac{2 \text{ mol Al}}{3 \text{ mol Cl}_2} \times \frac{26.98 \text{ g Al}}{1 \text{ mol Al}} = 3.71 \times 10^{-3} \text{ g Al}$$

f. Given: 500.00 kg Al; $T_2 = 15°C$; $P_2 = 83.0$ kPa
Unknown: V of Cl_2 in m^3

$$500.00 \text{ kg Al} \times \frac{1 \text{ mol Al}}{26.98 \text{ g Al}} \times \frac{3 \text{ mol Cl}_2}{2 \text{ mol Al}} \times \frac{1000 \text{ g}}{1 \text{ kg}} \times \frac{22.4 \text{ L}}{1 \text{ mol}} \times \frac{1 \text{ m}^3}{1000 \text{ L}}$$

$= 623 \text{ m}^3 \text{ Cl}_2$ at STP

$$V_2 = \frac{P_1V_1T_2}{T_1P_2} = \frac{101.3 \text{ kPa} \times 623 \text{ m}^3 \times (273+15) \text{ K}}{(273+0) \text{ K} \times 83.0 \text{ kPa}} = 802 \text{ m}^3$$

310. a. Given: balanced equation; STP; 57.0 mL H_2

Unknown: V of N_2

$$57.0 \text{ mL } H_2 \times \frac{1 \text{ mL } N_2}{3 \text{ mL } H_2} = 19.0 \text{ mL } N_2$$

b. Given: balanced equation; STP; 6.39×10^4 L H_2

Unknown: V of NH_3

$$6.39 \times 10^4 \text{ L } H_2 \times \frac{2 \text{ L } NH_3}{3 \text{ L } H_2} = 4.26 \times 10^4 \text{ L } NH_3$$

c. Given: balanced equation; 20.0 mol N_2; STP

Unknown: V of NH_3

$$20.0 \text{ mol } N_2 \times \frac{2 \text{ mol } NH_3}{1 \text{ mol } N_2} \times \frac{22.4 \text{ L } NH_3}{1 \text{ mol } NH_3} = 896 \text{ L } NH_3$$

d. Given: balanced equation; $V_1 = 800.$ L NH_3; $T_1 = 55°C$; $P_1 = 0.900$ atm

Unknown: V of H_2 at STP

V of NH_3 at STP =

$$V_2 = \frac{P_1 V_1 T_2}{T_1 P_2} = \frac{0.900 \text{ atm} \times 800. \text{ L } NH_3 \times (273 + 0) \text{ K}}{(273 + 55) \text{ K} \times 1.00 \text{ atm}} = 599 \text{ L } NH_3$$

$$599 \text{ L } NH_3 \times \frac{3 \text{ L } H_2}{2 \text{ L } NH_3} = 899 \text{ L } H_2$$

---

311. a. Given: balanced equation; $V_{C_3H_8} = 3$ L at STP; $T_2 = 250.°C$; $P_2 = 1.00$ atm

Unknown: V of H_2O

$$3.0 \text{ L } C_3H_8 \times \frac{4 \text{ L } H_2O}{1 \text{ L } C_3H_8} = 12 \text{ L } H_2O \text{ at STP}$$

$$V_2 = \frac{P_1 V_1 T_2}{T_1 P_2} = \frac{1.00 \text{ atm} \times 12 \text{ L} \times (273 + 250) \text{ K}}{(273 + 0) \text{ K} \times 1.00 \text{ atm}} = 23 \text{ L}$$

b. Given: balanced equation; 640. L CO_2

Unknown: V of O_2

$$640. \text{ L } CO_2 \times \frac{5 \text{ L } O_2}{3 \text{ L } CO_2} = 1070 \text{ L } O_2$$

c. Given: 465 mL O_2 at STP; balanced equation; $T_2 = 37°C$; $P_2 = 0.973$ atm

Unknown: V_2 of CO_2

$$465 \text{ mL } O_2 \times \frac{3 \text{ mL } CO_2}{5 \text{ mL } O_2} = 279 \text{ mL } CO_2 \text{ at STP}$$

$$V_2 = \frac{P_1 V_1 T_2}{T_1 P_2} = \frac{1.00 \text{ atm} \times 279 \text{ mL} \times (273 + 37) \text{ K}}{(273 + 0) \text{ K} \times 0.973 \text{ atm}} = 326 \text{ mL}$$

d. Given: 2.50 L of C_3H_8 at STP; balanced equation; $T_2 = 175°C$; $P_2 = 1.14$ atm

Unknown: V of products at T_2 and P_2 (V_2)

$$2.50 \text{ L } C_3H_8 \times \frac{3 \text{ L } CO_2}{1 \text{ L } C_3H_8} = 7.50 \text{ L } CO_2 \text{ at STP}$$

$$2.50 \text{ L } C_3H_8 \times \frac{4 \text{ L } H_2O}{1 \text{ L } C_3H_8} = 10.0 \text{ L } H_2O$$

V_1 at STP = 750 L + 10.0 L = 17.5 L

$$V_2 = \frac{P_1 V_1 T_2}{T_1 P_2} = \frac{1.00 \text{ atm} \times 17.5 \text{ L} \times (273 + 175) \text{ K}}{(273 + 0) \text{ K} \times 1.14 \text{ atm}} = 25.2 \text{ L}$$

312. Given: balanced equation; $V_1 = 3500.$ L CO; $T_1 = 20.°C$; $P_1 = 0.953$ atm

Unknown: V of O_2 at STP

$$V_2 = \frac{P_1V_1T_2}{T_1P_2} = \frac{0.953 \text{ atm} \times 3500. \text{ L} \times (273+0) \text{ K}}{(273+20) \text{ K} \times 1.00 \text{ atm}} = 3110 \text{ L CO at STP}$$

$$V \text{ of } O_2 = 3110 \text{ L CO} \times \frac{1 \text{ L } O_2}{2 \text{ L CO}} = 1550 \text{ L } O_2$$

313. Given: balanced equation; $V_1 = 1.00$ L HF; $P_1 = 3.48$ atm; $T_1 = 25°C$; $T_2 = 15°C$; $P_2 = 0.940$ atm

Unknown: V of SiF_4 at V_2 and P_2

$$V_2 = \frac{P_1V_1T_2}{T_1P_2} = \frac{3.48 \text{ atm} \times 1.00 \text{ L} \times (273+15) \text{ K}}{(273+25) \text{ K} \times 0.940 \text{ atm}} = 3.58 \text{ L HF}$$

$$3.58 \text{ L HF} \times \frac{1 \text{ L } SiF_4}{4 \text{ L HF}} = 0.894 \text{ L } SiF_4$$

314. Given: balanced equation

a. Given: 6.28 g Fe

Unknown: V of H_2 at STP

$$6.28 \text{ g Fe} \times \frac{1 \text{ mol Fe}}{55.85 \text{ g Fe}} \times \frac{4 \text{ mol } H_2}{3 \text{ mol Fe}} \times \frac{22.4 \text{ L}}{1 \text{ mol}} = 3.36 \text{ L } H_2$$

b. Given: $V_1 = 500.$ L H_2O; $T_1 = 250.°C$; $P_1 = 1.00$ atm

Unknown: mass Fe

Convert to STP:

$$V_2 = \frac{P_1V_1T_2}{T_1P_2} = \frac{1.00 \text{ atm} \times 500. \text{ L} \times (273+0) \text{ K}}{(273+250) \text{ K} \times 1.00 \text{ atm}} = 261 \text{ L}$$

$$261 \text{ L } H_2O \times \frac{1 \text{ mol}}{22.4 \text{ L}} \times \frac{3 \text{ mol Fe}}{4 \text{ mol } H_2O} \times \frac{55.85 \text{ g Fe}}{1 \text{ mol Fe}} = 488 \text{ g Fe}$$

c. Given: 285 g Fe_3O_4

Unknown: V of H_2 at 20.°C and 1.06 atm

$$285 \text{ g } Fe_3O_4 \times \frac{1 \text{ mol } Fe_3O_4}{231.55 \text{ g } Fe_3O_4} \times \frac{4 \text{ mol } H_2}{1 \text{ mol } Fe_3O_4} \times \frac{22.4 \text{ L}}{1 \text{ mol}} = 110. \text{ L } H_2 \text{ at STP}$$

$$V_2 = \frac{P_1V_1T_2}{T_1P_2} = \frac{1.00 \text{ atm} \times 110. \text{ L} \times (273+20) \text{ K}}{(273+0) \text{ K} \times 1.06 \text{ atm}} = 111 \text{ L}$$

315. Given: balanced equation; 0.027 g Na; excess H_2O

Unknown: V of H_2 at STP

$$0.027 \text{ g Na} \times \frac{1 \text{ mol Na}}{22.99 \text{ g Na}} \times \frac{1 \text{ mol } H_2}{2 \text{ mol Na}} \times \frac{22.4 \text{ L}}{1 \text{ mol}} = 0.013 \text{ L } H_2$$

316. Given: balanced equation; $V_1 = 7.15$ L CO_2; $T_1 = 125°C$; $P_1 = 1.02$ atm

Unknown: V of O_2 at STP; mass $C_4H_{10}O$

$$V_2 = \frac{P_1V_1T_2}{T_1P_2} = \frac{1.02 \text{ atm} \times 7.15 \text{ L} \times (273+0) \text{ K}}{(273+125) \text{ K} \times 1.00 \text{ atm}} = 5.00 \text{ L } CO_2 \text{ at STP}$$

$$5.00 \text{ L } CO_2 \times \frac{6 \text{ L } O_2}{4 \text{ L } CO_2} = 7.50 \text{ L } O_2$$

$$5.00 \text{ L } CO_2 \times \frac{1 \text{ mol}}{22.4 \text{ L}} \times \frac{1 \text{ mol } C_4H_{10}O}{4 \text{ mol } CO_2} \times \frac{74.14 \text{ g } C_4H_{10}O}{1 \text{ mol } C_4H_{10}O} = 4.14 \text{ g } C_4H_{10}O$$

317. Given: balanced equation

a. Given: 0.100 mol $C_3H_5N_3O_9$
Unknown: V of products at STP

$$0.100 \text{ mol } C_3H_5N_3O_9 \times \frac{6 \text{ mol } N_2}{4 \text{ mol } C_3H_5N_3O_9} \times \frac{22.4 \text{ L}}{1 \text{ mol}} = 3.36 \text{ L } N_2$$

$$0.100 \text{ mol } C_3H_5N_3O_9 \times \frac{12 \text{ mol } CO_2}{4 \text{ mol } C_3H_5N_3O_9} \times \frac{22.4 \text{ L}}{1 \text{ mol}} = 6.72 \text{ L } CO_2$$

$$0.100 \text{ mol } C_3H_5N_3O_9 \times \frac{10 \text{ mol } H_2O}{4 \text{ mol } C_3H_5N_3O_9} \times \frac{22.4 \text{ L}}{1 \text{ mol}} = 5.60 \text{ L } H_2O$$

$$0.100 \text{ mol } C_3H_5N_3O_9 \times \frac{1 \text{ mol } O_2}{4 \text{ mol } C_3H_5N_3O_9} \times \frac{22.4 \text{ L}}{1 \text{ mol}} = 0.560 \text{ L } O_2$$

b. Given: 10.0 g $C_3H_5N_3O_9$
Unknown: total V gases produced at 300.°C and 1.00 atm

V_T from **a** = 3.36 L + 6.72 L + 5.60 L + 0.560 L = 16.24 L

$$10.0 \text{ g } C_3H_5N_3O_9 \times \frac{1 \text{ mol } C_3H_5N_3O_9}{227.11 \text{ g } C_3H_5N_3O_9} \times \frac{16.24 \text{ L gases}}{0.100 \text{ mol } C_3H_5N_3O_9}$$

= 7.15 L gases at STP

$$V_2 = \frac{V_1 P_1 T_2}{T_1 P_2} = \frac{7.15 \text{ L} \times 1.00 \text{ atm} \times (273 + 300) \text{ K}}{(273 + 0) \text{ K} \times 1.00 \text{ atm}} = 15.0 \text{ L gases}$$

318. Given: balanced equation; 250. mL N_2O at STP
Unknown: mass NH_4NO_3

$$250. \text{ mL } N_2O \times \frac{1 \text{ L}}{1000 \text{ mL}} \times \frac{1 \text{ mol}}{22.4 \text{ L}} \times \frac{1 \text{ mol } NH_4NO_3}{1 \text{ mol } N_2O} \times \frac{80.06 \text{ g } NH_4NO_3}{1 \text{ mol } NH_4NO_3}$$

= 0.894 g NH_4NO_3

319. Given: balanced equation; 8.46 g Ca_3P_2
Unknown: V of PH_3 at 18°C and 102.4 kPa

$$8.46 \text{ g } Ca_3P_2 \times \frac{1 \text{ mol } Ca_3P_2}{182.18 \text{ g } Ca_3P_2} \times \frac{2 \text{ mol } PH_3}{1 \text{ mol } C_3P_2} \times \frac{22.4 \text{ L}}{1 \text{ mol}} = 2.08 \text{ L } PH_3 \text{ at STP}$$

$$V_2 = \frac{P_1 V_1 T_2}{T_1 P_2} = \frac{101.3 \text{ kPa} \times 2.08 \text{ L} \times (273 + 18) \text{ K}}{(273 + 0) \text{ K} \times 102.4 \text{ kPa}} = 2.19 \text{ PH}_3$$

320. Given: balanced equation; 6.0×10^3 kg $AlCl_3$
Unknown: mass Al; V of HCl at 4.71 atm + 43°C

$$6.0 \times 10^3 \text{ kg } AlCl_3 \times \frac{1 \text{ mol } AlCl_3}{133.33 \text{ g } AlCl_3} \times \frac{2 \text{ mol } Al}{2 \text{ mol } AlCl_3} \times \frac{26.98 \text{ g } Al}{1 \text{ mol } Al}$$

= 1.2×10^3 kg Al

$$6.0 \times 10^3 \text{ kg } AlCl_3 \times \frac{1 \text{ mol } AlCl_3}{133.33 \text{ g } AlCl_3} \times \frac{6 \text{ mol } HCl}{2 \text{ mol } AlCl_3} \times \frac{22.4 \text{ L}}{1 \text{ mol}} \times \frac{1000 \text{ g}}{1 \text{ kg}}$$

= 3.0×10^6 L HCl at STP

$$V_2 = \frac{P_1 V_1 T_2}{T_1 P_2} = \frac{1.00 \text{ atm} \times 3.0 \times 10^6 \text{ L} \times (273 + 43) \text{ K}}{(273 + 0) \text{ K} \times 4.71 \text{ atm}} = 7.4 \times 10^5 \text{ L HCl}$$

321. Given: balanced equation; actual yield = 8.50×10^4 kg $(NH_2)_2CO$; 89.5% yield
Unknown: V of NH_3

theoretical yield $(NH_2)_2CO = 8.50 \times 10^4$ kg $(NH_2)_2CO \times \frac{100\%}{89.5\%}$

= 9.50×10^4 kg $(NH_2)_2CO$

$$9.50 \times 10^4 \text{ kg } (NH_2)_2CO \times \frac{1000 \text{ g}}{1 \text{ kg}} \times \frac{1 \text{ mol } (NH_2)_2CO}{60.07 \text{ g } (NH_2)_2CO} \times \frac{2 \text{ mol } NH_3}{1 \text{ mol } (NH_2)_2CO}$$

$\times \frac{22.4 \text{ L}}{1 \text{ mol}} = 7.08 \times 10^7$ L NH_3

322. Given: $V_1 = 265$ mL O_2; $T_1 = 10.°C$; $P_1 = 0.975$ atm; balanced equation

Unknown: mass of BaO_2

$P_{O_2} = P_T - P_{H_2O} = 0.975 \text{ atm} - 1.23 \text{ kPa} \times \dfrac{1 \text{ atm}}{101.3 \text{ kPa}} = 0.963 \text{ atm}$

$V_2 = \dfrac{P_1 V_1 T_2}{T_1 P_2} = \dfrac{0.963 \text{ atm} \times 265 \text{ mL} \times (273 + 0) \text{ K}}{(273 + 10) \text{ K} \times 1.00 \text{ atm}} = 246 \text{ mL } O_2 \text{ at STP}$

$246 \text{ mL } O_2 \times \dfrac{1 \text{ L}}{1000 \text{ mL}} \times \dfrac{1 \text{ mol}}{22.4 \text{ L}} \times \dfrac{2 \text{ mol } BaO_2}{1 \text{ mol } O_2} \times \dfrac{169.33 \text{ g } BaO_2}{1 \text{ mol } BaO_2}$

$= 3.72 \text{ g } BaO_2$

323. Given: balanced equation; 15.0 g $KMnO_4$

Unknown: V of Cl_2 at 15°C and 0.959 atm

$15.0 \text{ g } KMnO_4 \times \dfrac{1 \text{ mol } KMnO_4}{158.04 \text{ g } KMnO_4} \times \dfrac{5 \text{ mol } Cl_2}{2 \text{ mol } KMnO_4} \times \dfrac{22.4 \text{ L}}{1 \text{ mol}}$

$= 5.32 \text{ L } Cl_2 \text{ at STP}$

$V_2 = \dfrac{P_1 V_1 T_2}{T_1 P_2} = \dfrac{1.00 \text{ atm} \times 5.32 \text{ L} \times (273 + 15) \text{ K}}{(273 + 0) \text{ K} \times 0.959 \text{ atm}} = 5.85 \text{ L } Cl_2$

324. Given: balanced equations; 35.0 kL O_2 at STP

Unknown: V of NH_3 at STP; V of NO_2 at STP

$35.0 \text{ kL } O_2 \times \dfrac{4 \text{ L } NH_3}{5 \text{ L } O_2} = 28.0 \text{ kL } NH_3$

$28.0 \text{ kL } NH_3 \times \dfrac{4 \text{ L } NO}{4 \text{ L } NH_3} \times \dfrac{2 \text{ L } NO_2}{2 \text{ L } NO} = 28.0 \text{ kL } NO_2$

325. Given: balanced equation; 5.00 L O_2 at STP

Unknown: mass of $KClO_3$

$5.00 \text{ L } O_2 \times \dfrac{1 \text{ mol } O_2}{22.4 \text{ L } O_2} \times \dfrac{2 \text{ mol } KClO_3}{3 \text{ mol } O_2} \times \dfrac{122.55 \text{ g } KClO_3}{1 \text{ mol } KClO_3} = 18.2 \text{ g } KClO_3$

326. Given: balanced equation

a. Given: 38 000 L CO_2

Unknown: V of NH_3 under the same conditions

$38\,000 \text{ L } CO_2 \times \dfrac{1 \text{ L } NH_3}{1 \text{ L } CO_2} = 38\,000 \text{ L } NH_3$

b. Given: 38 000 L CO_2 at 25°C and 1.00 atm

Unknown: mass of $NaHCO_3$

$V_2 = \dfrac{P_1 V_1 T_2}{T_1 P_2} = \dfrac{1.00 \text{ atm} \times 38\,000 \text{ L} \times (273 + 0) \text{ K}}{(273 + 25) \text{ K} \times 1.00 \text{ atm}}$

$= 34\,800 \text{ L } CO_2 \text{ at STP}$

$34\,800 \text{ L } CO_2 \times \dfrac{1 \text{ mol}}{22.4 \text{ L}} \times \dfrac{1 \text{ mol } NaHCO_3}{1 \text{ mol } CO_2} \times \dfrac{84.01 \text{ g } NaHCO_3}{1 \text{ mol } NaHCO_3}$

$= 1.30 \times 10^5 \text{ g } NaHCO_3$

c. Given: 46.0 kg $NaHCO_3$

Unknown: V of NH_3 at STP

$46.0 \text{ kg } NaHCO_3 \times \dfrac{1000 \text{ g}}{1 \text{ kg}} \times \dfrac{1 \text{ mol } NaHCO_3}{84.01 \text{ g } NaHCO_3} \times \dfrac{1 \text{ mol } NH_3}{1 \text{ mol } NaHCO_3} \times \dfrac{22.4 \text{ L}}{1 \text{ mol}}$

$= 1.23 \times 10^4 \text{ L } NH_3$

d. Given: 100.00 kg NaHCO$_3$
Unknown: V of CO$_2$ at 5.50 atm and 42°C

$$100.00 \text{ kg NaHCO}_3 \times \frac{1000 \text{ g}}{1 \text{ kg}} \times \frac{1 \text{ mol NaHCO}_3}{84.01 \text{ g NaHCO}_3} \times \frac{1 \text{ mol CO}_2}{1 \text{ mol NaHCO}_3} \times \frac{22.4 \text{ L}}{1 \text{ mol}}$$

$= 2.67 \times 10^4$ L CO$_2$ at STP

$$V_2 = \frac{P_1 V_1 T_2}{T_1 P_2} = \frac{1.00 \text{ atm} \times 2.67 \times 10^4 \text{ L} \times (273 + 42) \text{ K}}{(273 + 0) \text{ K} \times 5.50 \text{ atm}} = 5.60 \times 10^3 \text{ L CO}_2$$

327. Given: balanced equation

a. Given: 4.74 g C$_4$H$_{10}$; excess O$_2$
Unknown: V of CO$_2$ at 150.°C and 1.14 atm

$$4.74 \text{ g C}_4\text{H}_{10} \times \frac{1 \text{ mol C}_4\text{H}_{10}}{58.14 \text{ g C}_4\text{H}_{10}} \times \frac{8 \text{ mol CO}_2}{2 \text{ mol C}_4\text{H}_{10}} \times \frac{22.4 \text{ L}}{1 \text{ mol}} = 7.30 \text{ L CO}_2 \text{ at STP}$$

$$V_2 = \frac{P_1 V_1 T_2}{T_1 P_2} = \frac{1.00 \text{ atm} \times 7.30 \text{ L} \times (273 + 150) \text{ K}}{(273 + 0) \text{ K} \times 1.14 \text{ atm}} = 9.92 \text{ L CO}_2$$

b. Given: 0.500 g C$_4$H$_{10}$
Unknown: V of O$_2$ at 0.980 atm and 75°C

$$0.500 \text{ g C}_4\text{H}_{10} \times \frac{1 \text{ mol C}_4\text{H}_{10}}{58.14 \text{ g C}_4\text{H}_{10}} \times \frac{13 \text{ mol O}_2}{2 \text{ mol C}_4\text{H}_{10}} \times \frac{22.4 \text{ L}}{1 \text{ mol}} = 1.25 \text{ L O}_2 \text{ at STP}$$

$$V_2 = \frac{P_1 V_1 T_2}{T_1 P_2} = \frac{1.00 \text{ atm} \times 1.25 \text{ L} \times (273 + 75) \text{ K}}{(273 + 0) \text{ K} \times 0.980 \text{ atm}} = 1.63 \text{ L O}_2$$

c. Given: mass$_1$ (torch + fuel) = 876.2 g; mass$_2$ (torch + fuel) = 859.3 g
Unknown: V of CO$_2$ at STP

mass butane reacted = 876.2 g − 859.3 g = 16.9 g

$$16.9 \text{ g C}_4\text{H}_{10} \times \frac{1 \text{ mol C}_4\text{H}_{10}}{58.14 \text{ g C}_4\text{H}_{10}} \times \frac{8 \text{ mol CO}_2}{2 \text{ mol C}_4\text{H}_{10}} \times \frac{22.4 \text{ L}}{1 \text{ mol}} = 26.0 \text{ L CO}_2$$

d. Given: 3720 L of CO$_2$ at 35°C and 0.993 atm
Unknown: mass H$_2$O

$$V_2 = \frac{V_1 P_1 T_2}{T_1 P_2} = \frac{3720 \text{ L} \times 0.993 \text{ atm} \times (273 + 0) \text{ K}}{(273 + 35) \text{ K} \times 1.00 \text{ atm}} = 3270 \text{ L CO}_2 \text{ at STP}$$

$$3270 \text{ L CO}_2 \times \frac{1 \text{ mol}}{22.4 \text{ L}} \times \frac{10 \text{ mol H}_2\text{O}}{8 \text{ mol CO}_2} \times \frac{18.02 \text{ g H}_2\text{O}}{1 \text{ mol H}_2\text{O}} = 3290 \text{ g H}_2\text{O}$$

328. Given: 75.0 g ethanol; 500.0 g H$_2$O
Unknown: percentage concentration

$$\frac{75.0 \text{ g ethanol}}{(500.0 + 75.0) \text{ g}} \times 100 = 13.0\% \text{ ethanol}$$

329. Given: 3.50 g KIO$_3$; 6.23 g KOH; 805.05 g H$_2$O
Unknown: percentage concentration of KIO$_3$ and KOH

$$\frac{3.50 \text{ g KIO}_3}{(3.50 + 6.23 + 805.05) \text{ g}} \times 100 = 0.430\% \text{ KIO}_3$$

$$\frac{6.23 \text{ g KOH}}{(3.50 + 6.23 + 805.05) \text{ g}} \times 100 = 0.765\% \text{ KOH}$$

330. Given: 0.377 g RbCl to make a 5.00% solution

Unknown: mass of H_2O

$0.377 \text{ g RbCl} \times \dfrac{100 \text{ g solution}}{5 \text{ g RbCl}} = 7.54 \text{ g solution}$

$7.54 \text{ g solution} - 0.377 \text{ g RbCl} = 7.16 \text{ g } H_2O$

---

331. Given: 30.0 g H_2O; 18.0% $LiNO_3$ solution

Unknown: mass of $LiNO_3$

$\dfrac{x}{30.0 \text{ g} + x} = 0.18 \; ; \; x = 6.59 \text{ g } LiNO_3$

---

332. Given: 141.6 g $C_3H_5O(COOH)_3$; V of solution = 3500.0 mL

Unknown: molarity

$141.6 \text{ g } C_3H_5O(COOH)_3 \times \dfrac{1 \text{ mol } C_3H_5O(COOH)_3}{192.14 \text{ g } C_3H_5O(COOH)_3} = 0.7370 \text{ mol } C_3H_5O(COOH)_3$

$\dfrac{0.7370 \text{ mol}}{3500.0 \text{ mL}} \times \dfrac{1000 \text{ mL}}{1 \text{ L}} = 0.2106 \text{ M}$

---

333. Given: 280.0 mg NaCl; 2.00 mL H_2O

Unknown: molarity

$280.0 \text{ mg NaCl} \times \dfrac{1 \text{ g}}{1000 \text{ mg}} \times \dfrac{1 \text{ mol NaCl}}{58.44 \text{ g NaCl}} = 0.00479 \text{ mol NaCl}$

$\dfrac{0.00479 \text{ mol NaCl}}{2.00 \text{ mL solution}} \times \dfrac{1000 \text{ mL}}{1 \text{ L}} = 2.40 \text{ M}$

---

334. Given: 390.0 g CH_3COOH; 1000.0 mL solution

Unknown: molarity

$390.0 \text{ g } CH_3COOH \times \dfrac{1 \text{ mol } CH_3COOH}{60.06 \text{ g } CH_3COOH} = 6.494 \text{ mol } CH_3COOH$

$\dfrac{6.494 \text{ mol } CH_3COOH}{1000.0 \text{ mL solution}} \times \dfrac{1000 \text{ mL}}{1 \text{ L}} = 6.494 \text{ M}$

---

335. Given: 5.000×10^3 L; 0.215 M

Unknown: mass of $C_6H_{12}O_6$

$0.215 \text{ mol/L} \times (5.00 \times 10^3 \text{ L}) \times \dfrac{180.18 \text{ g } C_6H_{12}O_6}{1 \text{ mol } C_6H_{12}O_6} = 1.94 \times 10^5 \text{ g } C_6H_{12}O_6$

---

336. Given: 720. mL solution; 0.0939 M

Unknown: mass of $MgBr_2$

$\dfrac{0.0939 \text{ mol}}{L} \times 720. \text{ mL} \times \dfrac{1 \text{ L}}{1000 \text{ mL}} \times \dfrac{184.10 \text{ g } MgBr_2}{1 \text{ mol } MgBr_2} = 12.4 \text{ g } MgBr_2$

---

337. Given: 300. mL solution; 0.875 M

Unknown: mass of NH_4Cl

$\dfrac{0.875 \text{ mol}}{L} \times 300. \text{ mL} \times \dfrac{1 \text{ L}}{1000 \text{ mL}} \times \dfrac{53.50 \text{ g } NH_4Cl}{1 \text{ mol } NH_4Cl} = 14.0 \text{ g } NH_4Cl$

---

338. Given: 560 g CH_3COCH_3 solute; 620 g H_2O solvent

Unknown: molality

$560 \text{ g } CH_3COCH_3 \times \dfrac{1 \text{ mol } CH_3COCH_3}{58.09 \text{ g } CH_3COCH_3} = 9.64 \text{ mol } CH_3COCH_3$

$\dfrac{9.64 \text{ mol } CH_3COCH_3}{0.620 \text{ kg solvent}} = 16 \text{ m}$

339. Given: 12.9 g $C_6H_{12}O_6$ solute; 31.0 g H_2O solvent

Unknown: molality

$$12.9 \text{ g } C_6H_{12}O_6 \times \frac{1 \text{ mol } C_6H_{12}O_6}{180.18 \text{ g } C_6H_{12}O_6} = 0.0716 \text{ mol } C_6H_{12}O_6$$

$$\frac{0.0716 \text{ mol } C_6H_{12}O_6}{31.0 \text{ g solvent}} \times \frac{1000 \text{ g}}{1 \text{ kg}} = 2.31 \text{ m}$$

340. Given: 125 g solvent; 12.0 m solution

Unknown: moles of solute ($CH_3CHOHCH_2CH_3$); mass of solute

$$125 \text{ g solvent} \times \frac{12.0 \text{ mol solute}}{1 \text{ kg solvent}} \times \frac{1 \text{ kg}}{1000 \text{ g}} = 1.50 \text{ mol solute}$$

$$1.50 \text{ mol } CH_3CHOHCH_2CH_3 \times \frac{74.14 \text{ g } CH_3CHOHCH_2CH_3}{1 \text{ mol } CH_3CHOHCH_2CH_3}$$

$$= 111 \text{ g } CH_3CHOHCH_2CH_3$$

341. a. Given: 12.0% $KMnO_4$; 500.0 g solution

Unknown: mass solute; mass solvent (H_2O)

$500.0 \text{ g} \times 0.120 = 60.0 \text{ g } KMnO_4$

$500.0 \text{ g} - 60.0 \text{ g} = 440.0 \text{ g } H_2O$

b. Given: 0.60 M $BaCl_2$; 1.750 L solution

Unknown: mass solute

$$\frac{0.60 \text{ mol } BaCl_2}{1.00 \text{ L solution}} \times 1.750 \text{ L solution} = 1.1 \text{ mol } BaCl_2$$

$$1.1 \text{ mol } BaCl_2 \times \frac{208.23 \text{ g } BaCl_2}{1 \text{ mol } BaCl_2} = 230 \text{ g } BaCl_2$$

c. Given: 6.20 m glycerol ($HOCH_2CHOHCH_2OH$); 800.0 g H_2O solvent

Unknown: mass glycerol in g

$$\frac{6.20 \text{ mol glycerol}}{1 \text{ kg solvent}} \times 800.0 \text{ g solvent} \times \frac{1 \text{ kg}}{1000 \text{ g}} = 4.96 \text{ mol glycerol}$$

$$4.96 \text{ mol glycerol} \times \frac{92.11 \text{ glycerol}}{1 \text{ mol glycerol}} = 457 \text{ g glycerol}$$

d. Given: 12.27 g solute ($K_2Cr_2O_7$); 650. mL solution

Unknown: molarity

$$\frac{12.27 \text{ g } K_2Cr_2O_7}{650. \text{ mL solution}} \times \frac{1000 \text{ mL}}{1 \text{ L}} \times \frac{1 \text{ mol } K_2Cr_2O_7}{294.20 \text{ g } K_2Cr_2O_7} = 0.0642 \text{ M } K_2Cr_2O_7$$

e. Given: 288 g $CaCl_2$ solute; 2.04 kg H_2O solvent

Unknown: molality

$$\frac{288 \text{ g } CaCl_2}{2.04 \text{ kg } H_2O} \times \frac{1 \text{ mol } CaCl_2}{110.98 \text{ g } CaCl_2} = 1.27 \text{ m}$$

f. Given: 0.160 M NaCl solution; 25.0 mL solution

Unknown: mass solute in g

$$\frac{0.160 \text{ mol NaCl}}{1.00 \text{ L solution}} \times \frac{1 \text{ L}}{1000 \text{ mL}} \times 25.0 \text{ mL solution} = 0.00400 \text{ mol NaCl}$$

$$0.00400 \text{ mol NaCl} \times \frac{58.44 \text{ g NaCl}}{1 \text{ mol NaCl}} = 0.234 \text{ g NaCl}$$

g. Given: 2.00 m $C_6H_{12}O_6$; 1.50 kg H_2O solvent

Unknown: mass of solute and total mass of solution

$$\frac{2.00 \text{ mol } C_6H_{12}O_6}{1.00 \text{ kg } H_2O} \times 1.50 \text{ kg } H_2O = 3.00 \text{ mol } C_6H_{12}O_6$$

$$3.00 \text{ mol } C_6H_{12}O_6 \times \frac{180.18 \text{ g } C_6H_{12}O_6}{1 \text{ mol } C_6H_{12}O_6} = 541 \text{ g } C_6H_{12}O_6$$

$$1.50 \text{ kg} \times \frac{1000 \text{ g}}{1 \text{ kg}} + 541 \text{ g} = 2040 \text{ g total}$$

342. Given: 2.50 L of 4.25 M solution of H_2SO_4

Unknown: moles of H_2SO_4

$$2.50 \text{ L solution} \times \frac{4.25 \text{ mol } H_2SO_4}{1.00 \text{ L solution}} = 10.6 \text{ mol } H_2SO_4$$

343. Given: 71.5 g $C_{18}H_{32}O_2$ solute; 525 g solvent

Unknown: molality

$$71.5 \text{ g } C_{18}H_{32}O_2 \times \frac{1 \text{ mol } C_{18}H_{32}O_2}{280.50 \text{ g } C_{18}H_{32}O_2} = 0.255 \text{ mol } C_{18}H_{32}O_2$$

$$\frac{0.255 \text{ mol } C_{18}H_{32}O_2}{525 \text{ g solvent}} \times \frac{1000 \text{ g}}{1 \text{ kg}} = 0.486 \text{ } m$$

344. Given: 16.2% $Na_2S_2O_3$ solution by mass

a. Unknown: mass $Na_2S_2O_3$ in 80.0 g solution

$$80.0 \text{ g} \times \frac{16.2\%}{100\%} = 13.0 \text{ g } Na_2S_2O_3$$

b. Unknown: moles $Na_2S_2O_3$ in 80.0 g solution

$$80.0 \text{ g} \times \frac{16.2\%}{100\%} Na_2S_2O_3 \times \frac{1 \text{ mol } Na_2S_2O_3}{158.12 \text{ g } Na_2S_2O_3}$$

$$= 0.0820 \text{ mol } Na_2S_2O_3$$

c. Given: 80.0 g of 16.2% $Na_2S_2O_3$ solution (see **b**) diluted to 250.0 mL with H_2O

Unknown: molarity of final solution

$$\frac{0.0820 \text{ mol } Na_2S_2O_3}{250.0 \text{ mL solution}} \times \frac{1000 \text{ mL}}{1 \text{ L}} = 0.328 \text{ M}$$

345. Given: $CoCl_2$ solute; 650.00 mL of 4.00 M solution

Unknown: mass of $CoCl_2$

$$\frac{4.00 \text{ mol } CoCl_2}{1.00 \text{ L solution}} \times 650.00 \text{ mL solution} \times \frac{1 \text{ L}}{1000 \text{ mL}} = 2.60 \text{ mol } CoCl_2$$

$$2.60 \text{ mol } CoCl_2 \times \frac{129.83 \text{ g } CoCl_2}{1.00 \text{ mol } CoCl_2} = 338 \text{ g } CoCl_2$$

346. Given: 11.27 g $AgNO_3$; 0.150 M solution

Unknown: volume of solution

$$11.27 \text{ g } AgNO_3 \times \frac{1 \text{ mol } AgNO_3}{169.88 \text{ g } AgNO_3} = 0.0663 \text{ mol } AgNO_3$$

$$0.0663 \text{ mol } AgNO_3 \times \frac{1 \text{ L solution}}{0.150 \text{ mol } AgNO_3} = 0.442 \text{ L solution}$$

347. Given: 2250 g H_2O solvent; 1.50 m solution

Unknown: mass of NH_2CONH_2

$$\frac{1.50 \text{ mol } NH_2CONH_2}{1.00 \text{ kg solvent}} \times 2250 \text{ g solvent} \times \frac{1 \text{ kg}}{1000 \text{ g}} = 3.38 \text{ mol } NH_2CONH_2$$

$$3.38 \text{ mol } NH_2CONH_2 \times \frac{60.07 \text{ g } NH_2CONH_2}{1 \text{ mol } NH_2CONH_2} = 203 \text{ g } NH_2CONH_2$$

348. Given: 21.29 mL of a 3.38 M $Ba(NO_3)_2$ solution

Unknown: mass of $Ba(NO_3)_2$

$$21.29 \text{ mL} \times \frac{1 \text{ L}}{1000 \text{ mL}} \times \frac{3.38 \text{ mol } Ba(NO_3)_2}{1 \text{ L}} = 0.0720 \text{ mol } Ba(NO_3)_2$$

$$0.0720 \text{ mol } Ba(NO_3)_2 \times \frac{261.35 \text{ g } Ba(NO_3)_2}{1 \text{ mol } Ba(NO_3)_2} = 18.8 \text{ g } Ba(NO_3)_2$$

349. Given: 100.0 g of a 3.5% $(NH_4)_2SO_4$ solution

Unknown: description of preparation

$$100.0 \text{ g} \times \frac{3.5\%}{100\%} (NH_4)_2SO_4 = 3.5 \text{ g } (NH_4)_2SO_4$$

100.0 g solution − 3.5 g solute = 96.5 g solvent

Add 3.5 g $(NH_4)_2SO_4$ to 96.5 g H_2O

350. Given: 590.0 g water solvent; 0.82 m solution

Unknown: mass $CaCl_2$ solute

$$\frac{0.82 \text{ mol } CaCl_2}{1.00 \text{ kg } H_2O} \times 590.0 \text{ g } H_2O \times \frac{1 \text{ kg}}{1000 \text{ g}} = 0.48 \text{ mol } CaCl_2$$

$$0.48 \text{ mol } CaCl_2 \times \frac{110.98 \text{ g } CaCl_2}{1 \text{ mol } CaCl_2} = 53 \text{ g } CaCl_2$$

351. Given: 0.250 L of 5.00 M NH_3 diluted to 1.000 L

Unknown: moles NH_3; final molarity

$$0.250 \text{ L} \times \frac{5.00 \text{ mol } NH_3}{1 \text{ L}} = 1.25 \text{ mol } NH_3$$

$$\frac{1.25 \text{ mol } NH_3}{1.000 \text{ L}} = 1.25 \text{ M}$$

352. Given: 62.0 g solute; 125 g H_2O; 5.3 m solution

Unknown: molar mass of solute

$$\frac{5.3 \text{ mol solute}}{1.00 \text{ kg } H_2O} \times 125 \text{ g } H_2O \times \frac{1 \text{ kg}}{1000 \text{ g}} = 0.662 \text{ mol solute}$$

$$M = \frac{62.0 \text{ g solute}}{0.662 \text{ mol solute}} = 93.7 \text{ g/mol}$$

353. Given: 0.9% NaCl

Unknown: masses of NaCl and H_2O to prepare 50. L

$$50.\text{ L} \times \frac{1000 \text{ mL}}{1 \text{ L}} \times \frac{1.000 \text{ g}}{1 \text{ mL}} = 50\,000 \text{ g} = 50 \text{ kg}$$

$$\frac{0.9\%}{100\%} \times 50 \text{ kg} = 0.45 \text{ kg NaCl}; \quad 50 \text{ kg} − 0.45 = 49.6 \text{ kg } H_2O$$

354. Given: mass beaker = 68.60 g; mass beaker + H_2O = 115.12 g; 4.08 g glucose solute

Unknown: percentage concentration glucose

mass H_2O = 115.12 g − 68.60 g = 46.52 g

total mass of solution = 46.52 g + 4.08 g = 50.60 g

$$\frac{4.08 \text{ g glucose}}{50.60 \text{ g solution}} \times 100 = 8.06\% \text{ glucose}$$

---

355. Given: D = 0.902 g/mL at 20°C for ethyl acetate solvent; cellulose nitrate solute

Unknown: V of solvent to prepare a 2.0% solution using 25 g of solute

$$\frac{25 \text{ g}}{x + 25 \text{ g}} = 0.020$$

$$x = \frac{25 \text{ g}}{0.020} - 25 \text{ g} = 1200 \text{ g solvent}$$

$$1200 \text{ g solvent} \times \frac{1 \text{ mL}}{0.902 \text{ g}} = 1400 \text{ mL or 1.4 L solvent}$$

---

356. Given: reactants and products

Unknown: balanced equation

$CdCl_2 + Na_2S \rightarrow CdS + 2NaCl$

a. Given: 50.00 mL of a 3.91 M solution

Unknown: moles $CdCl_2$

$$\frac{3.91 \text{ mol } CdCl_2}{1 \text{ L}} \times 50.00 \text{ mL} \times \frac{1 \text{ L}}{1000 \text{ mL}} = 0.196 \text{ mol } CdCl_2$$

b. Given: 0.196 mol $CdCl_2$ from **a**; balanced equation; excess Na_2S

Unknown: moles CdS

$$0.196 \text{ mol } CdCl_2 \times \frac{1 \text{ mol CdS}}{1 \text{ mol } CdCl_2} = 0.196 \text{ mol CdS}$$

c. Given: 0.196 mol CdS from **b**

Unknown: mass CdS

$$0.196 \text{ mol CdS} \times \frac{144.48 \text{ g CdS}}{1 \text{ mol CdS}} = 28.3 \text{ g CdS}$$

---

357. Given: 60.00 mL of 5.85 M H_2SO_4 solution

Unknown: mass H_2SO_4

$$\frac{5.85 \text{ mol } H_2SO_4}{1.00 \text{ L}} \times 60.00 \text{ mL} \times \frac{1 \text{ L}}{1000 \text{ mL}} = 0.351 \text{ mol } H_2SO_4$$

$$0.351 \text{ mol } H_2SO_4 \times \frac{98.09 \text{ g } H_2SO_4}{1 \text{ mol } H_2SO_4} = 34.4 \text{ g } H_2SO_4$$

---

358. Given: 22.5 kL of 6.83 M HCl

Unknown: moles HCl

$$\frac{6.83 \text{ mol HCl}}{1.00 \text{ L}} \times 22.5 \text{ kL} \times \frac{1000 \text{ L}}{1 \text{ kL}} = 1.54 \times 10^5 \text{ mol HCl}$$

359. Given: balanced equation; excess H_2SO_4; 0.600 M $BaCl_2$ solution

Unknown: V $BaCl_2$ solution to produce 12.00 g $BaSO_4$

$$12.00 \text{ g } BaSO_4 \times \frac{1 \text{ mol } BaSO_4}{233.40 \text{ g } BaSO_4} = 0.05141 \text{ mol } BaSO_4$$

$$0.05141 \text{ mol } BaSO_4 \times \frac{1 \text{ mol } BaCl_2}{1 \text{ mol } BaSO_4} = 0.05141 \text{ mol } BaCl_2$$

$$0.05141 \text{ mol } BaCl_2 \times \frac{1.00 \text{ L}}{0.600 \text{ mol } BaCl_2} = 0.0857 \text{ L } BaCl_2 = 85.7 \text{ mL } BaCl_2$$

360. Given: $CuSO_4 \cdot 5H_2O$ solute

$$100. \text{ g} \times \frac{6.00\%}{100\%} CuSO_4 = 6.00 \text{ g } CuSO_4$$

a. Given: 100. g of a 6.00% $CuSO_4$ solution

Unknown: preparation of solution

$$6.00 \text{ g } CuSO_4 \times \frac{1 \text{ mol } CuSO_4}{159.62 \text{ g } CuSO_4} \times \frac{1 \text{ mol } CuSO_4 \cdot 5H_2O}{1 \text{ mol } CuSO_4}$$

$$\times \frac{249.72 \text{ g } CuSO_4 \cdot 5H_2O}{1 \text{ mol } CuSO_4 \cdot 5H_2O} = 9.39 \text{ g } CuSO_4 \cdot 5H_2O \text{ in } 100. \text{ g} - 9.39 \text{ g}$$

$$= 90.61 \text{ g } H_2O$$

b. Given: 1.00 L of a 0.800 M $CuSO_4$ solution

Unknown: preparation of solution

$$1.00 \text{ L} \times \frac{0.800 \text{ mol } CuSO_4}{1.00 \text{ L}} \times \frac{1 \text{ mol } CuSO_4 \cdot 5H_2O}{1 \text{ mol } CuSO_4}$$

$$\times \frac{249.72 \text{ g } CuSO_4 \cdot 5H_2O}{1 \text{ mol } CuSO_4 \cdot 5H_2O} = 200. \text{ g } CuSO_4 \cdot 5H_2O \text{ in water to make } 1.00 \text{ L of solution}$$

c. Given: 3.5 m solution of $CuSO_4$ in 1.0 kg H_2O

Unknown: preparation of solution

$$\frac{3.5 \text{ mol } CuSO_4}{1.0 \text{ kg } H_2O} \times 1.0 \text{ kg } H_2O \times \frac{1 \text{ mol } CuSO_4 \cdot 5H_2O}{1 \text{ mol } CuSO_4} \times \frac{249.72 \text{ g } CuSO_4 \cdot 5H_2O}{1 \text{ mol } CuSO_4 \cdot 5H_2O}$$

$$= 870 \text{ g } CuSO_4 \cdot 5H_2O$$

$$3.5 \text{ mol } CuSO_4 \times \frac{159.62 \text{ g } CuSO_4}{1 \text{ mol } CuSO_4} = 560 \text{ g } CuSO_4$$

$$870 \text{ g } CuSO_4 \cdot 5H_2O - 560 \text{ g } CuSO_4 = 310 \text{ g } H_2O$$

$$1.0 \text{ kg} \times \frac{1000 \text{ g}}{1 \text{ kg}} - 310 \text{ g} = 690 \text{ g added water}$$

361. Given: 700.0 mL of 2.50 M $CaCl_2$ solution

Unknown: mass $CaCl_2 \cdot 6H_2O$

$$700.0 \text{ mL} \times \frac{1 \text{ L}}{1000 \text{ mL}} \times \frac{2.50 \text{ mol } CaCl_2}{1 \text{ L}} \times \frac{1 \text{ mol } CaCl_2 \cdot 6H_2O}{1 \text{ mol } CaCl_2}$$

$$\times \frac{219.10 \text{ g } CaCl_2 \cdot 6H_2O}{1 \text{ mol } CaCl_2 \cdot 6H_2O} = 383 \text{ g } CaCl_2 \cdot 6H_2O$$

362. Given: 1.250 L of 0.00205 M $C_6H_{14}N_4O_2$

Unknown: mass of $C_6H_{14}N_4O_2$

$$1.250 \text{ L} \times \frac{0.00205 \text{ mol } C_6H_{14}N_4O_2}{1 \text{ L}} = 0.00256 \text{ mol } C_6H_{14}N_4O_2$$

$$0.00256 \text{ mol } C_6H_{14}N_4O_2 \times \frac{174.24 \text{ g } C_6H_{14}N_4O_2}{1 \text{ mol } C_6H_{14}N_4O_2} = 0.446 \text{ g } C_6H_{14}N_4O_2$$

363. Given: 2.402 kg $NiSO_4 \cdot 6H_2O$; 25% solution

Unknown: mass of water

$$2.402 \text{ kg } NiSO_4 \cdot 6H_2O \times \frac{1000 \text{ g}}{1 \text{ kg}} \times \frac{1 \text{ mol } NiSO_4 \cdot 6H_2O}{262.88 \text{ g } NiSO_4 \cdot 6H_2O}$$

$$\times \frac{1 \text{ mol } NiSO_4}{1 \text{ mol } NiSO_4 \cdot 6H_2O} \times \frac{154.76 \text{ g } NiSO_4}{1 \text{ mol } NiSO_4} = 1414 \text{ g } NiSO_4$$

$$\frac{1414 \text{ g}}{x + 1414 \text{ g}} = 0.25; \; x = \frac{1414 \text{ g}}{0.25} - 1414 \text{ g} = 4242 \text{ g } H_2O \text{ in solution}$$

$$4242 \text{ g } H_2O - (2402 \text{ g} - 1414 \text{ g}) = 3254 \text{ g } H_2O \text{ added}$$

364. Given: $KAl(SO_4)_2 \cdot 12H_2O$ solute; 35.0 g of a 15% $KAl(SO_4)_2$ solution

Unknown: mass of solute; mass of H_2O added

$$35.0 \text{ g} \times 0.15 \; KAl(SO_4)_2 = 5.25 \text{ g } KAl(SO_4)_2$$

$$5.25 \text{ g } KAl(SO_4)_2 \times \frac{1 \text{ mol } KAl(SO_4)_2}{258.22 \text{ g } KAl(SO_4)_2} \times \frac{1 \text{ mol } KAl(SO_4)_2 \cdot 12H_2O}{1 \text{ mol } KAl(SO_4)_2}$$

$$\times \frac{474.46 \text{ g } KAl(SO_4)_2 \cdot 12H_2O}{1 \text{ mol } KAl(SO_4)_2 \cdot 12H_2O} = 9.646 \text{ g } KAl(SO_4)_2 \cdot 12H_2O$$

$$35.0 \text{ g solution} - 9.646 \text{ g solute} = 25.35 \text{ g water added}$$

365. a. Given: $M_S = 0.500$ M KBr; $V_S = 20.00$ mL; $V_D = 100.00$ mL; $M_S V_S = M_D V_D$

Unknown: M_D

$$M_D = \frac{M_S V_S}{V_D} = \frac{0.500 \text{ M KBr} \times 20.00 \text{ mL}}{100.00 \text{ mL}} = 0.100 \text{ M KBr}$$

b. Given: $M_S = 1.00$ M LiOH; $M_D = 0.075$ M LiOH; $V_D = 500.00$ mL; $M_S V_S = M_D V_D$

Unknown: V_S

$$V_S = \frac{M_D V_D}{M_S} = \frac{0.075 \text{ M LiOH} \times 500.00 \text{ mL}}{1.00 \text{ M LiOH}} = 38 \text{ mL}$$

c. Given: $V_S = 5.00$ mL; $M_D = 0.0493$ M HI; $V_D = 100.00$ mL; $M_S V_S = M_D V_D$

Unknown: M_S

$$M_S = \frac{M_D V_D}{V_S} = \frac{0.0493 \text{ M HI} \times 100.00 \text{ mL}}{5.00 \text{ mL}} = 0.986 \text{ M HI}$$

d. Given: $M_S = 12.0$ M HCl; $V_S = 0.250$ L; $M_D = 1.8$ M HCl; $M_S V_S = M_D V_D$

Unknown: V_D

$$V_D = \frac{M_S V_S}{M_D} = \frac{12.0 \text{ M HCl} \times 0.250 \text{ L}}{1.8 \text{ M HCl}} = 1.7 \text{ L}$$

e. Given: $M_S = 7.44$ M NH_3; $M_D = 0.093$ M NH_3; $V_D = 4.00$ L; $M_SV_S = M_DV_D$

Unknown: V_S

$$V_S = \frac{M_DV_D}{M_S} = \frac{0.093 \text{ M } NH_3 \times 4.00 \text{ L}}{7.44 \text{ M } NH_3} = 0.050 \text{ L} = 50. \text{ mL}$$

366. Given: $M_S = 0.0813$ M; $V_S = 16.5$ mL; $M_D = 0.0200$ M; $M_SV_S = M_DV_D$

Unknown: V_D; V of water added

$$V_D = \frac{M_SV_S}{M_D} = \frac{0.0813 \text{ M} \times 16.5 \text{ mL}}{0.0200 \text{ M}} = 67.1 \text{ mL}$$

67.1 mL − 16.5 mL = 50.6 mL H_2O added

367. Given: $M_S = 3.79$ M NH_4Cl; $V_S = 50.00$ mL; $V_D = 2.00$ L; $M_SV_S = M_DV_D$

Unknown: M_D

$$M_D = \frac{M_SV_S}{V_D} = \frac{3.79 \text{ M } NH_4Cl \times 50.00 \text{ mL}}{2000 \text{ mL}} = 0.0948 \text{ M } NH_4Cl$$

368. Given: 100.00 mL H_2O added; $M_D = 0.046$ M KOH; $M_S = 2.09$ M KOH; $M_SV_S = M_DV_D$

Unknown: V_S

$V_D = V_S + 100.00$ mL

$M_SV_S = M_DV_D$

2.09 M KOH × V_S = 0.046 M KOH (V_S + 100.00 mL)

$V_S = 2.3$ mL

369. a. Given: $V_S = 20.00$ mL; $M_D = 0.50$ M; 100.00 mL H_2O added; $M_SV_S = M_DV_D$

Unknown: M_S

$$M_S = \frac{M_DV_D}{V_S} = \frac{0.50 \text{ M} \times (100.00 \text{ mL} + 20.00 \text{ mL})}{20.00 \text{ mL}} = 3.0 \text{ M}$$

b. Given: $V_D = 5.00$ L; $M_D = 3.0$ M; $M_S = 18.0$ M; $M_SV_S = M_DV_D$

Unknown: V_S

$$V_S = \frac{M_DV_D}{M_S} = \frac{3.0 \text{ M} \times 5.00 \text{ L}}{18.0 \text{ M}} = 0.83 \text{ L}$$

c. Given: V = 0.83 L, from **b**; D = 1.84 g/mL

Unknown: mass H_2SO_4

$$m = VD = 0.83 \text{ L} \times 1.84 \frac{\text{g}}{\text{mL}} \times \frac{1000 \text{ mL}}{1 \text{ L}} = 1.5 \times 10^3 \text{ g}$$

370. Given: $V_S = 1.19$ mL; $M_S = 8.00$ M; $M_D = 1.50$ M; $M_SV_S = M_DV_D$

Unknown: V_D

$$V_D = \frac{M_SV_S}{M_D} = \frac{8.00 \text{ M} \times 1.19 \text{ mL}}{1.50 \text{ M}} = 6.35 \text{ mL}$$

371. Given: $M_S = 5.75$ M; $V_D = 2.00$ L; $M_D = 1.00$ M; $M_S V_S = M_D V_D$

Unknown: V_S

$$V_S = \frac{M_D V_D}{M_S} = \frac{1.00 \text{ M} \times 2.00 \text{ L}}{5.75 \text{ M}} = 0.348 \text{ L or } 348 \text{ mL}$$

372. a. Given: $V_S = 25.00$ mL; $M_D = 0.186$ M; 50.00 mL H$_2$O added; $M_S V_S = M_D V_D$

Unknown: M_S

$$M_S = \frac{M_D V_D}{V_S} = \frac{0.186 \text{ M} \times (25.00 \text{ mL} + 50.00 \text{ mL})}{25.00 \text{ mL}} = 0.558 \text{ M}$$

373. a. Given: 36% HCl solution; $D = 1.18$ g/mL

Unknown: V_1 of 1.0 kg of HCl solution; V_2 that contains 1.0 g HCl; V_3 that contains 1.0 mol HCl

$$V_1 = \frac{m}{D} = 1.0 \text{ kg} \times \frac{1 \text{ mL}}{1.18 \text{ g}} \times \frac{1000 \text{ g}}{1 \text{ kg}} = 850 \text{ mL}$$

$$V_2 = \frac{100\% \times 1 \text{ g}}{36\%} \times \frac{1 \text{ mL}}{1.18 \text{ g}} = 2.4 \text{ mL}$$

$$V_3 = 36.46 \text{ g} \times \frac{100\%}{36\%} \times \frac{1 \text{ mL}}{1.18 \text{ g}} = 86 \text{ mL}$$

b. Given: $D = 1.42$ g/mL; 71% HNO$_3$ solution; $M_S V_S = M_D V_D$

Unknown: V of HNO$_3$ to prepare 10.0 L of 2.00 M HNO$_3$ (V_S)

$$\frac{71 \text{ g HNO}_3}{100 \text{ g}} \times \frac{1.42 \text{ g}}{1 \text{ mL}} \times \frac{1 \text{ mol HNO}_3}{63.02 \text{ g HNO}_3} \times \frac{1000 \text{ mL}}{1 \text{ L}} = 16 \frac{\text{mol HNO}_3}{\text{L}}$$
$$= 16 \text{ M HNO}_3$$

$$V_S = \frac{M_D V_D}{M_S} = \frac{2.00 \text{ M} \times 10.0 \text{ L}}{16 \text{ M}} = 1.2 \text{ L}$$

c. Given: 86 mL contains 1.0 mol (from **a**); $M_S V_S = M_D V_D$; $M_D = 3.0$ M; $V_D = 4.50$ L

Unknown: V_S

$$M_S = \frac{1 \text{ mol}}{86 \text{ mL}} \times \frac{1000 \text{ mL}}{1 \text{ L}} = 12 \text{ M}$$

$$V_S = \frac{M_D V_D}{M_S} = \frac{3.0 \text{ M} \times 4.50 \text{ L}}{12 \text{ M}} = 1.1 \text{ L}$$

374. Given: $M_S = 3.8$ M; $V_D = 8 V_S$; $M_S V_S = M_D V_D$

Unknown: M_D

$$M_D = \frac{M_S V_S}{V_D} = \frac{3.8 \text{ M} \times V_S}{8 V_S} = 0.48 \text{ M}$$

375. Given: $M_D = 2.50$ M;
$V_D = 480.$ mL;
39 mL H$_2$O evaporated;
$M_S V_S = M_D V_D$

Unknown: M_S

$M_S = \dfrac{M_D V_D}{V_S} = \dfrac{2.50 \text{ M} \times 480. \text{ mL}}{(480. - 39) \text{ mL}} = 2.72$ M

---

376. Given: $M_D = 1.22$ M;
$V_D = 25.00$ mL;
$M_S = 6.45$ M;
$M_S V_S = M_D V_D$

Unknown: V_S; procedure

$V_S = \dfrac{M_D V_D}{M_S} = \dfrac{1.22 \text{ M} \times 25.00 \text{ mL}}{6.45 \text{ M}} = 4.73$ mL

Dilute the 6.45 M acid by adding 4.73 mL of it to enough distilled water to make 25.00 mL of solution.

---

377. Given: 100.0 mL of a 2.41 M solution; 9.56 g solute

Unknown: molar mass of solute

$100.0 \text{ mL} \times \dfrac{2.41 \text{ mol}}{1 \text{ L}} \times \dfrac{1 \text{ L}}{1000 \text{ mL}} = 0.241$ mol solute

molar mass $= \dfrac{9.56 \text{ g}}{0.241 \text{ mol}} = 39.7$ g/mol

---

378. Given: 34 g I$_2$; $V_S = 25$ mL; $V_D = 500$ mL; $M_S V_S = M_D V_D$

Unknown: M_D

$M_S = \dfrac{34 \text{ g I}_2}{25 \text{ mL}} \times \dfrac{1 \text{ mol I}_2}{253.8 \text{ g I}_2} \times \dfrac{1000 \text{ mL}}{1 \text{ L}} = 5.36$ M

$M_D = \dfrac{M_S V_S}{V_D} = \dfrac{5.36 \text{ M} \times 25 \text{ mL}}{500 \text{ mL}} = 0.27$ M

---

379. Given: 85% H$_3$PO$_4$ solution;
$V = 600.0$ mL;
2.80 M

Unknown: mass of solution

$600.0 \text{ mL} \times \dfrac{2.80 \text{ mol}}{1 \text{ L}} \times \dfrac{1 \text{ L}}{1000 \text{ mL}} = 1.68$ mol H$_3$PO$_4$

$1.68 \text{ mol H}_3\text{PO}_4 \times \dfrac{98.00 \text{ g H}_3\text{PO}_4}{1 \text{ mol H}_3\text{PO}_4} = 165$ g H$_3$PO$_4$

mass of solution $= \dfrac{165 \text{ g H}_3\text{PO}_4 \times 100. \text{ g solution}}{85 \text{ g H}_3\text{PO}_4} = 190$ g solution

---

380. Given: $M_S = 18.0$ M;
$M_D = 4.0$ M;
$V_D = 3.00$ L;
$M_S V_S = M_D V_D$

Unknown: V_S

$V_S = \dfrac{M_D V_D}{M_S} = \dfrac{4.0 \text{ M} \times 3.00 \text{ L}}{18.0 \text{ M}} = 0.67$ L

---

381. Given: $V_D = 1.00$ L; $M_D = 0.495$ M;
$M_S = 3.07$ M;
$M_S V_S = M_D V_D$

Unknown: V_S

$V_S = \dfrac{M_D V_D}{M_S} = \dfrac{0.495 \text{ M} \times 1.00 \text{ L}}{3.07 \text{ M}} = 0.161$ L $= 161$ mL

Dilute 161 mL stock urea solution to 1.00 L.

---

382. a. Given: 76.2% C$_6$H$_{12}$O$_6$

Unknown: molality

$\dfrac{76.2 \text{ g C}_6\text{H}_{12}\text{O}_6}{(100. - 76.2) \text{ g H}_2\text{O}} \times \dfrac{1 \text{ mol C}_6\text{H}_{12}\text{O}_6}{180.18 \text{ g C}_6\text{H}_{12}\text{O}_6} \times \dfrac{1000 \text{ g}}{1 \text{ kg}} = 17.8$ m

b. Given: 76.2% $C_6H_{12}O_6$;
$D = 1.42$ g/mL;
$V = 1.00$ L

Unknown: mass of $C_6H_{12}O_6$; molarity

$$\frac{1.42 \text{ g}}{1 \text{ mL}} \times 1.00 \text{ L} \times \frac{1000 \text{ mL}}{1 \text{ L}} \times \frac{76.2\%}{100\%} = 1080 \text{ g}$$

$$\frac{1080 \text{ g}}{1 \text{ L}} \times \frac{1 \text{ mol } C_6H_{12}O_6}{180.18 \text{ g } C_6H_{12}O_6} = 5.99 \text{ M}$$

383. Given: $M_D = 0.0890$ M;
$V_S = 10.00$ mL;
$V_D = 50.00$ mL;
$M_S V_S = M_D V_D$;
molar mass $Na_2CO_3 = 105.99$ g

Unknown: M_S; percentage Na_2CO_3 in 50.00 g of a material used to make 1.000 L stock solution

$$M_S = \frac{M_D V_D}{V_S} = \frac{0.0890 \text{ M} \times 50.00 \text{ mL}}{10.00 \text{ mL}} = 0.445 \text{ M}$$

$$\frac{0.445 \text{ mol}}{1 \text{ L}} \times 1.000 \text{ L} \times \frac{105.99 \text{ g}}{1 \text{ mol}} = 47.17 \text{ g } Na_2CO_3$$

$$\frac{47.17 \text{ g } Na_2CO_3}{50.00 \text{ g sample}} \times 100 = 94.34\% \text{ } Na_2CO_3$$

384. Given: V_T of $CuCl_2$ stock solution $= 0.600$ L;
$V_S = 20.0$ mL;
$V_D = 150.0$ mL;
$M_D = 0.250$ M;
$M_S V_S = M_D V_D$

Unknown: mass $CuCl_2$ to make stock solution

$$M_S = \frac{M_D V_D}{V_S} = \frac{0.250 \text{ M} \times 150.0 \text{ mL}}{20.0 \text{ mL}} = 1.88 \text{ M}$$

$$\frac{1.88 \text{ mol } CuCl_2}{1 \text{ L}} \times 0.600 \text{ L} \times \frac{134.45 \text{ g } CuCl_2}{1 \text{ mol } CuCl_2} = 152 \text{ g } CuCl_2$$

385. Given: $M_S = 2.15$ M;
$M_D = 0.65$ M;
$M_S V_S = M_D V_D$

Unknown: dilution factor

$2.15 \text{ M} \times V = 0.65 \times kV$; $k = 3.3$; $V_D = 3.3 V_S$

$V_D - V_S = 3.3 V_S - 1.0 V_S = 2.3 V_S$; Add 2.3 volumes of H_2O per volume of stock solution.

386. a. Given: 18.2% $Sr(NO_3)_2$ solution;
$D = 1.02$ g/ml
$V = 80.00$ mL

Unknown: mass of $Sr(NO_3)_2$

$$80.00 \text{ mL} \times \frac{1.02 \text{ g}}{1 \text{ mL}} \times \frac{18.2 \text{ g } Sr(NO_3)_2}{100 \text{ g solution}} = 14.9 \text{ g } Sr(NO_3)_2$$

b. Given: 14.9 g $Sr(NO_3)_2$ from **a**

Unknown: moles $Sr(NO_3)_2$

$$14.9 \text{ g } Sr(NO_3)_2 \times \frac{1 \text{ mol } Sr(NO_3)_2}{211.64 \text{ g } Sr(NO_3)_2} = 7.04 \times 10^{-2} \text{ mol } Sr(NO_3)_2$$

c. Given: 7.04×10^{-2} mol Sr(NO$_3$)$_2$ from **b**; 420.0 mL H$_2$O added

Unknown: molarity (M_D)

$$M_S = \frac{7.04 \times 10^{-2} \text{ mol}}{80.0 \text{ mL}} \times \frac{1000 \text{ mL}}{1 \text{ L}} = 0.880 \text{ M}$$

$$M_D = \frac{M_S V_S}{V_D} = \frac{0.880 \text{ M} \times 80.0 \text{ mL}}{(80.0 + 420.0) \text{ mL}} = 0.141 \text{ M}$$

387. Given: 60.0 g C$_6$H$_{12}$O$_6$ in 80.0 g H$_2$O; K_f for water = $-1.86°$C/m; $\Delta t_f = K_f m$

Unknown: freezing point

$$60.0 \text{ g C}_6\text{H}_{12}\text{O}_6 \times \frac{1 \text{ mol C}_6\text{H}_{12}\text{O}_6}{180.18 \text{ g C}_6\text{H}_{12}\text{O}_6} = 0.333 \text{ mol C}_6\text{H}_{12}\text{O}_6$$

$$\frac{0.333 \text{ mol C}_6\text{H}_{12}\text{O}_6}{80.0 \text{ g H}_2\text{O}} \times \frac{1000 \text{ g}}{1 \text{ kg}} = 4.16 \ m$$

$$\Delta t_f = K_f m = \frac{-1.86°\text{C}}{m} \times 4.16 \ m = -7.74°\text{C}$$

fp (solution) = fp (solvent) + Δt_f = 0.00°C − 7.74°C = 7.74°C

388. Given: 645 g H$_2$NCONH$_2$; 980. g H$_2$O

Unknown: freezing point

$$645 \text{ g H}_2\text{NCONH}_2 \times \frac{1 \text{ mol H}_2\text{NCONH}_2}{60.07 \text{ g H}_2\text{NCONH}_2} = 10.7 \text{ mol H}_2\text{NCONH}_2$$

$$\frac{10.7 \text{ mol H}_2\text{NCONH}_2}{980. \text{ g H}_2\text{O}} \times \frac{1000 \text{ g}}{1 \text{ kg}} = 10.9 \ m$$

$$\Delta t_f = K_f m = -\frac{1.86°\text{C}}{m} \times 10.9 \ m = -20.3 \ °\text{C}$$

f_p = 0.0°C − 20.3°C = −20.3°C

389. Given: 30.00 g KBr; 100.00 g H$_2$O

Unknown: boiling point

$$30.00 \text{ g KBr} \times \frac{1 \text{ mol KBr}}{119.00 \text{ g KBr}} = 0.252 \text{ mol KBr}$$

$$\Delta t_b = \frac{0.252 \text{ mol KBr}}{100.00 \text{ g H}_2\text{O}} \times \frac{1000 \text{ g}}{1 \text{ kg}} \times \frac{2 \text{ mol ions}}{1 \text{ mol KBr}} \times \frac{0.51°\text{C}}{\text{mol}} = 2.6 \text{ C}$$

bp = 100.0°C + 2.6°C = 102.6°C

390. Given: 385 g CaCl$_2$; 1.230×10^3 g H$_2$O

Unknown: boiling point

$$385 \text{ g CaCl}_2 \times \frac{1 \text{ mol CaCl}_2}{110.97 \text{ g CaCl}_2} = 3.47 \text{ mol CaCl}_2$$

$$\Delta t_b = \frac{3.47 \text{ mol CaCl}_2}{1.230 \times 10^3 \text{ g H}_2\text{O}} \times \frac{10^3 \text{ g}}{1 \text{ kg}} \times \frac{3 \text{ mol ions}}{1 \text{ mol CaCl}_2} \times \frac{0.51°\text{C}}{m} = 4.3°\text{C}$$

bp = 100.0°C + 4.3°C = 104.3°C

391. Given: 0.827 g nonelectrolyte; 2.500 g H$_2$O; fp = −10.18°C

Unknown: molar mass

$$m = \frac{\Delta t_f}{K_f} = \frac{-10.18°\text{C}}{-1.86°\text{C}/m} = 5.47 \ m$$

$$\frac{5.47 \text{ mol}}{1.00 \text{ kg H}_2\text{O}} \times 2.500 \text{ g H}_2\text{O} \times \frac{1 \text{ kg}}{1000 \text{ g}} = 0.0137 \text{ mol}$$

$$\frac{0.827 \text{ g}}{0.0137 \text{ mol}} = 60.4 \text{ g/mol}$$

392. Given: 0.171 g nonelectrolyte; ether solvent; mass of solution = 2.470 g; bp = 36.43°C

Unknown: molar mass

mass of ether solvent = 2.470 g − 0.171 g = 2.299 g

Δt_b = bp − normal bp = 36.43°C − 34.6°C = 1.83°C

$$m = \frac{\Delta t_b}{K_b} = \frac{1.83°C}{2.02°C/m} = 0.906\ m$$

$$\frac{0.906\ \text{mol}}{1.00\ \text{kg ether}} \times 2.299\ \text{g ether} \times \frac{1\ \text{kg}}{1000\ \text{g}} = 0.00208\ \text{mol}$$

$$\frac{0.171\ \text{g}}{0.00208\ \text{mol}} = 82.2\ \text{g/mol}$$

---

393. Given: 383 g glucose; 400. g H$_2$O

Unknown: freezing point; boiling point

$$383\ \text{g C}_6\text{H}_{12}\text{O}_6 \times \frac{1\ \text{mol C}_6\text{H}_{12}\text{O}_6}{180.18\ \text{g C}_6\text{H}_{12}\text{O}_6} = 2.13\ \text{mol C}_6\text{H}_{12}\text{O}_6$$

$$\frac{2.13\ \text{mol}}{400.\ \text{g H}_2\text{O}} \times \frac{1000\ \text{g}}{1\ \text{kg}} = 5.32\ m$$

$$\text{fp} = \text{normal fp} + (K_f m) = 0.00°C + \left(\frac{-1.86°C}{m} \times 5.32\ m\right) = -9.90°C$$

$$\text{bp} = \text{normal bp} + (K_b m) = 100.00°C + \left(\frac{0.51°C}{m} \times 5.32\ m\right) = 102.7°C$$

---

394. Given: 72.4 g glycerol; 122.5 g H$_2$O

Unknown: boiling point

$$72.4\ \text{g glycerol} \times \frac{1\ \text{mol glycerol}}{92.08\ \text{g glycerol}} = 0.786\ \text{mol glycerol}$$

$$\frac{0.786\ \text{mol}}{122.5\ \text{g H}_2\text{O}} \times \frac{1000\ \text{g}}{1\ \text{kg}} = 6.42\ m$$

$$\text{bp} = \text{normal bp} + (K_b m) = 100.00°C + \left(\frac{0.51°C}{m} \times 6.42\ m\right) = 103.3°C$$

---

395. Given: 30.20 g HOCH$_2$CH$_2$OH solute; 88.40 g phenol

Unknown: boiling point

$$30.20\ \text{g HOCH}_2\text{CH}_2\text{OH} \times \frac{1\ \text{mol HOCH}_2\text{CH}_2\text{OH}}{62.08\ \text{g HOCH}_2\text{CH}_2\text{OH}}$$

$$= 0.4865\ \text{mol HOCH}_2\text{CH}_2\text{OH}$$

$$\frac{0.4865\ \text{mol}}{88.04\ \text{g}} \times \frac{1000\ \text{g}}{1\ \text{kg}} = 5.503\ m$$

$$\text{bp} = \text{normal bp} + (K_b m) = 181.8°C + \left(\frac{3.60°C}{m} \times 5.503\ m\right) = 201.6°C$$

---

396. Given: 450. g H$_2$O; fp = −4.5°C

Unknown: mass of ethanol solute

$$m = \frac{\text{fp}}{K_f} = \frac{-4.5°C}{-1.86°C/m} = 2.4\ m$$

$$\frac{2.4\ \text{mol}}{1\ \text{kg}} \times 450.\ \text{g} \times \frac{1\ \text{kg}}{1000\ \text{g}} = 1.1\ \text{mol}$$

$$1.1\ \text{mol ethanol} \times \frac{46.08\ \text{g ethanol}}{1\ \text{mol ethanol}} = 51\ \text{g ethanol}$$

397. Given: fp = −3.9°C; 25.00 g H_2O; 4.27 g solute

Unknown: molar mass of solute

$$m = \frac{\text{fp}}{K_f} = \frac{-3.9°C}{-1.86°C/m} = 2.1\ m$$

$$\frac{2.1\ \text{mol}}{1\ \text{kg}\ H_2O} \times 25.00\ \text{g}\ H_2O \times \frac{1\ \text{kg}}{1000\ \text{g}} = 0.052\ \text{mol}$$

$$\frac{4.27\ \text{g}}{0.052\ \text{mol}} = 82\ \text{g/mol}$$

---

398. Given: 1.17 g $C_{10}H_8O$; 2.00 mL benzene; $T = 20°C$; D of benzene = 0.876 g/mL; K_f for benzene = −5.12°C/m; normal fp of benzene = 5.53°C

Unknown: freezing point

$$\text{mass benzene} = 2.00\ \text{mL} \times \frac{0.8769}{1\ \text{mL}} = 1.75\ \text{g}$$

$$\text{mol}\ C_{10}H_8O = 1.17\ \text{g}\ C_{10}H_8O \times \frac{1\ \text{mol}\ C_{10}H_8O}{144.18\ \text{g}\ C_{10}H_8O} = 0.00811\ \text{mol}\ C_{10}H_8O$$

$$\frac{0.00811\ \text{mol}\ C_{10}H_8O}{1.75\ \text{g benzene}} \times \frac{1000\ \text{g}}{1\ \text{kg}} = 4.63\ m$$

$$\text{fp} = \text{normal fp} + (K_f m) = 5.53°C + \left(\frac{-5.12°C}{m} \times 4.63\ m\right) = -18.2°C$$

---

399. Given: 10.44 g solute; 50.00 g CH_3COOH solvent; bp = 159.2°C

Unknown: molar mass

$$m = \frac{\text{bp} - \text{normal bp}}{K_b} = \frac{159.2°C - 117.9°C}{3.07°C/m} = 13.5\ m$$

$$\frac{13.5\ \text{mol}}{1\ \text{kg solvent}} \times 50.00\ \text{g solvent} \times \frac{1\ \text{kg}}{1000\ \text{g}} = 0.675\ \text{mol}$$

$$\frac{10.44\ \text{g}}{0.675\ \text{mol}} = 15.5\ \text{g/mol}$$

---

400. Given: 0.0355 g solute; 1.000 g camphor solvent; $T = 200.0°C$; fp = 157.7°C

Unknown: molar mass

$$m = \frac{\text{fp} - \text{normal fp}}{K_f} = \frac{157.7°C - 178.8°C}{-39.7°C/m} = 0.531\ m$$

$$\frac{0.531\ \text{mol}}{1\ \text{kg solvent}} \times 1.000\ \text{g solvent} \times \frac{1\ \text{kg}}{1000\ \text{g}} = 5.31 \times 10^{-4}\ \text{mol}$$

$$\frac{0.0355\ \text{g}}{5.31 \times 10^{-4}\ \text{mol}} = 66.8\ \text{g/mol}$$

---

401. Given: 22.5 g $C_6H_{12}O_6$; 294 g phenol

Unknown: boiling point

$$22.5\ \text{g}\ C_6H_{12}O_6 \times \frac{1\ \text{mol}\ C_6H_{12}O_6}{180.18\ \text{g}\ C_6H_{12}O_6} = 0.125\ \text{mol}\ C_6H_{12}O_6$$

$$\frac{0.125\ \text{mol}}{294\ \text{g}} \times \frac{1000\ \text{g}}{1\ \text{kg}} = 0.425\ m$$

$$\text{bp} = 181.8°C + (3.60°C/m \times 0.425\ m) = 183.3°C$$

---

402. Given: 50.0% solution of ethylene glycol in H_2O

$$\frac{50.0\ \text{g ethylene glycol}}{50.0\ \text{g}\ H_2O} \times \frac{1000\ \text{g}}{1\ \text{kg}} \times \frac{1\ \text{mol ethylene glycol}}{62.08\ \text{g ethylene glycol}} = 16.1\ m$$

a. Unknown: freezing point

$$\text{fp} = \text{normal fp} + (K_f m) = 0.00°C + (-1.86°C/m \times 16.1\ m) = -29.9°C$$

b. Unknown: boiling point

$$\text{bp} = \text{normal bp} + (K_b m) = 100.00°C + (0.51°C/m \times 16.1\ m) = 108.2°C$$

403. Given: $K_f = -20.0°C/m$; normal fp = 6.6°C; cyclohexane solvent; 1.604 g solute; 10.000 g cyclohexane; fp = −4.4°C

Unknown: molar mass

$$m = \frac{\text{fp} - \text{normal fp}}{K_f} = \frac{-4.4°C - 6.6°C}{-20.0°C/m} = 0.550\ m$$

$$\frac{0.550\ \text{mol}}{1.00\ \text{kg solvent}} \times 10.000\ \text{g solvent} \times \frac{1\ \text{kg}}{1000\ \text{g}} = 0.00550\ \text{mol}$$

$$\frac{1.604\ \text{g}}{0.00550\ \text{mol}} = 292\ \text{g/mol}$$

404. Given: 2.62 kg HNO_3; H_2O solvent; mass of solution = 5.91 kg

Unknown: freezing point

mass H_2O = 5.91 kg − 2.62 kg = 3.29 kg

$$2.62\ \text{kg}\ HNO_3 \times \frac{1\ \text{mol}\ HNO_3}{63.02\ \text{g}\ HNO_3} \times \frac{1000\ \text{g}}{1\ \text{kg}} = 41.6\ \text{mol}\ HNO_3 = 83.2\ \text{mol ions}$$

$$\frac{83.2\ \text{mol ions}}{3.29\ \text{kg}\ H_2O} = 25.3\ m$$

fp = normal fp + ($K_f m$) = 0.00°C + (−1.86°C/m × 25.3 m) = −47.1°C

405. Given: 0.5190 g naphthalene solvent; mass solution = 0.5959 g; fp = 74.8°C

Unknown: molar mass

mass solute = 0.5959 g − 0.5190 g = 0.0769 g

$$m = \frac{\text{fp} - \text{normal fp}}{K_f} = \frac{74.8°C - 80.2°C}{-6.94°C/m} = 0.778\ m$$

$$\frac{0.778\ \text{mol}}{1.00\ \text{kg solvent}} \times 0.5190\ \text{g solvent} \times \frac{1\ \text{kg}}{1000\ \text{g}} = 4.04 \times 10^{-4}\ \text{mol}$$

$$\frac{0.0769\ \text{g}}{4.04 \times 10^{-4}\ \text{mol}} = 190.\ \text{g/mol}$$

406. Given: 8.69 g $NaCH_3COO$; 15.00 g H_2O

Unknown: boiling point

$$8.69\ \text{g}\ NaCH_3COO \times \frac{1\ \text{mol}\ NaCH_3COO}{82.04\ \text{g}\ NaCH_3COO} \times \frac{2\ \text{mol ions}}{1\ \text{mol}\ NaCH_3COO}$$
$$= 0.212\ \text{mol ions}$$

$$\frac{0.212\ \text{mol}}{15.00\ \text{g}\ H_2O} \times \frac{1000\ \text{g}}{1\ \text{kg}} = 14.1\ m$$

bp = normal bp + ($K_b m$) = 100.00°C + (0.51°C/m × 14.1 m) = 107.2°C

407. Given: 110.5 g H_2SO_4; 225 g H_2O

Unknown: freezing point

$$110.5\ \text{g}\ H_2SO_4 \times \frac{1\ \text{mol}\ H_2SO_4}{98.09\ \text{g}\ H_2SO_4} \times \frac{3\ \text{mol ions}}{1\ \text{mol}\ H_2SO_4} = 3.38\ \text{mol ions}$$

$$\frac{3.38\ \text{mol}}{225\ \text{g}\ H_2O} \times \frac{1000\ \text{g}}{1\ \text{kg}} = 15.0\ m$$

fp = normal fp + ($K_f m$) = 0.00°C + (−1.86°C/m × 15.0 m) = −27.9°C

408. Given: empirical formula is C_8H_5; 4.04 g pyrene; 10.00 g benzene solvent; bp = 85.1°C; K_b = 2.53°C/m; normal bp = 80.1°C

Unknown: molar mass; molecular formula

$m = \dfrac{\text{bp} - \text{normal bp}}{K_b} = \dfrac{85.1°C - 80.1°C}{2.53°C/m} = 1.98\ m$

$\dfrac{1.98\ \text{mol}}{1\ \text{Kg solvent}} \times 10.00\ \text{g solvent} \times \dfrac{1\ \text{kg}}{1000\ \text{g}} = 0.0198\ \text{mol}$

$\dfrac{4.04\ \text{g}}{0.0198\ \text{mol}} = 204\ \text{g/mol}$

molar mass C_8H_5 = 101.13 g/mol

$\dfrac{204\ \text{g}}{1\ \text{mol}} \times \dfrac{1\ \text{mol}}{101.13\ \text{g}} = 2.02 \approx 2$

molecular formula = $C_{8 \times 2}H_{5 \times 2} = C_{16}H_{10}$

---

409. Given: $CaCl_2$ solute; 100.00 g H_2O; fp = −5.0°C

Unknown: mass $CaCl_2$; mass $C_6H_{12}O_6$ for same fp

$m = \dfrac{\text{fp} - \text{normal fp}}{K_f} = \dfrac{-5.0°C - 0.0°C}{-1.86°C/m} = 2.7\ m$ (ions)

$\dfrac{2.7\ \text{mol ions}}{1.0\ \text{kg}} \times \dfrac{1\ \text{mol}\ CaCl_2}{3\ \text{mol ions}} = 0.90\ m\ CaCl_2$

$\dfrac{0.90\ \text{mol}\ CaCl_2}{1.0\ \text{kg}\ H_2O} \times 100.00\ \text{g}\ H_2O \times \dfrac{1\ \text{kg}}{1000\ \text{g}} = 0.090\ \text{mol}\ CaCl_2$

$0.090\ \text{mol}\ CaCl_2 \times \dfrac{110.98\ \text{g}\ CaCl_2}{1\ \text{mol}\ CaCl_2} = 10.\ \text{g}\ CaCl_2$

Because glucose is a nonelectrolyte, m of glucose = 2.7 m

$\dfrac{2.7\ \text{mol glucose}}{1.0\ \text{kg}\ H_2O} \times 100.00\ \text{g}\ H_2O \times \dfrac{1\ \text{kg}}{1000\ \text{g}} = 0.27\ \text{mol glucose}$

$0.27\ \text{mol}\ C_6H_{12}O_6 \times \dfrac{180.18\ \text{g}\ C_6H_{12}O_6}{1\ \text{mol}\ C_6H_{12}O_6} = 49\ \text{g glucose}$

---

410. Given: empirical formula is CH_2O; 0.0866 g solute; 1.000 g ether; bp = 36.5°C

Unknown: molecular formula

$m = \dfrac{\text{bp} - \text{normal bp}}{K_b} = \dfrac{36.5°C - 34.6°C}{2.02°C/m} = 0.941\ m$

$\dfrac{0.941\ \text{mol}}{1.00\ \text{kg solvent}} \times 1.000\ \text{g solvent} \times \dfrac{1\ \text{kg}}{1000\ \text{g}} = 9.41 \times 10^{-4}\ \text{mol}$

$\dfrac{0.0866\ \text{g}}{9.41 \times 10^{-4}\ \text{mol}} = 92.0\ \text{g/mol}$ for the molecule

molar mass CH_2O = 30.03 g/mol

$\dfrac{92.0\ \text{g/mol}}{30.03\ \text{g.mol}} = 3.06$

molecular formula = $C_{1 \times 3}H_{2 \times 3}O_{1 \times 3} = C_3H_6O_3$

---

411. Given: 28.6% HCl by mass

Unknown: freezing point

$\dfrac{28.6\ \text{g HCl}}{(100.0\ \text{g} - 28.6\ \text{g})\ H_2O} \times \dfrac{1\ \text{mol HCl}}{36.46\ \text{g HCl}} \times \dfrac{2\ \text{mol ions}}{1\ \text{mol HCl}} \times \dfrac{1000\ \text{g}}{1\ \text{kg}} = 22.0\ m$ ions

fp = normal fp + ($K_f m$) = 0.00°C + (−1.86°C/m × 22.0 m) = −40.9°C

412. Given: 4.510 kg H_2O;
fp = –18.0°C;
$HOCH_2CH_2OH$ solute

Unknown: mass of solute; boiling point

$$m = \frac{\text{fp} - \text{normal fp}}{K_f} = \frac{-18.0°C - 0.0°C}{-1.86°C/m} = 9.68\ m$$

$$\frac{9.68\ \text{mol}}{1.00\ \text{kg solvent}} \times 4.510\ \text{kg solvent} = 43.7\ \text{mol}$$

$$43.7\ \text{mol}\ HOCH_2CH_2OH \times \frac{62.08\ \text{g}\ HOCH_2CH_2OH}{1\ \text{mol}\ HOCH_2CH_2OH}$$

$$= 2710\ \text{g}\ HOCH_2CH_2OH = 2.71\ \text{kg}\ HOCH_2CH_2OH$$

bp = normal bp × ($K_b m$) = 100.00°C × (0.51°C/m × 9.68 m) = 104.9°C

---

413. a. Given: 2.00 g solute; 10.00 g H_2O; fp = –4.0°C

Unknown: molality

$$\frac{\text{fp} - \text{normal fp}}{K_f} = \frac{-4.0°C - 0.0°C}{-1.86°C/m} = 2.2\ m$$

b. Given: 2.00 g solute; acetone solvent; bp = 58.9°C; normal bp = 56.00°C; K_b = 1.71°C/m

Unknown: molality

$$\frac{\text{bp} - \text{normal bp}}{K_b} = \frac{58.9°C - 56.00°C}{1.71°C/m} = 1.7\ m$$

---

414. Given: fp = –22.0°C; 100.00 g glycerol solute

Unknown: mass of H_2O

$$m = \frac{\text{fp} - \text{normal fp}}{K_f} = \frac{-22.0°C - 0.0°C}{-1.86°C/m} = 11.8\ m$$

$$100.00\ \text{g}\ C_3H_5(OH)_3 \times \frac{1\ \text{mol}\ C_3H_5(OH)_3}{92.11\ \text{g}\ C_3H_5(OH)_3} = 1.086\ \text{mol}\ C_3H_5(OH)_3$$

$$1.110\ \text{mol}\ C_3H_5(OH)_3 \times \frac{1\ \text{kg}\ H_2O}{11.8\ \text{mol}\ C_3H_5(OH)_3} \times \frac{1000\ \text{g}}{1\ \text{kg}} = 92.0\ \text{g}\ H_2O$$

---

415. Given: empirical formula is CH_2O; 0.515 g solute; 1.717 g acetic acid; fp = 8.8°C

Unknown: molar mass; molecular formula

$$m = \frac{\text{fp} - \text{normal fp}}{K_f} = \frac{8.8°C - 16.6°C}{-3.90°C/m} = 2.0\ m$$

$$\frac{2.0\ \text{mol}}{1.00\ \text{kg acetic acid}} \times 1.717\ \text{g acetic acid} \times \frac{1\ \text{kg}}{1000\ \text{g}} = 0.0034\ \text{mol}$$

$$\frac{0.515\ \text{g}}{0.0034\ \text{mol}} = 150\ \text{g/mol}$$

molar mass CH_2O = 30.03 g/mol

$$\frac{150\ \text{g/mol}}{30.03\ \text{g/mol}} = 5.0$$

molecular formula = $C_{1 \times 5}H_{2 \times 5}O_{1 \times 5} = C_5H_{10}O_5$

416. Given: empirical formula is C_2H_2O;
3.775 g solute;
12.00 g H_2O;
fp = −4.72°C

Unknown: molar mass; molecular formula

$$m = \frac{\text{fp} - \text{normal fp}}{K_f} = \frac{-4.72°C - 0.00°C}{-1.86°C/m} = 2.54\ m$$

$$\frac{2.54\ \text{mol}}{1.00\ \text{kg}\ H_2O} \times 12.00\ \text{g}\ H_2O \times \frac{1\ \text{kg}}{1000\ \text{g}} = 0.0305\ \text{mol}$$

$$\frac{3.775\ \text{g}}{0.0305\ \text{mol}} = 124\ \text{g/mol}$$

molar mass C_2H_2O = 42.04 g/mol

$$\frac{124\ \text{g/mol}}{42.04\ \text{g/mol}} = 2.95 \approx 3$$

molecular formula = $C_{2 \times 3}H_{2 \times 3}O_{1 \times 3} = C_6H_6O_3$

---

417. Given: $[OH^-] = 6.4 \times 10^{-5}$ M

Unknown: $[H_3O^+]$

$[H_3O^+][OH^-] = 1.0 \times 10^{-14}\ M^2$

$$\frac{1.0 \times 10^{-14}\ M^2}{6.4 \times 10^{-5}\ M} = 1.6 \times 10^{-10}\ M$$

---

418. Given: 7.50×10^{-4} M HNO_3

Unknown: $[H_3O^+]$; $[OH^-]$

$[H_3O^+] = 7.50 \times 10^{-4}$ M $HNO_3 \times \dfrac{1\ M\ H_3O^+}{1\ M\ HNO_3} = 7.50 \times 10^{-4}$ M

$[H_3O^+][OH^-] = 1.00 \times 10^{-14}\ M^2$

$[OH^-] = \dfrac{1.00 \times 10^{-14}\ M^2}{7.50 \times 10^{-4}\ M} = 1.33 \times 10^{-11}\ M$

---

419. Given: 0.00118 M HBr

Unknown: pH

pH = $-\log[H_3O^+] = -\log(1.18 \times 10^{-3}) = 2.928$

---

420. a. Given: $[H_3O^+] = 1.0$ M

Unknown: pH

pH = $-\log(1.0) = 0.0$

b. Given: 2.0 M HCl solution

Unknown: pH

pH = $-\log(2.0) = -0.30$

c. Given: 10. M HCl solution

Unknown: pH

pH = $-\log(10.) = -1.00$

---

421. a. Given: $[OH^-] = 1 \times 10^{-5}$ M

Unknown: pH

$[H_3O^+][OH^-] = 1 \times 10^{-14}\ M^2$

$[H_3O^+] = \dfrac{1 \times 10^{-14}\ M^2}{[OH^-]} = \dfrac{1. \times 10^{-14}\ M^2}{1 \times 10^{-5}\ M} = 1 \times 10^{-9}$ M

pH = $-\log[H_3O^+] = -\log(1 \times 10^{-9}) = 9.0$

b. Given: $[OH^-] = 5 \times 10^{-8}$ M

Unknown: pH

$[H_3O^+][OH^-] = 1 \times 10^{-14}$ M^2

$[H_3O^+] = \dfrac{1 \times 10^{-14} \text{ M}^2}{[OH^-]} = \dfrac{1 \times 10^{-14} \text{ M}^2}{5 \times 10^{-8} \text{ M}} = 2 \times 10^{-7}$ M

pH $= -\log [H_3O^+] = -\log (2 \times 10^{-7}) = 7 - 0.3 = 6.7$

c. Given: $[OH^-] = 2.90 \times 10^{-11}$ M

Unknown: pH

$[H_3O^+][OH^-] = 1 \times 10^{-14}$ M^2

$[H_3O^+] = \dfrac{1 \times 10^{-14} \text{ M}^2}{[OH^-]} = \dfrac{1.00 \times 10^{-14} \text{ M}^2}{2.90 \times 10^{-11} \text{ M}} = 3.45 \times 10^{-4}$ M

pH $= -\log [H_3O^+] = -\log (3.45 \times 10^{-4}) = 3.462$

422. Given: pH = 8.92

Unknown: pOH; (OH^-)

pH + pOH = 14

pOH = 14.00 − pH = 14.00 − 8.92 = 5.08

$[OH^-]$ = antilog (−pOH) = $1 \times 10^{-5.08} = 8.3 \times 10^{-6}$ M

423. a. Given: $[H_3O^+] = 2.51 \times 10^{-13}$ M

Unknown: pOH

$[OH^-] = \dfrac{1.00 \times 10^{-14} \text{ M}^2}{[H_3O^+]} = \dfrac{1.00 \times 1.0^{-14} \text{ M}^2}{2.51 \times 10^{-13} \text{ M}} = 3.98 \times 10^{-2}$ M

pOH $= -\log [OH^-] = -\log (3.98 \times 10^{-2}) = 1.400$

b. Given: $[H_3O^+] = 4.3 \times 10^{-3}$ M

Unknown: pOH

$[OH^-] = \dfrac{1.0 \times 10^{-14} \text{ M}^2}{[H_3O^+]} = \dfrac{1.0 \times 10^{-14} \text{ M}^2}{4.3 \times 10^{-3} \text{ M}} = 2.3 \times 10^{-12}$ M

pOH $= -\log [OH^-] = -\log (2.3 \times 10^{-12}) = 11.64$

c. Given: $[H_3O^+] = 9.1 \times 10^{-6}$ M

Unknown: pOH

$[OH^-] = \dfrac{1.0 \times 10^{-14} \text{ M}^2}{[H_3O^+]} = \dfrac{1.0 \times 10^{-14} \text{ M}^2}{9.1 \times 10^{-6} \text{ M}} = 1.1 \times 10^{-9}$ M

pOH $= -\log [OH^-] = -\log (1.1 \times 10^{-9}) = 8.96$

d. Given: $[H_3O^+] = 0.070$ M

Unknown: pOH

$[OH^-] = \dfrac{1.0 \times 10^{-14} \text{ M}^2}{[H_3O^+]} = \dfrac{1.0 \times 10^{-14} \text{ M}^2}{7.0 \times 10^{-2} \text{ M}} = 1.4 \times 10^{-13}$ M

pOH $= -\log [OH^-] = -\log (1.4 \times 10^{-13}) = 12.85$

424. Given: 3.50 g NaOH solute; V = 2.50 L

Unknown: $[OH^-]$; $[H_3O^+]$

3.50 g NaOH $\times \dfrac{1 \text{ mol NaOH}}{40.00 \text{ g NaOH}} = 0.0875$ mol NaOH

$[OH^-] = \dfrac{0.0875 \text{ mol}}{2.50 \text{ L}} = 0.0350$ M

$[H_3O^+] = \dfrac{1.00 \times 10^{-14} \text{ M}^2}{[OH^-]} = \dfrac{1.00 \times 10^{-14} \text{ M}^2}{3.50 \times 10^{-2} \text{ M}} = 2.86 \times 10^{-13}$ M

425. Given: $V_1 = 1.00$ L; $pH_1 = 12.90$; $V_2 = 2.00$ L

Unknown: pH_2

$[H_3O^+]$ = antilog (−pH) = $1 \times 10^{-12.90} = 1.3 \times 10^{-13}$ M

$M_2 = \dfrac{M_1 V_1}{V_2} = \dfrac{1.3 \times 10^{-13} \text{ M} \times 1.00 \text{ L}}{2.00 \text{ L}} = 6.5 \times 10^{-14}$ M

$pH_2 = -\log [H_3O^+] = -\log (6.5 \times 10^{-14}) = 13.19$

426. Unknown: $[H_3O^+]$; $[OH^-]$

$[OH^-] = 5 \times 10^{-2}$ M

a. Given: 0.05 M NaOH

$[H_3O^+] = \dfrac{1 \times 10^{-14} \text{ M}^2}{[OH^-]} = \dfrac{1 \times 10^{-14} \text{ M}^2}{5 \times 10^{-2} \text{ M}} = 2 \times 10^{-13}$ M

b. Given: 0.0025 M H_2SO_4

$[H_3O^+] = 0.0025 \text{ M } H_2SO_4 \times \dfrac{2 \text{ M } H_3O^+}{1 \text{ M } H_2SO_4} = 0.0050 \text{ M} = 5.0 \times 10^{-3}$ M

$[OH^-] = \dfrac{1.0 \times 10^{-14} \text{ M}^2}{[H_3O^+]} = \dfrac{1.0 \times 10^{-14} \text{ M}^2}{5.0 \times 10^{-13} \text{ M}} = 2.0 \times 10^{-12}$ M

c. Given: 0.013 M LiOH

$[OH^-] = 1.3 \times 10^{-2}$ M

$[H_3O^+] = \dfrac{1.0 \times 10^{-14} \text{ M}^2}{1.3 \times 10^{-2} \text{ M}} = 7.7 \times 10^{-13}$ M

d. Given: 0.150 M HNO_3

$[H_3O^+] = 1.50 \times 10^{-1}$ M

$[OH^-] = \dfrac{1.00 \times 10^{-14} \text{ M}^2}{1.50 \times 10^{-1} \text{ M}} = 6.67 \times 10^{-14}$ M

e. Given: 0.0200 M Ca(OH)$_2$

$[OH^-] = 2[Ca(OH)_2] = 4.00 \times 10^{-2}$ M

$[H_3O^+] = \dfrac{1.00 \times 10^{-14} \text{ M}^2}{4.00 \times 10^{-2} \text{ M}} = 2.50 \times 10^{-13}$ M

f. Given: 0.390 M $HClO_4$

$[H_3O^+] = 3.90 \times 10^{-1}$ M

$[OH^-] = \dfrac{1.00 \times 10^{-14} \text{ M}^2}{3.90 \times 10^{-1} \text{ M}} = 2.56 \times 10^{-14}$ M

427. Unknown: pH

a. Given: $[H_3O^+] = 2 \times 10^{-13}$ M

$pH = -\log(2 \times 10^{-13}) = 12.7$

b. Given: $[H_3O^+] = 5.0 \times 10^{-3}$ M

$pH = -\log(5.0 \times 10^{-3}) = 2.30$

c. Given: $[H_3O^+] = 7.7 \times 10^{-13}$ M

$pH = -\log(7.7 \times 10^{-13}) = 12.11$

d. Given: $[H_3O^+] = 1.50 \times 10^{-1}$ M

$pH = -\log(1.50 \times 10^{-1}) = 0.824$

e. Given: $[H_3O^+] = 2.50 \times 10^{-13}$ M

$pH = -\log(2.50 \times 10^{-13}) = 12.602$

f. Given: $[H_3O^+] = 3.90 \times 10^{-1}$ M

$pH = -\log(3.90 \times 10^{-1}) = 0.409$

428. Given: 0.160 M KOH

Unknown: $[H_3O^+]$; $[OH^-]$

$[OH^-] = [KOH] = 1.60 \times 10^{-1}$ M

$[H_3O^+] = \dfrac{1.00 \times 10^{-14} \text{ M}^2}{1.60 \times 10^{-1} \text{ M}} = 6.25 \times 10^{-14}$ M

429. Given: pH = 12.9; NaOH solution

Unknown: molarity

pOH = 14.0 − 12.9 = 1.1

[NaOH] = [OH⁻] = antilog (− pOH) = $1.0 \times 10^{-1.1}$ = 0.08 M

---

430. Given: 0.001 25 M HBr; V_1 = 175 mL; V_2 = 3.00 L; $M_1V_1 = M_2V_2$

Unknown: pH before and after dilution

$M_2 = \dfrac{M_1V_1}{V_2} = \dfrac{0.001\ 25\ \text{M} \times 175\ \text{mL}}{3.00\ \text{L}} \times \dfrac{1\ \text{L}}{1000\ \text{mL}} = 7.29 \times 10^{-5}$ M

[HBr] = [H₃O⁺]

pH₁ = − log [H₃O⁺] = − log (1.25×10^{-3}) = 2.903

pH₂ = − log [H₃O⁺] = − log (7.29×10^{-5}) = 4.137

---

431. Given: NaOH solutions of 0.0001 M and 0.0005 M

Unknown: pH for both solutions

[NaOH] = [OH⁻] = 1×10^{-4} M; pOH = − log (1×10^{-4}) = 4.0

pH = 14.0 − 4.0 = 10.0

[NaOH] = [OH⁻] = 5×10^{-4} M; pOH = − log (5×10^{-4}) = 3.3

pH = 14.0 − 3.3 = 10.7

---

432. Given: V_1 = 15.0 mL; M_1 = 1.0 M HCl; V_2 = 20.0 mL; M_2 = 0.50 M HNO₃; V_F = 1.25 L

a. Unknown: [H₃O⁺] and [OH⁻] of final solution

Before dilution: 15.0 mL × $\dfrac{1.0\ \text{mol}}{1\ \text{L}}$ × $\dfrac{1\ \text{L}}{1000\ \text{mL}}$ + 20.0 mL × $\dfrac{0.50\ \text{mol}}{1\ \text{L}}$ × $\dfrac{1\ \text{L}}{1000\ \text{mL}}$ = 0.0150 mol + 0.0100 mol = 0.0250 mol

[H₃O⁺] = $\dfrac{0.0250\ \text{mol}}{(15.0 + 20.0)\ \text{mL}} \times \dfrac{1000\ \text{mL}}{1\ \text{L}}$ = 0.714 M

After dilution: [H₃O⁺] = $M_F = \dfrac{0.714\ \text{M} \times 0.035\ \text{L}}{1.25\ \text{L}}$ = 0.020 M

[OH⁻] = $\dfrac{1.0 \times 10^{-14}\ \text{M}^2}{2.0 \times 10^{-2}\ \text{M}} = 5.0 \times 10^{-13}$ M

b. Unknown: pH of final solution

pH = − log [H₃O⁺] = − log (2.0×10^{-2}) = 1.70

---

433. a. Given: 0.001 57 M HNO₃

Unknown: pH

[H₃O⁺] = [HNO₃] = 1.57×10^{-3} M

pH = − log [H₃O⁺] = − log (1.57×10^{-3}) = 2.804

b. Given: V_1 = 500.0 mL; M_1 = 0.001 57 M; V_2 = 447.0 mL

Unknown: pH at V_2

$M_2 = \dfrac{M_1V_1}{V_2} = \dfrac{1.57 \times 10^{-3}\ \text{M} \times 500.0\ \text{mL}}{447.0\ \text{mL}} = 1.76 \times 10^{-3}$ M

pH = − log [H₃O⁺] = − log (1.76×10^{-3}) = 2.754

---

434. Given: [H₃O⁺] = 0.00035 M

Unknown: [OH⁻]

[OH⁻] = $\dfrac{1.0 \times 10^{-14}\ \text{M}^2}{3.5 \times 10^{-4}\ \text{M}} = 2.9 \times 10^{-11}$ M

435. Given: NaOH solute;
$pH_1 = 12.14$;
$V_1 = 50.00$ mL;
$V_2 = 2.000$ L

Unknown: pH_2

$pOH_1 = 14.00 - pH = 14.00 - 12.14 = 1.86$

$[OH^-]_1 = $ antilog $(-pOH) = 1.0 \times 10^{-1.86} = 1.4 \times 10^{-2}$ M

$[OH^-]_2 = \dfrac{[OH^-]_1 V_1}{V_2} = \dfrac{1.4 \times 10^{-2} \text{ M} \times 50.00 \text{ mL}}{2.000 \text{ L}} \times \dfrac{1 \text{ L}}{1000 \text{ mL}} = 3.5 \times 10^{-4}$ M

$pOH_2 = -\log(3.5 \times 10^{-4}) = 3.46$

$pH_2 = 14.00 - 3.46 = 10.54$

---

436. Given: pH = 4.0

Unknown: $[H_3O^+]$; $[OH^-]$

$[H_3O^+] = $ antilog $(-pH) = $ antilog $(-4.0) = 1 \times 10^{-4}$ M

$[OH^-] = \dfrac{1 \times 10^{-14} \text{ M}^2}{1 \times 10^{-4} \text{ M}} = 1 \times 10^{-10}$ M

---

437. Given: 0.000 460 M $Ca(OH)_2$ solution

Unknown: pH

$[OH^-] = 4.60 \times 10^{-4}$ M $Ca(OH)_2 \times \dfrac{2 \text{ M OH}^-}{1 \text{ M Ca(OH)}_2} = 9.20 \times 10^{-4}$ M

$[H_3O^+] = \dfrac{1.00 \times 10^{-14} \text{ M}^2}{9.20 \times 10^{-4} \text{ M}} = 1.09 \times 10^{-11}$ M

$pH = -\log(1.09 \times 10^{-11}) = 10.963$

---

438. Given: $Sr(OH)_2$ solute; pH = 11.4; 1.00 L

Unknown: mass $Sr(OH)_2$

$pOH = 14.0 - 11.4 = 2.6$

$[OH^-] = $ antilog $(-pOH) = $ antilog $(-2.6) = 10^{-2.6} = 3 \times 10^{-3}$ M

$[Sr(OH)_2] = 3 \times 10^{-3}$ M $OH^- \times \dfrac{1 \text{ M Sr(OH)}_2}{2 \text{ M OH}^-} = 2 \times 10^{-3}$ M

$\dfrac{2 \times 10^{-3} \text{ mol}}{1 \text{ L}} \times 1.00 \text{ L} = 2 \times 10^{-3}$ mol

2×10^{-3} mol $Sr(OH)_2 \times \dfrac{121.64 \text{ g Sr(OH)}_2}{1 \text{ mol Sr(OH)}_2} = 0.2$ g $Sr(OH)_2$

---

439. Given: NH_3 solute; pH = 11.00

Unknown: $[H_3O^+]$; $[OH^-]$

$[H_3O^+] = $ antilog $(-11.00) = 1.0 \times 10^{-11}$ M

$[OH^-] = \dfrac{1.0 \times 10^{-14} \text{ M}^2}{1.0 \times 10^{-11} \text{ M}} = 1.0 \times 10^{-3}$ M

---

440. a. Given: 1.0 M CH_3COOH; pH = 2.40

Unknown: percent ionization

$[H_3O^+] = $ antilog $(-2.40) = 1.00 \times 10^{-2.40} = 3.98 \times 10^{-3}$ M

$\dfrac{3.98 \times 10^{-3} \text{ M}}{1.0 \text{ M}} \times 100 = 0.40\%$ ionized

b. Given: 0.10 M CH_3COOH; pH = 2.90

Unknown: percent ionization

$[H_3O^+] = $ antilog $(-2.90) = 1.00 \times 10^{-2.90} = 1.26 \times 10^{-3}$ M

$\dfrac{1.26 \times 10^{-3} \text{ M}}{0.10 \text{ M}} \times 100 = 1.3\%$ ionized

c. Given: 0.010 M CH_3COOH; pH = 3.40

Unknown: percent ionization

$[H_3O^+] = $ antilog $(-3.40) = 1.00 \times 10^{-3.40} = 3.98 \times 10^{-4}$ M

$\dfrac{3.98 \times 10^{-4} \text{ M}}{1.0 \times 10^{-2} \text{ M}} \times 100 = 4.0\%$

441. Given: 5.00 g HNO_3; 2.00 L

Unknown: pH

$5.00 \text{ g } HNO_3 \times \dfrac{1 \text{ mol } HNO_3}{63.02 \text{ g } HNO_3} = 0.0793 \text{ mol } HNO_3$

$[H_3O^+] = [HNO_3] = \dfrac{0.0793 \text{ mol}}{2.00 \text{ L}} = 0.0396 \text{ M}$

$pH = -\log(0.0396) = 1.402$

---

442. Given: stock pH = 1.50; HCl solute

Unknown: pH of diluted solution

$M_S = [H_3O^+] = [HCl] = \text{antilog}(-1.50) = 1.0 \times 10^{-1.50} \text{ M} = 3.16 \times 10^{-2} \text{ M}$

a. Given: $V_S = 1.00$ mL; $V_D = 1000.$ mL

$M_D = \dfrac{M_S V_S}{V_D} = \dfrac{3.16 \times 10^{-2} \text{ M} \times 1.00 \text{ mL}}{1000. \text{ mL}} = 3.16 \times 10^{-5} \text{ M}$

$pH = -\log(3.16 \times 10^{-5}) = 4.50$

b. Given: $V_S = 25.00$ mL; $V_D = 200.$ mL

$M_D = \dfrac{M_S V_S}{V_D} = \dfrac{3.16 \times 10^{-2} \text{ M} \times 25.00 \text{ mL}}{200. \text{ mL}} = 3.95 \times 10^{-3} \text{ M}$

$pH = -\log(3.95 \times 10^{-3}) = 2.40$

c. Given: $V_S = 18.83$ mL; $V_D = 4.000$ L

$M_D = \dfrac{M_S V_S}{V_D} = \dfrac{3.16 \times 10^{-2} \text{ M} \times 18.83 \text{ mL}}{4.000 \text{ L}} \times \dfrac{1 \text{ L}}{1000 \text{ mL}} = 1.49 \times 10^{-4} \text{ M}$

$pH = -\log(1.49 \times 10^{-4}) = 3.83$

d. Given: $V_S = 1.50$ L; $V_D = 20.0$ kL

$M_D = \dfrac{M_S V_S}{V_D} = \dfrac{3.16 \times 10^{-2} \text{ M} \times 1.50 \text{ L}}{20.0 \text{ kL}} \times \dfrac{1 \text{ kL}}{1000 \text{ L}} = 2.37 \times 10^{-6} \text{ M}$

$pH = -\log(2.37 \times 10^{-6}) = 5.63$

---

443. Given: $[H_3O^+] = 10\,000\,[OH^-]$; aqueous solution

Unknown: $[H_3O^+]$; $[OH^-]$

$[H_3O^+][OH^-] = 10\,000\,[OH^-][OH^-] = 1 \times 10^{-14}\,M^2$

$[OH^-]^2 = \dfrac{1 \times 10^{-14}\,M^2}{10^4} = 1 \times 10^{-18}\,M^2$

$[OH^-] = 1 \times 10^{-9}\,M$

$[H_3O^+] = 10^4\,[OH^-] = 10^4 \times 1 \times 10^{-9}\,M = 1 \times 10^{-5}$

---

444. Given: KOH solute; pH = 12.90; acid reacts with half of OH^-

Unknown: resulting pH

$pOH = 14.00 - pH = 14.00 - 12.90 = 1.10$

$[OH^-] = \text{antilog}(-1.10) = 1.0 \times 10^{-1.10} = 0.079\,M$

after reaction:

$[OH^-] = \dfrac{0.079\,M}{2} = 0.040\,M$

$[H_3O^+] = \dfrac{1.0 \times 10^{-14}\,M^2}{4.0 \times 10^2\,M} = 2.5 \times 10^{-13}\,M$

$pH = -\log(2.5 \times 10^{-13}) = 12.60$

---

445. Given: HCl solute; pH = 1.70

Unknown: $[H_3O^+]$; $[HCl]$

$[H_3O^+] = \text{antilog}(-1.70) = 1.0 \times 10^{-1.70} = 0.020\,M$

$[HCl] = [H_3O^+] = 0.020\,M$

446. Given: $Ca(OH)_2$ solute:
pH = 10.80

Unknown: molarity

pOH = 14.00 − pH = 14.00 − 10.80 = 3.20

$[OH^-]$ = antilog (−3.20) = $1.0 \times 10^{-3.20}$ = 6.3×10^{-4} M

$[Ca(OH)_2] = 6.3 \times 10^{-4}$ M $OH^- \times \dfrac{1 \text{ M Ca(OH)}_2}{2 \text{ M OH}^-} = 3.2 \times 10^{-4}$ M

447. Given: 1.00 M stock HCl

Unknown: pH

pH = − log (1.00×10^0) = 0.000

Given: pH = 4.00; 1.00 L stock HCl

Unknown: V_D

$[H_3O^+]$ = antilog (−4.00) = $1.0 \times 10^{-4.00}$ = 1.0×10^{-4} M

$V_D = \dfrac{M_S V_S}{M_D} = \dfrac{1.00 \text{ M} \times 1.00 \text{ L}}{1.0 \times 10^{-4} \text{ M}} = 1.0 \times 10^{4.}$ L = 10. kL

Given: 1.00 L of pH 4.00

Unknown: V at pH 6.00

$[H_3O^+]$ at pH 4.00 = 1.0×10^{-4} M

$[H_3O^+]$ at pH 6.00 = antilog (−6.00) = $1.0 \times 10^{-6.00}$ = 1.0×10^{-6} M

$V = \dfrac{1.0 \times 10^{-4} \text{ M} \times 1.00 \text{ L}}{1.0 \times 10^{-6} \text{ M}} = 1.0 \times 10^2$ L

Given: 1.00 L of pH 4.00

Unknown: V at pH 8.00

$[H_3O^+]$ at pH 4.00 = 1.00×10^{-4} M

$[H_3O^+]$ at pH 8.00 = antilog (−8.00) = $1.0 \times 10^{-8.00}$ = 1.0×10^{-8} M

$V = \dfrac{1.0 \times 10^{-4} \text{ M} \times 1.00 \text{ L}}{1.0 \times 10^{-8} \text{ M}} = 1.0 \times 10^4$ L = 10. kL

448. Given: pH = 1.28; 1.00 L $HClO_3$

Unknown: moles NaOH to react; mass NaOH

$HClO_3 + NaOH \rightarrow NaClO_3 + H_2O$

$[H_3O^+]$ = antilog (−1.28) = $1.0 \times 10^{-1.28}$ = 5.2×10^{-2} M

$\dfrac{5.2 \times 10^{-2} \text{ mol}}{1.0 \text{ L}} \times 1.00$ L $HClO_3 = 5.2 \times 10^{-2}$ mol $HClO_3$

= 5.2×10^{-2} mol NaOH

0.052 mol NaOH $\times \dfrac{40.00 \text{ g NaOH}}{1.0 \text{ mol NaOH}}$ = 2.1 g NaOH

449. Given: NH_3 solute; pH = 11.90; 1.00 L NH_3 solution

Unknown: moles HCl to react

pOH = 14.00 − 11.90 = 2.10

$[OH^-]$ = antilog (−2.10) = $1.0 \times 10^{-2.10}$ = 7.9×10^{-3} M

$\dfrac{7.9 \times 10^{-3} \text{ mol}}{1.0 \text{ L}} \times 1.00$ L = 7.9×10^{-3} mol OH^- = 7.9×10^{-3} mol HCl

450. Given: pH = 3.15

Unknown: $[H_3O^+]$; $[OH^-]$

$[H_3O^+]$ = antilog (−3.15) = $1.0 \times 10^{-3.15}$ = 7.1×10^{-4} M

$[OH^-] = \dfrac{1.0 \times 10^{-14} \text{ M}^2}{7.1 \times 10^{-4} \text{ M}} = 1.4 \times 10^{-11}$ M

451. Given: 20.00 mL HBr;
20.05 mL of
0.1819 M NaOH

Unknown: molarity HBr

HBr + NaOH → NaBr + H$_2$O

$$\frac{0.1819 \text{ mol NaOH}}{\text{L}} \times 20.05 \text{ mL} \times \frac{1 \text{ L}}{1000 \text{ mL}} = 3.647 \times 10^{-3} \text{ mol NaOH}$$

$$3.647 \times 10^{-3} \text{ mol NaOH} \times \frac{1 \text{ mol HBr}}{1 \text{ mol NaOH}} = 3.647 \times 10^{-3} \text{ mol HBr}$$

$$\frac{3.647 \times 10^{-3} \text{ mol HBr}}{20.00 \text{ mL}} \times \frac{1000 \text{ mL}}{1 \text{ L}} = 0.1824 \text{ M HBr}$$

452. Given: 15.00 mL CH$_3$COOH;
22.70 mL of 0.550 M NaOH

Unknown: molarity of CH$_3$COOH

CH$_3$COOH + NaOH → NaCH$_3$COO + H$_2$O

$$\frac{0.550 \text{ mol NaOH}}{\text{L}} \times 22.70 \text{ mL} \times \frac{1 \text{ L}}{1000 \text{ mL}} = 1.25 \times 10^{-2} \text{ mol NaOH}$$

$$1.25 \times 10^{-2} \text{ mol NaOH} \times \frac{1 \text{ mol CH}_3\text{COOH}}{1 \text{ mol NaOH}} = 1.25 \times 10^{-2} \text{ mol CH}_3\text{COOH}$$

$$\frac{1.25 \times 10^{-2} \text{ mol CH}_3\text{COOH}}{15.00 \text{ mL}} \times \frac{1000 \text{ mL}}{1 \text{ L}} = 0.833 \text{ M CH}_3\text{COOH}$$

453. Given: 20.00 mL Sr(OH)$_2$;
43.03 mL of 0.1159 M HCl

Unknown: molarity of Sr(OH)$_2$ solution

Sr(OH)$_2$ + 2HCl → SrCl$_2$ + 2H$_2$O

$$\frac{0.1159 \text{ mol HCl}}{\text{L}} \times 43.03 \text{ mL} \times \frac{1 \text{ L}}{1000 \text{ mL}} = 4.987 \times 10^{-3} \text{ mol HCl}$$

$$4.987 \times 10^{-3} \text{ mol HCl} \times \frac{1 \text{ mol Sr(OH)}_2}{2 \text{ mol HCl}} = 2.494 \times 10^{-3} \text{ mol Sr(OH)}_2$$

$$\frac{2.494 \times 10^{-3} \text{ mol Sr(OH)}_2}{20.00 \text{ mL}} \times \frac{1000 \text{ mL}}{1 \text{ L}} = 0.1247 \text{ M Sr(OH)}_2$$

454. Given: 35.00 mL NH$_3$ solution;
54.95 mL of 0.400 M H$_2$SO$_4$

Unknown: molarity of NH$_3$ solution

2NH$_3$ + H$_2$SO$_4$ → (NH$_4$)$_2$SO$_4$

$$\frac{0.400 \text{ mol H}_2\text{SO}_4}{\text{L}} \times 54.95 \text{ mL} \times \frac{1 \text{ L}}{1000 \text{ mL}} = 2.20 \times 10^{-2} \text{ mol H}_2\text{SO}_4$$

$$2.20 \times 10^{-2} \text{ mol H}_2\text{SO}_4 \times \frac{2 \text{ mol NH}_3}{1 \text{ mol H}_2\text{SO}_4} = 4.40 \times 10^{-2} \text{ mol NH}_3$$

$$\frac{4.40 \times 10^{-2} \text{ mol NH}_3}{35.00 \text{ mL}} \times \frac{1000 \text{ mL}}{1 \text{ L}} = 1.26 \text{ M NH}_3$$

455. Given: 28.25 mL of 0.218 M NaOH; 2.000 g acetic acid diluted to 100.00 mL; 20.00 mL acetic acid

Unknown: % acetic acid in stock solution

$CH_3COOH + NaOH \rightarrow NaCH_3COO + H_2O$

$$\frac{0.218 \text{ mol NaOH}}{L} \times 28.25 \text{ mL} \times \frac{1 \text{ L}}{1000 \text{ mL}} = 6.16 \times 10^{-3} \text{ mol NaOH}$$

$$6.16 \times 10^{-3} \text{ mol NaOH} \times \frac{1 \text{ mol } CH_3COOH}{1 \text{ mol NaOH}} = 6.16 \times 10^{-3} \text{ mol } CH_3COOH$$

$$\frac{6.16 \times 10^{-3} \text{ mol } CH_3COOH}{20.00 \text{ mL}} \times \frac{1000 \text{ mL}}{1 \text{ L}} = 0.308 \text{ M } CH_3COOH$$

$$\frac{2.000 \text{ g } CH_3COOH}{100.00 \text{ mL}} \times \frac{1000 \text{ mL}}{1 \text{ L}} \times \frac{1 \text{ mol } CH_3COOH}{60.06 \text{ g } CH_3COOH}$$

$$= 0.3333 \text{ M } CH_3COOH$$

$$\frac{0.308 \text{ M}}{0.3333 \text{ M}} \times 100 = 92.5\% \text{ } CH_3COOH$$

---

456. Given: 9.709 g Na_2CO_3 diluted to 1.0000 L; 10.00 mL Na_2CO_3 solution; 16.90 mL of 0.1022 M HCl

Unknown: percentage of Na_2CO_3

$Na_2CO_3 + 2HCl \rightarrow 2NaCl + H_2O + CO_2$

$$\frac{0.1022 \text{ mol HCl}}{L} \times 16.90 \text{ mL} \times \frac{1 \text{ L}}{1000 \text{ mL}} = 1.727 \times 10^{-3} \text{ mol HCl}$$

$$1.727 \times 10^{-3} \text{ mol HCl} \times \frac{1 \text{ mol } Na_2CO_3}{2 \text{ mol HCl}} = 8.635 \times 10^{-4} \text{ mol } Na_2CO_3$$

$$\frac{8.635 \times 10^{-4} \text{ mol } Na_2CO_3}{10.00 \text{ mL}} \times \frac{1000 \text{ mL}}{1 \text{ L}} = 8.635 \times 10^{-2} \text{ M } Na_2CO_3$$

$$\frac{9.709 \text{ g } Na_2CO_3}{1.0000 \text{ L}} \times \frac{1 \text{ mol } Na_2CO_3}{105.99 \text{ g } Na_2CO_3} = 9.160 \times 10^{-2} \text{ M } Na_2CO_3$$

$$\frac{8.635 \times 10^{-2} \text{ M } Na_2CO_3}{9.160 \times 10^{-2} \text{ M } Na_2CO_3} \times 100 = 94.27\% \text{ } Na_2CO_3$$

---

457. Given: 50.00 mL KOH; 27.87 mL of 0.8186 M HCl

Unknown: molarity of KOH

$KOH + HCl \rightarrow KCl + H_2O$

$$\frac{0.8186 \text{ mol HCl}}{L} \times 27.87 \text{ mL} \times \frac{1 \text{ L}}{1000 \text{ mL}} = 2.281 \times 10^{-2} \text{ mol HCl}$$

$$2.281 \times 10^{-2} \text{ mol HCl} \times \frac{1 \text{ mol KOH}}{1 \text{ mol HCl}} = 2.281 \times 10^{-2} \text{ mol KOH}$$

$$\frac{2.281 \times 10^{-2} \text{ mol KOH}}{50.00 \text{ mL}} \times \frac{1000 \text{ mL}}{1 \text{ L}} = 0.4562 \text{ M KOH}$$

---

458. Given: 15.00 mL CH_3COOH; 34.13 mL of 0.9940 M NaOH

Unknown: molarity of CH_3COOH

$CH_3COOH + NaOH \rightarrow NaCH_3COO + H_2O$

$$\frac{0.9940 \text{ mol NaOH}}{L} \times 34.13 \text{ mL} \times \frac{1 \text{ L}}{1000 \text{ mL}} = 3.393 \times 10^{-2} \text{ mol NaOH}$$

$$3.393 \times 10^{-2} \text{ mol NaOH} \times \frac{1 \text{ mol } CH_3COOH}{1 \text{ mol NaOH}}$$

$$= 3.393 \times 10^{-2} \text{ mol } CH_3COOH$$

$$\frac{3.393 \times 10^{-2} \text{ mol } CH_3COOH}{15.00 \text{ mL}} \times \frac{1000 \text{ mL}}{1 \text{ L}} = 2.262 \text{ M } CH_3COOH$$

459. Given: 12.00 mL NH_3 solution; 19.48 mL of 1.499 M HNO_3

Unknown: molarity of NH_3 solution

$NH_3 + HNO_3 \rightarrow NH_4NO_3$

$$\frac{1.499 \text{ mol } HNO_3}{L} \times 19.48 \text{ mL} \times \frac{1 \text{ L}}{1000 \text{ mL}} = 2.920 \times 10^{-2} \text{ mol } HNO_3$$

$$2.920 \times 10^{-2} \text{ mol } HNO_3 \times \frac{1 \text{ mol } NH_3}{1 \text{ mol } HNO_3} = 2.920 \times 10^{-2} \text{ mol } NH_3$$

$$\frac{2.920 \times 10^{-2} \text{ mol } NH_3}{12.00 \text{ mL}} \times \frac{1000 \text{ mL}}{1 \text{ L}} = 2.433 \text{ M } NH_3$$

460. a. Given: 1 mol acid : 1 mol base; $M_A = M_B$; 20.00 mL base

Unknown: V of acid

20.00 mL base because $M \times V$ provides number of moles, which are in a 1 : 1 ratio

b. Given: $M_A = 2M_B$; 20.00 mL base

Unknown: V of acid

$M_A \times V_A = M_B \times 20.00 \text{ mL}$

$2M_B \times V_A = M_B \times 20.00 \text{ mL}$

$$V_A = \frac{M_B \times 20.00 \text{ mL base}}{2M_B} = 10.00 \text{ mL}$$

c. Given: $M_B = 4M_A$; 20.00 mL base

Unknown: V of acid

$M_A \times V_A = M_B \times 20.00 \text{ mL}$

$M_A \times V_A = 4M_A \times 20.00 \text{ mL}$

$$V_A = \frac{4M_A \times 20.00 \text{ mL}}{M_A} = 80.00 \text{ mL}$$

461. Given: 10.00 mL stock HF diluted to 500.00 mL; 20.00 mL HF; 13.51 mL of 0.1500 M NaOH

Unknown: molarity of stock HF

$HF + NaOH \rightarrow NaF + H_2O$

$$\frac{0.1500 \text{ mol NaOH}}{L} \times 13.51 \text{ mL} \times \frac{1 \text{ L}}{1000 \text{ mL}} = 2.026 \times 10^{-3} \text{ mol NaOH}$$

$$2.026 \times 10^{-3} \text{ mol NaOH} \times \frac{1 \text{ mol HF}}{1 \text{ mol NaOH}} = 2.026 \times 10^{-3} \text{ mol HF}$$

$$\frac{2.026 \times 10^{-3} \text{ mol HF}}{20.00 \text{ mL}} \times \frac{1000 \text{ mL}}{1 \text{ L}} = 0.1013 \text{ M HF}$$

$$\frac{0.1013 \text{ mol HF}}{L} \times 500.00 \text{ mL} \times \frac{1 \text{ L}}{1000 \text{ mL}} = 0.05065 \text{ mol HF}$$

$$\frac{0.05065 \text{ mol HF}}{10.00 \text{ mL}} \times \frac{1000 \text{ mL}}{L} = 5.065 \text{ M HF}$$

462. Given: 16.22 mL of 0.5030 M KOH; 18.41 mL diprotic acid

Unknown: molarity of acid

$$\frac{0.5030 \text{ mol KOH}}{L} \times 16.22 \text{ mL} \times \frac{1 \text{ L}}{1000 \text{ mL}} = 8.159 \times 10^{-3} \text{ mol KOH}$$

$$8.159 \times 10^{-3} \text{ mol KOH} \times \frac{1 \text{ mol acid}}{2 \text{ mol KOH}} = 4.080 \times 10^{-3} \text{ mol acid}$$

$$\frac{4.080 \times 10^{-3} \text{ mol acid}}{18.41 \text{ mL}} \times \frac{1000 \text{ mL}}{1 \text{ L}} = 0.2216 \text{ M acid}$$

463. Given: 42.27 mL of 1.209 M NaOH; 25.00 mL H_2SO_4

Unknown: molarity of the H_2SO_4

$H_2SO_4 + 2NaOH \rightarrow Na_2SO_4 + 2H_2O$

$$\frac{1.209 \text{ mol NaOH}}{L} \times 42.27 \text{ mL} \times \frac{1 \text{ L}}{1000 \text{ mL}} = 5.110 \times 10^{-2} \text{ mol NaOH}$$

$$5.110 \times 10^{-2} \text{ mol NaOH} \times \frac{1 \text{ mol } H_2SO_4}{2 \text{ mol NaOH}} = 2.555 \times 10^{-2} \text{ mol } H_2SO_4$$

$$\frac{2.555 \times 10^{-2} \text{ mol } H_2SO_4}{25.00 \text{ mL}} \times \frac{1000 \text{ mL}}{1 \text{ L}} = 1.022 \text{ M } H_2SO_4$$

464. Given: 1 mol acid: 1 mol base; 0.7025 g $KHC_8H_4O_4$; 20.18 mL KOH

Unknown: molarity of KOH

$$0.7025 \text{ g } KHC_8H_4O_4 \times \frac{1 \text{ mol } KHC_8H_4O_4}{204.23 \text{ g } KHC_8H_4O_4}$$

$$= 3.440 \times 10^{-3} \text{ mol } KHC_8H_4O_4$$

$$3.440 \times 10^{-3} \text{ mol } KHC_8H_4O_4 \times \frac{1 \text{ mol KOH}}{1 \text{ mol } KHC_8H_4O_4} = 3.440 \times 10^{-3} \text{ mol KOH}$$

$$\frac{3.440 \times 10^{-3} \text{ mol KOH}}{20.18 \text{ mL}} \times \frac{1000 \text{ mL}}{1 \text{ L}} = 0.1705 \text{ M KOH}$$

465. Given: 20.00 mL triprotic acid; 17.03 mL of 2.025 M NaOH

Unknown: molarity of acid

$$\frac{2.025 \text{ mol NaOH}}{L} \times 17.03 \text{ mL} \times \frac{1 \text{ L}}{1000 \text{ mL}} = 3.449 \times 10^{-2} \text{ mol NaOH}$$

$$3.449 \times 10^{-2} \text{ mol NaOH} \times \frac{1 \text{ mol acid}}{3 \text{ mol NaOH}} = 1.150 \times 10^{-2} \text{ mol acid}$$

$$\frac{1.150 \times 10^{-2} \text{ mol acid}}{20.00 \text{ mL}} \times \frac{1000 \text{ mL}}{1 \text{ L}} = 0.5750 \text{ M acid}$$

466. Given: 41.04 mL KOH; 21.65 mL of 0.6515 M HNO_3

Unknown: molarity of KOH

$KOH + HNO_3 \rightarrow KNO_3 + H_2O$

$$\frac{0.6515 \text{ mol } HNO_3}{L} \times 21.65 \text{ mL} \times \frac{1 \text{ L}}{1000 \text{ mL}} = 1.410 \times 10^{-2} \text{ mol } HNO_3$$

$$1.410 \times 10^{-2} \text{ mol } HNO_3 \times \frac{1 \text{ mol KOH}}{1 \text{ mol } HNO_3} = 1.410 \times 10^{-2} \text{ mol KOH}$$

$$\frac{1.410 \times 10^{-2} \text{ mol KOH}}{41.04 \text{ mL}} \times \frac{1000 \text{ mL}}{1 \text{ L}} = 0.3436 \text{ M KOH}$$

467. Given: 20.00 mL of 2.00 M H_2SO_4; 1.85 M NaOH

Unknown: V of NaOH

$H_2SO_4 + 2NaOH \rightarrow Na_2SO_4 + 2H_2O$

$$\frac{2.00 \text{ mol } H_2SO_4}{L} \times 20.00 \text{ mL} \times \frac{1 \text{ L}}{1000 \text{ mL}} = 4.00 \times 10^{-2} \text{ mol } H_2SO_4$$

$$4.00 \times 10^{-2} \text{ mol } H_2SO_4 \times \frac{2 \text{ mol NaOH}}{1 \text{ mol } H_2SO_4} = 8.00 \times 10^{-2} \text{ mol NaOH}$$

$$\frac{8.00 \times 10^{-2} \text{ mol NaOH}}{1.85 \text{ mol/L}} = 0.0432 \text{ L NaOH} = 43.2 \text{ mL NaOH}$$

468. Given: 0.5200 M H_2SO_4; 100.00 mL of 0.1225 M $Sr(OH)_2$

Unknown: V of H_2SO_4

$H_2SO_4 + Sr(OH)_2 \rightarrow SrSO_4 + 2H_2O$

$$\frac{0.1225 \text{ mol Sr(OH)}_2}{\text{L}} \times 100.00 \text{ mL} \times \frac{1 \text{ L}}{1000 \text{ mL}} = 1.225 \times 10^{-2} \text{ mol Sr(OH)}_2$$

$$1.225 \times 10^{-2} \text{ mol Sr(OH)}_2 \times \frac{1 \text{ mol } H_2SO_4}{1 \text{ mol Sr(OH)}_2} = 1.225 \times 10^{-2} \text{ mol } H_2SO_4$$

$$\frac{1.225 \times 10^{-2} \text{ mol } H_2SO_4}{0.5200 \text{ mol/L}} = 0.02356 \text{ L } H_2SO_4 = 23.56 \text{ mL } H_2SO_4$$

469. Given: 4.005 g KOH in 200.00 mL solution; 25.00 mL KOH; 19.93 mL of 0.4388 M HCl

Unknown: moles KOH in 4.005 g; mass KOH; percent KOH

$KOH + HCl \rightarrow KCl + H_2O$

$$\frac{0.4388 \text{ mol HCl}}{\text{L}} \times 19.93 \text{ mL} \times \frac{1 \text{ L}}{1000 \text{ mL}} = 8.745 \times 10^{-3} \text{ mol HCl}$$

$$8.745 \times 10^{-3} \text{ mol HCl} \times \frac{1 \text{ mol KOH}}{1 \text{ mol HCl}} = 8.745 \times 10^{-3} \text{ mol KOH in 25.00 mL}$$

$$8.745 \times 10^{-3} \text{ mol KOH} \times \frac{200.00 \text{ mL}}{25.00 \text{ mL}} = 6.996 \times 10^{-2} \text{ mol KOH}$$

in 4.005 g KOH

$$6.996 \times 10^{-2} \text{ mol KOH} \times \frac{56.11 \text{ g KOH}}{1 \text{ mol KOH}} = 3.925 \text{ g KOH}$$

$$\frac{3.925 \text{ g}}{4.005 \text{ g}} \times 100 = 98.00\% \text{ KOH}$$

470. Given: 558 mL of 3.18 M HCl

Unknown: mass of $Mg(OH)_2$ to react

$Mg(OH)_2 + 2HCl \rightarrow MgCl_2 + 2H_2O$

$$\frac{3.18 \text{ mol HCl}}{\text{L}} \times 558 \text{ mL} \times \frac{1 \text{ L}}{1000 \text{ mL}} = 1.77 \text{ mol HCl}$$

$$1.77 \text{ mol HCl} \times \frac{1 \text{ mol Mg(OH)}_2}{2 \text{ mol HCl}} = 0.885 \text{ mol Mg(OH)}_2$$

$$0.885 \text{ mol Mg(OH)}_2 \times \frac{58.32 \text{ g Mg(OH)}_2}{1 \text{ mol Mg(OH)}_2} = 51.6 \text{ g Mg(OH)}_2$$

471. Given: 12.61 mL NH_3 solution; 5.19 mL of 1.25 M HCl

Unknown: molarity of NH_3 solution

$$\frac{1.25 \text{ mol HCl}}{\text{L}} \times 5.19 \text{ mL} \times \frac{1 \text{ L}}{1000 \text{ mL}} = 6.49 \times 10^{-3} \text{ mol HCl}$$

$$6.49 \times 10^{-3} \text{ mol HCl} \times \frac{1 \text{ mol } NH_3}{1 \text{ mol HCl}} = 6.49 \times 10^{-3} \text{ mol } NH_3$$

$$\frac{6.49 \times 10^{-3} \text{ mol } NH_3}{12.61 \text{ mL}} \times \frac{1000 \text{ mL}}{1 \text{ L}} = 0.515 \text{ M } NH_3$$

472. Given: 5.090 g sample of 92.10% NaOH; 2.811 M diprotic acid

Unknown: V of acid

$5.090 \text{ g NaOH} \times 0.9210 = 4.688 \text{ g NaOH}$

$$4.688 \text{ g NaOH} \times \frac{1 \text{ mol NaOH}}{40.00 \text{ g NaOH}} \times \frac{1 \text{ mol acid}}{2 \text{ mol NaOH}} = 0.05860 \text{ mol acid}$$

$$\frac{0.05860 \text{ mol acid}}{2.811 \text{ mol/L}} = 0.02085 \text{ L acid} = 20.85 \text{ mL acid}$$

473. Given: 43.09 mL of 0.1529 M Ba(OH)$_2$; 26.06 mL HCl for Ba(OH)$_2$; 27.05 mL HCl for 15.00 mL RbOH

a. Unknown: molarity of HCl

$$Ba(OH)_2 + 2HCl \rightarrow BaCl_2 + 2H_2O$$

$$\frac{0.1529 \text{ mol Ba(OH)}_2}{L} \times 43.09 \text{ mL} \times \frac{1 \text{ L}}{1000 \text{ mL}} = 6.588 \times 10^{-3} \text{ mol Ba(OH)}_2$$

$$6.588 \times 10^{-3} \text{ mol Ba(OH)}_2 \times \frac{2 \text{ mol HCl}}{1 \text{ mol Ba(OH)}_2} = 1.318 \times 10^{-2} \text{ mol HCl}$$

$$\frac{1.318 \times 10^{-2} \text{ mol HCl}}{26.06 \text{ mL}} \times \frac{1000 \text{ mL}}{1 \text{ L}} = 0.5058 \text{ M HCl}$$

b. Unknown: molarity of RbOH

$$HCl + RbOH \rightarrow RbCl + H_2O$$

$$\frac{0.5058 \text{ mol HCl}}{L} \times 27.05 \text{ mL} \times \frac{1 \text{ L}}{1000 \text{ mL}} = 1.368 \times 10^{-2} \text{ mol HCl}$$

$$1.368 \times 10^{-2} \text{ mol HCl} \times \frac{1 \text{ mol RbOH}}{1 \text{ mol HCl}} = 1.368 \times 10^{-2} \text{ mol RbOH}$$

$$\frac{1.368 \times 10^{-2} \text{ mol RbOH}}{15.00 \text{ mL}} \times \frac{1000 \text{ mL}}{1 \text{ L}} = 0.9120 \text{ M RbOH}$$

474. Given: 2800 kg of 6.0 M HCl; $D_{HCl} = 1.10$ g/mL

Unknown: mass Ca(OH)$_2$

$$Ca(OH)_2 + 2HCl \rightarrow CaCl_2 + 2H_2O$$

$$2800 \text{ kg HCl} \times \frac{1 \text{ mL}}{1.10 \text{ g}} \times \frac{1000 \text{ g}}{1 \text{ kg}} \times \frac{1 \text{ L}}{1000 \text{ mL}} = 2500 \text{ L HCl}$$

$$\frac{6.0 \text{ mol HCl}}{L} \times 2500 \text{ L} = 1.5 \times 10^4 \text{ mol HCl}$$

$$1.5 \times 10^4 \text{ mol HCl} \times \frac{1 \text{ mol Ca(OH)}_2}{2 \text{ mol HCl}} = 7.5 \times 10^3 \text{ mol Ca(OH)}_2$$

$$7.5 \times 10^3 \text{ mol Ca(OH)}_2 \times \frac{74.10 \text{ g Ca(OH)}_2}{1 \text{ mol Ca(OH)}_2} = 5.6 \times 10^5 \text{ g Ca(OH)}_2$$

$$= 560 \text{ kg Ca(OH)}_2$$

475. Given: 1.00 mL HNO$_3$ diluted to 200.00 mL; 10.00 mL of diluted HNO$_3$; 23.94 mL of 0.0177 M Ba(OH)$_2$

Unknown: molarity of original HNO$_3$

$$Ba(OH)_2 + 2HNO_3 \rightarrow Ba(NO_3)_2 + 2H_2O$$

$$\frac{0.0177 \text{ mol Ba(OH)}_2}{L} \times 23.94 \text{ mL} \times \frac{1 \text{ L}}{1000 \text{ mL}} = 4.24 \times 10^{-4} \text{ mol Ba(OH)}_2$$

$$4.24 \times 10^{-4} \text{ mol Ba(OH)}_2 \times \frac{2 \text{ mol HNO}_3}{1 \text{ mol Ba(OH)}_2} = 8.48 \times 10^{-4} \text{ mol HNO}_3$$

$$\frac{8.48 \times 10^{-4} \text{ mol HNO}_3}{10.00 \text{ mL}} \times \frac{1000 \text{ mL}}{1 \text{ L}} = 8.48 \times 10^{-2} \text{ M HNO}_3$$

$$\frac{8.48 \times 10^{-2} \text{ mol HNO}_3}{1 \text{ L}} \times 200.00 \text{ mL} \times \frac{1 \text{ L}}{1000 \text{ mL}} = 1.70 \times 10^{-2} \text{ mol HNO}_3$$

$$\frac{1.70 \times 10^{-2} \text{ mol HNO}_3}{1.00 \text{ mL}} \times \frac{1000 \text{ mL}}{L} = 17.0 \text{ M HNO}_3$$

476. Given: 4.494 M H_2SO_4; 7.2280 g LiOH

Unknown: V of H_2SO_4

$H_2SO_4 + 2LiOH \rightarrow Li_2SO_4 + 2H_2O$

$7.2280 \text{ g LiOH} \times \dfrac{1 \text{ mol LiOH}}{23.95 \text{ g LiOH}} = 0.3018 \text{ mol LiOH}$

$0.3018 \text{ mol LiOH} \times \dfrac{1 \text{ mol } H_2SO_4}{2 \text{ mol LiOH}} = 0.1509 \text{ mol } H_2SO_4$

$\dfrac{0.1509 \text{ mol } H_2SO_4}{4.494 \text{ mol/L}} = 0.03358 \text{ L } H_2SO_4 = 33.58 \text{ mL } H_2SO_4$

---

477. Given: $5CO_2(g) + Si_3N_4(s) \rightarrow 3SiO(s) + 2N_2O(g) + 5CO(g)$

Unknown: reaction enthalpy for the given reaction

(1) $CO(g) + SiO_2(s) \rightarrow SiO(g) + CO_2(g)$

(2) $8CO_2(g) + Si_3N_4(s) \rightarrow 3SiO_2(s) + 2N_2O(g) + 8CO(g)$

$\Delta H_{\text{reaction 1}} = +520.9 \text{ kJ}$

$\Delta H_{\text{reaction 2}} = +461.05 \text{ kJ}$

$3[CO(g) + SiO_2(s) \rightarrow SiO(g) + CO_2(g)] \quad \Delta H = 3(+520.9) = +1562.7 \text{ kJ}$

$8CO_2(g) + Si_3N_4(s) \rightarrow 3SiO_2(g) + 2N_2O(g) + 8CO(g) \quad \Delta H = +461.05 \text{ kJ}$

$5CO_2(g) + Si_3N_4(s) \rightarrow 3SiO(s) + 2N_2O(g) + 5CO(g)$

$\Delta H = 2024 \text{ kJ}$

---

478. Given: $CaCO_3(s) \rightarrow CaO(s) + CO_2(g)$

Unknown: ΔH

$CaCO_3(s) \rightarrow Ca(s) + C(s) + \dfrac{3}{2}O_2(g) \quad \Delta H = 1207.6 \text{ kJ/mol}$

$Ca(s) + \dfrac{1}{2}O_2(g) \rightarrow CaO(s) \quad \Delta H = -634.9 \text{ kJ/mol}$

$C(s) + O_2(g) \rightarrow CO_2(g) \quad \Delta H = -393.5 \text{ kJ/mol}$

$CaCO_3(s) \rightarrow CaO(s) + CO_2(g) \quad \Delta H = 179.2 \text{ kJ/mol}$

---

479. Given: $2FeO(s) + O_2(g) \rightarrow Fe_2O_3(s)$

Unknown: ΔH

$4Fe(s) + 3O_2(g) \rightarrow 2Fe_2O_3(s) \quad \Delta H = -1118.4 \text{ kJ/mol}$

$2[2FeO(s) \rightarrow 2Fe(s) + O_2(g)] \quad \Delta H = 2(+825.5) = +1651.0$

$4FeO(s) + O_2(g) \rightarrow 2Fe_2O_3(s) \quad \Delta H = 533 \text{ kJ/mol}$

---

480. Given: $NH_3(g) + HF(g) \rightarrow NH_4F(s)$

Unknown: ΔH

$NH_3(g) \rightarrow N(g) + \dfrac{3}{2}H_2(g) \quad \Delta H = 45.9 \text{ kJ/mol}$

$HF(g) \rightarrow \dfrac{1}{2}H_2(g) + F(s) \quad \Delta H = 273.3 \text{ kJ/mol}$

$N(g) + 2H_2(g) + F(s) \rightarrow NH_4F(s) \quad \Delta H = -125 \text{ kJ/mol}$

$NH_3(g) + HF(g) \rightarrow NH_4F(s) \quad \Delta H = 194 \text{ kJ/mol}$

---

481. Given: $H_2S(g) + O_2(g) \rightarrow H_2O(l) + SO_2(g)$

$\Delta H_{\text{reaction}} = -562.1 \text{ kJ/mol}$

$\Delta S_{\text{reaction}} = -0.092\,78 \text{ kJ/mol} \cdot \text{K}$

$T = 25°C = 298 \text{ K}$

Unknown: ΔG

$\Delta G = \Delta H - T\Delta S$

$= -562.1 \text{ kJ/mol} - [(298 \text{ K})(-0.092\,78 \text{ kJ/mol} \cdot \text{K})]$

$= -534.5 \text{ kJ/mol}$

482. Given: $NaClO_3(s) \rightarrow NaCl(s) + O_2(g)$
$\Delta H_{reaction} = -19.1$ kJ/mol
$\Delta S_{reaction} = 0.1768$ kJ/mol · K
$T = 25°C = 298$ K

Unknown: ΔG

$\Delta G = \Delta H - T\Delta S$

$= -19.1$ kJ/mol $- [(298$ K$)(0.1768$ kJ/mol · K$)]$

$= -71.8$ kJ/mol

---

483. Given: $C_2H_6(g) + O_2(g) \rightarrow 2CO_2(g) + 3 H_2O(l)$
$\Delta H_{reaction} = -1561$ kJ/mol
$\Delta S_{reaction} = -1.4084$ kJ/mol · K
$T = 25°C = 298$ K

Unknown: ΔG for combustion of 1 mole of C_2H_6

$\Delta G = \Delta H - T\Delta S$

$= -1561$ kJ/mol $- [(298$ K$)(-0.4084$ kJ/mol · K$)]$

$= -1683$ kJ/mol

---

484. Given: $F_2(g) + H_2O(l) \rightarrow 2HF(g) + O_2(g)$

Unknown: ΔH

$2[H_2 + F_2 \rightarrow 2$ HF$]$ $\Delta H = 2(-273.3) = -546.6$ kJ/mol

$2H_2O \rightarrow 2H_2 + O_2$ $\Delta H = +285.8$

$2F_2(g) + 2H_2O(l) \rightarrow 4HF(g) + O_2(g)$ $\Delta H = -260.8$ kJ/mol

---

485. Given: $CaO(s) + SO_3(g) \rightarrow CaSO_4(s)$
$H_2O(l) + SO_3(g) \rightarrow H_2SO_4(l)$
$\Delta H = -132.5$ kJ/mol
$H_2SO_4(l) + Ca(g) \rightarrow CaSO_4(s) + H_2(g)$ $\Delta H = -602.5$ kJ/mol
$Ca(s) + O_2(g) \rightarrow CaO(s)$ $\Delta H = -634.9$ kJ/mol
$H_2(g) + O_2(g) \rightarrow H_2O(l)$ $\Delta H = -285.8$ kJ/mol

Unknown: ΔH for reaction of CaO + SO_3

$CaO(s) \rightarrow Ca(s) + \frac{1}{2}O_2(g)$ $\Delta H = +634.9$ kJ/mol

$H_2O(l) + SO_3(g) \rightarrow H_2SO_4(l)$ $\Delta H = -132.5$ kJ/mol

$H_2SO_4(l) + Ca(s) \rightarrow CaSO_4(s) + H_2(g)$ $\Delta H = -602.5$ kJ/mol

$H_2 + \frac{1}{2}O_2 \rightarrow H_2O(l)$ $\Delta H = -285.8$ kJ/mol

$CaO(s) + SO_3(g) \rightarrow CaSO_4(s)$ $\Delta H = -385.9$ kJ/mol

---

486. Given: $Na_2O(s) + SO_2(g) \rightarrow Na_2SO_3(s)$

Unknown: ΔH

$Na_2O(s) \rightarrow 2Na(s) + \frac{1}{2}O_2(g)$ $\Delta H = +414.2$ kJ/mol

$SO_2(g) \rightarrow S(s) + O_2(g)$ $\Delta H = +296.8$ kJ/mol

$2Na(s) + S(s) + \frac{3}{2}O_2(g) \rightarrow Na_2SO_3(s)$ $\Delta H = -1101$ kJ/mol

$Na_2O(s) + SO_2(g) \rightarrow Na_2SO_3(s)$ $\Delta H = -390.$ kJ/mol

487. Given: $C_4H_9OH(l)$ + $O_2(g) \rightarrow$ $C_3H_7COOH(l)$ + $H_2O(l)$;
$C_4H_9OH(l)$ + $6O_2(g) \rightarrow 4CO_2(g)$ + $5H_2O(l)$ $\Delta H_c =$ −2675.9 kJ/mol;
$C_3H_7COOH(l)$ + $5O_2(g) \rightarrow$ $4CO_2(g)$ + $4H_2O(l)$ $\Delta H_c =$ −2183.6 kJ/mol

Unknown: ΔH for oxidation of C_4H_9OH to make C_3H_7COOH

$C_4H_9OH(l)$ + $6O_2(g) \rightarrow 4CO_2(g)$ + $5H_2O(l)$ $\Delta H_c =$ −2675.9 kJ/mol

$4CO_2(g)$ + $4H_2O(l) \rightarrow C_3H_7COOH(l)$ + $5O_2(g)$ $\Delta H_c =$ 2183.6 kJ/mol

$C_4H_9OH(l)$ + $O_2(g) \rightarrow C_3H_7COOH(l)$ + $H_2O(l)$ $\Delta H =$ −492.3 kJ/mol

---

488. Given: $CuO(s) + H_2(g)$ $\rightarrow Cu(s) + H_2O(l)$
$\Delta H =$ −128.5 kJ/mol
$\Delta S =$ −70.1 J/mol·K
$T = 25°C = 298$ K

Unknown ΔG

$\Delta G = \Delta H - T\Delta S$

$\Delta S = (-70.1 \text{ J/mol·K}) \left(\dfrac{\text{kJ}}{1000 \text{ J}} \right) = -0.0701$ kJ/mol·K

$\Delta G = -128.5$ kJ/mol − [(298 K)(−0.0701 kJ/mol·K)]
        = −107.6 kJ/mol

---

489. Given: $NaI(s) + Cl_2(g)$ $\rightarrow NaCl(s) + I_2(l)$
$\Delta S =$ −79.9 J/mol·K
$\Delta G =$ −98.0 kJ/mol
$T = 25°C = 298$ K

Unknown: ΔH

$\Delta G = \Delta H - T\Delta S$

$\Delta H + \Delta G + T\Delta S$

$\Delta S = (-79.9 \text{ J/mol·K}) \left(\dfrac{\text{kJ}}{1000 \text{ J}} \right) = -0.0799$ kJ/mol·K

$\Delta H = -98.0$ kJ/mol + [(298 K)(−0.0799 kJ/mol·K)] = −121.8 kJ/mol

---

490. Given: $4HBr(g) + MnO_2(s) \rightarrow$ $MnBr_2(s) +$ $2H_2O(l) + Br_2(l)$
$\Delta H = -291.3$ kJ/mol;
$T = 25°C = 298$ K
$\Delta H^o_{f\,HBr} = -36.29$ kJ/mol
$\Delta H^o_{f\,MnO_2} = -520.0$ kJ/mol
$\Delta H^o_{f\,H_2O} = -285.8$ kJ/mol
$\Delta H^o_{f\,Br_2} = 0.00$ kJ/mol

Unknown: ΔH^o_f of $MnBr_2(s) = x$

Net ΔH = [Sum of ΔH_f of products] − [Sum of ΔH_f of reactants]

$-291.3 = \left[\Delta H_{f\,MnBr_2} + 2\,\Delta H_{f\,H_2O} + \Delta H_{f\,Br_2} \right] -$
$\left[4\,\Delta H_{f\,HBr} + \Delta H_{f\,MnO_2} \right]$

$-291.3 = [x + (2)(-258.8) + (0)] - [(4)(-36.29) + (-520)]$

$-291.3 = x + 93.56$

$-291.3 - 93.56 = x = -384.9$ kJ/mol

491. Given: $CaC_2(s) + 2H_2O(l) \rightarrow C_2H_2(g) + Ca(OH)_2(s)$
$\Delta G = -147.7$ kJ/mol
$\Delta H = -125.6$ kJ/mol
$T = 25°C = 298$ K
Unknown: ΔS

$\Delta G = \Delta H - T\Delta S$
$\Delta S = (\Delta H - \Delta G)/T$
$= [-125.6 \text{ kJ/mol} - (-147.7 \text{ kJ/mol})]/298 \text{ K}$
$= 0.0742$ kJ/mol

---

492. Given: $NH_4NO_3(s) \rightarrow N_2O(g) + 2H_2O(g)$
$\Delta S = 446.4$ J/mol·K
$T = 25°C = 298$ K
Unknown: ΔG

$2N(g) + \frac{1}{2}O_2(g) \rightarrow N_2O(g)$ $\Delta H = 82.1$ kJ/mol

$2[H_2(g) + \frac{1}{2}O_2(g) \rightarrow H_2O(g)]$ $\Delta H = (2)(-241.82) = -483.64$ kJ/mol

$2[H_2O(l) \rightarrow H_2(g) + \frac{1}{2}O_2(g)]$ $\Delta H = (2)(285.8) = 571.6$ kJ/mol

$NH_4NO_3(s) \rightarrow N_2(g) + 2H_2O(l) + \frac{1}{2}O_2$ $\Delta H = 365.56$ kJ/mol

$NH_4NO_3(s) \rightarrow N_2O(g) + 2H_2O(g)$ $\Delta H = -35.98$ kJ/mol

$\Delta S = (446.4$ J/mol·K$)($kJ/1000 J$) = 0.4464$ kJ/mol·K

$\Delta G = \Delta H - T\Delta S$
$= -35.98$ kJ/mol $- [(298$ K$)(0.4464$ kJ/mol·K$)] = -169.0$ kJ/mol

---

493. a. Unknown: Chemical equations for combustion of (1) 1 mol of methane (CH_4) and (2) 1 mol of propane (C_3H_8)

1. *methane*: $CH_4(g) + 2O_2(g) \rightarrow CO_2(g) + 2H_2O(l)$
2. *propane*: $C_3H_8(g) + 5O_2(g) \rightarrow 3CO_2(g) + 4H_2O(g)$

b. Unknown: Enthalpy change (ΔH) for each reaction

1. *methane*: $CH_4(g) \rightarrow C(s) + 2H_2(g)$ $\Delta H = 74.9$ kJ/mol

$C(s) + O_2(g) \rightarrow CO_2(g)$ $\Delta H = -393.5$ kJ/mol

$2[H_2(g) + \frac{1}{2}O_2(g) \rightarrow H_2O(g)]$ $\Delta H = (2)(-241.82) = -483.64$ kJ/mol

$CH_4(g) + 2O_2(g) \rightarrow CO_2(g) + 2H_2O(g)$ $\Delta H = -802.2$ kJ/mol

2. *propane*: $3[C(s) + O_2(g) \rightarrow CO_2(g)]$ $\Delta H = (3)(-393.5) = -1180.5$ kJ/mol

$4[H_2(g) + \frac{1}{2}O_2(g) \rightarrow H_2O(g)]$ $\Delta H = (4)(-241.82) = -967.28$ kJ/mol

$C_3H_8(g) \rightarrow 3C(s) + 4H_2(g)$ $\Delta H = 104.7$ kJ/mol

$C_3H_8(g) + 5O_2(g) \rightarrow 3CO_2(g) + 4H_2O(g)$ $\Delta H = -2043$ kJ/mol

c. Unknown: Heat output per kilogram of each fuel

1. *methane*: $(1000 \text{ g } CH_4)\left(\dfrac{\text{mol } CH_4}{16.05 \text{ g } CH_4}\right) = 62.3$ mol CH_4

$(62.3 \text{ mol } CH_4)(-802.2 \text{ kJ/mol}) = -4.998 \times 10^{-4}$ kJ/kg

2. *propane*: $(1000 \text{ g}/C_3H_8)\left(\dfrac{\text{mol } C_3H_8}{44.11 \text{ g } C_3H_8}\right) = 22.67$ mol C_3H_8

$(22.67 \text{ mol } C_3H_8)(-2043 \text{ kJ/mol}) = -4.632 \times 10^4$ kJ/kg

494. Given: $C_2H_2(g) + H_2O(l) \rightarrow CH_3CHO(l)$;
$C_2H_2(g) + 2O_2(g) \rightarrow 2CO_2(g) + H_2O(l)$ $\Delta H = -1299.6$ kJ/mol;
$CH_3CHO(l) + 2O_2(g) \rightarrow 2CO_2(g) + 2H_2O(l)$ $\Delta H = -1166.9$ kJ/mol

Unknown: Enthalpy (ΔH) for reaction of acetylene with water

$C_2H_2(g) + 2O_2(g) \rightarrow 2CO_2(g) + H_2O(l)$ $\quad \Delta H = -1299.6$ kJ/mol

$2CO_2(g) + 2H_2O(l) \rightarrow CH_3CHO(l) + 2O_2(g)$ $\quad \Delta H = 1166.9$ kJ/mol

$C_2H_2(g) + H_2O(l) \rightarrow CH_3CHO(l)$ $\quad \Delta H = -132.7$ kJ/mol

---

495. Given: $C_{10}H_{22}(l) + 15O_2(g) \rightarrow 10CO_2(g) + 11H_2O(l)$
ΔH_f° for liquid decane ($C_{10}H_{22}$) $= -300.9$ kJ/mol

Unknown: Enthalpy (ΔH) for combustion of decane

$10[C(s) + O_2(g) \rightarrow CO_2(g)]$ $\quad \Delta H = (10)(-393.5) = -3935$ kJ/mol

$11[H_2(g) + \tfrac{1}{2}O_2(g) \rightarrow H_2O(l)]$ $\quad \Delta H = (11)(-285.8) = -3143.8$ kJ/mol

$C_{10}H_{22}(l) \rightarrow 10C(s) + 11H_2(g)$ $\quad \Delta H = 300.9$ kJ/mol

$C_{10}H_{22}(l) + \tfrac{31}{2}O_2(g) \rightarrow 10CO_2(g) + 11H_2O(l)$ $\quad \Delta H = -6777.9$ kJ/mol

---

496. Given: $MgO(s) + 2HCl(g) \rightarrow MgCl_2(s) + H_2O(l)$;
$Mg(s) + 2HCl(g) \rightarrow MgCl_2(s) + H_2(g)$ $\Delta H = -456.9$ kJ/mol;
$Mg(s) + O_2(g) \rightarrow MgO(s)$ $\Delta H = -601.6$ kJ/mol;
$H_2O(l) \rightarrow H_2(g) + O_2(g)$ $\Delta H = +285.8$ kJ/mol

Unknown: Enthalpy (ΔH) of reaction of MgO with HCl

$Mg(s) + 2HCl(g) \rightarrow MgCl_2(s) + H_2(g)$ $\quad \Delta H = -456.9$ kJ/mol

$MgO(s) \rightarrow Mg(s) + \tfrac{1}{2}O_2(g)$ $\quad \Delta H = 601.6$ kJ/mol

$H_2(g) + \tfrac{1}{2}O_2(g) \rightarrow H_2O(l)$ $\quad \Delta H = -285.8$ kJ/mol

$MgO(s) + 2HCl(g) \rightarrow MgCl_2(s) + H_2O(l)$ $\quad \Delta H = -141.1$ kJ/mol

---

497. Given: $2NaOH(s) + 2Na(s) \xrightarrow{\Delta} 2Na_2O(s) + H_2(g)$
$\Delta S = 10.6$ J/mol·K
$\Delta H^\circ_{f\,NaOH} = -425.9$ kJ/mol
$T = 25°C = 298$ K

Unknown: ΔG

$2[2Na(s) + \tfrac{1}{2}O_2(g) \rightarrow Na_2O(s)]$ $\quad \Delta H = (2)(-414.2) = -828.4$ kJ/mol

$2[NaOH(s) \rightarrow 2Na(s) + O_2(g) + H_2(g)]$ $\quad \Delta H = (2)(425.9) = 851.8$ kJ/mol

$2NaOH(s) + 2Na(s) \rightarrow 2Na_2O(s) + H_2(g)$ $\quad \Delta H = 23.4$ kJ/mol

$\Delta S = (10.6$ J/mol·K$)($kJ/1000 J$) = 0.0106$ kJ/mol

$\Delta G = \Delta H - T\Delta S$
$= 23.4$ kJ/mol $- [(298$ K$)(0.0106$ kJ/mol·K$)]$
$= 20.2$ kJ/mol

498. Given: $NH_3(g) + HCl(g)$
$\rightarrow NH_4Cl(s)$
$T = 25°C = 298$ K
$\Delta G = -91.2$ kJ/mol

Unknown: Entropy change (ΔS) in J/mol·K

$\frac{1}{2}N_2(g) + 2H_2(g) + \frac{1}{2}Cl_2(g) \rightarrow NH_4Cl(s)$ $\Delta H = -314.4$ kJ/mol

$NH_3(g) \rightarrow \frac{1}{2}N_2(g) + \frac{3}{2}H_2(g)$ $\Delta H = 45.9$ kJ/mol

$HCl(s) \rightarrow \frac{1}{2}H_2(g) + \frac{1}{2}Cl_2(g)$ $\Delta H = 92.3$ kJ/mol

$NH_3(g) + HCl(g) \rightarrow NH_4Cl(s)$ $\Delta H = -176.2$ kJ/mol

$\Delta G = \Delta H - T\Delta S$

$\Delta S = \Delta H - \Delta G/T$

$= (-176.2 \text{ kJ/mol}) - (-91.2 \text{ kJ/mol})/298 \text{ K}$

$= (-0.285 \text{ kJ/mol} \cdot \text{K})\left(\dfrac{1000 \text{ J}}{\text{kJ}}\right) = -285$ J/mol·K

---

499. a. Given: $3C(s) + Fe_2O_3(s) \rightarrow 3CO(g) + Fe(s)$
$\Delta H^o_{fCO(g)} = -110.53$ kJ/mol

Unknown: Enthalpy (ΔH)

$3C(s) + \frac{3}{2}O_2(g) \rightarrow 3CO(g)$ $\Delta H = (3)(-110.53) = -331.59$ kJ/mol

$Fe_2O_3(s) \rightarrow 2Fe(s) + \frac{3}{2}O_2(g)$ $\Delta H = +1118.4$ kJ/mol

$3C(s) + Fe_2O_3(s) \rightarrow 3CO(g) + 2Fe(s)$ $\Delta H = 786.8$ kJ/mol

b. Given: $3Mn(s) + Fe_2O_3(s) \rightarrow 3MnO(s) + 2Fe(s)$
$\Delta H^o_{fMnO(s)} = -384.9$ kJ/mol

Unknown: ΔH

$3[Mn(s) + \frac{1}{2}O_2(g) \rightarrow MnO(s)]$ $\Delta H = (3)(-384.9) = -1154.7$ kJ/mol

$Fe_2O_3(s) \rightarrow 2Fe(s) + \frac{3}{2}O_2(g)$ $\Delta H = +1118.4$ kJ/mol

$3Mn(s) + Fe_2O_3(s) \rightarrow 3MnO(s) + 2Fe(s)$ $\Delta H = -36$ kJ/mol

c. Given: $12P(s) + 10Fe_2O_3(s) \rightarrow 3P_4O_{10}(s) + 20Fe(s)$
$\Delta H^o_{fP_4O_{10}(s)} = -3009.9$ kJ/mol

Unknown: ΔH

$3[4P(s) + 5O_2(g) \rightarrow P_4O_{10}(s)]$ $\Delta H = (3)(-3009.9) = -9029.7$ kJ/mol

$10[Fe_2O_3(s) \rightarrow 2Fe(s) + \frac{3}{2}O_2(g)]$ $\Delta H = (10)(1118.4) = 11\,184.$ kJ/mol

$12P(s) + 10Fe_2O_3(s) \rightarrow 3P_4O_{10}(s) + 20Fe(s)$ $\Delta H = +2154$ kJ/mol

d. Given: $3Si(s) + 2Fe_2O_3(s) \rightarrow 3SiO_2(s) + 4Fe(s)$
$\Delta H^o_{fSiO_2(s)} = -910.9$ kJ/mol

Unknown: ΔH

$3[Si(s) + O_2(g) \rightarrow SiO_2(s)]$ $\Delta H = (3)(-910.9) = 2732.7$ kJ/mol

$2[Fe_2O_3(s) \rightarrow 2Fe(s) + \frac{3}{2}O_2(g)]$ $\Delta H = (2)(1118.4) = 2236.8$ kJ/mol

$3Si(s) + 2Fe_2O_3(s) \rightarrow 3SiO_2(s) + 4Fe(s)$ $\Delta H = -496$ kJ/mol

e. Given: $3S(s) + 2Fe_2O_3(s) \rightarrow 3SO_2(g) + 4Fe(s)$

Unknown: ΔH

$3[S(s) + O_2(g) \rightarrow SO_2(g)]$ $\Delta H = (3)(-296.8) = -890.4$ kJ/mol

$2[Fe_2O_3(s) \rightarrow 2Fe(s) + \frac{3}{2}O_2(g)]$ $\Delta H = (2)(1118.4) = 2236.8$ kJ/mol

$3S(s) + 2Fe_2O_3(s) \rightarrow 3SO_2(g) + 4Fe(s)$ $\Delta H\ +1346.4$ kJ/mol

500. a. Given: $A \rightleftharpoons C + D$
$[A] = 2.24 \times 10^{-2}$ M
$[C] = 6.41 \times 10^{-3}$ M
$[D] = 6.41 \times 10^{-3}$ M

Unknown: K

$$K = \frac{[C][D]}{[A]} = \frac{(6.41 \times 10^{-3} \text{ M})^2}{2.24 \times 10^{-2} \text{ M}} = 1.83 \times 10^{-3}$$

b. Given: $A + B \rightleftharpoons C + D$
$[A] = 3.23 \times 10^{-5}$ M = [B]
$[C] = 1.27 \times 10^{-2}$ M = [D]

Unknown: K

$$K = \frac{[C][D]}{[A][B]} = \frac{(1.27 \times 10^{-2} \text{ M})^2}{(3.23 \times 10^{-5} \text{M})^2} = 1.55 \times 10^5$$

c. Given: $A + B \rightleftharpoons 2C$
$[A] = 7.02 \times 10^{-3}$ M = [B]
$[C] = 2.16 \times 10^{-2}$ M

Unknown: K

$$K = \frac{[C]^2}{[A][B]} = \frac{(2.16 \times 10^{-2} \text{ M})^2}{(7.02 \times 10^{-3} \text{ M})^2} = 9.47$$

d. Given: $2A \rightleftharpoons 2C + D$
$[A] = 6.59 \times 10^{-4}$ M
$[C] = 4.06 \times 10^{-3}$ M
$[D] = 2.03 \times 10^{-3}$ M

Unknown: K

$$K = \frac{[C]^2[D]}{[A]^2} = \frac{(4.06 \times 10^{-3} \text{ M})^2(2.03 \times 10^{-3} \text{ M})}{(6.59 \times 10^{-4} \text{ M})^2} = 7.71 \times 10^{-2}$$

e. Given: $A + B \rightleftharpoons C + D + E$
$[A] = 3.73 \times 10^{-4}$ M = [B]
$[C] = 9.35 \times 10^{-4}$ M = [D] = [E]

Unknown: K

$$K = \frac{[C][D][E]}{[A][B]} = \frac{(9.35 \times 10^{-4} \text{ M})^3}{(3.73 \times 10^{-4} \text{ M})^2} = 5.88 \times 10^{-3}$$

f. Given: $2A + B \rightleftharpoons 2C$
$[A] = 5.50 \times 10^{-3}$ M
$[B] = 2.25 \times 10^{-3}$ M
$[C] = 1.02 \times 10^{-2}$ M

Unknown: K

$$K = \frac{[C]^2}{[A]^2[B]} = \frac{(1.02 \times 10^{-2} \text{ M})^2}{(5.50 \times 10^{-3} \text{ M})^2(2.25 \times 10^{-3} \text{ M})} = 1.53 \times 10^3$$

501. Given: $2A(g) \rightleftharpoons 2C(g) + D(g)$
[A] = 1.88×10^{-1} M
[C] = 6.56 M
$K = 2.403 \times 10^2$
Unknown: [D]

$K = \dfrac{[C]^2[D]}{[A]^2}$

[D] = (K) [A]^2/[C]^2

$= (2.403 \times 10^2)(1.88 \times 10^{-1}\text{ M})^2/(6.56\text{ M})^2$

= 0.197 M

502. Given: $T = 700$ K
$K = 3.164 \times 10^3$
$C_2H_4(g) + H_2(g) \rightleftharpoons C_2H_6(g)$
Unknown: [C_2H_4] if [H_2] = 0.0619 M and [C_2H_6] = 1.055 M

$K = \dfrac{[C_2H_6]}{[C_2H_4][H_2]}$

$[C_2H_4] = \dfrac{[C_2H_6]}{[H_2](K)} = \dfrac{1.055\text{ M}}{(0.0619\text{ M})(3.164 \times 10^3)} = 5.39 \times 10^{-3}$ M

503. Given: $A + 2B \rightleftharpoons C + 2D$
[A] = 0.0567 M
[B] = 0.1171 M
[C] = 0.000 3378 M
[D] = 0.000 6756 M
Unknown: K

$K = \dfrac{[C][D]^2}{[A][B]^2} = \dfrac{(0.000\,3378\text{ M})(0.000\,6756\text{ M})^2}{(0.0567\text{ M})(0.1171\text{ M})^2} = 1.98 \times 10^{-7}$

504. Given: $2A \rightleftharpoons 2C + 2D$
[A] = 0.1077 M
[C] = 0.000 4104 M
[D] = 0.000 4104 M
Unknown: K

$K = \dfrac{[C]^2[D]^2}{[A]^2} = \dfrac{(0.000\,4104\text{ M})^2(0.000\,4104\text{ M})^2}{0.1077\text{ M}^2} = 2.446 \times 10^{-12}$

506. a. Given: $COCl_2(g) \rightleftharpoons CO(g) + Cl_2(g)$
$T = 25°C$
$K = 4.282 \times 10^{-2}$
[CO] = 5.90×10^{-3} M = [Cl_2]
Unknown: [$COCl_2$]

$K = \dfrac{[CO][Cl_2]}{[COCl_2]}$

$[COCl_2] = \dfrac{[CO][Cl_2]}{K} = \dfrac{(5.90 \times 10^{-3}\text{ M})^2}{4.282 \times 10^{-2}} = 8.13 \times 10^{-4}$ M

b. Given: $COCl_2(g) \rightleftharpoons CO(g) + Cl_2(g)$
$K = 4.282 \times 10^{-2}$
[$COCl^2$] = 0.003 70 M
[CO] = [Cl_2]
Unknown: [CO] and [Cl_2]

$K = \dfrac{[CO][Cl_2]}{[COCl_2]}$

$K = \dfrac{x^2}{[COCl_2]}$

$x^2 = (K)[COCl_2]$

$x = \sqrt{(K)[COCl_2]} = \sqrt{(4.282 \times 10^{-2})(0.003\,70\text{ M})} = 0.0126$ M

507. Given: $A(g) + B(s) \rightleftharpoons C(g) + D(s)$
$K = 1$
$T = 500$ K

  a. Unknown: [A] and [C]

B and D are solids; therefore their concentrations = 1.

$$K = \frac{[C][1]}{[A][1]}$$

If $K = 1$ [C][D] = [A][B]

$$K = 1 = \frac{[C][1]}{[A][1]} = \frac{[C]}{[A]}$$

Therefore, [A] = [C]

---

508. Given: $C(s) + H_2O(g) \rightleftharpoons CO(g) + H_2(g)$
$K = 4.251 \times 10^{-2}$
$T = 800$ K
$[H_2O] = 0.1990$ M

Unknown: [CO] and $[H_2]$

C is a solid; [C] = 1

$$K = \frac{[CO][H_2]}{[1][H_2O]}$$

$[CO][H_2] = (K)[H_2O]$

$x^2 = (4.251 \times 10^{-2})(0.1990 \text{ M})$

$x = \sqrt{8.459 \times 10^{-3} \text{ M}}$

$\quad = 0.0918$ M

---

509. a. Given: $2NO(g) + O_2(g) \rightleftharpoons 2NO_2(g)$
$T = 500$ K
$K = 1.671 \times 10^4$
$[NO] = 6.200 \times 10^{-2}$ M
$[O_2] = 8.305 \times 10^{-3}$ M

Unknown: $[NO_2]$

$$K = \frac{[NO_2]^2}{[NO]^2[O_2]}$$

$[NO_2]^2 = K[NO]^2[O_2]$

$[NO_2] = \sqrt{K[NO]^2[O_2]}$

$\quad = \sqrt{(1.671 \times 10^4)(6.200 \times 10^{-2} \text{ M})^2(8.305 \times 10^{-3} \text{ M})}$

$\quad = 0.7304$ M

  b. Given: $T = 1000$ K
$K = 1.315 \times 10^{-2}$
$[NO] = 6.200 \times 10^{-2}$ M
$[O_2] = 8.305 \times 10^{-3}$ M

Unknown: $[NO_2]$

$[NO_2] \sqrt{K[NO]^2[O_2]}$

$\quad = \sqrt{(1.315 \times 10^{-2})(6.200 \times 10^{-2} \text{ M})^2(8.305 \times 10^{-3} \text{ M})}$

$\quad = 6.479 \times 10^{-4}$ M

---

511. b. Given: $H_2(g) + Br_2(g) \rightleftharpoons 2HBr(g)$
$K = 5.628 \times 10^{18}$
$[H_2]_{initial} = [Br_2]_{initial}$
$[HBr] = 0.500$ M

Unknown: $[H_2]$

$$K = \frac{[HBr]^2}{[H_2][Br_2]}$$

$[H_2] = [Br_2] = x$

$$[H_2] = \frac{[HBr]^2}{(K)(x)}$$

$$x^2 = \frac{(0.500 \text{ M})^2}{(5.628 \times 10^{18})}$$

$$x = \sqrt{\frac{(0.500 \text{ M})^2}{(5.628 \times 10^{18})}} = 2.11 \times 10^{-10} \text{ M}$$

512. Given: $N_2F_4(g) \rightleftharpoons 2 NF_2(g)$
$T = 25°C$
$[N_2F_4] = 0.9989$ M
$[NF_2] = 1.131 \times 10^{-3}$ M
Unknown: K

$$K = \frac{[NF_2]^2}{[N_2F_4]} = \frac{(1.131 \times 10^{-3} \text{ M})^2}{0.9989 \text{ M}} = 1.281 \times 10^{-6}$$

513. Given: $N_2O_4(g) \rightleftharpoons 2NO_2(g)$
$T = 25°C$
$[N_2O_4] = 5.95 \times 10^{-1}$ M
$[NO_2] = 5.24 \times 10^{-2}$ M
Unknown: K

$$K = \frac{[NO_2]^2}{[N_2O_4]} = \frac{(5.24 \times 10^{-2} \text{ M})^2}{(5.95 \times 10^{-1} \text{ M})} = 4.61 \times 10^{-3}$$

514. a. Given: $NaCN(s) + HCl(g) \rightleftharpoons HCN(g) + NaCl(s)$
Unknown: Expression for equilibrium constant (K)

$$K = \frac{[HCN][NaCl]}{[NaCN][HCl]}$$

NaCN and NaCl are solids; therefore their concentrations equal 1.

$$K = \frac{[HCN]}{[HCl]}$$

b. Given: $K = 2.405 \times 10^6$
$[HCN] = 0.8959$ M
Unknown: $[HCl]$

$$K = \frac{[HCN]}{[HCl]}$$

$$[HCl] = \frac{[HCN]}{K} = \frac{0.8959 \text{ M}}{2.405 \times 10^6} = 3.725 \times 10^{-7} \text{ M}$$

515. a. Given: $CH_4(g) + H_2O(g) \rightleftharpoons CO(g) + 3 H_2(g)$
At 1100 K, $K = 3.112 \times 10^2$

b. Temperature = 110 K, $K = 3.112 \times 10^2$
$[H_2] = 1.56$ M
$[CH_4] = 3.70 \times 10^{-2}$ M
$[H_2O] = 8.27 \times 10^{-1}$ M
Unknown: $[CO]$

$$K = \frac{[CO][H_2]^3}{[CH_4][H_2O]}$$

$$[CO] = (K)\frac{[CH_4][H_2O]}{[H_2]} = \frac{(3.112 \times 10^2)(3.70 \times 10^{-2} \text{ M})(8.27 \times 10^{-1} \text{ M})}{(1.56 \text{ M})^3}$$

$$= 2.51 \text{ M}$$

516. Given: $N_2O_4 \rightleftharpoons NO_2$
$T = 20°C = 293$ K
$[N_2O_4] = 2.55 \times 10^{-3}$ M
$[NO_2] = 10.4 \times 10^{-3}$ M
Unknown: K

$N_2O_4 \rightleftharpoons 2NO_2$

$K = \dfrac{[NO_2]^2}{[N_2O_4]} = \dfrac{(10.4 \times 10^{-3} \text{ M})^2}{(2.55 \times 10^{-3} \text{ M})} = 0.0424$

517. Given: $N_2O_4 \rightleftharpoons NO_2$
$T = 20°C = 293$ K
$[N_2O_4] = 2.67 \times 10^{-3}$ M
$[NO_2] = 10.2 \times 10^{-3}$ M
Unknown: K

$N_2O_4 \rightleftharpoons 2NO_2$

$K = \dfrac{[NO_2]^2}{[N_2O_4]} = \dfrac{(10.2 \times 10^{-3} \text{ M})^2}{2.67 \times 10^{-3} \text{ M}} = 0.0390$

518. Given: $T = 25°C$
$[HCOOH] = 0.025$ M
$[H_3O^+] = 2.03 \times 10^{-3}$ M
Unknown: K_a

$HCOOH + H_2O \rightleftharpoons H_3O^+ + HCOO^-$

$K_a = \dfrac{[H_3O^+][HCOO^-]}{[HCOOH]}$

$[H_3O^+] = [HCOO^-] = 2.03 \times 10^{-3}$ M

$[HCOOH] = 0.025 - 2.03 \times 10^{-3} = 0.022\,97$ M

$K_a = \dfrac{(2.03 \times 10^{-3})^2}{0.022\,97} = 1.8 \times 10^{-4}$

519. Given: $[HIO_3] = 0.400$ M
$pH = 0.726$
$T = 25°C$
Unknown: K_a

$pH = -\log[H_3O^+]$

$\log[H_3O^+] = -pH$

$[H_3O^+] = \text{antilog}(-pH)$

$[H_3O^+] = 1 \times 10^{-pH}$

$\quad = 1 \times 10^{-0.726}$

$\quad = 0.1879$ M

$HIO_3 + H_2O \rightleftharpoons H_3O^+ + IO_3^-$

$[H_3O^+] = [IO_3^-] = 0.1879$ M

$[HIO_3] = 0.400 - 0.1879 = 0.212$ M

$K_a = \dfrac{[H_3O^+][IO_3^-]}{[HIO_3]} = \dfrac{(0.1879)^2}{0.212} = 0.167$

520. Given: [HClO] = 0.150 M
pH = 4.55
T = 25°C
Unknown: K_a

$[H_3O^+]$ = antilog (–pH)
$[H_3O^+] = 1 \times 10^{-pH} = 1 \times 10^{-4.55} = 2.8 \times 10^{-5}$

$HClO + H_2O \rightleftharpoons H_3O^+ + ClO^-$

$[H_3O^+] = [ClO^-] = 2.8 \times 10^{-5}$ M

$[HClO] = 0.150 - 2.8 \times 10^{-5} = 0.15$ M

$K_a = \dfrac{[H_3O^+][ClO^-]}{[HClO]} = \dfrac{(2.8 \times 10^{-5})^2}{0.15} = 5.2 \times 10^{-9}$

---

521. Given: $[CH_3CH_2CH_2NH_2]$ = 0.039 M
$[OH^-] = 3.74 \times 10^{-3}$ M
Unknown: (a) pH of solution;
(b) K_b for propylamine

a. pH = –log $[H_3O^+]$
$[H_3O^+][OH^-] = 1 \times 10^{-14}$ M^2

$[H_3O^+] = \dfrac{1.0 \times 10^{-14} \text{ M}^2}{3.74 \times 10^{-3} \text{ M}} = 2.67 \times 10^{-12}$ M

pH = –log (2.67×10^{-12}) = 11.573

b. $CH_3CH_2CH_2NH_2 + H_2O \rightleftharpoons CH_3CH_2CH_2NH_3^+ + OH^-$

$K_b = \dfrac{[CH_3CH_2CH_2NH_3^+][OH^-]}{[CH_3CH_2CH_2NH_2]}$

$[OH^-] = [CH_3CH_2CH_2NH_3^+] = 3.74 \times 10^{-3}$ M

$[CH_3CH_2CH_2NH_2] = 0.039 - 3.74 \times 10^{-3}$ M = 0.03526

$K_b = \dfrac{(3.74 \times 10^{-3})^2}{0.03526} = 4.0 \times 10^{-4}$

---

522. Given: K_a of HNO_2 = 4.6×10^{-4}
T = 25°C
$[HNO_2]$ = 0.0450 M
Unknown: $[H_3O^+]$

$HNO_2 + H_2O \rightleftharpoons H_3O^+ + NO_2^-$

$K_a = \dfrac{[H_3O^+][NO_2^-]}{[HNO_2]}$

$[H_3O^+] = [NO_2^-] = x$

$x^2 = (K_a)[HNO_2] = (4.6 \times 10^{-4})(0.0450 \text{ M}) = 2.07 \times 10^{-5}$ M

$x = \sqrt{2.07 \times 10^{-5} \text{ M}} = 4.5 \times 10^{-3}$ M

523. Given: $[HN_3] = 0.102$ M
$[H_3O^+] = 1.39 \times 10^{-3}$ M
$T = 25°C$
Unknown: **a.** pH
b. K_a

a. pH $= -\log[H_3O^+] = -\log(1.39 \times 10^{-3}) = 2.857$

b. $HN_3 + H_2O \rightleftharpoons H_3O^+ + N_3^-$

$$K_a = \frac{[H_3O^+][N_3^-]}{[HN_3]}$$

$[H_3O^+] = [N_3^-] = 1.39 \times 10^{-3}$ M

$[HN_3] = 0.102$ M $- 1.39 \times 10^{-3}$ M $= 0.10061$ M

$$K_a = \frac{(1.39 \times 10^{-3})^2}{0.10061} = 1.92 \times 10^{-5}$$

---

524. Given: $[BrCH_2COOH] = 0.200$ M
$[H_3O^+] = 0.0192$ M
$T = 25°C$
Unknown: **a.** pH
b. K_a

a. pH $= -\log[H_3O^+] = -\log(0.0192) = 1.717$

b. $BrCH_2COOH + H_2O \rightleftharpoons H_3O^+ + BrCH_2COO^-$

$$K_a = \frac{[H_3O^+][BrCH_2COO^-]}{[BrCH_2COOH]}$$

$[H_3O^+] = [BrCH_2COO^-] = 0.0192$ M

$[BrCH_2COOH] = 0.200$ M $- 0.0192$ M $= 0.1808$ M

$$K_a = \frac{(0.0192)^2}{0.1808} = 2.04 \times 10^{-3}$$

---

525. a. Given: $B + H_2O \rightleftharpoons BH^+ + OH^-$
$[B]_{initial} = 0.400$ M
$[OH^-] = 2.70 \times 10^{-4}$ M
Unknown: (1) $[H_3O^+]$
(2) pH
(3) K_b

(1) $[H_3O^+][OH^-] = 1.00 \times 10^{-14}$ M^2

$$[H_3O^+] = \frac{1.00 \times 10^{-14} \text{ M}^2}{2.70 \times 10^{-4} \text{ M}} = 3.70 \times 10^{-11} \text{ M}$$

(2) pH $= -\log(3.70 \times 10^{-11}) = 10.431$

(3) $B + H_2O \rightleftharpoons BH^+ + OH^-$

$$K_b = \frac{[BH^+][OH^-]}{[B]}$$

$[OH^-] = [BH^+] = 2.70 \times 10^{-4}$ M

$[B] = 0.400$ M $- 2.70 \times 10^{-4}$ M $= 0.399\,73$ M

$$K_b = \frac{(2.70 \times 10^{-4})^2}{0.399\,73} = 1.82 \times 10^{-7}$$

b. Given: $B_{initial}$ = 0.005 50 M
$[OH^-] = 8.45 \times 10^{-4}$ M
Unknown: (1) [B] at equilibrium
(2) K_b
(3) pH

(1) [B] at equilibrium = $0.005\ 50$ M $- 8.45 \times 10^{-4}$ M $= 4.66 \times 10^{-3}$ M

(2) $B + H_2O \rightleftharpoons BH^+ + OH^-$

$$K_b = \frac{[BH^+][OH^-]}{[B]}$$

$[OH^-] = [BH^+] = 8.45 \times 10^{-4}$ M

$$K_b = \frac{(8.45 \times 10^{-4}\text{ M})^2}{4.66 \times 10^{-3}\text{ M}} = 1.53 \times 10^{-4}$$

(3) pH = $-\log [H_3O^+]$

$[H_3O^+][OH^-] = 1 \times 10^{-14}$ M^2

$$[H_3O^+] = \frac{1 \times 10^{-14}\text{ M}^2}{8.45 \times 10^{-4}\text{ M}} = 1.18 \times 10^{-11}\text{ M}$$

pH = $-\log (1.18 \times 10^{-11}) = 10.93$

c. Given: $[B]_{initial}$ = 0.0350 M
pH = 11.29
Unknown: (1) $[H_3O^+]$
(2) $[OH^-]$
(3) [B] at equilibrium
(4) K_b

(1) $[H_3O^+]$ = antilog $(-\text{pH}) = 1 \times 10^{-\text{pH}} = 1 \times 10^{-11.29} = 5.13 \times 10^{-12}$ M

(2) $[H_3O^+][OH^-] = 1 \times 10^{-14}$ M^2

$$[OH^-] = \frac{1 \times 10^{-14}\text{ M}^2}{[H_3O^+]} = \frac{1 \times 10^{-14}\text{ M}^2}{5.13 \times 10^{-12}\text{ M}} = 1.9 \times 10^{-3}\text{ M}$$

(3) [B] at equilibrium = 0.0350 M $- 1.9 \times 10^{-3}$ M = 0.0331 M

(4) $K_b = \dfrac{[BH^+][OH^-]}{[B]}$

$[OH^-] = [BH^+] = 1.9 \times 10^{-3}$ M

$$K_b = \frac{(1.9 \times 10^{-3})^2}{0.0331} = 1.1 \times 10^{-4}$$

d. Given: [B] at equilibrium = 0.006 28 M
$[OH^-]$ = 0.000 92 M
Unknown: (1) $[B]_{initial}$
(2) K_b
(3) pH

(1) $[B]_{eq} = [B]_{initial} - [OH^-]$

$[B]_{initial} = [B]_{eq} + [OH^-] = 0.006\ 28$ M $+ 0.000\ 92$ M $= 7.2 \times 10^{-3}$ M

(2) $K_b = \dfrac{[BH^+][OH^-]}{[B]}$

$[OH^-] = [BH^+] = 0.000\ 92$ M

$$K_b = \frac{(0.000\ 92\text{ M})^2}{0.006\ 28\text{ M}} = 1.35 \times 10^{-4}$$

(3) pH = $-\log [H_3O^+]$

$[H_3O^+][OH^-] = 1 \times 10^{-14}$ M^2

$$[H_3O^+] = \frac{1 \times 10^{14}\text{ M}^2}{0.000\ 92\text{ M}} = 1.1 \times 10^{-11}$$

pH = $-\log (1.1 \times 10^{-11}) = 10.96$

526. Given: Solubility of C_6H_5COOH in water = 2.9 g/L
pH = 2.92
$T = 25°C$

Unknown: K_a

$C_6H_5COOH + H_2O \rightleftharpoons H_3O^+ + C_6H_5COO^-$

$pH = -\log[H_3O^+]$

$\log[H_3O^+] = -pH$

$[H_3O^+] = \text{antilog}(-pH)$

$[H_3O^+] = 1 \times 10^{-pH} = 1 \times 10^{-2.92} = 1.2 \times 10^{-3}$ M

$[H_3O^+] = [C_6H_5COO^-] = 1.2 \times 10^{-3}$ M

$\left(\dfrac{2.9 \text{ g } C_6H_5COOH}{L}\right)\left(\dfrac{1 \text{ mol } C_6H_5COOH}{122.11 \text{ g } C_6H_5COOH}\right) = 0.02375$ M

$= [C_6H_5COOH]_{initial}$

$[C_6H_5COOH] = 0.02375 - 1.2 \times 10^{-3} = 0.02255$

$K_a = \dfrac{[H_3O^+][C_6H_5COO^-]}{[C_6H_5COOH]} = \dfrac{(1.2 \times 10^{-2} \text{ M})^2}{0.022\,55 \text{ M}} = 6.4 \times 10^{-5}$

527. Given: $[H_2NCH_2CH_2OH]_{initial} = 0.006\,50$ M
pH = 10.64
$T = 25°C$

Unknown: **a.** $[H_2NCH_2CH_2OH]$ at equilibrium
b. K_b

a. $H_2NCH_2CH_2OH + H_2O \rightleftharpoons H_3NCH_2CH_2OH^+ + OH^-$

$pH = -\log[H_3O^+]$

$\log[H_3O^+] = -pH$

$[H_3O^+] = \text{antilog}(-pH)$

$[H_3O^+] = 1 \times 10^{-pH} = 1 \times 10^{-10.64} = 2.3 \times 10^{-11}$ M

$[H_3O^+][OH^-] = 1 \times 10^{-14}$ M²

$[OH^-] = \dfrac{1 \times 10^{-14} \text{ M}^2}{[H_3O^+]} = \dfrac{1 \times 10^{-14} \text{ M}^2}{2.3 \times 10^{-11} \text{ M}} = 4.35 \times 10^{-4}$ M

$[H_2NCH_2CH_2OH] = 0.006\,50 \text{ M} - 4.35 \times 10^{-4} \text{ M} = 6.06 \times 10^{-3}$ M

$[H_3NCH_2CH_2OH^+] = [OH^-] = 4.35 \times 10^{-4}$ M

b. $K_b = \dfrac{[H_3NCH_2CH_2OH^+][OH^-]}{[H_2NCH_2CH_2OH]} = \dfrac{(4.35 \times 10^{-4} \text{ M})^2}{6.06 \times 10^{-3} \text{ M}} = 3.1 \times 10^{-5}$

528. Given: $[H_2Se] = 0.060$ M
$[H_3O^+] = 2.72 \times 10^{-3}$ M
$T = 25°C$

Unknown: K_a

$H_2Se + H_2O \rightleftharpoons H_3O^+ + HSe^-$

$K_a = \dfrac{[H_3O^+][HSe^-]}{H_2Se}$

$[HSe^-] = [H_3O^+] = 2.72 \times 10^{-3}$ M

$H_2Se = 0.060 \text{ M} - 2.72 \times 10^{-3} \text{ M} = 0.05728$ M

$K_a = \dfrac{(2.72 \times 10^{-3} \text{ M})^2}{0.05728 \text{ M}} = 1.3 \times 10^{-4}$

529. Given: K_b of C_5H_5N
$= 1.78 \times 10^{-9}$
$[C_5H_5N] = 0.140$ M

Unknown: **a.** $[OH^-]$
b. pH

a. $C_5H_5N + H_2O \rightleftharpoons C_5H_5NH^+ + OH^-$

$$K_b = \frac{[C_5H_5NH^+][OH^-]}{C_5H_5N}$$

$[C_5H_5NH^+] = [OH^-] = x$

$x^2 = K_b[C_5H_5N] = (1.78 \times 10^{-9})(0.140 \text{ M}) = 2.5 \times 10^{-10}$

$x = \sqrt{2.5 \times 10^{-10}} = 1.58 \times 10^{-5}$ M $= [OH^-]$

b. pH $= -\log[H_3O^+]$

$[H_3O^+][OH^-] = 1 \times 10^{-14}$ M^2

$[H_3O^+] = \dfrac{1 \times 10^{-14} \text{ M}^2}{[OH^-]}$

$= \dfrac{1 \times 10^{-14} \text{ M}^2}{1.58 \times 10^{-5} \text{ M}} = 6.329 \times 10^{-10}$ M

pH $= -\log[6.329 \times 10^{-10} \text{ M}] = 9.20$

---

530. Given: $[HA] = 0.0208$ M
pH $= 2.17$

Unknown: **a.** $[HA]_{initial}$
b. K_a

a. $[HA] + [H_2O] \rightleftharpoons H_3O^+ + A^-$

pH $= -\log[H_3O^+]$

$\log[H_3O^+] = -$pH

$[H_3O^+] =$ antilog $(-$pH$) = 1 \times 10^{-pH} = 1 \times 10^{-2.17} = 6.76 \times 10^{-3}$ M

$[HA] = [HA]_{initial} - [H_3O^+]$

$[HA]_{initial} = [HA] + [H_3O^+]$

$= 0.0208$ M $+ 6.76 \times 10^{-3}$ M $= 0.02756$ M

b. $K_a = \dfrac{[H_3O^+][A^-]}{[HA]}$

$[H_3O^+] = [A^-] = 6.76 \times 10^{-3}$ M

$K_a = \dfrac{(6.76 \times 10^{-3} \text{ M})^2}{0.0208 \text{ M}} = 2.2 \times 10^{-3}$

531. Given: Mass of solute
($CH_3COCOOH$)
= 438 mg
volume of
solvent (H_2O)
= 10.00 mL
pH = 1.34

Unknown: K_a

$CH_3COCOOH + H_2O \rightleftharpoons H_3O^+ + CH_3COCOO^-$

pH = $-\log [H_3O^+]$

$\log [H_3O^+] = -pH$

$[H_3O^+]$ = antilog ($-pH$)

$[H_3O^+] = 1 \times 10^{-pH} = 1 \times 10^{-1.34} = 0.0457$ M

$[H_3O^+] = [CH_3COCOO^-] = 0.0457$ M

$\left(\dfrac{438 \text{ mg } CH_3COCOOH}{10.00 \text{ mL}} \right) \left(\dfrac{g}{1000 \text{ mg}} \right) \left(\dfrac{1000 \text{ mL}}{L} \right) \left(\dfrac{1 \text{ mol } CH_3COCOOH}{88.04 \text{ g } CH_3COCOOH} \right)$

= 0.4975 M = $[CH_3COCOOH]_{initial}$

$[CH_3COCOOH] = 0.4975 - .0457 = 0.4518$ M

$K_a = \dfrac{[H_3O^+][CH_3COCOO^-]}{[CH^3COCOOH]} = \dfrac{(0.0457 \text{ M})^2}{0.4518 \text{ M}} = 4.63 \times 10^{-3}$

532. Given: $[H_3O^+]$ of
solution of
acetoacetic acid
(CH_3COCH_2CO
OH) = 4.38
$\times 10^{-3}$ M
$[CH_3COCH_2CO$
OH] (nonionized)
= 0.0731 M

Unknown: K_a

$CH_3COCH_2COOH + H_2O \rightleftharpoons H_3O^+ \ CH_3COCH_2COO^-$

$K_a = \dfrac{[H_3O^+][CH_3COCH_2COO^-]}{[CH_3COCH_2COOH]}$

$[CH_3COCH_2COO^-] = [H_3O^+] = 4.38 \times 10^{-3}$ M

$[CH_3COCH_2COOH] = 0.0731$ M $- 4.38 \times 10^{-3}$ M $= 0.068 \ 72$ M

$K_a = \dfrac{(4.38 \times 10^{-3} \text{ M})^2}{0.0731 \text{ M}} = 2.62 \times 10^{-4}$

533. Given: K_a of
$CH_3CHClCOOH$
= 1.48×10^{-3}
$[CH_3CHClCOO$
H]$_{initial}$ =
0.116 M

Unknown: **a.** $[H_3O^+]$
b. pH

a. $CH_3CHClCOOH + H_2O \rightleftharpoons H_3O^+ + CH_3CHClCOO^-$

$K_a = \dfrac{[H_3O^+][CH_3CHClCOO^-]}{[CH_3CHClCOOH]}$

$[H_3O^+]$ = x = $[CH_3CHClCOO^-]$

$[CH_3CHClCOOH] = 0.116 - x$

$K_a = \dfrac{(x)(x)}{0.116 - x} = \dfrac{x^2}{0.116 - x}$

$K_a = 1.48 \times 10^{-3}$

$1.48 \times 10^{-3} = \dfrac{x^2}{0.116 - x}$

By the quadratic equation, x = 0.0124 M.

b. pH = $-\log [H_3O^+] = -\log (0.0124) = 1.907$

534. Given: First ionization:
$H_2SO_4 + H_2O \rightarrow H_3O^+ + HSO_4^-$
Ionization = 100%
Second ionization: $HSO_4^- + H_2O \rightleftharpoons H_3O^+ + SO_4^{2-}$
$K_{a_2} = 1.3 \times 10^{-2}$
$[H_2SO_4] = 0.0788$ M

Unknown: **a.** total $[H_3O^+]$
b. pH of H_2SO_4 solution

a. $K_{a_2} = 1.3 \times 10^{-2} = \dfrac{[H_3O^+][SO_4^{2-}]}{HSO_4^-}$

$[H_3O^+] = 0.0788 + x$

$[SO_4^{2-}] = x$

$[HSO_4^-] = 0.0788 - x$

$1.3 \times 10^{-2} = \dfrac{(0.0788 + x)(x)}{(0.0788 - x)}$

$(1.3 \times 10^{-2})(0.0788 - x) = (0.0788 + x)(x)$

$(1.0244 \times 10^{-3}) - (1.3 \times 10^{-2})x = (7.88 \times 10^{-2})x + x^2$

$x^2 + (9.18 \times 10^{-2})x - (1.0244 \times 10^{-3}) = 0$

By the quadratic formula, x = 0.010 024.

$[H_3O^+] = 0.0788 + 0.01 = 0.0888$

b. pH = $-\log [H_3O^+] = -\log (0.0888) = 1.05$

---

535. Given: [HOCN] = 0.100 M
$[H_3O^+] = 5.74 \times 10^{-3}$ M

Unknown: **a.** K_a
b. pH

a. $HOCN + H_2O \rightleftharpoons H_3O^+ + OCN^-$

$K_a = \dfrac{[H_3O^+][OCN^-]}{[HOCN]}$

$[H_3O^+] = [OCN^-] = 5.74 \times 10^{-3}$ M

$[HOCN] = 0.100 \text{ M} - 5.74 \times 10^{-3} \text{ M} = 0.094$ M

$K_a = \dfrac{(5.74 \times 10^{-3})^2}{0.094} = 3.5 \times 10^{-4}$

b. pH = $-\log [H_3O^+] = -\log (5.74 \times 10^{-3}) = 2.241$

---

536. Given: [HCN] = 0.025 M
$[CN^-] = 3.16 \times 10^{-6}$ M

$HCN + H_2O \rightleftharpoons H_3O^+ + CN^-$

a. Unknown: $[H_3O^+]$

$[CN^-] = [H_3O^+] = 3.16 \times 10^{-6}$ M

b. Unknown: pH

pH = $-\log [H_3O^+] = -\log (3.16 \times 10^{-6}) = 5.500$

d. Unknown: K_a

$K_a = \dfrac{[H_3O^+][CN^-]}{[HCN]}$

$[HCN] = 0.025 \text{ M} - 3.16 \times 10^{-6} \text{ M} = 0.0249$ M

$K_a = \dfrac{(3.16 \times 10^{-6} \text{ M})^2}{0.0249 \text{ M}} = 4.0 \times 10^{-10}$

f. Given: [HCN] = 0.085 M

Unknown: $[H_3O^+]$

$K_a = \dfrac{x^2}{0.085}$

$x^2 = (K_a)(0.085)$

$x = \sqrt{(K_a)(0.085)} = \sqrt{(4.0 \times 10^{-10})(0.085)} = \sqrt{3.4 \times 10^{-11}} = 5.8 \times 10^{-6}$

537. Given: $[CCl_2HCOOH] = 1.20$ M
$[H_3O^+] = 0.182$ M

a. Unknown: pH

pH $= -\log [H_3O^+] = -\log (0.182) = 0.740$

b. Unknown: K_a

$CCl_2HCOOH + H_2O \rightleftharpoons H_3O^+ + CCl_2HCOO^-$

$K_a = \dfrac{[H_3O^+][CCl_2HCOO^-]}{[CCl_2HCOOH]}$

$[H_3O^+] = [CCl_2HCOO^-] = 0.182$ M

$[CCl_2HCOOH] = 1.20$ M $- 0.182$ M $= 1.018$ M

$K_a = \dfrac{(0.182)^2}{1.018} = 0.0325$

---

538. Given: $[C_6H_5OH] = 0.215$ M
pH $= 5.61$
Unknown: K_a

$C_6H_5OH + H_2O \rightleftharpoons H_3O^+ + C_6H_5O^-$

$K_a = \dfrac{[H_3O^+][C_6H_5O^-]}{[C_6H_5OH]}$

pH $= -\log [H_3O^+]$

$\log [H_3O^+] = -$pH

$[H_3O^+] =$ antilog $(-$pH$)$

$[H_3O^+] = 1 \times 10^{-pH} = 1 \times 10^{-5.61} = 2.45 \times 10^{-6}$ M

$[H_3O^+] = [C_6H_5O^-] = 2.45 \times 10^{-6}$ M

$[C_6H_5OH] = 0.215 - 2.45 \times 10^{-6} = 0.215$ M

$K_a = \dfrac{(2.45 \times 10^{-6})^2}{0.215} = 2.80 \times 10^{-11}$

---

539. Given: Solution of NH_2CH_2COOH is 3.75 g in 250.0 mL H_2O
pH $= 0.890$

a. Unknown: molarity of NH_2CH_2COOH

$\left(\dfrac{3.75 \text{ g } NH_2CH_2COOH}{250 \text{ mL}}\right)\left(\dfrac{1000 \text{ mL}}{\text{L}}\right)\left(\dfrac{1 \text{ mol } NH_2CH_2COOH}{75.06 \text{ g } NH_2CH_2COOH}\right)$

$= 0.200$ M NH_2CH_2COOH

b. Unknown: K_a

$$NH_2CH_2COOH + H_2O \rightleftharpoons H_3O^+ + NH_2CH_2COO^-$$

$$K_a = \frac{[H_3O^+][NH_2CH_2COO^-]}{[NH_2CH_2COOH]}$$

$$pH = -\log[H_3O^+]$$

$$\log[H_3O^+] = -pH$$

$$[H_3O^+] = \text{antilog}(-pH)$$

$$[H_3O^+] = 1 \times 10^{-pH} = 1 \times 10^{-0.890} = 0.1288 \text{ M}$$

$$[H_3O^+] = [NH_2CH_2COO^-] = 0.1288 \text{ M}$$

$$[NH_2CH_2COOH] = 0.200 \text{ M} - 0.1288 \text{ M} = 0.0712 \text{ M}$$

$$K_a = \frac{(0.1288)^2}{0.0712} = 0.233$$

---

540. Given: $(CH_3)_3N + H_2O \rightleftharpoons CH_3NH^+ + OH^-$
$[(CH_3)_3N] = 0.0750$ M
$[OH^-] = 2.32 \times 10^{-3}$ M
Unknown: **a.** pH **b.** K_b

a. $[H_3O^+][OH^-] = 1 \times 10^{-14} \text{ M}^2$

$$[H_3O^+] = \frac{1 \times 10^{-14} \text{ M}^2}{[OH^-]} = \frac{1 \times 10^{-14} \text{ M}^2}{2.32 \times 10^{-3} \text{ M}} = 4.31 \times 10^{-12} \text{ M}$$

$$pH = -\log[H_3O^+] = -\log(4.31 \times 10^{-12}) = 11.37$$

b. $K_b = \frac{[CH_3NH^+][OH^-]}{[(CH_3)_3N]}$

$$[OH^-] = [CH_3NH^+] = 2.32 \times 10^{-3} \text{ M}$$

$$[(CH_3)_3N] = 0.0750 \text{ M} - 2.32 \times 10^{-3} \text{ M} = 0.07268 \text{ M}$$

$$K_b = \frac{(2.32 \times 10^{-3})^2}{0.07268} = 7.41 \times 10^{-5}$$

---

541. Given: $[(CH_3)_2NH] = 5.00 \times 10^{-3}$ M
pH = 11.20
Unknown: **a.** K_b
b. Which base is stronger: $(CH_3)_2NH$ or $(CH_3)_3N$?

a. $(CH_3)_2NH + H_2O \rightleftharpoons (CH_3)_2NH_2^+ + OH^-$

$$K_b = \frac{[(CH_3)_2NH_2^+][OH^-]}{[(CH_3)_2NH]}$$

$$pH = -\log[H_3O^+]$$

$$\log[H_3O^+] = -pH$$

$$[H_3O^+] = \text{antilog}(-pH) = 1 \times 10^{-pH} = 1 \times 10^{-11.20} = 6.31 \times 10^{-12} \text{ M}$$

$$[H_3O^+][OH^-] = 1 \times 10^{-14} \text{ M}^2$$

$$[OH^-] = \frac{1 \times 10^{-14} \text{ M}^2}{[H_3O^+]} = \frac{1 \times 10^{-14} \text{ M}^2}{6.31 \times 10^{-12} \text{ M}} = 1.585 \times 10^{-3} \text{ M}$$

$$[OH^-] = [(CH_3)_2NH_2^+] = 1.585 \times 10^{-3} \text{ M}$$

$$[(CH_3)_2NH] = 5.00 \times 10^{-3} \text{ M} - 1.585 \times 10^{-3} \text{ M} = 3.415 \times 10^{-3} \text{ M}$$

$$K_b = \frac{(1.585 \times 10^{-3})^2}{3.415 \times 10^{-3}} = 7.36 \times 10^{-4}$$

b. K_b of $(CH_3)_3N = 7.41 \times 10^{-5}$

K_b of $(CH_3)_2NH = 7.36 \times 10^{-4}$

$7.36 \times 10^{-4} > 7.41 \times 10^{-5}$

Therefore $(CH_3)_2NH$ (dimethylamine) is the stronger base.

542. Given: $H_2NNH_2 + H_2O(l) \rightleftharpoons H_2NNH_3^+ (aq) + OH^- (aq)$
$H_2NNH_3^+ (aq) + H_2O(l) \rightleftharpoons H_3NNH_3^{2+} (aq) + OH^- (aq)$
$K_{b_2} = 8.9 \times 10^{-16}$

a. $[H_2NNH_2] = 0.120$ M
pH = 10.50
Unknown: K_{b_1}
(Assume that $[H_2NNH_2]_{initial}$ does not change.)

$K_{b_1} = \dfrac{[H_2NNH_3^+][OH^-]}{[H_2NNH_2]}$

$[H_3O^+] = $ antilog $(-pH)$

$[H_3O^+] = 1 \times 10^{-pH}$

$= 1 \times 10^{-10.50}$

$= 3.16 \times 10^{-11}$ M

$[H_3O^+][OH^-] = 1 \times 10^{-14}$ M^2

$[OH^-] = \dfrac{1 \times 10^{-14} \text{ M}^2}{[H_3O^+]} = \dfrac{1 \times 10^{-14} \text{ M}^2}{3.16 \times 10^{-11} \text{ M}}$

$= 3.16 \times 10^{-4}$ M

$[OH^-] = 3.16 \times 10^{-4}$ M $= [H_2NNH_3^+]$

$K_{b_1} = \dfrac{(3.16 \times 10^{-4})^2}{0.120 \text{ M}} = 8.3 \times 10^{-7}$

b. Given: $[H_2NNH_2] = 0.020$ M
Unknown: $[OH^-]$

$K_{b_1} = \dfrac{[H_2NNH_3^+][OH^-]}{[H_2NNH_2]}$

$8.3 \times 10^{-7} = \dfrac{x^2}{0.020}$

$x = \sqrt{(8.3 \times 10^{-7})(0.020)} = 1.3 \times 10^{-4}$ M $= [OH^-]$

c. Unknown: pH of 0.020 M $[H_2NNH_2]$ solution

$[H_3O^+][OH^-] = 1 \times 10^{-14}$ M^2

$[H_3O^+] = \dfrac{1 \times 10^{-14} \text{ M}^2}{[OH^-]} = \dfrac{1 \times 10^{-14} \text{ M}^2}{1.3 \times 10^{-4} \text{ M}} = 7.7 \times 10^{-11}$ M

pH $= -\log [H_3O^+] = -\log (7.7 \times 10^{-11}) = 10.11$

---

543. Given: Saturated solution: 0.276 g $AgBrO_3$ in 150.0 mL H_2O
Unknown: K_{sp}

solubility $= \left(\dfrac{0.276 \text{ g AgBrO}_3}{150.0 \text{ mL H}_2\text{O}}\right)\left(\dfrac{1000 \text{ mL}}{\text{L}}\right)\left(\dfrac{1 \text{ mol AgBrO}_3}{235.74 \text{ g AgBrO}_3}\right)$

$= 7.80 \times 10^{-3}$ M $AgBrO_3$

$AgBrO_3(s) \rightleftharpoons Ag^+(aq) + BrO_3^-(aq)$

$K_{sp} = [Ag^+][BrO_3^-]$

$[Ag^+] = 7.80 \times 10^{-3}$ M

$[BrO_3^-] = 7.80 \times 10^{-3}$ M

$K_{sp} = (7.80 \times 10^{-3})^2 = 6.08 \times 10^{-5}$

544. Given: solubility of CaF_2
= 0.0427 g/2.50 L
Unknown: K_{sp}

$$\left(\frac{0.0427 \text{ g CaF}_2}{2.50 \text{ L}}\right)\left(\frac{1 \text{ mol CaF}_2}{78.06 \text{ g CaF}_2}\right) = 2.19 \times 10^{-4} \text{ M CaF}_2$$

$CaF_2(s) \rightleftharpoons Ca^{2+}(aq) + 2F^-(aq)$

$K_{sp} = [Ca^{2+}][F^-]^2$

$[Ca^{2+}] = 2.19 \times 10^{-4}$ M

$[F^-] = (2)(2.19 \times 10^{-4}) = 4.38 \times 10^{-4}$ M

$K_{sp} = (2.19 \times 10^{-4})(4.38 \times 10^{-4})^2$

$= 4.20 \times 10^{-11}$

545. Given: K_{sp} of $CaSO_4$
= 9.1×10^{-6}
Unknown: $[CaSO_4]$ in a saturated solution

$CaSO_4(s) \rightleftharpoons Ca^{2+}(aq) + SO_4^{2-}(aq)$

$K_{sp} = [Ca^{2+}][SO_4^{2-}]$

$[Ca^{2+}] = [SO_4^{2-}] = x$

$K_{sp} = x^2 = 9.1 \times 10^{-6}$

$x = \sqrt{9.1 \times 10^{-6}}$

$= 3.0 \times 10^{-3}$ M = $[CaSO_4]$

546. Given: A salt = X_2Y
$K_{sp} = 4.25 \times 10^{-7}$
Unknown: **a.** molarity of a saturated solution of the salt
b. molarity of a solution of AZ with the same K_{sp}

a. $X_2Y(s) \rightleftharpoons 2X(aq) + Y(aq)$

$K_{sp} = [X]^2[Y]$

$[X] = 2x$

$[Y] = x$

$K_{sp} = 4.25 \times 10^{-7} = (2x)^2(x) = 4x^3$

$x = 4.74 \times 10^{-3}$ M = $[X_2Y]$

b. $AZ(s) \rightleftharpoons A(aq) + Z(aq)$

$K_{sp} = [A][Z] = x^2$

$x = \sqrt{4.25 \times 10^{-7}} = 6.52 \times 10^{-4}$ M

547. Given: V NaOH = 0.320 L
[NaOH] = 0.046 M
V CaCl$_2$ = 0.400 L
[CaCl$_2$] = 0.085 M
K_{sp} Ca(OH)$_2$
= 5.5 × 10^{-6}

Unknown: whether a precipitate of Ca(OH)$_2$ forms

$2\text{NaOH} + \text{CaCl}_2 \rightarrow 2\text{NaCl} + \text{Ca(OH)}_2$

$\text{Ca(OH)}_2(s) \rightleftharpoons \text{Ca}^{2+}(aq) + 2\text{OH}^-(aq)$

$K_{sp} = [\text{Ca}^{2+}][\text{OH}^-]^2 = 5.5 \times 10^{-6}$

$(0.400 \text{ L})\left(\dfrac{0.085 \text{ mol Ca}^{2+}}{\text{L}}\right) = 0.034 \text{ mol Ca}^{2+}$

$(0.320 \text{ L})\left(\dfrac{0.046 \text{ mol OH}^-}{\text{L}}\right) = 0.0147 \text{ mol OH}^-$

Total volume = 0.320 L + 0.400 L = 0.720 L

$\dfrac{0.034 \text{ mol Ca}^{2+}}{0.720 \text{ L}} = 0.0472 \text{ mol/L Ca}^{2+}$

$\dfrac{0.0147 \text{ mol OH}^-}{0.720 \text{ L}} = 0.0204 \text{ mol/L OH}^-$

$[\text{Ca}^{2+}][\text{OH}^-]^2 = (0.0472)(0.0204)^2$
$= 1.9 \times 10^{-5} = $ ion product

$1.9 \times 10^{-5} > K_{sp}$; precipitation occurs

---

548. Given: V AgNO$_3$ = 0.020 L
[AgNO$_3$] = 0.077 M
V NaC$_2$H$_3$O$_2$ = 0.030 L
[NaC$_2$H$_3$O$_2$] = 0.043 M
K_{sp} AgC$_2$H$_3$O$_2$ = 2.5 × 10^{-3}

Unknown: whether a precipitate forms

$\text{AgNO}_3 + \text{NaC}_2\text{H}_3\text{O}_2 \rightarrow \text{AgC}_2\text{H}_3\text{O}_2 + \text{NaNO}_3$

$\text{AgC}_2\text{H}_3\text{O}_2(s) \rightleftharpoons \text{Ag}^+(aq) + \text{C}_2\text{H}_3\text{O}_2^-(aq)$

$K_{sp} = [\text{Ag}^+][\text{C}_2\text{H}_3\text{O}_2^-] = 2.5 \times 10^{-3}$

$(0.020 \text{ L})\left(\dfrac{0.077 \text{ mol Ag}^+}{\text{L}}\right) = 1.54 \times 10^{-3} \text{ mol Ag}^+$

$(0.030 \text{ L})\left(\dfrac{0.043 \text{ mol C}_2\text{H}_3\text{O}_2}{\text{L}}\right) = 1.29 \times 10^{-3} \text{ mol C}_2\text{H}_3\text{O}_2$

Total volume = 0.020 L + 0.030 L = 0.050 L

$\dfrac{1.54 \times 10^{-3} \text{ mol Ag}^+}{0.050 \text{ L}} = 0.031 \text{ mol/L Ag}^+$

$\dfrac{1.29 \times 10^{-3} \text{ mol C}_2\text{H}_3\text{O}_2^-}{0.050 \text{ L}} = 0.026 \text{ mol/L C}_2\text{H}_3\text{O}_2^-$

$[\text{Ag}^+][\text{C}_2\text{H}_3\text{O}_2^-] = (0.031)(0.026) = 8.1 \times 10^{-4} = $ ion product

$8.1 \times 10^{-4} < 2.5 \times 10^{-3}$; no precipitation

549. Given: V $Pb(C_2H_3O_2)_2$
= 0.100 L
$[Pb(C_2H_3O_2)]$
= 0.036 M
V NaCl = 0.050 L
[NaCl] = 0.074 M
K_{sp} $PbCl_2$
= 1.9×10^{-4}

Unknown: whether a precipitate forms

$Pb(C_2H_3O_2)_2 + 2NaCl \rightarrow PbCl_2 + 2Na(C_2H_3O_2)$

$PbCl_2(s) \rightleftharpoons Pb^+(aq) + 2Cl^-(aq)$

$K_{sp} = [Pb^+][Cl^-]^2 = 1.9 \times 10^{-4}$

$(0.100 \text{ L})\left(\dfrac{0.036 \text{ mol Pb}^+}{\text{L}}\right) = 3.6 \times 10^{-3}$ mol Pb^+

$(0.050 \text{ L})\left(\dfrac{0.074 \text{ mol Cl}^-}{\text{L}}\right) = 3.7 \times 10^{-3}$ mol Cl^-

Total volume = 0.1 L + 0.05 L = 0.15 L

$\dfrac{3.6 \times 10^{-3} \text{ mol Pb}^+}{0.15 \text{ L}} = 0.024$ mol/L Pb^+

$\dfrac{3.7 \times 10^{-3} \text{ mol Cl}^-}{0.15 \text{ L}} = 0.025$ mol/L Cl^-

$[Pb^+][Cl^-]^2 = (0.024)(0.025)^2 = 1.5 \times 10^{-5}$ = ion product

$1.5 \times 10^{-5} < 1.9 \times 10^{-4}$; no precipitation

550. Given: V $(NH_4)_2S$
= 0.020 L
$[(NH_4)_2S]$
= 0.0090 M
V $Al(NO_3)_3$
= 0.120 L
$[Al(NO_3)_3]$
= 0.0082 M
K_{sp} Al_2S_3
= 2.00×10^{-7}

Unknown: whether a precipitate forms

$3(NH_4)_2S + 2Al(NO_3)_3 \rightarrow Al_2S_3 + 6NH_4NO_3$

$Al_2S_3(s) \rightleftharpoons 2Al^{3+}(aq) + 3S^{2-}(aq)$

$K_{sp} = [Al^{3+}]^2[S^{2-}]^3 = 2.00 \times 10^{-7}$

$(0.120 \text{ L})\left(\dfrac{0.0082 \text{ mol Al}^{3+}}{\text{L}}\right) = 9.8 \times 10^{-4}$ mol Al^{3+}

$(0.020 \text{ L})\left(\dfrac{0.0090 \text{ mol S}^{2-}}{\text{L}}\right) = 1.8 \times 10^{-4}$ mol S^{2-}

Total volume = 0.120 L + 0.020 L = 0.140 L

$\dfrac{9.8 \times 10^{-4} \text{ mol Al}^{3+}}{0.140 \text{ L}} = 7.0 \times 10^{-3}$ mol/L Al^{3+}

$\dfrac{1.8 \times 10^{-4} \text{ mol S}^{2-}}{0.140 \text{ L}} = 1.3 \times 10^{-3}$ mol/L S^{2-}

$[Al^{3+}]^2[S^{2-}]^3 = (7.0 \times 10^{-3})^2(1.3 \times 10^{-3})^3 = 1.1 \times 10^{-13}$ = ion product

$1.1 \times 10^{-13} < 2.00 \times 10^{-7}$; no precipitation

551. Given: $[CaCrO_4]$ = 0.010 M

Unknown: K_{sp}

$CaCrO_4(s) \rightleftharpoons Ca^{2+} + CrO_4^{2-}$

$K_{sp} = [Ca^{2+}][CrO_4^{2-}]$

$[Ca^{2+}] = 0.010$ M

$[CrO_4^{2-}] = 0.010$ M

$K_{sp} = (0.010)^2 = 1.0 \times 10^{-4}$

552. Given: Solubility of
PbSeO$_4$ = 0.001 36
g/10.00 mL

Unknown: K_{sp}

$$\left(\frac{0.001\ 36\ \text{g PbSeO}_4}{10.00\ \text{mL}}\right)\left(\frac{1000\ \text{mL}}{\text{L}}\right)\left(\frac{1\ \text{mol PbSeO}_4}{350.12\ \text{g PbSeO}_4}\right) = 3.88 \times 10^{-4}\ \text{M PbSeO}_4$$

PbSeO$_4$(s) ⇌ Pb^{2+}(aq) + SeO$_4^{2-}$(aq)

K_{sp} = [Pb^{2+}][SeO$_4^{2-}$]

[Pb^{2+}] = [SeO$_4^{2-}$] = 3.88 × 10^{-4} M

K_{sp} = (3.88 × 10^{-4})^2 = 1.51 × 10^{-7}

---

553. Given: V of CuSCN =
0.0225 L
[CuSCN] =
4.0 × 10^{-6} M

Unknown: **a.** K_{sp}
b. mass of CuSCN dissolved in 1 × 10^{-3} L of solution

a. CuSCN(s) ⇌ Cu^{1+}(aq) + SCN^{1-}(aq)

K_{sp} = [Cu^{1+}][SCN^{1-}]

[Cu^{1+}] = [SCN^{1-}] = 4.0 × 10^{-6} M

K_{sp} = (4.0 × 10^{-6})^2 = 1.6 × 10^{-11}

b. solubility =
$4.0 \times 10^{-6}\ \text{M} = \left(\dfrac{x\ \text{g CuSCN}}{1 \times 10^3\ \text{L}}\right)\left(\dfrac{1\ \text{mol CuSCN}}{121.64\ \text{g CuSCN}}\right)$

x = (4.0 × 10^{-6} M)(1000 L)(121.64 g)/1 mol = 0.49 g

---

554. Given: [Ag$_2$Cr$_2$O$_7$]
= 3.684 × 10^{-3} M

Unknown: K_{sp}

Ag$_2$Cr$_2$O$_7$(s) ⇌ 2Ag$^+$(aq) + Cr$_2$O$_7^{2-}$(aq)

K_{sp} = [Ag$^+$]^2[Cr$_2$O$_7^{2-}$]

[Ag$^+$] = (2)(3.684 × 10^{-3} M) = 7.368 × 10^{-3} M

[Cr$_2$O$_7^-$] = 3.684 × 10^{-3} M

K_{sp} = (7.368 × 10^{-3})^2(3.684 × 10^{-3}) = 2.000 × 10^{-7}

---

555. Given: K_{sp} BaSO$_3$ =
8.0 × 10^{-7}

Unknown: **a.** [BaSO$_3$]
b. mass of BaSO$_3$ dissolved in 500. mL H$_2$O

a. BaSO$_3$(s) ⇌ Ba^{2+}(aq) + SO$_3^{2-}$(aq)

K_{sp} = [Ba^{2+}][SO$_3^{2-}$]

[Ba^{2+}] = [SO$_3^{2-}$] = x

K_{sp} = x^2 = 8.0 × 10^{-7}

x = $\sqrt{8.0 \times 10^{-7}}$ = 8.9 × 10^{-4} M = [BaSO$_3$]

b. solubility =
$8.9 \times 10^{-4}\ \text{M} = \left(\dfrac{x\ \text{g BaSO}_3}{500\ \text{mL}}\right)\left(\dfrac{1000\ \text{mL}}{1\ \text{L}}\right)\left(\dfrac{1\ \text{mol BaSO}_3}{217.37\ \text{g BaSO}_3}\right)$

$x = \dfrac{(8.9 \times 10^{-4}\ \text{M})(500\ \text{mL})(1\ \text{L})(217.37\ \text{g})}{(1000\ \text{mL})(1\ \text{mol})}$

= 0.097 g

---

556. Given: K_{sp} PbCl$_2$ =
1.9 × 10^{-4}

Unknown: [PbCl$_2$]

PbCl$_2$(s) ⇌ Pb^{2+}(aq) + 2Cl$^-$

K_{sp} = [Pb^{2+}][Cl$^-$]^2

[Pb^{2+}] = x

[Cl$^-$] = 2x

K_{sp} = 1.9 × 10^{-4} = (x)(2x)^2 = 4x^3

x = $\sqrt[3]{1.9 \times 10^{-4}/4}$ = 0.036 M = [PbCl$_2$]

557. Given: K_{sp} BaCO$_3$ = 1.2×10^{-8}

Unknown: **a.** [BaCO$_3$]
b. volume of water needed to dissolve 0.10 g BaCO$_3$

a. BaCO$_3$(s) $\rightleftharpoons$ Ba^{2+}(aq) + CO$_3^{2-}$(aq)

K_{sp} = [Ba^{2+}][CO$_3^{2-}$] = 1.2×10^{-8}

$1.2 \times 10^{-8} = x^2$

$x = 1.1 \times 10^{-4}$ M = [BaCO$_3$]

b. solubility =
1.1×10^{-4} M = $\left(\dfrac{0.10 \text{ g BaCO}_3}{x}\right)\left(\dfrac{1 \text{ mol BaCO}_3}{197.31 \text{ g BaCO}_3}\right)$

$x = (0.10 \text{ g BaCO}_3)\left(\dfrac{1 \text{ mol BaCO}_3}{197.31 \text{ g BaCO}_3}\right)\left(\dfrac{1}{1.1 \times 10^{-4} \text{ M}}\right)$

$x = 4.6$ L

---

558. Given: K_{sp} SrSO$_4$ = 3.2×10^{-7}

Unknown: **a.** [SrSO$_4$]
b. mass of SrSO$_4$ remaining after 20.0 L saturated solution is evaporated

a. SrSO$_4$(s) $\rightleftharpoons$ Sr^{2+}(aq) + SO$_4^{2-}$(aq)

K_{sp} = [Sr^{2+}][SO$_4^{2-}$] = 3.2×10^{-7}

$3.2 \times 10^{-7} = x^2$

$x = 5.7 \times 10^{-4}$ M = [SrSO$_4$]

b. solubility =
5.7×10^{-4} M = $\left(\dfrac{x \text{ g SrSO}_4}{20.0 \text{ L}}\right)\left(\dfrac{1 \text{ mol SrSO}_4}{183.65 \text{ g SrSO}_4}\right)$

$x = (5.7 \times 10^{-4} \text{ M})(20.0 \text{ L})(183.65 \text{ g SrSO}_4/\text{mol})$

$= 2.1$ g

---

559. Given: K_{sp} SrSO$_3$ = 4.0×10^{-8}
solubility = 1.0000 g/5.0 L H$_2$O

Unknown: mass of SrSO$_3$ remaining after saturated solution is filtered

SrSO$_3$(s) $\rightleftharpoons$ Sr^{2+}(aq) + SO$_3^{2-}$(aq)

K_{sp} = [Sr^{2+}][SO$_3^{2-}$] = 4.0×10^{-8}

$4.0 \times 10^{-8} = x^2$

$x = 2.0 \times 10^{-4}$ = [SrSO$_3$]

solubility =
2.0×10^{-4} M = $\left(\dfrac{x}{5 \text{ L}}\right)\left(\dfrac{1 \text{ mol}}{167.66 \text{ g SrSO}_3}\right)$

$x = (2.0 \times 10^{-4} \text{ M})(5 \text{ L})\left(\dfrac{167.66 \text{ g SrSO}_3}{1 \text{ mol}}\right)$

$= 0.17$ g SrSO$_3$

1.000 g $- 0.17$ g $= 0.83$ g SrSO$_3$

---

560. Given: K_{sp} Mn$_3$(AsO$_4$)$_2$ = 1.9×10^{-11}

Unknown: [Mn$_3$(AsO$_4$)$_2$] in a saturated solution

Mn$_3$(AsO$_4$)$_2$(s) $\rightleftharpoons$ 3Mn^{2+}(aq) + 2AsO$_4^{3-}$(aq)

K_{sp} = [Mn^{2+}]^3[AsO$_4^{3-}$]^2

[Mn^{2+}] = 3x

[AsO$_4^{3-}$] = 2x

K_{sp} = 1.9×10^{-11} = $(3x)^3(2x)^2 = 108x^5$

$x = 2.8 \times 10^{-3}$ M = [Mn$_3$(AsO$_4$)$_2$]

561. Given: V $Sr(NO_3)_2$ = 0.030 L
$[Sr(NO_3)_2]$ = 0.0050 M
V K_2SO_4 = 0.020 L
$[K_2SO_4]$ = 0.010 M
K_{sp} $SrSO_4$ = 3.2 × 10^{-7}

Unknown: **a.** ion product of the ions that can form a precipitate
(b) whether a precipitate forms

a. $Sr(NO_3)_2 + K_2SO_4 \rightarrow SrSO_4 + 2KNO_3$
$SrSO_4(s)$ $Sr^{2+}(aq) + SO_4^{2-}(aq)$

$K_{sp} = [Sr^{2+}][SO_4^{2-}] = 3.2 \times 10^{-7}$

$(0.030 \text{ L})\left(\dfrac{0.0050 \text{ mol Sr}^{2+}}{\text{L}}\right) = 1.5 \times 10^{-4} \text{ mol Sr}^{2+}$

$(0.020 \text{ L})\left(\dfrac{0.010 \text{ mol SO}_4^{2-}}{\text{L}}\right) = 2.0 \times 10^{-4} \text{ mol SO}_4^{2-}$

Total volume = 0.030 L + 0.020 L = 0.050 L

$\dfrac{1.5 \times 10^{-4} \text{ mol Sr}^{2+}}{0.050 \text{ L}} = 3.0 \times 10^{-3}$ mol/L Sr^{2+}

$\dfrac{2.0 \times 10^{-4} \text{ mol SO}_4^{2-}}{0.050 \text{ L}} = 4.0 \times 10^{-3}$ mol/L SO$_4^{2-}$

$[Sr^{2+}][SO_4^{2-}] = (3.0 \times 10^{-3})(4.0 \times 10^{-3})$

$= 1.2 \times 10^{-5}$ = ion product

b. $1.2 \times 10^{-5} > 3.2 \times 10^{-7}$; precipitation occurs

562. Given: K_{sp} $PbBr_2$ = 6.3 × 10^{-6}
V $MgBr_2$ = 0.120 L
$[MgBr_2]$ = 0.0035 M
V $Pb(C_2H_3O_2)_2$ = 0.180 L
$[Pb(C_2H_3O_2)_2]$ = 0.0024 M

Unknown: **a.** ion product of Br$^-$ and Pb^{2+} in the mixed solution
b. whether a precipitate forms

a. $MgBr_2 + Pb(C_2H_3O_2)_2 \rightarrow PbBr_2 + Mg(C_2H_3O_2)_2$
$PbBr_2(s)$ $Pb^{2+}(aq) + 2Br^-(aq)$

$K_{sp} = [Pb^{2+}][Br^-]^2 = 6.3 \times 10^{-6}$

$(0.180 \text{ L})\left(\dfrac{0.0024 \text{ mol Pb}^{2+}}{\text{L}}\right) = 4.3 \times 10^{-4} \text{ mol Pb}^{2+}$

$(0.120 \text{ L})\left(\dfrac{0.0035 \text{ mol Br}^-}{\text{L}}\right) = 4.2 \times 10^{-4} \text{ mol Br}^-$

Total volume = 0.180 L + 0.120 L = 0.3 L

$\dfrac{4.3 \times 10^{-4} \text{ mol Pb}^{2+}}{0.3 \text{ L}} = 1.4 \times 10^{-3}$ mol/L Pb^{2+}

$\dfrac{4.2 \times 10^{-4} \text{ mol Br}^-}{0.3 \text{ L}} = 1.4 \times 10^{-3}$ mol/L Br$^-$

$[Br^-] = (2)(1.4 \times 10^{-3}) = 2.8 \times 10^{-3}$

ion product = $[Pb^{2+}][Br^-]^2 = (1.4 \times 10^{-3})(2.8 \times 10^{-3})^2$

$= 1.1 \times 10^{-8}$

b. $1.1 \times 10^{-8} < 6.3 \times 10^{-6}$; no precipitation occurs

563. Given: K_{sp} Mg(OH)$_2$ = 1.5×10^{-11}

Unknown: **b.** volume of H$_2$O required to dissolve 0.10 g Mg(OH)$_2$

b. $K_{sp} = [\text{Mg}^{2+}][\text{OH}^-]^2$

$[\text{Mg}^{2+}] = x$

$[\text{OH}^-] = 2x$

$K_{sp} = 1.5 \times 10^{-11} = (x)(2x)^2 = 4x^3$

$x = 1.6 \times 10^{-4} \text{ M} = [\text{Mg(OH)}_2]$

solubility =

$1.6 \times 10^{-4} \text{ M} = \left(\dfrac{0.10 \text{ g Mg(OH)}_2}{x}\right)\left(\dfrac{1 \text{ mol Mg(OH)}_2}{58.3 \text{ g Mg(OH)}_2}\right)$

$x = (0.10 \text{ g Mg(OH)}_2)\left(\dfrac{1 \text{ mol Mg(OH)}_2}{58.3 \text{ g Mg(OH)}_2}\right)\left(\dfrac{1}{1.6 \times 10^{-4} \text{ M}}\right)$

= 11 L

---

564. Given: K_{sp} Li$_2$CO$_3$ = 2.51×10^{-2}

Unknown: **a.** [Li$_2$CO$_3$]
b. mass of Li$_2$CO$_3$ dissolved to make 3440 mL of saturated solution

a. Li$_2$CO$_3$ 2Li$^+$ + CO$_3^{2-}$

$K_{sp} = [\text{Li}^+]^2[\text{CO}_3^{2-}]$

$[\text{Li}^+] = 2x$

$[\text{CO}_3^{2-}] = x$

$K_{sp} = 2.51 \times 10^{-2} = (2x)^2(x) = 4x^3$

$x = 0.184 \text{ M} = [\text{Li}_2\text{CO}_3]$

b. solubility =

$0.184 \text{ M} = \left(\dfrac{x}{3440 \text{ mL}}\right)\left(\dfrac{1000 \text{ mL}}{\text{L}}\right)\left(\dfrac{1 \text{ mol Li}_2\text{CO}_3}{73.86 \text{ g Li}_2\text{CO}_3}\right)$

$x = (0.184 \text{ M})(3440 \text{ mL})\left(\dfrac{1 \text{ L}}{1000 \text{ mL}}\right)\left(\dfrac{73.86 \text{ g Li}_2\text{CO}_3}{1 \text{ mol}}\right)$

= 46.8 g

---

565. Given: V Ba(OH)$_2$ = 0.050 L
V HCl = 0.03161 L
[HCl] = 0.3417 M

Unknown: K_{sp} Ba(OH)$_2$

Ba(OH)$_2$ + 2HCl → BaCl$_2$ + 2H$_2$O

BaCl$_2$ Ba^{2+} + 2Cl$^-$

$(0.03161 \text{ L})\left(\dfrac{0.3417 \text{ mol HCl}}{\text{L}}\right) = 0.01080 \text{ mol HCl}$

$\left(\dfrac{1 \text{ mol Ba(OH)}_2}{2 \text{ mol HCl}}\right)(0.01080 \text{ mol HCl}) = 5.400 \times 10^{-3} \text{ mol Ba(OH)}_2$

$\dfrac{5.400 \times 10^{-3} \text{ mol Ba(OH)}_2}{0.050 \text{ L}} = 0.1080 \text{ M Ba(OH)}_2$

$K_{sp} = [\text{Ba}^{2+}][\text{Cl}^-]^2$

[Ba(OH)$_2$] = 0.1080 M

[Ba^{2+}] = 0.1080 M

[Cl$^-$] = (2)(0.1080) = 0.2160 M

$K_{sp} = (0.1080)(0.2160)^2$

$= 5.040 \times 10^{-3}$

566. $QR \rightarrow Q^+ + R^-$

    a. Given: $[QR] = 1.0$ M
    Unknown: K_{sp}

$K_{sp} = [Q^+][R^-]$
$= (1.0)(1.0) = 1.0$

    b. Given: $[QR] = 0.50$ M
    Unknown: K_{sp}

$K_{sp} = [Q^+][R^-]$
$= (0.50)(0.50) = 0.25$

    c. Given: $[QR] = 0.1$ M
    Unknown: K_{sp}

$K_{sp} = (0.1)(0.1) = 0.01$

    d. Given: $[QR] = 0.001$ M
    Unknown: K_{sp}

$K_{sp} = (0.001)(0.001) = 1 \times 10^{-6}$

---

567. Given: Saturated solutions of the salts QR, X_2Y, KL_2, A_3Z, and D_2E_3 are 0.02 M

Unknown: K_{sp} for each salt

    a. $QR \rightarrow Q^+ + R^-$

$K_{sp} = [Q^+][R^-] = (0.02)^2 = 4 \times 10^{-4}$

    b. $X_2Y \rightarrow 2X^+ + Y^-$

$K_{sp} = [X^+]^2[Y^-]$

$[X^+] = (2)(0.02) = 0.04$ M

$[Y^-] = 0.02$ M

$K_{sp} = (0.04)^2(0.02) = 3 \times 10^{-5}$

    c. $KL_2 \rightarrow K^+ + 2L^-$

$K_{sp} = [K^+][L^-]^2$

$[K^+] = 0.02$ M

$[L^-] = (2)(0.02) = 0.04$ M

$K_{sp} = (0.02)(0.04)^2 = 3 \times 10^{-5}$

    d. $A_3Z \rightarrow 3A^+ + Z^-$

$K_{sp} = [A^+]^3[Z^-]$

$[A^+] = (3)(0.02) = 0.06$ M

$[Z^-] = 0.02$ M

$K_{sp} = (0.06)^3(0.02) = 4 \times 10^{-6}$

    e. $D_2E_3 \rightarrow 2D^+ + 3E^-$

$K_{sp} = [D^+]^2[E^-]^3$

$[D^+] = (2)(0.02) = 0.04$ M

$[E^-] = (3)(0.02) = 0.06$ M

$K_{sp} = (0.04)^2(0.06)^3 = 3 \times 10^{-7}$

568. Given: K_{sp} of AgBr = 5.0×10^{-13}

Unknown: **a.** [AgBr]
b. mass of AgBr in 10.0 L of saturated solution

a. AgBr(s) ⇌ Ag$^+$(aq) + Br$^-$(aq)

K_{sp} = [Ag$^+$][Br$^-$]

[Ag$^+$] = [Br$^-$] = x

$K_{sp} = x^2 = 5.0 \times 10^{-13}$

x = 7.1×10^{-7} M = [AgBr]

b. solubility =

7.1×10^{-7} M = $\left(\dfrac{x \text{ AgBr}}{10.0 \text{ L}}\right)\left(\dfrac{1 \text{ mol AgBr}}{187.87 \text{ g AgBr}}\right)$

x = $(7.1 \times 10^{-7}$ M$)(10.0$ L$)(187.87$ g$)/1$ mol

= 1.3×10^{-3} g

---

569. Given: K_{sp} of Ca(OH)$_2$ = 5.5×10^{-6}

Unknown: **a.** molarity of saturated Ca(OH)$_2$ solution
b. [OH$^-$]
c. pH

a. Ca(OH)$_2$(s) ⇌ Ca^{2+}(aq) + 2OH$^-$(aq)

K_{sp} = [Ca^{2+}][OH$^-$]^2

[Ca^{2+}] = x

[OH$^-$] = 2x

$K_{sp} = 5.5 \times 10^{-6} = (x)(2x)^2 = 4x^3$

x = 0.011 M = [Ca(OH)$_2$]

b. [OH$^-$] = (2)(x) = (2)(0.011 M) = 0.022 M

c. [H$_3$O$^+$][OH$^-$] = 1×10^{-14} M^2

[H$_3$O$^+$] = $\dfrac{1 \times 10^{-14} \text{ M}^2}{0.022 \text{ M}}$

= 4.5×10^{-13}

pH = –log [H$_3$O$^+$]

= –log (4.5×10^{-13})

= 12.35

---

570. Given: K_{sp} of MgCO$_3$ = 3.5×10^{-8}

Unknown: mass of MgCO$_3$ dissolved in 4.00 L of water

MgCO$_3$(s) ⇌ Mg^{2+}(aq) + CO$_3^{2-}$(aq)

K_{sp} = [Mg^{2+}][CO$_3^{2-}$]

[Mg^{2+}] = [CO$_3^{2-}$] = x

$K_{sp} = x^2 = 3.5 \times 10^{-8}$

x = 1.9×10^{-4} M = [MgCO$_3$]

solubility =

1.9×10^{-4} M = $\left(\dfrac{x}{4.00 \text{ L}}\right)\left(\dfrac{1 \text{ mol MgCO}_3}{84.29 \text{ g MgCO}_3}\right)$

x = $(1.9 \times 10^{-4}$ M$)(4.00$ L$)(84.29$ g$)/1$ mol

= 0.064 g

571. *Formula equation:*

$$Fe + SnCl_4 \rightarrow FeCl_3 + SnCl_2$$

Ionic equation:

$$\overset{0}{Fe} + \overset{+4}{Sn^{4+}} + 4\overset{-1}{Cl^-} \rightarrow \overset{+3}{Fe^{3+}} + 3\overset{-1}{Cl^-} + \overset{+2}{Sn^{2+}} + 2\overset{-1}{Cl^-}$$

Oxidation half-reaction:

$$2[\overset{0}{Fe} \rightarrow \overset{+3}{Fe^{3+}} + 3e^-] = 2Fe \rightarrow 2Fe^{3+} + 6e^-$$

Reduction half-reaction:

$$3[\overset{+4}{SnCl_4} + 2e^- \rightarrow \overset{+2}{SnCl_2} + 2Cl^-] = 3SnCl_4 + 6e^- \rightarrow 3SnCl_2 + 6Cl^-$$

Combine half-reactions:

$$2Fe \rightarrow 2Fe^{3+} + 6e^-$$
$$\underline{3SnCl_4 + 6e^- \rightarrow 3SnCl_2 + 6Cl^-}$$
$$3SnCl_4 + 2Fe \rightarrow 3SnCl_2 + 2Fe^{3+} + 6Cl^-$$

Combine ions to form balanced equation:

$$2Fe + 3SnCl_4 \rightarrow 2FeCl_3 + 3SnCl_2$$

---

572. *Formula equation:*

$$H_2O_2 + FeSO_4 + H_2SO_4 \rightarrow Fe_2(SO_4)_3 + H_2O$$

Ionic equation:

$$\overset{+1\ -1}{H_2O_2} + \overset{+2}{Fe^{2+}} + (\overset{+6\ -2}{SO_4^{2-}}) + 2\overset{+1}{H^+} + (\overset{+6\ -2}{SO_4^{2-}}) \rightarrow 2\overset{+3}{Fe^{3+}} + 3(\overset{+6\ -2}{SO_4^{2-}}) + \overset{+1\ -2}{H_2O}$$

Oxidation half-reaction:

$$2\overset{+2}{Fe^{2+}} \rightarrow 2\overset{+3}{Fe^{3+}} + 2e^-$$

Reduction half-reaction:

$$\overset{+1\ -1}{H_2O_2} + 2H^+ + 2e^- \rightarrow 2\overset{+1\ -2}{H_2O}$$

Combine half-reactions:

$$2Fe^{2+} \rightarrow 2F^{3+} + 2e^-$$
$$\underline{H_2O_2 + 2H^+ + 2e^- \rightarrow H_2O + H_2O}$$
$$H_2O_2 + 2H^+ + 2Fe^{2+} \rightarrow 2Fe^{3+} + 2H_2O$$

Combine ions to form balanced equation:

$$H_2O_2 + 2FeSO_4 + H_2SO_4 \rightarrow Fe_2(SO_4)_3 + 2H_2O$$

573. *Formula equation:*

$CuS + HNO_3 \rightarrow Cu(NO_3)_2 + NO + S + H_2O$

Ionic equation:

$Cu^{2+} + S^{2-} + H^+ + \overset{+5-2}{NO_3^-} \rightarrow Cu^{2+} + 2\overset{+5-2}{NO_3^-} + \overset{+2-2}{NO} + \overset{0}{S} + \overset{+1-2}{H_2O}$

Oxidation half-reaction:

$3[\overset{-2}{CuS} \rightarrow 2e^- + Cu^{2+} + \overset{0}{S}] = 3CuS \rightarrow 6e^- + 3Cu^{2+} + 3S$

Reduction half-reaction:

$2[4\overset{+5}{HNO_3} + 3e^- \rightarrow 3NO_3^- + \overset{+2}{NO} + 2H_2O]$

$= 8HNO_3 + 6e^- \rightarrow 6NO_3^- + 2NO + 4H_2O$

Combine half-reactions:

$3CuS \rightarrow 6e^- + 3Cu^{2+} + 3S$

$\underline{8HNO_3 + 6e^- \rightarrow 6NO_3^- + 2NO + 4H_2O}$

$3CuS + 8HNO_3 \rightarrow 3Cu^{2+} + 3S + 6NO_3^- + 2NO + 4H_2O$

Combine ions to form balanced equation:

$3CuS + 8HNO_3 \rightarrow 3Cu(NO_3)_2 + 2NO + 3S + 4H_2O$

---

574. *Formula equation:*

$K_2Cr_2O_7 + HI \rightarrow CrI_3 + KI + I_2 + H_2O$

Ionic equation:

$2\overset{+1}{K^+} + \overset{+6\ -2}{Cr_2O_7^{2-}} + \overset{+1-1}{H\ I} \rightarrow \overset{+3}{Cr^{3+}} + 3\overset{-1}{I^-} + \overset{+1-1}{K\ I} + \overset{0}{I_2} + \overset{+1-2}{H_2O}$

Oxidation half-reaction:

$3[2\overset{-1}{I^-} \rightarrow \overset{0}{I_2} + 2e^-] = 6I^- \rightarrow 3I_2 + 6e^-$

Reduction half-reaction:

$\overset{+6}{K_2Cr_2O_7} + 14H^+ + 8I^- + 6e^- \rightarrow 2\overset{+3}{CrI_3} + 2KI + 7H_2O$

Combine half-reactions:

$6I^- \rightarrow 3I_2 + 6e^-$

$\underline{K_2Cr_2O_7 + 14H^+ + 8I^- + 6e^- \rightarrow 2CrI_3 + 2KI + 7H_2O}$

$K_2Cr_2O_7 + 14H^+ + 14I^- \rightarrow 2CrI_3 + 2KI + 3I_2 + 7H_2O$

Combine ions to form balanced equation:

$K_2Cr_2O_7 + 14HI \rightarrow 2CrI_3 + 2KI + 3I_2 + 7H_2O$

575. *Formula equation:*

$$CO_2 + NH_2OH \rightarrow CO + N_2 + H_2O$$

Ionic equation:

$$\overset{+4-2}{CO_2} + \overset{-1+1-2+1}{NH_2OH} \rightarrow \overset{+2-2}{CO} + \overset{0}{N_2} + \overset{+1-2}{H_2O}$$

Oxidation half-reaction:

$$2\overset{-1}{N}H_2OH + 2OH^- \rightarrow \overset{0}{N_2} + 4H_2O + 2e^-$$

Reduction half-reaction:

$$\overset{+4}{C}O_2 + H_2O + 2e^- \rightarrow \overset{+2}{C}O + 2OH^-$$

Combine half-reactions:

$$2NH_2OH + 2OH^- \rightarrow N_2 + 4H_2O + 2e^-$$

$$\underline{CO_2 + H_2O + 2e^- \rightarrow CO + 2OH^-}$$

$$CO_2 + 2NH_2OH \rightarrow CO + N_2 + 3H_2O$$

Combine ions to form balanced equation:

$$CO_2 + 2NH_2OH \rightarrow CO + N_2 + 3H_2O$$

---

576. *Formula equation:*

$$Bi(OH)_3 + K_2SnO_2 \rightarrow Bi + K_2SnO_3$$

Ionic equation:

$$\overset{+3}{Bi^{3+}} + 3\overset{-2+1}{OH^-} + 2\overset{+1}{K^+} + \overset{+2-2}{SnO_2^{2-}} \rightarrow \overset{0}{Bi} + 2\overset{+1}{K^+} + \overset{+4-2}{SnO_3^{2-}}$$

Oxidation half-reaction:

$$3[2K^+ + \overset{+2}{Sn}O_2^{2-} + H_2O \rightarrow 2K^+ + \overset{+4}{Sn}O_3^{2-} + 2H^+ + 2e^-]$$

$$= 6K^+ + 3SnO_2^{2-} + 3H_2O \rightarrow 6K^+ + 3SnO_3^{2-} + 6H^+ + 6e^-$$

Reduction half-reaction:

$$2[\overset{+3}{Bi^{3+}} + 3OH^- + 3H^+ + 3e^- \rightarrow \overset{0}{Bi} + 3H_2O]$$

$$= 2Bi^{3+} + 6OH^- + 6H^+ + 6e^- \rightarrow 2Bi + 6H_2O$$

Combine half-reactions:

$$6K^+ + 3SnO_2^{2-} + 3H_2O \rightarrow 6K^+ + 3SnO_3^{2-} + 6H^+ + 6e^-$$

$$\underline{2Bi^{3+} + 6OH^- + 6H^+ + 6e^- \rightarrow 2Bi + 6H_2O}$$

$$6K^+ + 2Bi^{3+} + 6OH^- + 3SnO_2^{2-} \rightarrow 6K^+ + 2Bi + H_2O + 3SnO_3^{2-}$$

Combine ions to form balanced equation:

$$2Bi(OH)_3 + 3K_2SnO_2 \rightarrow 2Bi + 3K_2SnO_3 + 3H_2O$$

577. *Formula equation:*

$Mg + N_2 \rightarrow Mg_3N_2$

Ionic equation:

$\overset{0}{Mg} + \overset{0}{N_2} \rightarrow 3\overset{+2}{Mg}{}^{2+} + 2\overset{-3}{N}{}^{3-}$

Oxidation half-reaction:

$3\overset{0}{Mg} \rightarrow 3\overset{+2}{Mg}{}^{2+} + 6e^-$

Reduction half-reaction:

$\overset{0}{N_2} + 6e^- \rightarrow 2\overset{-3}{N}{}^{3-}$

Combine half-reactions:

$3Mg \rightarrow 3Mg^{2+} + 6e^-$

$\underline{N_2 + 6e^- \rightarrow 2N^{3-}}$

$3Mg + N_2 \rightarrow 3Mg^{2+} + 2N^{3+}$

Combine ions to form balanced equation:

$3Mg + N_2 \rightarrow Mg_3N_2$

---

578. *Formula equation:*

$SO_2 + Br_2 + H_2O \rightarrow HBr + H_2SO_4$

Ionic equation:

$\overset{+4-2}{SO_2} + \overset{0}{Br_2} + \overset{+1-2}{H_2O} \rightarrow \overset{+1}{H^+} + \overset{-1}{Br^-} + 2\overset{+1}{H^+} + \overset{+6-2}{SO_4^{2-}}$

Oxidation half-reaction:

$\overset{+4}{SO_2} + 2H_2O \rightarrow \overset{+6}{SO_4^{2-}} + 4H^+ + 2e^-$

Reduction half-reaction:

$\overset{0}{Br_2} + 2e^- \rightarrow 2\overset{-1}{Br^-}$

Combine half-reactions:

$SO_2 + 2H_2O \rightarrow SO_4^{2-} + 4H^+ + 2e^-$

$\underline{Br_2 + 2e^- \rightarrow 2Br^-}$

$SO_2 + 2H_2O + Br_2 \rightarrow 2Br^- + SO_4^{2-} + 4H^+$

Combine ions to form balanced equation:

$SO_2 + Br_2 + 2H_2O \rightarrow 2HBr + H_2SO_4$

579. *Formula equation:*

$$H_2S + Cl_2 \rightarrow S + HCl$$

Ionic equation:

$$2\overset{+1}{H_2^+} + \overset{-2}{S^{2-}} + \overset{0}{Cl_2} \rightarrow \overset{0}{S} + \overset{+1}{H^+} + \overset{-1}{Cl^-}$$

Oxidation half-reaction:

$$\overset{-2}{S^{2-}} + 2H^+ \rightarrow \overset{0}{S} + 2H^+ + 2e^-$$

Reduction half-reaction:

$$\overset{0}{Cl_2} + 2e^- \rightarrow 2\overset{-1}{Cl^-}$$

Combine half-reactions:

$$S^{2-} + 2H^+ \rightarrow S + 2H^+ + 2e^-$$

$$\underline{Cl_2 + 2e^- \rightarrow 2Cl^-}$$

$$S^{2-} + 2H^+ + Cl_2 \rightarrow S + 2Cl^- + 2H^+$$

Combine ions to form balanced equation:

$$H_2S + Cl_2 \rightarrow S + 2HCl$$

---

580. *Formula equation:*

$$PbO_2 + HBr \rightarrow PbBr_2 + Br_2 + H_2\overset{-2}{O}$$

Ionic equation:

$$\overset{+4\ -2}{PbO_2} + \overset{+1}{H^+} + \overset{-1}{Br^-} \rightarrow \overset{+2\ -1}{PbBr_2} + \overset{0}{Br_2} + \overset{+1\ -2}{H_2O}$$

Oxidation half-reaction:

$$2\overset{-1}{Br^-} \rightarrow \overset{0}{Br_2} + 2e^-$$

Reduction half-reaction:

$$2Br^- + \overset{+4}{PbO_2} + 4H^+ + 2e^- \rightarrow \overset{+2}{PbBr_2} + 2H_2O$$

Combine half-reactions:

$$2Br^- \rightarrow Br_2 + 2e^-$$

$$\underline{2Br^- + PbO_2 + 4H^+ + 2e^- \rightarrow PbBr_2 + 2H_2O}$$

$$4Br^- + PbO_2 + 4H^+ \rightarrow PbBr_2 + Br_2 + 2H_2O$$

Combine ions to form balanced equation:

$$PbO_2 + 4HBr \rightarrow PbBr_2 + Br_2 + 2H_2O$$

581. *Formula equation:*

$$S + HNO_3 \rightarrow NO_2 + H_2SO_4 + H_2O$$

Ionic equation:

$$\overset{0}{S} + \overset{+1+5-2}{HNO_3} \rightarrow \overset{+4-2}{NO_2} + \overset{+1}{2H^+} + \overset{+6-2}{SO_4^{2-}} + \overset{+1-2}{H_2O}$$

Oxidation half-reaction:

$$\overset{0}{S} + 4H_2O \rightarrow \overset{+6}{SO_4^{2-}} + 8H^+ + 6e^-$$

Reduction half-reaction:

$$6[\overset{+5}{HNO_3^-} + H^+ + 1e^- \rightarrow \overset{+4}{NO_2} + H_2O] = 6HNO_3 + 6H^+ + 6e^- \rightarrow 6NO_2 + 6H_2O$$

Combine half-reactions:

$$S + 4H_2O \rightarrow SO_4^{2-} + 8H^+ + 6e^-$$
$$\underline{6HNO_3 + 6H^+ + 6e^- \rightarrow 6NO_2 + 6H_2O}$$
$$S + 6HNO_3 \rightarrow SO_4^{2-} + 2H^+ + 6NO_2 + 2H_2O$$

Combine ions to form balanced equation:

$$S + 6HNO_3 \rightarrow 6NO_2 + H_2SO_4 + 2H_2O$$

---

582. *Formula equation:*

$$NaIO_3 + N_2H_4 + HCl \rightarrow N_2 + NaICl_2 + H_2O$$

Ionic equation:

$$\overset{+1+5-2}{NaIO_3} + \overset{-2+1}{N_2H_4} + \overset{+1}{H^+} + \overset{-1}{Cl^-} \rightarrow \overset{0}{N_2} + \overset{+1+1-1}{NaICl_2} + \overset{+1-2}{H_2O}$$

Oxidation half-reaction:

$$\overset{-2}{N_2H_4} \rightarrow \overset{0}{N_2} + 4H^+ + 4e^-$$

Reduction half-reaction:

$$\overset{+5}{NaIO_3} + 6H^+ + 2Cl^- + 4e^- \rightarrow \overset{+1}{NaICl_2} + 3H_2O$$

Combine half-reactions:

$$N_2H_4 \rightarrow N_2 + 4H^+ + 4e^-$$
$$\underline{NaIO_3 + 6H^+ + 2Cl^- + 4e^- \rightarrow NaICl_2 + 3H_2O}$$
$$NaIO_3 + 2H^+ + 2Cl^- + N_2H_4 \rightarrow NaICl_2 + 3H_2O + N_2$$

Combine ions to form balanced equation:

$$NaIO_3 + N_2H_4 + 2HCl \rightarrow N_2 + NaICl_2 + 3H_2O$$

583. *Formula equation:*

$$MnO_2 + H_2O_2 + HCl \rightarrow MnCl_2 + O_2 + H_2O$$

Ionic equation:

$$\overset{+4\ -2}{MnO_2} + \overset{+1\ -1}{H_2O_2} + \overset{+1}{H^+} + \overset{-1}{Cl^-} \rightarrow \overset{+2}{Mn^{2+}} + 2\overset{-1}{Cl^-} + \overset{0}{O_2} + \overset{+1\ -2}{H_2O}$$

Oxidation half-reaction:

$$\overset{-1}{H_2O_2} \rightarrow \overset{0}{O_2} + 2H^+ + 2e^-$$

Reduction half-reaction:

$$\overset{+4}{MnO_2} + 4H^+ + 2Cl^- + 2e^- \rightarrow \overset{+2}{Mn^{2+}} + 2H_2O + 2Cl^-$$

Combine half-reactions:

$$H_2O_2 \rightarrow O_2 + 2H^+ + 2e^-$$
$$\underline{MnO_2 + 4H^+ + 2Cl^- + 2e^- \rightarrow Mn^{2+} + 2H_2O + 2Cl^-}$$
$$MnO_2 + 2H^+ + H_2O_2 + 2Cl^- \rightarrow Mn^{2+} + 2H_2O + O_2 + 2Cl^-$$

Combine ions to form balanced equation:

$$MnO_2 + H_2O_2 + 2HCl \rightarrow MnCl_2 + O_2 + 2H_2O$$

---

584. *Formula equation:*

$$AsH_3 + NaClO_3 \rightarrow H_3AsO_4 + NaCl$$

Ionic equation:

$$\overset{-3\ +1}{AsH_3} + \overset{+1}{Na^+} + \overset{+5\ -2}{ClO_3^-} \rightarrow 3\overset{+1}{H^+} + \overset{+5\ -2}{AsO_4^{3-}} + \overset{+1}{Na^+} + \overset{-1}{Cl^-}$$

Oxidation half-reaction:

$$3[\overset{-3}{AsH_3} + 4H_2O \rightarrow \overset{+5}{AsO_4^{3-}} + 11H^+ + 8e^-]$$
$$= 3AsH_3 + 12H_2O \rightarrow 3AsO_4 + 33H^+ + 24e^-$$

Reduction half-reaction:

$$4[\overset{+5}{ClO_3^-} + 6H^+ + Na^+ + 6e^- \rightarrow \overset{-1}{Cl^-} + 3H_2O + Na^+]$$
$$= 4ClO_3^- + 24H^+ + 4Na^+ + 24e^- \rightarrow 4Cl^- + 12H_2O + 4Na^+$$

Combine half-reactions:

$$3AsH_3 + 12H_2O \rightarrow 3AsO_4 + 33H^+ + 24e^-$$
$$\underline{4ClO_3^- + 24H^+ + 4Na^+ + 24e^- \rightarrow 4Cl^- + 12H_2O + 4Na^+}$$
$$3AsH_3 + 4ClO_3^- + 4Na^+ \rightarrow 3AsO_4 + 9H^+ + 4Na^+ + 4Cl^-$$

Combine ions to form balanced equation:

$$3AsH_3 + 4NaClO_3 \rightarrow 3H_3AsO_4 + 4NaCl$$

585. *Formula equation:*

$$K_2Cr_2O_7 + H_2C_2O_4 + HCl \rightarrow CrCl_3 + CO_2 + KCl + H_2O$$

Ionic equation:

$$2\overset{+1}{K^+} + \overset{+6\ -2}{Cr_2O_7^{2-}} + 2\overset{+1}{H^+} + \overset{+3\ -2}{C_2O_4^{2-}} + \overset{+1}{H^+} + \overset{-1}{Cl^-} \rightarrow$$

$$\overset{+3}{Cr^{3+}} + 3\overset{-1}{Cl^-} + \overset{+4-2}{CO_2} + \overset{+1}{K^+} + \overset{-1}{Cl^-} + \overset{+1\ -2}{H_2O}$$

Oxidation half-reaction:

$$3[\overset{+3}{C_2O_4} \rightarrow 2\overset{+4}{CO_2} + 2e^-] = 3C_2O_4 \rightarrow 6CO_2 + 6e^-$$

Reduction half-reaction:

$$\overset{+6}{Cr_2O_7} + 14H^+ + 6e^- \rightarrow 2\overset{+3}{Cr^{3+}} + 7H_2O$$

Combine half-reactions:

$$3C_2O_4 \rightarrow 6CO_2 + 6e^-$$
$$\underline{Cr_2O_7 + 14H^+ + 6e^- \rightarrow 2Cr^{3+} + 7H_2O}$$
$$Cr_2O_7 + 14H^+ + 3C_2O_4 \rightarrow 2Cr^{3+} + 6CO_2 + 7H_2O$$

Combine ions to form balanced equation:

$$K_2Cr_2O_7 + 3H_2C_2O_4 + 8HCl \rightarrow 2CrCl_3 + 6CO_2 + 2KCl + 7H_2O$$

---

586. *Formula equation:*

$$\overset{+2\ +5-2}{Hg(NO_3)_2} \rightarrow \overset{+2\ -2}{HgO} + \overset{+4-2}{NO_2} + \overset{0}{O_2}$$

Oxidation half-reaction:

$$2\overset{-2}{O} \rightarrow \overset{0}{O_2} + 4e^-$$

Reduction half-reaction:

$$4[\overset{+5}{N} + 1e^- \rightarrow \overset{+4}{N}] = 4N + 4e^- \rightarrow 4N$$

Balanced equation:

$$2Hg(NO_3)_2 \rightarrow 2HgO + 4NO_2 + O_2$$

587. *Formula equation:*

$$HAuCl_4 + N_2H_4 \rightarrow Au + N_2 + HCl$$

Ionic equation:

$$\overset{+1}{H^+} + \overset{+3\ -1}{AuCl_4^-} + \overset{-2\ +1}{N_2H_4} \rightarrow \overset{0}{Au} + \overset{0}{N_2} + \overset{+1}{H^+} + \overset{-1}{Cl^-}$$

Oxidation half-reaction:

$$3[\overset{-2}{N_2H_4} \rightarrow \overset{0}{N_2} + 4H^+ + 4e^-] = 3N_2H_4 \rightarrow 3N_2 + 12H^+ + 12e^-$$

Reduction half-reaction:

$$4[\overset{+3}{HAuCl_4} + 3e^- \rightarrow \overset{0}{Au} + 4H^+ + 4Cl^-] = 4HAuCl_4 + 12e^- \rightarrow 4Au + 16Cl + 4H^+$$

Combine half-reactions:

$$3N_2H_4 \rightarrow 3N_2 + 12H^+ + 12e^-$$
$$\underline{4HAuCl_4^- + 12e^- \rightarrow 4Au + 16Cl^- + 4H^+}$$
$$4AuCl_4^- + 3N_2H_4 \rightarrow 4Au + 16Cl^- + 3N_2 + 16H^+$$

Combine ions to form balanced equation:

$$4HAuCl_4 + 3N_2H_4 \rightarrow 4Au + 3N_2 + 16HCl$$

---

588. *Formula equation:*

$$Sb_2(SO_4)_3 + KMnO_4 + H_2O \rightarrow H_3SbO_4 + K_2SO_4 + MnSO_4 + H_2SO_4$$

Ionic equation:

$$2\overset{+3}{Sb^{3+}} + 3\overset{+6\ -2}{SO_4^{2-}} + \overset{+1}{K^+} + \overset{+7\ -2}{MnO_4^-} + \overset{+1\ -2}{H_2O} \rightarrow$$

$$3\overset{+1}{H^+} + \overset{+5\ -2}{SbO_4^{3-}} + 2\overset{+1}{K^+} + \overset{+6\ -2}{SO_4^{2-}} + \overset{+2}{Mn^{2+}} + \overset{+6\ -2}{SO_4^{2-}} + 2\overset{+1}{H^+} + \overset{+6\ -2}{SO_4^{2-}}$$

Oxidation half-reaction:

$$5[2\overset{+3}{Sb^{3+}} + 3SO_4^{2-} + 8H_2O \rightarrow 2\overset{+5}{SbO_4^{3-}} + 3SO_4^{2-} + 16H^+ + 4e^-]$$
$$= 10Sb^{3+} + 15SO_4^{2-} + 40H_2O \rightarrow 10SbO_4^{3-} + 15SO_4^{2-} + 80H^+ + 20e^-$$

Reduction half-reaction:

$$4[K^+ + \overset{+7}{MnO_4^-} + 8H^+ + 5e^- \rightarrow K^+ + \overset{+2}{Mn^{2+}} + 4H_2O]$$
$$= 4K^+ + 4MnO_4^- + 32H^+ + 20e^- \rightarrow 4K^+ + 4Mn^{2+} + 16H_2O$$

Combine half-reactions:

$$10Sb^{3+} + 15SO_4^{2-} + 40H_2O \rightarrow 10SbO_4^{3-} + 15SO_4^{2-} + 80H^+ + 20e^-$$
$$\underline{4MnO_4^- + 4K^+ + 32H^+ + 20e^- \rightarrow 4Mn^{2+} + 4K^+ + 16H_2O}$$
$$10Sb^{3+} + 24H_2O + 4MnO_4^- + 15SO_4^{2-} + 4K^+ \rightarrow$$
$$10SbO_4^{3-} + 48H^+ + 4Mn^{2+} + 15SO_4^{2-} + 4K^+$$

Combine ions to form balanced equation:

$$5Sb_2(SO_4)_3 + 4KMnO_4 + 24H_2O \rightarrow 10H_3SbO_4 + 2K_2SO_4 + 4MnSO_4 + 9H_2SO_4$$

589. *Formula equation:*

$Mn(NO_3)_2 + NaBiO_3 + HNO_3 \rightarrow Bi(NO_3)_2 + HMnO_4 + NaNO_3 + H_2O$

Ionic equation:

$\overset{+2}{Mn}{}^{2+} + 2\overset{+5-2}{NO_3^-} + Na\overset{+1}{Bi}\overset{+5-2}{O_3} + \overset{+1}{H^+} + \overset{+5-2}{NO_3^-}$

$\rightarrow \overset{+2}{Bi}{}^{2+} + 2\overset{+5-2}{NO_3^-} + \overset{+1+7\ -2}{HMnO_4} + \overset{+1}{Na^+} + \overset{+5-2}{NO_3^-} + \overset{+1-2}{H_2O}$

Oxidation half-reaction:

$3[\overset{+2}{Mn}{}^{2+} + 4H_2O \rightarrow \overset{+7}{H}MnO_4 + 7H^+ + 5e^-]$

$= 3Mn^{2+} + 12H_2O \rightarrow 3HMnO_4 + 21H^+ + 15e^-$

Reduction half-reaction:

$5[Na\overset{+5}{Bi}O_3 + 6H^+ + 3e^- \rightarrow \overset{+2}{Bi}{}^{2+} + Na^+ + 3H_2O]$

$= 5NaBiO_3 + 30H^+ + 15e^- \rightarrow 5Bi^{2+} + 5Na^+ + 15H_2O$

Combine half-reactions:

$3Mn^{2+} + 12H_2O \rightarrow 3HMnO_4 + 21H^+ + 15e^-$

$\underline{5NaBiO_3 + 30H^+ + 15e^- \rightarrow 5Bi^{2+} + 5Na^+ + 15H_2O}$

$3Mn^{2+} + 5NaBiO_3 + 9H^+ \rightarrow 3MnO_4 + 5Na^+ + 5Bi^{2+} + 3H_2O$

Combine ions to form balanced equation:

$3Mn(NO_3)_2 + 5NaBiO_3 + 9HNO_3 \rightarrow 5Bi(NO_3)_2 + 3HMnO_4 + 5NaNO_3 + 3H_2O$

---

590. *Formula equation:*

$H_3AsO_4 + Zn + HCl \rightarrow AsH_3 + ZnCl_2 + H_2O$

Ionic equation:

$3\overset{+1}{H^+} + \overset{+5\ -2}{AsO_4^{3-}} + \overset{0}{Zn} + \overset{+1}{H^+} + \overset{-1}{Cl^-} \rightarrow \overset{-3+1}{AsH_3} + \overset{+2}{Zn^{2+}} + 2\overset{-1}{Cl^-} + \overset{+1-2}{H_2O}$

Oxidation half-reaction:

$4[\overset{0}{Zn} \rightarrow \overset{+2}{Zn}{}^{2+} + 2e^-] = 4Zn \rightarrow 4Zn^{2+} + 8e^-$

Reduction half-reaction:

$\overset{+5}{As}O_4 + 11H^+ + 8Cl^- + 8e^- \rightarrow \overset{-3}{As}H_3 + 4H_2O + 8Cl^-$

Combine half-reactions:

$4Zn \rightarrow 4Zn^{2+} + 8e^-$

$\underline{AsO_4 + 11H^+ + 8Cl^- + 8e^- \rightarrow AsH_3 + 4H_2O + 8Cl^-}$

$AsO_4 + 11H^+ + 4Zn + 8Cl^- \rightarrow AsH_3 + 4H_2O + 4Zn^{2+} + 8Cl^-$

Combine ions to form balanced equation:

$H_3AsO_4 + 4Zn + 8HCl \rightarrow AsH_3 + 4ZnCl_2 + 4H_2O$

591. *Formula equation:*

$$KClO_3 + HCl \rightarrow Cl_2 + H_2O + KCl$$

Ionic equation:

$$K^+ + \overset{+5\ -2}{ClO_3^-} + \overset{+1}{H^+} + \overset{-1}{Cl^-} \rightarrow \overset{0}{Cl_2} + \overset{+1\ -2}{H_2O} + \overset{+1}{K^+} + \overset{-1}{Cl^-}$$

Oxidation half-reaction:

$$3[2\overset{-1}{Cl^-} \rightarrow \overset{0}{Cl_2} + 2e^-] = 6Cl^- \rightarrow 3Cl_2 + 6e^-$$

Reduction half-reaction:

$$K^+ + \overset{+5}{ClO_3^-} + 6H^+ + 6e^- \rightarrow K^+ + \overset{-1}{Cl^-} + 3H_2O$$

Combine half-reactions:

$$6Cl^- \rightarrow 3Cl_2 + 6e^-$$
$$\underline{K^+ + ClO_3^- + 6H^+ + 6e^- \rightarrow K^+ + Cl^- + 3H_2O}$$
$$K^+ + ClO_3^- + 6H^+ + 6Cl^- \rightarrow K^+ + Cl^- + 3H_2O + 3Cl_2$$

Combine ions to form balanced equation:

$$KClO_3 + 6HCl \rightarrow 3Cl_2 + 3H_2O + KCl$$

---

592. *Formula equation:*

$$KClO_3 + HCl \rightarrow Cl_2 + ClO_2 + H_2O + KCl$$

Ionic equation:

$$\overset{+1\ +5\ -2}{KClO_3} + \overset{+1\ -1}{HCl^-} \rightarrow \overset{0}{Cl_2} + \overset{+4\ -2}{ClO_2} + \overset{+1\ -2}{H_2O} + \overset{+1}{K^+} + \overset{-1}{Cl^-}$$

Oxidation half-reaction:

$$2[2\overset{-1}{HCl} + 2H_2O \rightarrow \overset{+4}{ClO_2} + 6H^+ + Cl^- + 5e^-]$$
$$= 4HCl + 4H_2O \rightarrow 2ClO_2 + 12H^+ + 2Cl^- + 10e^-$$

Reduction half-reaction:

$$2K\overset{+5}{ClO_3} + 12H^+ + 10e^- \rightarrow \overset{0}{Cl_2} + 2K^+ + 6H_2O$$

Combine half-reactions:

$$4HCl + 4H_2O \rightarrow 2ClO_2 + 12H^+ + 2Cl^- + 10e^-$$
$$\underline{2KClO_3 + 12H^+ + 10e^- \rightarrow Cl_2 + 2K^+ + 6H_2O}$$
$$4HCl + 2KClO_3 \rightarrow 2ClO_2 + 2K^+ + Cl_2 + 2H_2O + 2Cl^-$$

Combine ions to form balanced equation:

$$2KClO_3 + 4HCl \rightarrow Cl_2 + 2ClO_2 + 2H_2O + 2KCl$$

593. *Formula equation:*

$$MnCl_3 + H_2O \rightarrow MnCl_2 + MnO_2 + HCl$$

Ionic equation:

$$\overset{+3}{Mn}{}^{3+} + 3\overset{-1}{Cl}{}^- + \overset{+1\ -2}{H_2O} \rightarrow \overset{+2}{Mn}{}^{2+} + 2\overset{-1}{Cl}{}^- + \overset{+4\ -2}{MnO_2} + \overset{+1}{H}{}^+ + \overset{-1}{Cl}{}^-$$

Oxidation half-reaction:

$$\overset{+3}{Mn}{}^{3+} + 3Cl^- + 2H_2O \rightarrow \overset{+4}{MnO_2} + 3Cl^- + 4H^+ + 1e^-$$

Reduction half-reaction:

$$\overset{+3}{Mn}{}^{3+} + 3Cl^- + 1e^- \rightarrow \overset{+2}{Mn}{}^{2+} + 3Cl^-$$

Combine half-reactions:

$$Mn^{3+} + 3Cl^- + 2H_2O \rightarrow MnO_2 + 3Cl^- + 4H^+ + 1e^-$$
$$Mn^{3+} + 3Cl^- + 1e^- \rightarrow Mn^{2+} + 3Cl^-$$
$$\overline{2Mn^{3+} + 6Cl^- + 2H_2O \rightarrow Mn^{2+} + 6Cl^- + MnO_2 + 4H^+}$$

Combine ions to form balanced equation:

$$2MnCl_3 + 2H_2O \rightarrow MnCl_2 + MnO_2 + 4HCl$$

---

594. *Formula equation:*

$$NaOH + H_2O + Al \rightarrow NaAl(OH)_4 + H_2 \text{ (in basic solution)}$$

Ionic equation:

$$\overset{+1}{Na}{}^+ + \overset{-2\ +1}{OH}{}^- + \overset{+1\ -2}{H_2O} + \overset{0}{Al} \rightarrow \overset{+1}{Na}{}^+\overset{+3\ -2+1}{Al(OH)_4^-} + \overset{0}{H_2}$$

Oxidation half-reaction:

$$2[\overset{0}{Al} + Na^+ + 4OH^- \rightarrow \overset{+3}{Al(OH)_4^-} + Na^+ + 3e^-]$$
$$= 2Al + 2Na^+ + 8OH^- \rightarrow 2Al(OH)_4^- + 2Na^+ + 6e^-$$

Reduction half-reaction:

$$3[2\overset{+1}{H_2O} + 2e^- \rightarrow \overset{0}{H_2} + 2OH^-] = 6H_2O + 6e^- \rightarrow 3H_2 + 6OH^-$$

Combine half-reactions:

$$2Al + 2Na^+ + 8OH^- \rightarrow 2Al(OH)_4^- + 2Na^+ + 6e^-$$
$$6H_2O + 6e^- \rightarrow 3H_2 + 6OH^-$$
$$\overline{2Al + 2Na^+ + 2OH^- + 6H_2O \rightarrow 2Al(OH)_4^- + 2Na^+ + 3H_2}$$

Combine ions to form balanced equation:

$$2NaOH + 6H_2O + 2Al \rightarrow 2NaAl(OH)_4 + 3H_2$$

595. *Formula equation:*

$$Br_2 + Ca(OH)_2 \to CaBr_2 + Ca(BrO_3)_2 + H_2O \text{ (in basic solution)}$$

Ionic equation:

$$\overset{0}{Br_2} + \overset{+2}{Ca^{2+}} + 2\overset{-2+1}{OH^-} \to \overset{+2}{Ca^{2+}} + 2\overset{-1}{Br^-} + \overset{+2}{Ca^{2+}} + 2\overset{+5\,-2}{BrO_3^-} + \overset{+1\,-2}{H_2O}$$

Oxidation half-reaction:

$$\overset{0}{Br_2} + 6Ca^{2+} + 12OH^- \to 2\overset{+5}{BrO_3^-} + 6Ca^{2+} + 6H_2O + 10e^-$$

Reduction half-reaction:

$$5[\overset{0}{Br_2} + 2e^- \to 2\overset{-1}{Br^-}] = 5Br_2 + 10e \to 10Br^-$$

Combine half-reactions:

$$Br_2 + Ca^{2+} + 12OH^- \to 2BrO_3 + Ca^{2+} + 6H_2O + 10e^-$$
$$\underline{5Br_2 + 10e^- \to 10Br^-}$$
$$6Br_2 + 12OH^- \to 10Br^- + 2BrO_3 + 6H_2O$$

Combine ions to form balanced equation:

$$6Br_2 + 6Ca(OH)_2 \to 5CaBr_2 + Ca(BrO_3)_2 + 6H_2O$$

---

596. *Formula equation:*

$$N_2O + NaClO + NaOH \to NaCl + NaNO_2 + H_2O \text{ (in basic solution)}$$

Ionic equation:

$$\overset{+1\,-2}{N_2O} + \overset{+1\,+1-2}{NaClO} + \overset{+1}{Na^+}\overset{-2+1}{OH^-} \to \overset{+1}{Na^+} + \overset{-1}{Cl^-} + \overset{+1\,+3-2}{NaNO_2} + \overset{+1\,-2}{H_2O}$$

Oxidation half-reaction:

$$2\overset{+1}{Na} + \overset{}{N_2O} + 6OH^- \to 2Na\overset{+3}{N}O_2 + 3H_2O + 4e^-$$

Reduction half-reaction:

$$2[Na\overset{+1}{C}lO + H_2O + 2e^- \to Cl^- + 2OH^- + Na^+]$$
$$= 2NaClO + 2H_2O + 4e^- \to 2Cl^- + 4OH^- + 2Na^+$$

Combine half-reactions:

$$2Na^+ + N_2O + 6OH^- \to 2NaNO_2 + 3H_2O + 4e^-$$
$$\underline{2NaClO + 2H_2O + 4e^- \to 2Cl^- + 4OH^- + 2Na^+}$$
$$2Na^+ + N_2O + 2OH^- + 2NaClO \to 2NaNO_2 + H_2O + 2Cl^- + 2Na^+$$

Combine ions to form balanced equation:

$$N_2O + 2NaClO + 2NaOH \to 2NaCl + 2NaNO_2 + H_2O$$

597. *Formula equation:*

$$HBr + MnO_2 \rightarrow MnBr_2 + H_2O + Br_2$$

Ionic equation:

$$\overset{+1}{H^+} + \overset{-1}{Br^-} + \overset{+4\ -2}{MnO_2} \rightarrow \overset{+2\ -1}{MnBr_2} + \overset{+1\ -2}{H_2O} + \overset{0}{Br_2}$$

Oxidation half-reaction:

$$2\overset{-1}{Br^-} \rightarrow \overset{0}{Br_2} + 2e^-$$

Reduction half-reaction:

$$2Br^- + \overset{+4}{MnO_2} + 4H^+ + 2e^- \rightarrow \overset{+2}{MnBr_2} + 2H_2O$$

Combine half-reactions:

$$2Br^- \rightarrow Br_2 + 2e^-$$
$$\underline{2Br^- + MnO_2 + 4H^+ + 2e^- \rightarrow MnBr_2 + 2H_2O}$$
$$4Br^- + MnO_2 + 4H^+ \rightarrow Br_2 + MnBr_2 + 2H_2O$$

Combine ions to form balanced equation:

$$4HBr + MnO_2 \rightarrow MnBr_2 + 2H_2O + Br_2$$

---

598. *Formula equation:*

$$Au + HCl + HNO_3 \rightarrow HAuCl_4 + NO + H_2O \text{ (in } aqua\ regia\text{*)}$$

Ionic equation:

$$\overset{0}{Au} + \overset{+1}{H^+} + \overset{-1}{Cl^-} + \overset{+1}{H^+} + \overset{+5\ -2}{NO_3^-} \rightarrow \overset{+1+3\ -1}{HAuCl_4} + \overset{+2-2}{NO} + \overset{+1\ -2}{H_2O}$$

Oxidation half-reaction:

$$\overset{0}{Au} + 4Cl^- + H^+ \rightarrow \overset{+3}{HAuCl_4} + 3e^-$$

Reduction half-reaction:

$$\overset{+5}{NO_3^-} + 4H^+ + 3e^- \rightarrow \overset{+2}{NO} + 2H_2O$$

Combine half-reactions:

$$Au + 4Cl^- + H^+ \rightarrow HAuCl_4 + 3e^-$$
$$\underline{NO_3^- + 4H^+ + 3e^- \rightarrow NO + 2H_2O}$$
$$Au + 4Cl^- + 5H^+ + NO_3^- \rightarrow HAuCl_4 + NO + 2H_2O$$

Combine ions to form balanced equation:

$$Au + 4HCl + HNO_3 \rightarrow HAuCl_4 + NO + 2H_2O$$

---

*Aqua regia = concentrated HCl + concentrated HNO_3.

599. $\overset{+2}{Cu^{2+}} + \overset{0}{Fe} \rightarrow \overset{+2}{Fe^{2+}} + \overset{0}{Cu}$

Anode = Fe (oxidation)
Cathode = Cu (reduction)

Half-reactions:

$Fe \rightleftharpoons Fe^{2+} + 2e^-$	$E^0 = +0.45$ V
$Cu^{2+} + 2e^- \rightleftharpoons Cu$	$E^0 = +0.34$ V
	$E^0_{cell} = +0.79$ V; spontaneous

600. $\overset{+2}{Pb^{2+}} + \overset{+2}{Fe^{2+}} \rightarrow \overset{+3}{Fe^{3+}} + \overset{0}{Pb}$

Anode = Fe (oxidation)
Cathode = Pb (reduction)

Half-reactions:

$Fe^{2+} \rightleftharpoons Fe^{3+} + 1e^-$	$E^0 = -0.77$ V
$Pb^{2+} + 2e^- \rightleftharpoons Pb$	$E^0 = -0.13$ V
	$E^0_{cell} = -0.90$ V; not spontaneous

601. $\overset{+2}{Mn^{2+}} + 3H_2O + \overset{+2}{Sn^{2+}} \rightarrow \overset{+7}{MnO_4^-} + 8H^+ + \overset{0}{Sn}$

Anode = Mn (oxidation)
Cathode = Sn (reduction)

Half-reactions:

$Mn^{2+} + 4H_2O \; MnO_4^- + 8H^+ + 5e^-$	$E^0 = -1.50$ V
$Sn^{2+} + 2e^- \rightleftharpoons Sn$	$E^0 = -0.14$ V
	$E^0_{cell} = -1.64$ V; not spontaneous

602. $\overset{+6}{MnO_4^{2-}} + \overset{0}{Cl_2} \rightarrow \overset{+7}{MnO_4^-} + 2\overset{-1}{Cl^-}$

Anode = Mn (oxidation)
Cathode = Cl (reduction)

Half-reactions:

$MnO_4^{2-} \rightleftharpoons MnO_4^- + e^-$	$E^0 = -0.56$ V
$Cl_2 + 2e^- \rightleftharpoons 2Cl^-$	$E^0 = +1.36$ V
	$E^0_{cell} = +0.80$ V; spontaneous

603. $\overset{+1}{Hg_2^{2+}} + 2\overset{+6}{MnO_4^{2-}} \rightarrow 2\overset{0}{Hg} + 2\overset{+7}{MnO_4^-}$

Anode = Mn (oxidation)
Cathode = Hg (reduction)

Half-reactions:

$MnO_4^{2-} \rightleftharpoons MnO_4^- + e^-$	$E^0 = -0.56$ V
$Hg_2^{2+} + 2e^- \rightleftharpoons 2Hg$	$E^0 = +0.80$ V
	$E^0_{cell} = +0.24$ V; spontaneous

---

604. $2\overset{+1}{Li^+} + \overset{0}{Pb} \rightarrow 2\overset{0}{Li} + \overset{+2}{Pb^{2+}}$

Anode = Pb (oxidation)
Cathode = Li (reduction)

Half-reactions:

$Pb \rightleftharpoons Pb^{2+} + 2e^-$	$E^0 = +0.13$ V
$Li^+ + e^- \rightleftharpoons Li$	$E^0 = -3.04$ V
	$E^0_{cell} = -2.91$ V; not spontaneous

---

605. $\overset{0}{Br_2} + 2\overset{-1}{Cl^-} \rightarrow 2\overset{-1}{Br^-} + \overset{0}{Cl_2}$

Anode = Cl (oxidation)
Cathode = Br (reduction)

Half-reactions:

$2Cl^- \rightleftharpoons Cl_2 + 2e^-$	$E^0 = -1.36$ V
$Br_2 + 2e^- \rightleftharpoons 2Br^-$	$E^0 = +1.07$ V
	$E^0_{cell} = -0.29$ V; not spontaneous

---

606. $\overset{0}{S} + 2\overset{-1}{I^-} \rightarrow \overset{-2}{S^{2-}} + \overset{0}{I_2}$

Anode = I (oxidation)
Cathode = S (reduction)

Half-reactions:

$2I^- \rightleftharpoons I_2 + 2e^-$	$E^0 = -0.54$ V
$S + 2e^- \rightleftharpoons S^{2-}$	$E^0 = -0.48$ V
	$E^0_{cell} = -1.02$ V; not spontaneous

607. $Ca^{2+} + 2e^- \rightleftharpoons Ca$ $E^0 = -2.87$ V

$Fe^{3+} + 3e^- \rightleftharpoons Fe$ $E^0 = -0.04$ V

Anode = Ca (oxidation)
Cathode = Fe (reduction)

Anode reaction:

$Ca \rightleftharpoons Ca^{2+} + 2e^-$

Cathode reaction:

$Fe^{3+} + 3e^- \rightleftharpoons Fe$

$E^0_{cell} = E^0_{cathode} - E^0_{anode} = -0.04 - (-2.87) = +2.83$ V

---

608. $Ag^+ + e^- \rightleftharpoons Ag$ $E^0 = +0.80$ V

$S + 2H^+ + 2e^- \rightleftharpoons H_2S$ $E^0 = +0.14$ V

Anode = S (oxidation)
Cathode = Ag (reduction)

Anode reaction:

$H_2S \rightleftharpoons S + 2H^+ + 2e^-$

Cathode reaction:

$Ag^+ + e^- \rightleftharpoons Ag$

$E^0_{cell} = E^0_{cathode} - E^0_{anode} = +0.80 - (+0.14) = +0.66$ V

---

609. $Fe^{3+} + e^- \rightleftharpoons Fe^{2+}$ $E^0 = +0.77$ V

$Sn^{2+} + 2e^- \rightleftharpoons Sn$ $E^0 = -0.14$ V

Anode = Sn (oxidation)
Cathode = Fe (reduction)

Anode reaction:

$Sn \rightleftharpoons Sn^{2+} + 2e^-$

Cathode reaction:

$Fe^{3+} + e^- \rightleftharpoons Fe^{2+}$

$E^0_{cell} = E^0_{cathode} - E^0_{anode} = +0.77 - (-0.14) = +0.91$ V

610. $Cu^{2+} + 2e^- \rightleftharpoons Cu$ $\hspace{3cm}$ $E^0 = +0.34$ V

$Au^{3+} + 3e^- \rightleftharpoons Au$ $\hspace{3cm}$ $E^0 = +1.50$ V

Anode = Cu (oxidation)
Cathode = Au (reduction)

Anode reaction:

$Cu \rightleftharpoons Cu^{2+} + 2e^-$

Cathode reaction:

$Au^{3+} + 3e^- \rightleftharpoons Au$

$E^0{}_{cell} = E^0{}_{cathode} - E^0{}_{anode} = +1.50 - (+0.34) = +1.16$ V

---

611. $\overset{0}{Ba} + \overset{+2}{Sn^{2+}} \rightarrow \overset{+2}{Ba^{2+}} + \overset{0}{Sn}$

Anode = Ba (oxidation)
Cathode = Sn (reduction)

Half-reactions:

$Ba \rightleftharpoons Ba^{2+} + 2e^-$ $\hspace{2cm}$ $E^0 = +2.91$ V

$Sn^{2+} + 2e^- \rightleftharpoons Sn$ $\hspace{2cm}$ $E^0 = -0.14$ V

$\hspace{5cm}$ $E^0{}_{cell} = +2.77$ V; spontaneous

---

612. $\overset{0}{Ni} + \overset{+2}{Hg^{2+}} \rightarrow \overset{+2}{Ni^{2+}} + \overset{0}{Hg}$

Anode = Ni (oxidation)
Cathode = Hg (reduction)

Half-reactions:

$Ni \rightleftharpoons Ni^{2+} + 2e^-$ $\hspace{2cm}$ $E^0 = +0.26$ V

$Hg^{2+} + 2e^- \rightleftharpoons Hg$ $\hspace{2cm}$ $E^0 = +0.85$ V

$\hspace{5cm}$ $E^0{}_{cell} = +1.11$ V; spontaneous

---

613. $2\overset{+3}{Cr^{3+}} + 7H_2O + 6\overset{+3}{Fe^{3+}} \rightarrow \overset{+6}{Cr_2O_7^{2-}} + 14H^+ + 6\overset{+2}{Fe^{2+}}$

Anode = Cr (oxidation)
Cathode = Fe (reduction)

Half-reactions:

$2Cr^{3+} + 7H_2O \rightleftharpoons Cr_2O_7^{2-} + 14H^+ + 6e^-$ $\hspace{1cm}$ $E^0 = -1.23$ V

$Fe^{3+} + e^- \rightleftharpoons Fe^{2+}$ $\hspace{4cm}$ $E^0 = +0.77$ V

$\hspace{7cm}$ $E^0{}_{cell} = -0.46$ V; not spontaneous

614. $\overset{0}{Cl_2} + \overset{0}{Sn} \rightarrow 2\overset{-1}{Cl^-} + \overset{+2}{Sn^{2+}}$

Anode = Sn (oxidation)
Cathode = Cl (reduction)

Half-reactions:

$Sn \rightleftharpoons Sn^{2+} + 2e^-$	$E^0 = +0.14$ V
$Cl_2 + 2e^- \rightleftharpoons 2Cl^-$	$E^0 = +1.36$ V
	$E^0_{cell} = +1.50$ V; spontaneous

615. $\overset{0}{Al} + 3\overset{+3}{Ag^+} \rightarrow \overset{+3}{Al^{3+}} + 3\overset{0}{Ag}$

Anode = Al (oxidation)
Cathode = Ag (reduction)

Half-reactions:

$Al \rightleftharpoons Al^{3+} + 3e^-$	$E^0 = +1.66$ V
$Ag^+ + e^- \rightleftharpoons Ag$	$E^0 = +0.80$ V
	$E^0_{cell} = +2.46$ V; spontaneous

616. $\overset{+1}{Hg_2^{2+}} + \overset{-2}{S^{2-}} \rightarrow 2\overset{0}{Hg} + \overset{0}{S}$

Anode = S (oxidation)
Cathode = Hg (reduction)

Half-reactions:

$S^{2-} \rightleftharpoons S + 2e^-$	$E^0 = +0.48$ V
$Hg_2^{2+} + 2e^- \rightleftharpoons 2Hg$	$E^0 = +0.80$ V
	$E^0_{cell} = +1.28$ V; spontaneous

617. $\overset{0}{Ba} + 2\overset{+1}{Ag^+} \rightarrow \overset{+2}{Ba^{2+}} + 2\overset{0}{Ag}$

Anode = Ba (oxidation)
Cathode = Ag (reduction)

Half-reactions:

$Ba \rightleftharpoons Ba^{2+} + 2e^-$	$E^0 = +2.91$ V
$Ag^+ + e^- \rightleftharpoons Ag$	$E^0 = +0.80$ V
	$E^0_{cell} = +3.71$ V; spontaneous

618. $\overset{-1}{2I^-} + \overset{+2}{Ca^{2+}} \rightarrow \overset{0}{I_2} + \overset{0}{Ca}$

Anode = I (oxidation)
Cathode = Ca (reduction)

Half-reactions:

$2I^- \rightleftharpoons I_2 + 2e^-$ $\qquad E^0 = -0.54$ V

$Ca^{2+} + 2e^- \rightleftharpoons Ca$ $\qquad E^0 = -2.87$ V

$E^0_{cell} = -3.41$ V; not spontaneous

619. $\overset{0}{Zn} + 2\overset{+7}{MnO_4^-} \rightarrow \overset{+2}{Zn^{2+}} + 2\overset{+6}{MnO_4^{2-}}$

Anode = Zn (oxidation)
Cathode = Mn (reduction)

Half-reactions:

$Zn \rightleftharpoons Zn^{2+} + 2e^-$ $\qquad E^0 = +0.76$ V

$MnO_4^- + e^- \rightleftharpoons MnO_4^{2-}$ $\qquad E^0 = +0.56$ V

$E^0_{cell} = +1.32$ V; spontaneous

620. $2\overset{+3}{Cr^{3+}} + 3\overset{+2}{Mg^{2+}} + 7H_2O \rightarrow \overset{+6}{Cr_2O_7^{2-}} + 14H^+ + 3\overset{0}{Mg}$

Anode = Cr (oxidation)
Cathode = Mg (reduction)

Half-reactions:

$2Cr^{3+} + 7H_2O \rightleftharpoons Cr_2O_7^{2-} + 14H^+ + 6e^-$ $\qquad E^0 = -1.23$ V

$Mg^{2+} + 2e^- \rightleftharpoons Mg$ $\qquad E^0 = -2.37$ V

$E^0_{cell} = -3.60$ V; not spontaneous

621. $Cl_2 + 2e^- \rightleftharpoons 2Cl^-$ $\qquad E^0 = +1.36$ V

$Ni^{2+} + 2e^- \rightleftharpoons Ni$ $\qquad E^0 = -0.26$ V

Anode = Ni (oxidation)
Cathode = Cl (reduction)

Anode reaction:

$Ni \rightleftharpoons Ni^{2+} + 2e^-$

Cathode reaction:

$Cl_2 + 2e^- \rightleftharpoons 2Cl^-$

$E^0_{cell} = E^0_{cathode} - E^0_{anode} = +1.36 - (-0.26) = +1.62$ V

Combine half-reactions:

$Ni \rightleftharpoons Ni^{2+} + 2e^-$

$Cl_2 + 2e^- \rightleftharpoons 2Cl^-$

$Cl_2 + Ni \rightleftharpoons Ni^{2+} + 2Cl^-$

622. $Fe^{3+} + 3e^- \rightleftharpoons Fe$ $\quad\quad E^0 = -0.04$ V

$Hg^{2+} + 2e^- \rightleftharpoons Hg$ $\quad\quad E^0 = +0.85$ V

Anode = Fe (oxidation)
Cathode = Hg (reduction)

Anode reaction:

$2[Fe \rightleftharpoons Fe^{3+} + 3e^-] = 2Fe \rightleftharpoons 2Fe^{3+} + 6e^-$

Cathode reaction:

$3[Hg^{2+} + 2e^- \rightleftharpoons Hg] = 3Hg^{2+} + 6e^- \rightleftharpoons 3Hg$

$E^0_{cell} = E^0_{cathode} - E^0_{anode} = +0.85 - (-0.04) = +0.89$ V

Combine half-reactions:

$2Fe \rightleftharpoons 2Fe^{3+} + 6e^-$

$3Hg^{2+} + 6e^- \rightleftharpoons 3Hg$

$3Hg^{2+} + 2Fe \rightarrow 3Hg + 2Fe^{3+}$

---

623. $MnO_4^- + e^- \rightleftharpoons MnO_4^{2-}$ $\quad\quad E^0 = +0.56$ V

$Al^{3+} + 3e^- \rightleftharpoons Al$ $\quad\quad E^0 = -1.66$ V

Anode = Al (oxidation)
Cathode = Mn (reduction)

Anode reaction:

$Al \rightleftharpoons Al^{3+} + 3e^-$

Cathode reaction:

$3[MnO_4^- + e^- \rightleftharpoons MnO_4^{2-}] = 3MnO_4^- + 3e^- \rightleftharpoons 3MnO_4^{2-}$

$E^0_{cell} = E^0_{cathode} - E^0_{anode} = +0.56 - (-1.66) = +2.22$ V

Combine half-reactions:

$Al \rightleftharpoons Al^{3+} + 3e^-$

$3MnO_4^- + 3e^- \rightleftharpoons 3MnO_4^{2-}$

$3MnO_4^- + Al \rightarrow 3MnO_4^{2-} + Al^{3+}$

624. $MnO_4^- + 8H^+ + 5e^- \rightleftharpoons Mn^{2+} + 4H_2O$ $\quad E^0 = +1.50$ V

$S + 2H^+ + 2e^- \rightleftharpoons H_2S$ $\quad E^0 = +0.14$ V

Anode = S (oxidation)
Cathode = Mn (reduction)

Anode reaction:

$5[H_2S \rightleftharpoons S + 2H^+ + 2e^-] = 5H_2S \rightleftharpoons 5S + 10H^+ + 10e^-$

Cathode reaction:

$2[MnO_4^- + 8H^+ + 5e^- \rightleftharpoons Mn^{2+} + 4H_2O]$
$= 2MnO_4^- + 16H^+ + 10e^- \rightleftharpoons 2Mn^{2+} + 8H_2O$

$E^0_{cell} = E^0_{cathode} - E^0_{anode} = +1.50 - (+0.14) = +1.36$ V

Combine half-reactions:

$5H_2S \rightleftharpoons 5S + 10H^+ + 10e^-$
$\underline{2MnO_4^- + 16H^+ + 10e^- \rightleftharpoons 2Mn^{2+} + 8H_2O}$

$2MnO_4^- + 6H^+ + 5H_2S \rightarrow 2Mn^{2+} + 8H_2O + 5S$

---

625. $Ca^{2+} + 2e^- \rightleftharpoons Ca$ $\quad E^0 = -2.87$ V

$Li^+ + e^- \rightleftharpoons Li$ $\quad E^0 = -3.04$ V

Anode = Li (oxidation)
Cathode = Ca (reduction)

Anode reaction:

$2[Li \rightleftharpoons Li^+ + e^-] = 2Li \rightleftharpoons 2Li^+ + 2e^-$

Cathode reaction:

$Ca^{2+} + 2e^- \rightleftharpoons Ca$

$E^0_{cell} = E^0_{cathode} - E^0_{anode} = -2.87 - (-3.04) = +0.17$ V

Combine half-reactions:

$2Li \rightleftharpoons 2Li^+ + 2e^-$
$\underline{Ca^{2+} + 2e^- \rightleftharpoons Ca}$

$Ca^{2+} + 2Li \rightarrow Ca + 2Li^+$

626. $Br_2 + 2e^- \rightleftharpoons 2Br^-$ $\quad\quad\quad\quad\quad\quad\quad E^0 = +1.07$ V

$MnO_4^- + 8H^+ + 5e^- \rightleftharpoons Mn^{2+} + 4H_2O \quad E^0 = +1.50$ V

Anode = Br (oxidation)
Cathode = Mn (reduction)

Anode reaction:

$5[2Br^- \rightleftharpoons Br_2 + 2e^-] = 10Br^- \rightleftharpoons 5Br_2 + 10e^-$

Cathode reaction:

$2[MnO_4^- + 8H^+ + 5e^- \rightleftharpoons Mn^{2+} + 4H_2O]$
$= 2MnO_4^- + 16H^+ + 10e^- \rightleftharpoons 2Mn^{2+} + 8H_2O$

$E^0_{cell} = E^0_{cathode} - E^0_{anode} = +1.50 - (+1.07) = +0.43$ V

Combine half-reactions:

$\quad\quad\quad 10Br^- \rightleftharpoons 5Br_2 + 10e^-$
$\underline{2MnO_4^- + 16H^+ + 10e^- \rightleftharpoons 2Mn^{2+} + 8H_2O}$

$2MnO_4^- + 16H^+ + 10Br^- \rightarrow 2Mn^{2+} + 8H_2O + 5Br_2$

---

627. $Sn^{2+} + 2e^- \rightleftharpoons Sn$ $\quad\quad\quad E^0 = -0.14$ V

$Fe^{3+} + e^- \rightleftharpoons Fe^{2+}$ $\quad\quad\quad\quad E^0 = +0.77$ V

Anode = Sn (oxidation)
Cathode = Fe (reduction)

Anode reaction:

$Sn \rightleftharpoons Sn^{2+} + 2e^-$

Cathode reaction:

$2[Fe^{3+} + e^- \rightleftharpoons Fe^{2+}] = 2Fe^{3+} + 2e^- \rightleftharpoons 2Fe^{2+}$

$E^0_{cell} = E^0_{cathode} - E^0_{anode} = +0.77 - (-0.14) = +0.91$ V

Combine half-reactions:

$\quad\quad\quad Sn \rightleftharpoons Sn^{2+} + 2e^-$
$\underline{2Fe^{3+} + 2e^- \rightleftharpoons 2Fe^{2+}}$

$2Fe^{3+} + Sn \rightarrow 2Fe^{2+} + Sn^{2+}$

628. $Zn^{2+} + 2e^- \rightleftharpoons Zn$ $\qquad E^0 = -0.76$ V

$Cr_2O_7^{2-} + 14H^+ + 6e^- \rightleftharpoons 2Cr^{3+} + 7H_2O \qquad E^0 = +1.23$ V

Anode = Zn (oxidation)
Cathode = Cr (reduction)

Anode reaction:

$3[Zn \rightleftharpoons Zn^{2+} + 2e^-] = 3Zn \rightleftharpoons 3Zn^{2+} + 6e^-$

Cathode reaction:

$Cr_2O_7^{2-} + 14H^+ + 6e^- \rightleftharpoons 2Cr^{3+} + 7H_2O$

$E^0_{cell} = E^0_{cathode} - E^0_{anode} = +1.23 - (-0.76) = +1.99$ V

Combine half-reactions:

$3Zn \rightleftharpoons 3Zn^{2+} + 6e^-$
$Cr_2O_7^{2-} + 14H^+ + 6e^- \rightleftharpoons 2Cr^{3+} + 7H_2O$

$Cr_2O_7^{2-} + 14H^+ + 3Zn \rightarrow 2Cr^{3+} + 7H_2O + 3Zn^{2+}$

---

629. $Ba^{2+} + 2e^- \rightleftharpoons Ba$ $\qquad E^0 = -2.91$ V

$Ca^{2+} + 2e^- \rightleftharpoons Ca$ $\qquad E^0 = -2.87$ V

Anode = Ba (oxidation)
Cathode = Ca (reduction)

Anode reaction:

$Ba \rightleftharpoons Ba^{2+} + 2e^-$

Cathode reaction:

$Ca^{2+} + 2e^- \rightleftharpoons Ca$

$E^0_{cell} = E^0_{cathode} - E^0_{anode} = -2.87 - (-2.91) = +0.04$ V

Combine half-reactions:

$Ba \rightleftharpoons Ba^{2+} + 2e^-$
$Ca^{2+} + 2e^- \rightleftharpoons Ca$

$Ca^{2+} + Ba \rightarrow Ca + Ba^{2+}$

630. $Hg_2^{2+} + 2e^- \rightleftharpoons 2Hg$ $\qquad E^0 = +0.80$ V

$Cd^{2+} + 2e^- \rightleftharpoons Cd$ $\qquad E^0 = -0.40$ V

Anode = Cd (oxidation)
Cathode = Hg (reduction)

Anode reaction:

$Cd \rightleftharpoons Cd^{2+} + 2e^-$

Cathode reaction:

$Hg_2^{2+} + 2e^- \rightleftharpoons 2Hg$

$E^0_{cell} = E^0_{cathode} - E^0_{anode} = +0.80 - (-0.40) = +1.20$ V

Combine half-reactions:

$\qquad Cd \rightleftharpoons Cd^{2+} + 2e^-$

$\underline{Hg_2^{2+} + 2e^- \rightleftharpoons 2Hg}$

$Hg_2^{2+} + Cd \rightarrow 2Hg + Cd^{2+}$